Tumors of the Eye and Ocular Adnexa

AFIP Atlases of Tumor and Non-Tumor Pathology

AMERICAN REGISTRY OF PATHOLOGY
Publication & Education

ARP PRESS

Arlington, Virginia

Senior Director of Publications: Mirlinda Q. Caton
Production Editor: Dian S. Thomas
Technical Editor and Subscription Manager: Magdalena C. Silva
Copyeditor: Audrey Kahn
Editorial Assistant: Elizabeth Tomlinson

Available from the American Registry of Pathology
Arlington, Virginia 22209
www.arppress.org
ISBN 1-933477-92-X
978-1-933477-92-3

Printed in Korea

AFIP ATLASES OF TUMOR AND NON-TUMOR PATHOLOGY

Fifth Series
Fascicle 3

TUMORS OF THE EYE AND OCULAR ADNEXA

by

Robert Folberg, MD
Departments of Ophthalmology and Pathology
Beaumont Health Royal Oak
Royal Oak, Michigan

Patricia Chévez-Barrios, MD
Department of Pathology and Genomic Medicine
Houston Methodist Hospital
Departments of Pathology and Laboratory Medicine, and Opthalmology
Weill Medical College of Cornell University
Ithaca, New York

Amy Y. Lin, MD
Department of Pathology
Department of Opthalmology and Visual Sciences
University of Illinois Hospital and Clinics
University of Illinois College of Medicine at Chicago
Chicago, Illinois

Tatyana Milman, MD
Associate Professor, Ophthalmology and Pathology
Wills Eye Hospital, Sidney Kimmel Medical College
Thomas Jefferson University
Philadelphia, Pennsylvania

Published by the
American Registry of Pathology
Arlington, Virginia
2020

AFIP ATLASES OF TUMOR AND NON-TUMOR PATHOLOGY

EDITOR
Jason L. Hornick, MD, PhD
Professor of Pathology, Harvard Medical School
Director of Surgical Pathology and Immunohistochemistry
Brigham and Women's Hospital
Boston, Massachusetts

EDITORIAL ADVISORY BOARD

Manuscript reviewed by:
Charles Eberhart, MD
Fausto Rodriguez, MD

EDITOR'S NOTE

The Atlases of Tumor Pathology have a long and distinguished history. They were first conceived at a cancer research meeting held in St. Louis in September 1947, as an attempt to standardize the nomenclature of neoplastic diseases. The first series was sponsored by the National Academy of Sciences-National Research Council. The organization of this formidable effort was entrusted to the Subcommittee on Oncology of the Committee on Pathology, and Dr. Arthur Purdy Stout was the first editor-in-chief. Many of the illustrations were provided by the Medical Illustration Service of the Armed Forces Institute of Pathology (AFIP), the type was set by the Government Printing Office, and the final printing was done at the Armed Forces Institute of Pathology. The American Registry of Pathology (ARP) purchased the Fascicles from the Government Printing Office and sold them virtually at cost. Over a period of 20 years, approximately 15,000 copies each of nearly 40 Fascicles were produced. The worldwide impact of these publications over the years has largely surpassed the original goal. They quickly became among the most influential publications on tumor pathology, primarily because of their overall high quality, but also because their low cost made them easily accessible the world over to pathologists and other students of oncology.

Upon completion of the first series, the National Academy of Sciences-National Research Council handed further pursuit of the project over to the newly created Universities Associated for Research and Education in Pathology (UAREP). The Second Series was started, generously supported by grants from the AFIP, the National Cancer Institute, and the American Cancer Society. Dr. Harlan I. Firminger became the editor-in-chief and was succeeded by Dr. William H. Hartmann. The Second Series Fascicles were produced as bound volumes instead of loose leaflets. They featured a more comprehensive coverage of the subjects, to the extent that the Fascicles could no longer be regarded as "atlases" but rather as monographs describing and illustrating in detail the tumors and tumor-like conditions of the various organs and systems.

Dr. Juan Rosai was appointed as editor-in-chief of the Third Series, and Dr. Leslie Sobin became associate editor. A distinguished Editorial Advisory Board was also convened, and these outstanding pathologists and educators played a major role in the success of this series, the first publication of which appeared in 1991 and the last (number 32) in 2003.

The same organizational framework applied to the Fourth Series, meticulously edited by Dr. Steven Silverberg with Dr. Ronald DeLellis as the associate editor. With UAREP and AFIP no longer functioning, ARP remained the responsible organization. The Fourth Series volumes were hardbound with illustrations almost exclusively in color. There was also an increased emphasis on the cytopathologic (intraoperative, exfoliative, or fine needle aspiration) and molecular features that are important in diagnosis and prognosis. At the time of the Fourth Series, ARP also produced

Atlases of Non-Tumor Pathology; these volumes were numbered separately from the tumor volumes.

As in the prior series, the goal of the Fifth Series includes a continuous attempt to correlate, whenever possible, the nomenclature used in the Fascicles with that proposed by the World Health Organization Classification of Tumors, as well as to ensure a consistency of style. Including molecular diagnostics is more important than ever, as is the availability of an online component, a more nimble website (www.arppress.org) and even a social media presence linked to the website. Now in this series, the tumor and non-tumor volumes are combined and consecutively numbered as the Atlases of Tumor and Non-Tumor Pathology. Close cooperation between the various authors and their respective liaisons from the Editorial Board will continue to be emphasized in order to minimize unnecessary repetition and discrepancies in the text and illustrations.

Particular thanks are due to the members of the Editorial Advisory Board, our reviewers, the editorial and production staff, and the individual Fascicle authors for their ongoing efforts to ensure that this series is a worthy successor to the previous four.

Jason L. Hornick, MD, PhD

PREFACE AND ACKNOWLEDGMENTS

We, pathologists, understand how much we can impact the lives of patients entrusted to our care. When we encounter neoplasms in our pathology practices, we often assume the burden of making life-altering decisions.

When we encounter ophthalmic tumors, there is an added burden: the disease or even the treatment of the disease may cause the patient to lose vision. Each of us, the authors of this book, is trained in both ophthalmology and pathology. As ophthalmologists, each of us has heard patients say, "Doctor, I would rather be dead than be blind!"

A book on the pathology of ophthalmic tumors must address not only how to make a correct diagnosis, but also how to guide the ophthalmic surgeon to a treatment plan for the disease while simultaneously preserving vision, if at all possible.

It has been a privilege for us to collaborate in bringing to you, our honored readers, an opportunity for us to be consultants as you study a case or come to terms with entities and concepts that are unique to ophthalmic pathology. We hope that this text will help you to collaborate with ophthalmic surgeons and oncologists to preserve life and sight.

Robert Folberg, MD
Patricia Chévez-Barrios, MD
Amy Y. Lin, MD
Tatyana Milman, MD

DEDICATIONS

Dedicated to the memory of Dr. Wallace H. Clark, Jr., who inspired me to love pathology, and dedicated in honor of my wife, Amy, and my children, Abby and Effy, who inspire me with love.

Robert Folberg, MD

In loving memory of an esteemed pathologist, my father, Dr. Agustín Chévez, who taught me how to use the microscope, to love research and to follow my dreams; and to another esteemed pathologist, my husband, Dr. Roberto Barrios, who has encouraged me with love, respect, and unconditional support. Dedicated to my daughter Gaby who inspires me with her wisdom, kindness, and love.

Patricia Chévez-Barrios, MD

Dedicated to my parents, Dr. Ann Lin (an ophthalmologist), and Dr. Ryan Lin (a pathologist), who inspired me to follow in their footsteps and forge my own path; to my husband, Nate Burgos, for his unconditional love and support; and to my mentor, Dr. Robert Folberg, for sharing his time, wisdom, and friendship.

Amy Y. Lin, MD

Dedicated to my mentor, colleague, and friend, Ralph C. Eagle, Jr., for his generosity with sharing his knowledge, inspiration, and unfailing support. Dedicated to my family—my husband, Alex, and my children, Mark and Samuel—without their love and support this work would not be possible.

Tatyana Milman, MD

Permission to use copyrighted illustrations has been granted by:

American Academy of Ophthalmology, Inc.
Jones LT. The lacrimal apparatus. In: Jones LT, Reeh MJ, Wirtschapter JD, et al. Ophthalmic anatomy: a manual with some clinical application. Rochester: American Academy of Ophthalmology and Otolaryngology; 1970:225. For figure 6-1.

Springer
AJCC Cancer Staging Manual, 8th ed. Chicago: Springer International Publishing; 2017. For appendices.

CONTENTS

1 PATHOLOGIC EXAMINATION OF OCULAR SPECIMENS

This chapter describes the most common approaches for the handling of ophthalmic pathology specimens. It is not intended to be a comprehensive document for grossing, but rather to highlight the essential techniques and features of each type of ophthalmic specimen removed for the diagnosis and treatment of neoplasia.

GLOBE

Enucleation

In an enucleation procedure, the eye (globe) is usually removed intact, with an attached segment of an optic nerve and small amounts of periocular soft tissue (fig. 1-1). In a procedure known as evisceration, the surgeon curettes the intraocular contents and removes corneoscleral tissue leaving the shell of the sclera behind. The evisceration technique is not appropriate in the context of a suspected neoplasm. Evisceration carries the risk of tumor spread into the orbital tissues when an intraocular neoplasm is present.

Enucleation is the preferred approach to remove an eye with a tumor. Fresh tumor retrieval techniques for genetics and molecular testing are described in other chapters. Most eyes come to the pathology laboratory fixed in 10 percent formalin. Fixation is recommended for at least 48 hours to obtain complete penetration by the fixative of the sclera and intraocular tissues, including the tumor. Injection of formalin or opening of the anterior chamber is discouraged because these procedures may alter the interpretation of the relationship between the tumor and the intraocular tissues and ocular coats. The recommended approach is to place the eye in at least five times its volume of formalin for adequate fixation.

The specimen is first identified as a right or left eye. The extraocular muscles include the four recti and the two oblique muscles. The recti muscles insert a few millimeters behind the limbus. The distance from the limbus to the muscle insertion progressively increases from the medial rectus muscle, to the inferior rectus muscle, to the lateral rectus muscle, to the superior rectus muscle (farthest from the limbus), forming an opening spiral (fig. 1-2). The differences between muscles are minute, however, making this approach impractical for confirming laterality of the eye (right or left eye). Also, various ocular pathologic conditions may distort the anatomic placement of rectus muscles relative to the corneoscleral limbus.

The most reliable way to confirm the laterality of the eye is to identify the oblique muscle insertions. The oblique muscles insert in precise locations, with the inferior oblique inserting directly into the temporal side of the optic nerve behind the location of the macula in the retina. The superior oblique inserts slightly temporal (lateral) to the center of the eye (12:00

Figure 1-1

ENUCLEATED EYE

An enucleated eye with an attached segment of the optic nerve (to the left). No significant amount of orbital tissue is present.

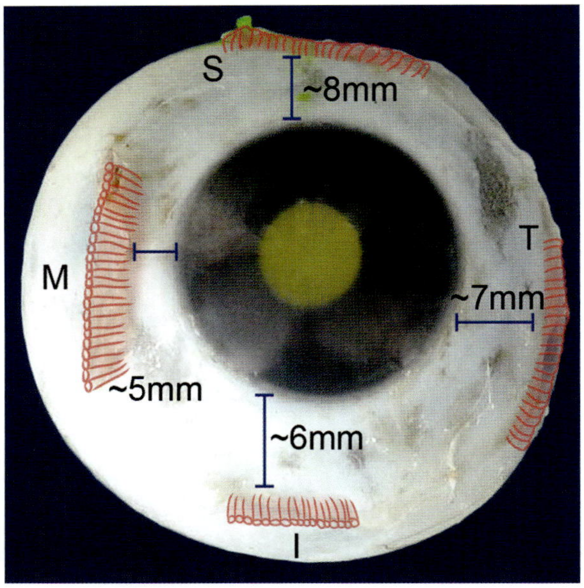

Figure 1-2

ENUCLEATED EYE

Frontal view of an enucleated eye with the distance of the superior (S), temporal/lateral (T), inferior (I), and medial/nasal (M) recti extraocular muscle insertions to the limbus. The insertion openings from medial to superior rectus muscle insertions are spiral shaped.

o'clock position), anterior to the equator, and the insertion is through a long tendon.

In many eyes, the track of the long ciliary nerves/vessels is seen as a light blue/gray line that runs horizontally, a few millimeters from the optic nerve to the equator. This feature is helpful in orienting the eye for the localization of the horizontal plane and finding the insertion of the inferior oblique muscle (fig. 1-3).

After orientation, it is recommended to place a mark on the sclera at the 12:00 o'clock position to maintain orientation during grossing. The eye is then measured: first, anterior-posterior (from the center of the cornea to the edge of the sclera meeting the optic nerve); then horizontal (at equator); and finally, vertical (at equator) preferably using measuring calipers. Also recorded are the horizontal and vertical measurements of the cornea from one edge of the limbus to the other, iris color, iris appearance, and status of the anterior chamber and pupil (fig. 1-4).

The length of the optic nerve is measured from the edge of the sclera to the tip of the cut margin while extending the nerve to correct the natural fixation shrinkage of the tissue (this is particularly useful in eyes harboring retinoblastoma). In some laboratories, the maximum width of the nerve near the scleral exit of the nerve is measured.

There are two approaches to taking cross sections of the optic nerve. In some laboratories, a cross-section is obtained 2 to 3 mm posterior to the insertion of the nerve into the sclera because tissue at the actual surgical margin may be crushed. In cases of retinoblastoma, a cross section at the 2- to 3-mm point that is free of a tumor means that the resection margin is also free.

Many laboratories take a cross section of the nerve near or at the surgical resection margin. Ink is applied to the optic nerve margin, and a transsection perpendicular to the long axis is performed to obtain the margin for histopathologic examination.

If a laboratory receives the optic nerve as a separate piece of tissue, the length and width should be measured. The surgical resection margin is then inked (fig. 1-5).

For eyes suspected of harboring uveal melanoma, some pathologists sample the vortex veins before opening the eye. These veins are believed to be a pathway for metastasis. The vortex veins exit the sclera posteriorly, and these structures can be cut with scissors and submitted individually in separate cassettes (fig. 1-6). Other laboratories do not submit this tissue for histologic sectioning because there is no evidence that involvement of the veins by uveal melanoma is associated independently with metastasis. Practices even vary across the laboratories of the contributors to this text.

Transmission of light through the eye may be helpful in locating an intraocular tumor before sectioning. In a darkened room, an intensive light source is focused onto the center of the cornea, and the eye is slowly rotated to find the edges of the shadow if present. A marker is used to trace the edge of the shadow on the sclera. The orientation of the sectioning of the eye is centered on the tumor shadow to obtain the central anterior-posterior section, the pupil-optic nerve (PO) section, through the tumor (fig. 1-7).

Using a sharp blade, the eye is opened starting from the back at less than 1 mm from the edge of the optic nerve into the sclera while firmly holding the eye with the fingers on top of

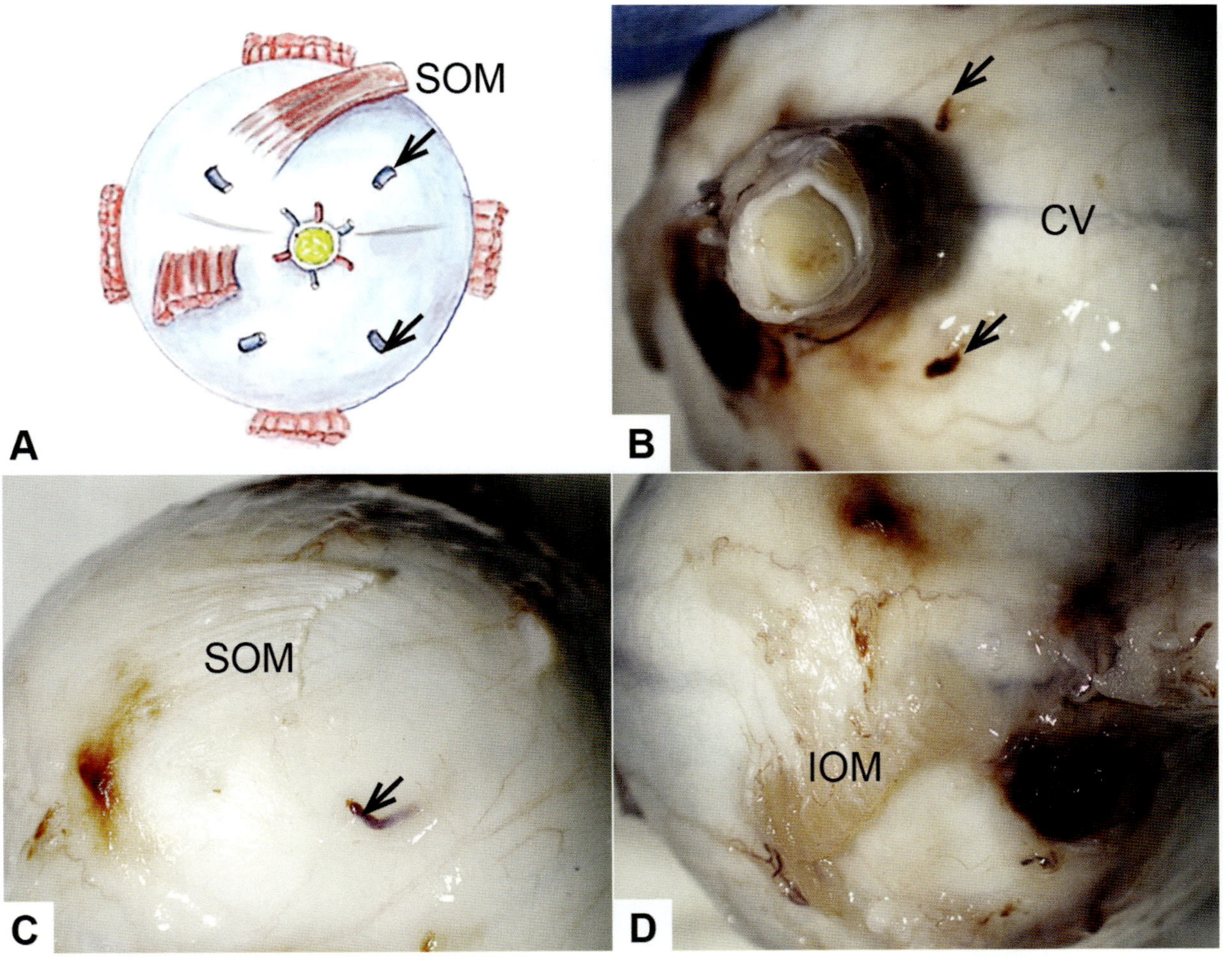

Figure 1-3

POSTERIOR VIEW OF THE EYE

A: Schematic representation of the back of a left eye. The tendinous insertion of the superior oblique muscle (SOM) superior and temporal to the optic nerve (in yellow). The four vortex veins (arrows) exit at the center of the quadrants and posterior to the equator. Slightly superior to the optic nerve in the sclera is the long ciliary vessels in the horizontal plane, seen as gray lines. The insertion of the inferior oblique muscle is seen. This muscle has a direct insertion into the sclera without a long tendon (seen in this illustration just left of the optic nerve.

B: Arrows point to the superior-nasal and inferior-nasal vortex veins in this enucleated eye.

C: The shiny white tendinous insertion of the SOM and the superior-nasal vortex vein (arrow) are seen.

D: In contrast with the SOM, the inferior oblique muscle (IOM) inserts directly into the sclera, without a tendon.

the equator on a nonslippery surface. The blade is advanced with a lateral zigzag motion toward the anterior segment, aiming to enter the anterior chamber at about 2-mm internal to the limbus.

When the blade reaches the anterior portion of the globe, after having crossed the equator, the blade is kept in place while the eye is rotated onto the corneal surface. Then, using both hands to hold the blade, it is possible to cut through the remainder of the eye with a swift guillotine-like movement toward the cornea, passing through the lens and cornea without dragging intraocular tissues (fig. 1-8).

The interior of the eye is examined after removing the first calotte (cap), and the tumor is described (size at the base in contact with the choroid or sclera and maximum height on the cut surface), noting any morphologic effects on the intraocular tissues. The eye is then turned down on the cut surface, and the same cutting method is used to remove the opposite calotte. The central PO nerve section is about 7.0 mm

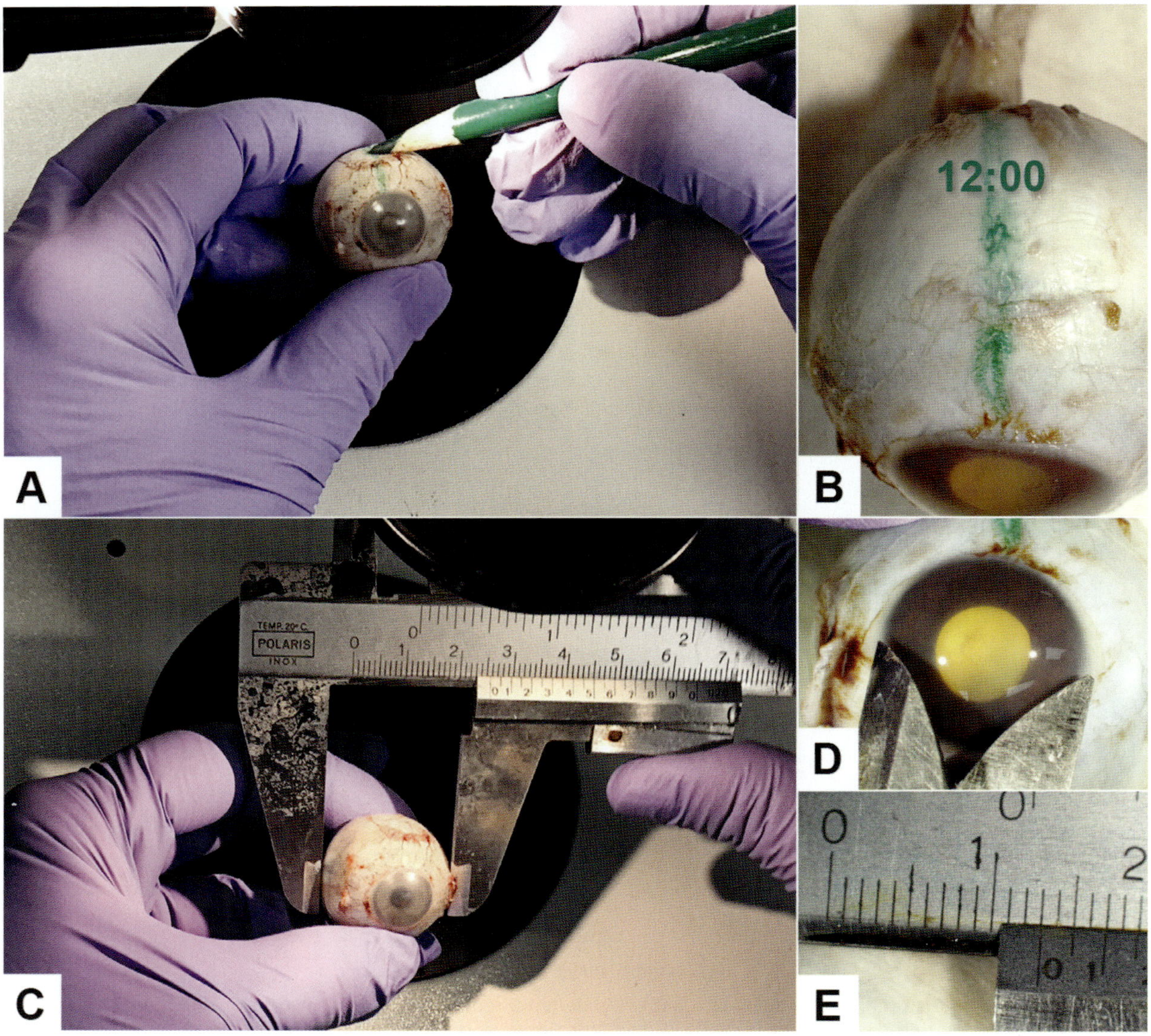

Figure 1-4

ORIENTATION OF THE EYE DURING GROSSING

A: After conferring laterality to the eye, the 12:00 o'clock position starting at the limbus is marked with a colored pencil or a similar implement.

B: The line extends to the back of the sclera until meeting the optic nerve.

C: Using a vernier, the eye is measured in the horizontal dimension (seen here) and also anterior posteriorly and vertically.

D: The diameter of the cornea is measured with the vernier from one aspect of the limbus to the other.

E: The measuring scale shows 13 mm for the measurement of this cornea in the horizontal plane.

in thickness at the center, wider in the cornea (7.5 to 8.0 mm) and narrower by the optic nerve (about 5.5 mm). Findings on the two calottes are reported after separating them.

The eye can be submitted entirely in three cassettes. The calottes may be further sectioned in anterior-posterior segments to increase the surface of tumor and intraocular structures studied in the histologic sections (see figs. 1-8 and 9-20). It may be helpful to obtain photographs of the cut surface of the eye (1,2).

Alternatively, for suspected uveal melanomas, the eye may be opened in a plane that permits the pathologist to visualize the tumor from the same perspective as the ophthalmologist. In the case of uveal melanoma in the posterior pole,

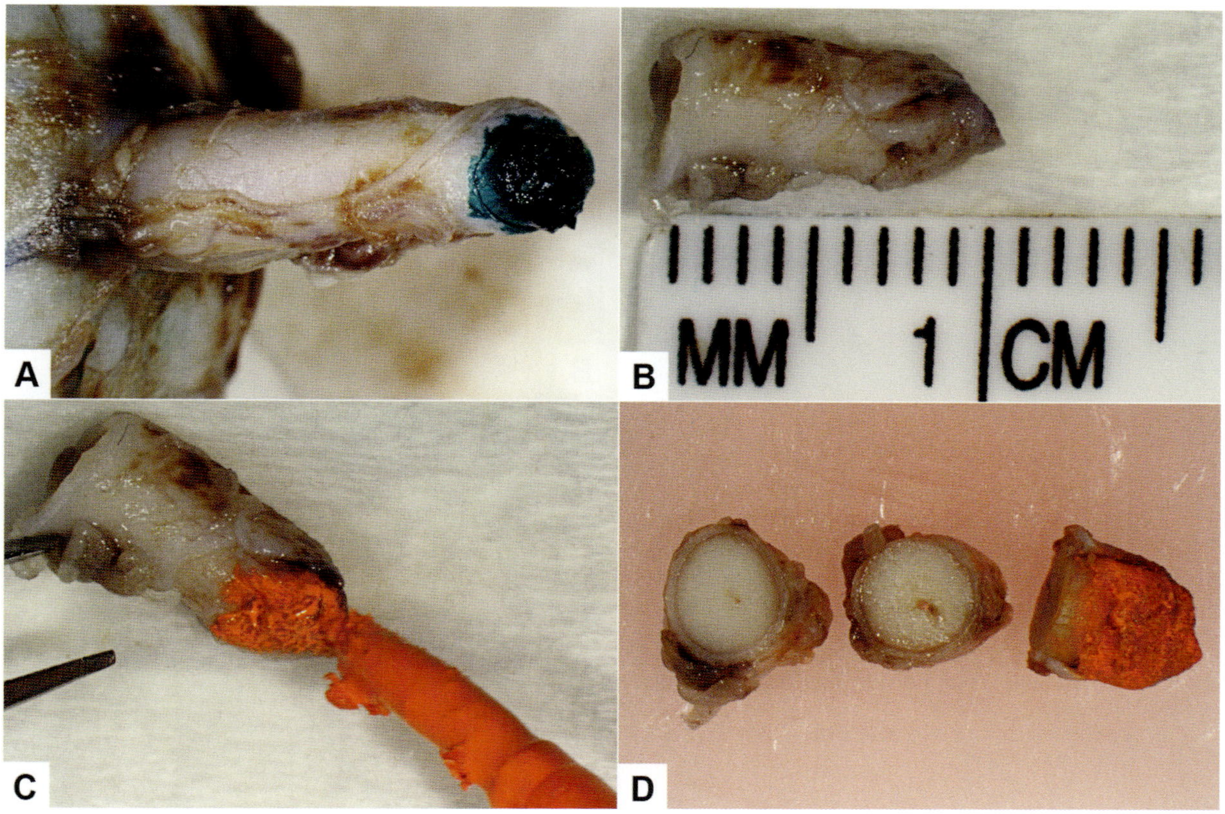

Figure 1-5

OPTIC NERVE MARGIN

A: The optic nerve margin is inked and sectioned to be submitted separately.
B: A separately received optic nerve in an enucleation for retinoblastoma is measured to the tip of the margin.
C: Ink is applied to the real margin (note the slant rough cut surface).
D: The nerve is serially sectioned and submitted for examination in one cassette.

it is possible to take a coronal section (parallel to the surface of the cornea) through the pars plana, revealing the tumor in the posterior pole, just as the ophthalmologist visualized the tumor. This maneuver permits the pathologist to take sections through the tumor to address questions that the ophthalmologist may have about the clinical appearance (for example, to sample a particular margin precisely). If an alternative protocol is used for tumors anterior to the equator of the eye, then the eye is opened in a plane that is parallel to the apex of the tumor, allowing for the generation of precise section planes through areas of interest by direct tumor visualization (3).

Each paraffin block is sectioned to obtain representative sections of all the intraocular structures of importance for providing prognostic information from the histopathologic review. For example, in retinoblastoma, it is crucial that the PO nerve section demonstrates the optic nerve in its entirety, with prelaminar, laminar, post-laminar components. For retinoblastoma, experts recommend trisecting the calottes (caps) and submitting them for histopathologic review (see chapter 9).

Although the hematoxylin-eosin (H&E) stain is used conventionally for uveal melanoma and retinoblastoma, most laboratories also generate slides stained by periodic acid–Schiff (PAS) because this stain reveals pathology in the various basement membranes of the ocular interior. Most experts also favor the bleaching of melanin with H&E counterstaining when studying uveal melanomas and other highly pigmented intraocular tumors.

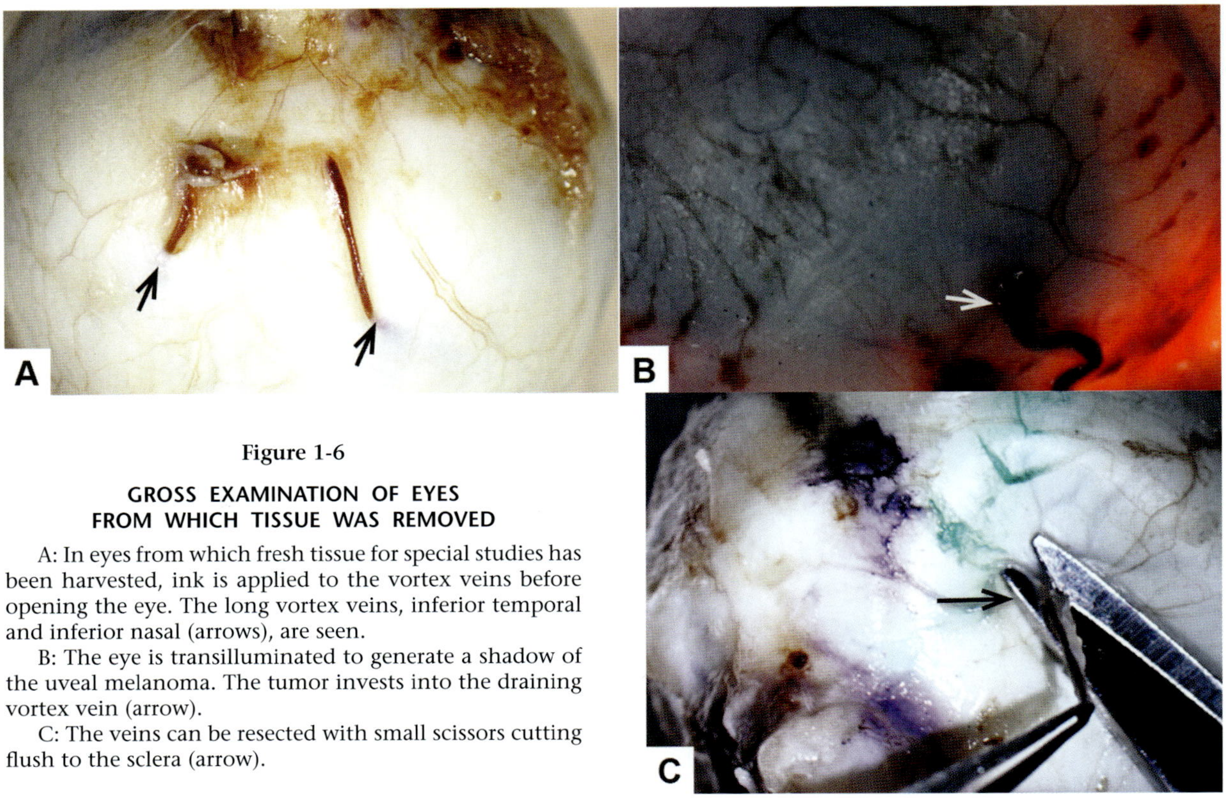

Figure 1-6

GROSS EXAMINATION OF EYES FROM WHICH TISSUE WAS REMOVED

A: In eyes from which fresh tissue for special studies has been harvested, ink is applied to the vortex veins before opening the eye. The long vortex veins, inferior temporal and inferior nasal (arrows), are seen.

B: The eye is transilluminated to generate a shadow of the uveal melanoma. The tumor invests into the draining vortex vein (arrow).

C: The veins can be resected with small scissors cutting flush to the sclera (arrow).

Segmental Excision

Segmental excisions of iris or iridociliary tumors are seldom performed. Orientation by the surgeon for laterality and identification of the clock position occupied by the lesion contribute significantly to a thorough evaluation of margins.

It is helpful for surgeons to submit iridectomy specimens on a piece of thick paper or other flat surface material (a piece of wooden tongue depressor facilitates the adhesion of the fresh tissue to the wood before placing into formalin) that preserves the shape of the specimen during fixation for adequate orientation and sectioning. The use of a stereoscopic microscope is encouraged when grossing these specimens, since the lesions may be less than 5 mm in maximum diameter and the localization and orientation of the margins may be challenging without the magnification. Handling of these tiny and flimsy specimens can benefit from the use of gel-tissue embedding before processing at the time of the grossing. This protocol may facilitate the maintenance of specimen orientation (see Pretissue Processing Embedding Technique below; fig. 1-15).

Iridectomy Specimen. Excision of iris tumors usually is accompanied by a small amount of iris surrounding the lesion and the pupil margin. The lesion may be inked for the peripheral margins (toward the sclera, and either inferior and superior, temporal and nasal, or preferably called by the clock hour to identify the margin). The pupillary margin and the deep and superficial margins are not attached to any structure. Thus, these are not margins unless there are synechias to the lens or the cornea. In these cases, the surface may be inked and referred to as cornea or lens surface.

After inking, the specimen is serially sectioned using a sharp thin blade (razor blades are better for small specimens). They are then embedded on edge (in pretissue processing gel if available), preferably no more than two pieces in each cassette.

Iridocyclectomy Specimens. Tumors in the ciliary body or iris and ciliary body are sometimes excised "en block" with small amounts of adjacent iris, ciliary body pars plana, and sclera. In these excisions, similar to the iridectomies, it is

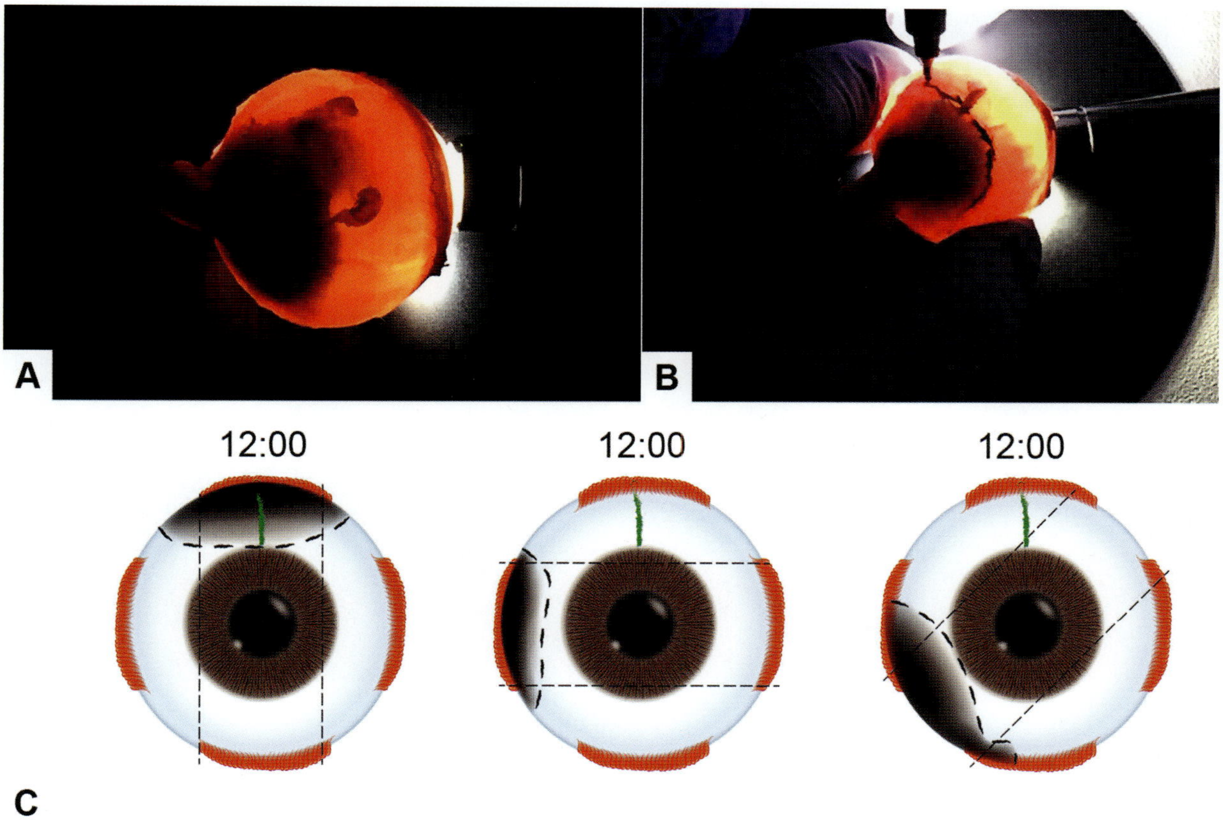

Figure 1-7

TRANSILLUMINATION OF THE EYE

A: In a darkened room, an intense light is focused onto the cornea, rotating the globe to show the shadow and its limits.

B: A marker is used to delineate the shadow.

C: After the shadow is delineated, a decision of how to cut open the eye is made based on the location of the tumor. The goal is to create a central pupil-optic nerve section to have all essential structures in one section, including the tumor. The diagram shows the possibilities depending on the location of the tumor. The globe on the left has a tumor in the superior quadrant, and a vertical section is selected. The globe in the middle has tumor in the medial or nasal quadrant, and the best section is horizontal. If the tumor is in the middle of the quadrants, then the section should be oblique, as seen in the globe on the right.

helpful to know the clock position of the excision to orient the margins. The posterior margin in this excision is the margin with the remainder of the ciliary body, usually the pars plana. The pathologist should ink the scleral margin in an iridocyclectomy specimen because the scleral margin is a critical landmark. After the specimen is inked, it is serially sectioned into pieces (as in a pie) with each section taken with a pupillary margin and a scleral margin (fig. 1-9).

Biopsy

Rarely, ophthalmologists perform biopsies of the interior of the eye: intraocular biopsies. Retina and chorioretinal biopsies are useful in rare tumors that do not shed cells to the vitreous. A transvitreal or transscleral approach may be used to obtain the biopsies. The transvitreal approach is preferred because the surgeon can visualize the lesion. This technique requires specialized instruments (laser, special scissors and forceps, and a vitrectomy instrument) which are used to minimize trauma to the retina and vitreous, and avoid bleeding. The resulting biopsy is delicate and small. Ideally, it is placed on filter paper to maintain orientation and flat architecture. Some of these biopsies, however, are received in fluid because they are suctioned through the vitrectomy instrument, to avoid further retinal damage or bleeding.

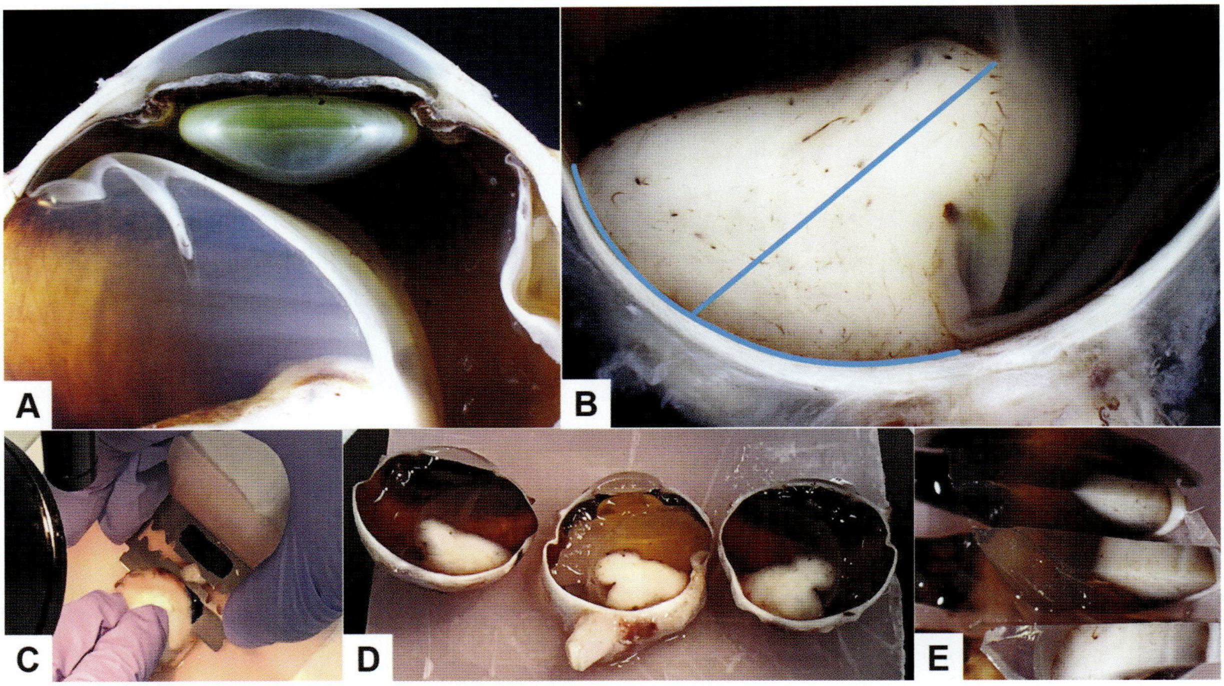

Figure 1-8

INTERIOR OF THE EYE

A: Cutting the first calotte allows the pathologist to visualize the interior of the eye. At the top, the cornea, anterior chamber, and lens are without pathologic alteration. The pathology appears in the posterior portion of the eye with a detached retina (white dome-shaped membrane).

B: The detached retina overlies the amelanotic mushroom-shaped uveal melanoma. The tumor is measured at the base in contact with the sclera as a tangent to the scleral curve and the tumor's maximum height at the cut surface is also recorded.

C: The second calotte is obtained by a section parallel to the first section, starting in the sclera adjacent to the optic nerve. The two fingers of the hand not holding the blade (inferior left) are firmly securing the eye on the equator and away from the track of the blade.

D: After obtaining the two calottes (left and right), a central pupil optic nerve section is obtained (at center) containing the tumor.

E: The calottes are further sectioned in a bread-loaf manner to achieve larger surface of tumor and ocular tissue for histopathologic evaluation. Here, segments were generated from one calotte.

Ophthalmologists who typically obtain biopsies from the choroid through the sclera commonly place tissue in formalin. If the clinical diagnosis points to intraocular lymphoma, then the tissue sample may be submitted in RPMI medium.

In any of these scenarios, the small tissue sample may be grossed under a dissecting microscope. It may be beneficial to consider using gel embedding to facilitate orientation (choroid or retina).

Fine-Needle Aspiration Biopsy. Most diagnoses of intraocular tumors depend on clinical and imaging findings. The reasons to avoid biopsies include the risk of complications of the biopsy techniques that may compromise visual acuity when the naturally clear media of the eye (cornea, anterior chamber fluid, lens, and vitreous) become cloudy with hemorrhage, scarring of the intraocular tissues, or retinal detachment.

In retinoblastoma, the possibility of seeding tumor during the biopsy procedure and converting an intraocular tumor into an extraocular and possibly metastatic tumor is one of the main reasons to avoid taking biopsies. In uveal melanoma, the most common indication for intraocular fine-needle aspiration biopsy (FNAB) is to obtain tumor cells for molecular prognostic testing. Some ocular oncologists only obtain the sample and send the entire specimen for molecular testing; others submit some sample to cytology to confirm the diagnosis, and most importantly, adequacy of the sample. Few practices

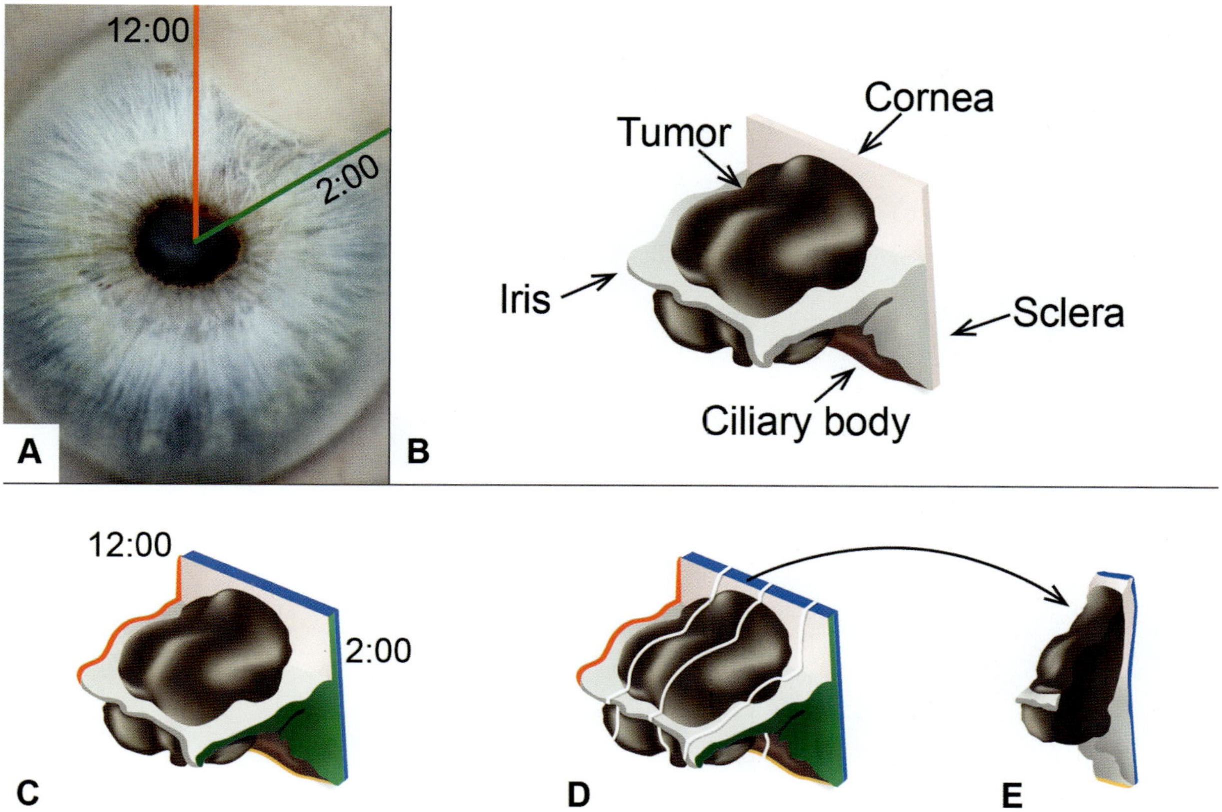

Figure 1-9

IRIDOCYCLECTOMY SPECIMENS

A: The tumor at the root of the iris extends into the iris at the 12:00 to 2:00 o'clock position.

B: Schematic representation of a segmental excision (iridocyclectomy) of this lesion includes a small amount of cornea, sclera, ciliary body, iris, and tumor.

C: The margins are inked with different colors. Apart from the corneoscleral margin and the posterior pars plana margin (yellow), the lateral margins are inked and represent the hour of the clock that limits the lesion (12:00 in red, 2:00 in green).

D: The specimen is then sectioned anterior posteriorly to obtain the margins and the center of the specimen. The sections contain all the essential anatomic structures and the margins.

E: The section generated by the step illustrated in "D".

include an on-site cytologic evaluation of the specimen at the time of the biopsy for adequacy of tumor sampling for molecular testing.

There are two approaches to FNAB in uveal melanomas: transvitreal or transscleral. Transvitreal biopsies require a surgical microscope and vitreoretinal instrument to visualize the tumor and minimize complications. Although some surgeons use a 25-gauge vitrectomy instrument to obtain a core of tissue, this technique increases the likelihood of bleeding.

Most surgeons use a 25- to 30-gauge needle. The needle is connected to long tubing and suction is created by a syringe in a biopsy gun. Usually, one or two passes are done (fig. 1-10). The transscleral approach requires of transillumination of the eye to localize the tumor, and the needle is introduced through the sclera after the conjunctiva is dissected (fig. 1-10). If adequacy evaluation is required, a small drop of the sample is shared on a slide to create a smear for Diff-Quick staining and on-site cytologic evaluation. The needle is rinsed after collecting samples for molecular testing and fixed for cytologic evaluation, which may include a cell block (fig. 1-10). Samples are usually scant but diagnostic, since the normal melanocytes lack atypical prominent nucleoli, abundant cytoplasm, epithelioid appearance, and discohesion, features seen in melanoma (fig. 1-10)

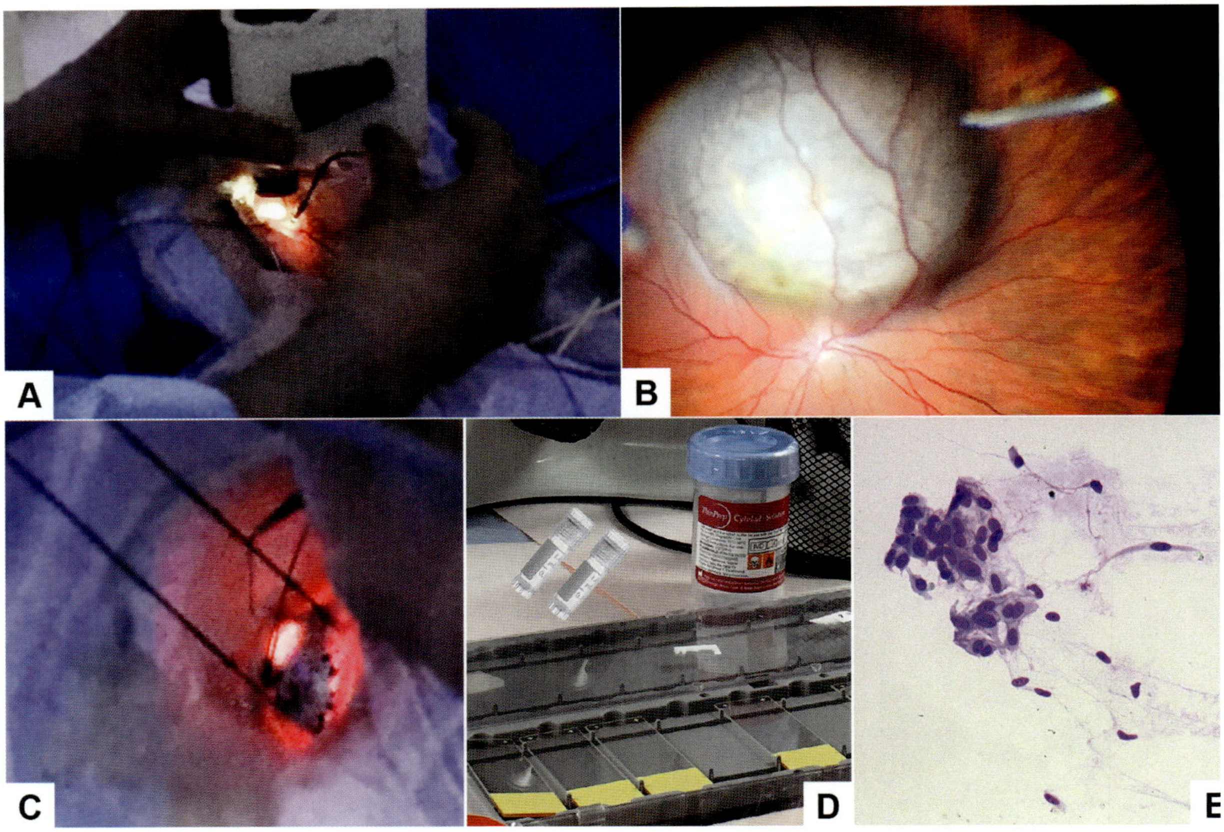

Figure 1-10

FINE-NEEDLE ASPIRATION BIOPSY (FNAB) OF INTRAOCULAR TUMORS

A: Transvitreal biopsy performed using an operating microscope allows visualization of the intraocular structures.

B: A 25-gauge needle (left) is entering the partially amelanotic choroidal tumor and seen through the operating microscope.

C: In a transscleral approach to the choroidal tumors, transillumination is used to localize the tumor, as seen here where the tumor cast a shadow. Traction sutures (to the muscles) are used to position and handle the eye during the procedure.

D: In the rapid on-site evaluation of the FNAB of uveal melanoma, the biopsy is evaluated by creating a slide with one drop of the needle and allocating the rest of the specimen in cryovials for molecular analysis and in cytology fixative for possible cell block and liquid-based cytology from the needle rinses.

E: Groups of spindle melanoma cells are seen in this Diff-Quick-stained smear prepared during an adequacy check to confirm the presence of tumor in the sample.

(4,5). Vitreous biopsies or vitrectomy specimens are used to diagnose intraocular vitreoretinal lymphomas (see chapter 11).

EYELIDS

Full-Thickness Wedge Resection

Segmental resections of eyelids for tumor are usually full-thickness wedge (pentagonal) resections that include all layers. Knowing the precise origin of a biopsy sample assists the pathologist in orienting the tissue to evaluate the margins. For example, the location of the eyelashes (eyelid margin-conjunctival/epidermal junction) assists in orienting the tissue because, in lower eyelids, the eyelid margin is on the superior area of the specimen while in the upper eyelids it is located inferiorly.

After the pathologist has determined the laterality (right or left) and the origin of the specimen (upper or lower eyelid), the temporal (lateral) and nasal (medial) margins are assessed. In lower eyelid wedge biopsies, the inferior margin is assessed, and in upper eyelid biopsies, the superior margin is assessed. There is no "deep margin," because in a full-thickness eyelid resection, the surface of the specimen opposite the cutaneous surface of the eyelid is conjunctival mucosa. The

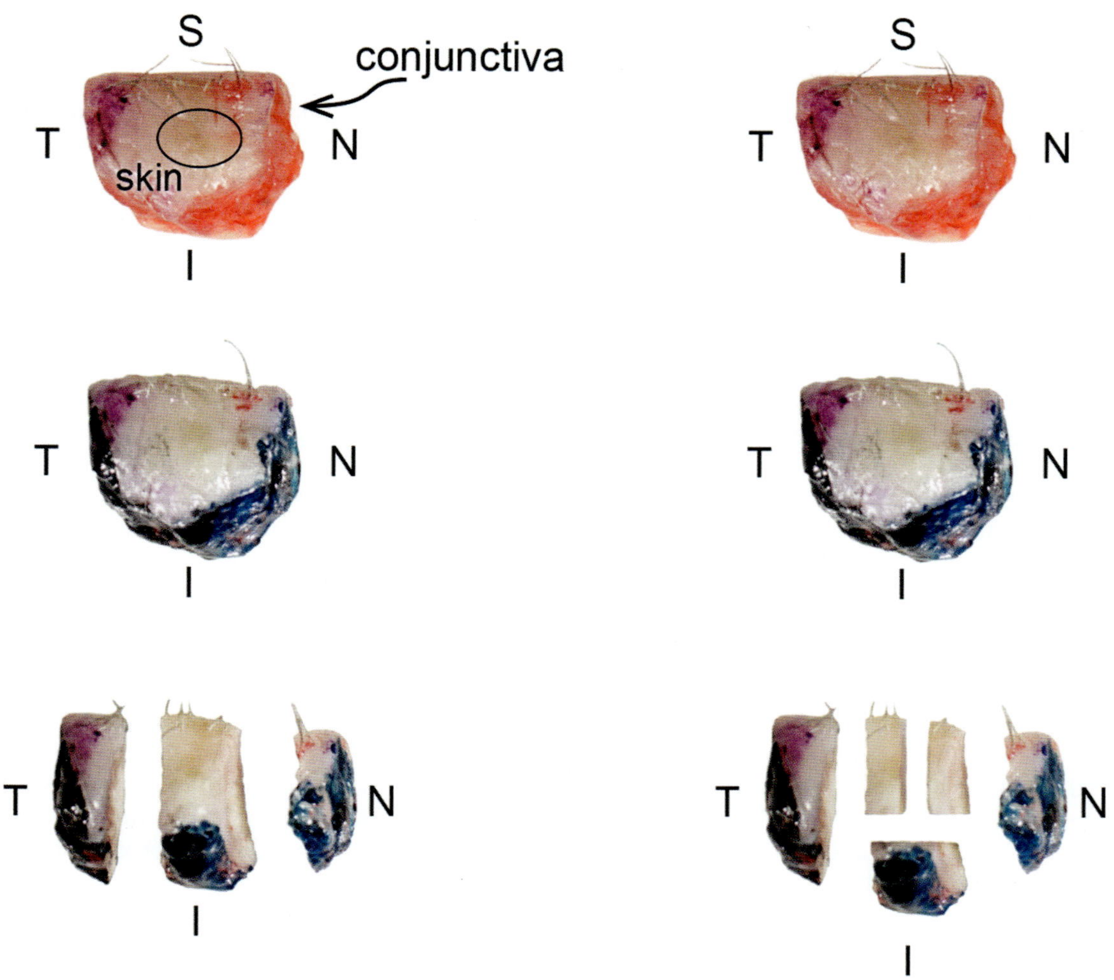

Figure 1-11

WEDGE EXCISION OF AN EYELID

Left: It is crucial for the surgeon to provide clinical information about laterality (right eye or left eye) and whether the specimen originates from a superior or inferior eyelid. The eyelashes are at the free border of the eyelid, and in this right lower eyelid, the temporal/lateral margin is on the left side. The anterior surface is the skin and the posterior surface the tarsal conjunctiva. The margins are inked as temporal and nasal (or medial), extending the color to the center of the inferior margin. The specimen is then trisected to obtain a temporal (T) section, including the temporal margin, a central section, including the inferior margin (I), and the nasal/medial margin (M), including the medial margin.

Right: Margins are taken separately from the central portion of the resected tissue that contains the tumor.

conjunctiva lining the undersurface of the eyelid is known as the palpebral conjunctiva.

There are two main dissecting approaches for full-thickness wedge resections (fig. 1-11) to examine the three margins (temporal, nasal, and either superior or inferior) while preserving the anatomic structures in the sections. In both approaches, the surgical margins are inked, temporal half one color and nasal half a different color, extending the colors into the center of the superior or inferior margin. If the specimen is large enough, ink is applied to the inferior or superior margin with a third color.

In small specimens, the eyelid is sectioned in three segments: nasal, central, temporal (fig. 1-11 left). In larger specimens, there is the option of obtaining the nasal and temporal margins and then obtaining the inferior margin separate from the central area where the tumor is usually present. The central portion may be divided into two pieces (fig. 1-11 right). This approach is usually preferred for frozen sections to freeze only the three margins. It is customary to submit the central portion with the tumor for permanent sections, embedding all sections on edge (1,6). Some ophthalmic surgeons take sentinel node biopsies, and the technique is similar to that used in other sites (7).

Partial-Thickness (Lamellar) Wedge Resection

Partial-thickness (lamellar) wedge resection includes the tarsus with conjunctiva, or rarely, the skin with muscle. In these cases, it is essential to ink the margin either anterior/superficial (surface of the tarsus/muscle) or deep (muscle in contact with tarsus) in addition to the superior or inferior margin and the temporal and nasal margins. These types of excisions are performed to preserve the function of the eyelid or for well-localized lesions in the conjunctiva or skin. They usually include the eyelid margin.

Biopsy

An eyelid biopsy is obtained from the palpebral (also known as the tarsal) conjunctiva or the skin, and sometimes from the eyelid margin. Skin lesions may be biopsied similarly to skin biopsies elsewhere, through punch biopsy, shave biopsy, or excisional biopsy. These are oriented and handled in the same manner as those elsewhere. Biopsies from the eyelid margin should be embedded on edge to obtain sections demonstrating the skin, tarsus, and conjunctiva. Biopsies from the tarsal conjunctiva also should be included on one edge for orientation of the epithelium and the tarsus.

CONJUNCTIVA AND CARUNCLE

Conventional Biopsies

An excellent gross evaluation of a conjunctival biopsy begins with the surgeon. Conjunctival tissue is very floppy, somewhat like plastic wrap used in a kitchen. Unless precautions are taken, conjunctival tissue will fold during fixation. Folds in the tissue can distort margins and generate tangential sections through lesions that may require precise measurement of thickness, such as conjunctival melanomas. Therefore, pathologists must play an active role in educating the surgeon on how to handle conjunctival biopsies.

The surgeon must take steps to ensure that the conjunctival mucosal biopsy remains flat before and during fixation. Frequently, ophthalmologists take these biopsies in outpatient settings, even in their offices. Placing the conjunctival biopsy on filter paper for about 10 seconds before immersion into formalin is an effective method. If filter paper is not available, any absorbable surface will suffice, and we have seen ophthalmologists use fragments of their business cards effectively for this purpose.

Pathologists should understand that the treatment of conjunctival malignancies requires a minimal loss of tissue. Even in cases of conjunctival squamous cell carcinoma and melanoma, ophthalmic surgeons do not take wide margins of the type used for cutaneous malignancies. It is essential for surgeons to conserve conjunctival tissue in order to maintain the integrity of the ocular surface. This is an essential objective for minimizing the likelihood of dry eye (xerosis) after the removal of a malignancy. Xerosis can be painful and lead to visual loss. Ophthalmic surgeons understand that surgical margins are not extensive in conjunctival biopsies, and cryotherapy or topical chemotherapy may be delivered to the affected site to augment surgical extirpation.

Some ophthalmologists mark the surgical margins with dye to provide orientation (for example, the margin closest to the limbus). They may use markers that are water soluble, such as methylene blue, and when they place the tiny fragment into formalin, the blue dye spreads across the entire specimen, and the ability to identify a critical margin is lost.

Some ophthalmologists mark margins with sutures. The suture must be thin, because the passage of a thick suture through the flimsy mucosal tissue disrupts the microanatomy and pathology. Drawings of the biopsy sample in the context of the ocular surface may be an effective, nondisruptive method of indicating margins.

Many lesions of the conjunctiva arise at the limbus. The sections should be taken perpendicular to the limbus to obtain histologic sections with one end at the cornea and the opposite end the conjunctiva toward the fornix or caruncle to evaluate invasion into the episclera/sclera, cornea, or conjunctival margins (fig. 1-12).

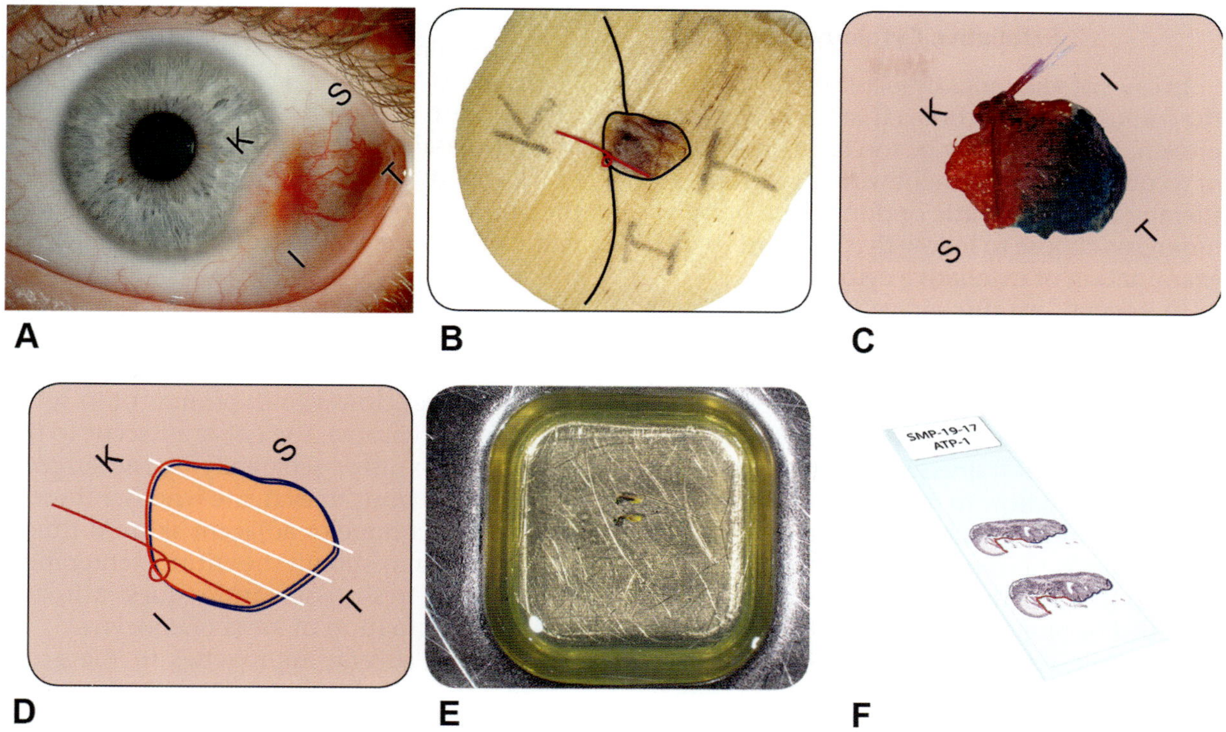

Figure 1-12

CONJUNCTIVAL EXCISIONS

A: The conjunctival tumor involves the cornea (K) at the temporal side of the left eye. Therefore, the superior (S), inferior (I), and temporal (T) margins must be sampled.

B: The excised conjunctival tumor is on a piece of tongue depressor with an orienting drawing and a suture placed in the inferior corneal tip. The specimen remains flat and in position after fixation using this approach.

C: The specimen is turned with the conjunctival surface down to ink the deep and peripheral margins: the temporal half is (T) blue and the corneal half (K) red.

D: After inking, the specimen is turned back to the original orientation with the deep margin down, and serial sections are obtained from the cornea to the temporal margin to obtain a section of the superior margin, a section of the inferior margin, and the central sections, each with corneal and temporal margins.

E: Each of the superior and inferior margins is in a separate cassette, and the central section may be submitted two in one cassette. Pretissue processing gel embedding was performed to orient the two central sections on edge.

F: Diagram of expected sections in the glass slide, with margins inked (temporal, corneal, and deep).

We cannot overemphasize the importance of the pathologist as an educator. Surgeons must be taught in an ongoing bidirectional dialog about the best methods for preparing conjunctival tissue samples for processing and interpretation. Frozen sections of the conjunctival lesion should be discouraged, if possible. Permanent sections are more precise and informative in these limited specimens (6,7).

Conjunctival "Map" Biopsies

The term conjunctival map biopsy describes the taking of multiple small incisional biopsies to determine the extent of ocular surface involvement by the neoplasm. Intraepithelial squamous neoplasia, conjunctival involvement by sebaceous carcinoma, and what is described clinically as primary acquired melanosis (a conjunctival melanoma precursor) may all affect the conjunctival surface diffusely.

In these clinical settings, the ophthalmologist takes multiple small biopsies, most commonly accompanied by a sketch of the surface of the eye. The site of each biopsy is matched to a number or letter indicating the label on the jar (for example, "1" or "A"). In these cases, it is not necessary for the pathologist to comment on the adequacy of the margins. The pathologist's goal for each biopsy is to identify the presence or absence of pathology.

Exfoliative Cytology

In cases of squamous dysplasia or carcinoma of the conjunctiva and cornea that has been biopsy proven, the lesion may be too large, multifocal, or recurrent, and treatment with topical chemotherapy is a therapeutic option. Some ophthalmologists choose to follow these patients by the examination of exfoliative cytology.

In general, we discourage ophthalmologists from using exfoliative cytology to follow patients with ocular surface neoplasms by the aphorism, "just because it is possible to do something does not mean that one should do it." Ophthalmologists are not well trained in how to obtain such specimens and how to fix the samples in the office. Furthermore, to obtain exfoliative cytology, topical anesthesia must be applied to the eye, and if this is the case, then more information can be obtained by biopsy than by exfoliative cytology. For example, it is easier to obtain immunohistochemistry on biopsy tissue than on exfoliative cytology specimens. The use of exfoliative cytology in the management of ocular surface neoplasia is yet another area of pathology practice where we believe that pathologists must educate their referring ophthalmologists on the best ways to obtain high-quality material so that the pathologist can render a timely and accurate diagnosis.

ORBIT AND OPTIC NERVE

Exenteration

When malignant orbital tumors infiltrate the soft tissues and possibly orbital bones, the ophthalmologist may elect to remove the contents of the orbital socket. The tissue removed includes the eye and all orbital soft tissue, with resection of adjacent bone if indicated. In many or most cases, the eyelids are removed with the orbital content, although, as described below, the procedure may be modified to spare eyelid tissue.

Exenteration is a surgical procedure for any malignancy that invades the soft tissues of the orbit, including squamous or basal cell carcinoma of the eyelid and malignancies of the conjunctiva. Other tumors of the orbit often necessitating exenteration are those from the lacrimal gland (e.g., adenoid cystic carcinoma), optic nerve (e.g., infiltrating meningiomas), or intraocular tumors that extend through the sclera and into the orbit (e.g., uveal melanoma, an extraocular extension of retinoblastoma).

The most common indications for radical exenteration (including orbital bones) include invasive basal cell or squamous cell carcinomas of the eyelids or malignancies originating from the sinuses into the orbit. Some cases of adenoid cystic carcinoma of the lacrimal gland require this surgery.

It is common practice for surgeons to submit tissue for a frozen section during an orbital exenteration. The margin that is most likely to be positive (closest to the tumor or in squamous cell carcinoma, the nerve adjacent to the tumor) is submitted separately by the surgeon for frozen sections to evaluate whether there is need to resect bone or adjacent skin. At times, the surgeon may submit the entire exenteration specimen as fresh tissue and request a frozen section examination of areas of concern.

There are several approaches to the gross examination of orbital exenteration specimens, and each approach begins with the proper orientation of the specimen. The orbit has a cuboidal shape with six walls. The eyelids and associated skin represent the anterior surface. The other surfaces include the nasal (the side where the eyelids have the lacrimal puncta, the conjunctiva, and caruncle), superior, temporal (the lacrimal gland is at the junction of the superior and temporal walls), inferior, and posterior (optic nerve cut margin) surfaces (fig. 1-13A).

The specimen should be fixed in formalin for about 3 to 5 days to allow the formalin to penetrate the tumor and intraocular tissues. The specimen is measured from front to back, from side to side, and from top to bottom. Areas of the eye are described and measured (globe, cornea, and segment of the optic nerve) and the dimensions and location of tumor tissue measured and identified in the report.

The soft tissue margins are inked, one color for each (each wall of the cube) margin (fig. 1-13B–D). The ink should extend to the anterior surface for the inferior and superior margins if the skin is absent from the eyelids (fig. 1-13B,D). The posterior margin includes the optic nerve margin, and the same color ink is used to designate the posterior soft tissue (fig. 1-13C).

After the margins are inked, a vertical section is taken through the entire specimen with the

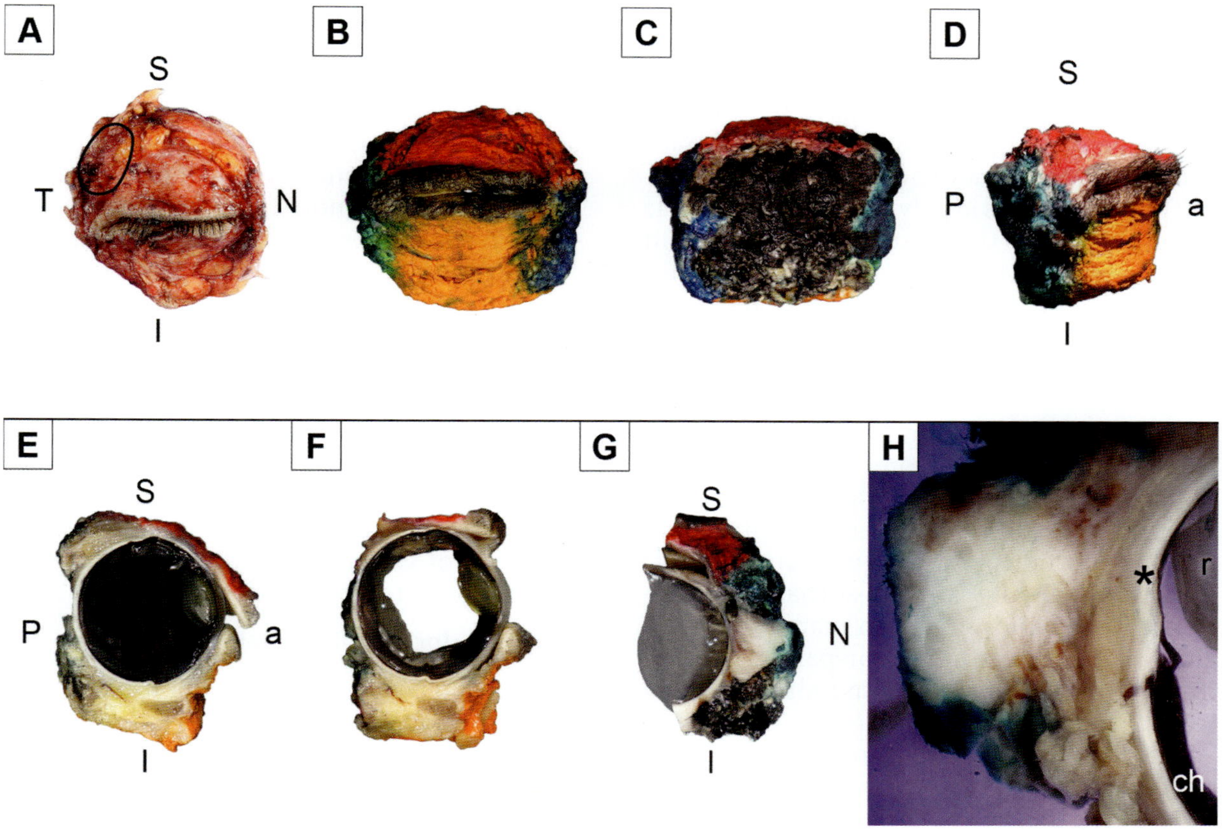

Figure 1-13

EXENTERATION SPECIMEN

A: Right exenteration specimen with lid skin sparing approach received fresh. Eyelid skin is absent. The exenteration contains a superior temporal tumor (circle) from the lacrimal gland.

B: View of the specimen after fixation and inking of the margins: temporal-green; superior-red; nasal-blue; inferior-yellow. The red and orange inks extend to the edge of the eyelid margin.

C: View of the posterior portion of the specimen with the posterior margin inked in black.

D: The orbit has a cuboidal shape in this lateral view.

E: View of the cut section of the specimen after the first vertical section is made to remove the temporal calotte. The upper eyelid is on top, and the soft tissues of the orbit appear in relationship with the eye.

F: View of the central pupil-optic nerve section after removing the nasal calotte. This section can be submitted as a whole mount.

G: Temporal calotte after bread-loaf sections. Green ink surrounds the superior temporal tumor (white lesion) in contact with the sclera of the eye.

H: Temporal calotte segment under the stereoscopic microscope reveals the adhesion of the tumor to the sclera (*). The choroid (ch) and the retina (r) are tumor free.

specimen positioned on the cutting board on the anterior surface (eyelids in contact with the cutting board surface). In this way, it is possible to obtain a section through the PO, with the surrounding orbital tissue and both eyelids in the central section. In order to generate this section, the optic nerve is first identified, so the sectioning starts at the edge of the nerve, with an anterior-downward movement to enter the edge of the cornea. This maneuver results in a section that shows the interior of the eye, the coats of the eye, and the relationships of these with the orbital and eyelid tissues (fig. 1-13E).

After describing the findings, especially if the tumor infiltrates the eye, the other side (calotte) of the eye is removed. After placing the specimen on the previous cut surface, the other side of the optic nerve is cut parallel to the cutting board with an anterior movement. Shortly after penetrating the globe, the specimen is turned down on the anterior surface to finish the cut through the cornea and eyelids, obtaining a central PO section (fig. 1-13 F). The temporal and nasal calottes are then cut in an anterior-posterior direction in segments (bread loaf) to obtain sections that have the eye and the orbital tissues, maximizing the surface for histopathologic examination (fig. 1-13G,H).

The PO section is ideally submitted as a whole mount (similar to the prostate whole mount) in one large cassette or wrapped in a porous bag for long processing (including longer times in formalin, alcohols, and xylenes). The calottes are also ideally submitted in these large cassettes, so the specimen is not further sectioned. However, if this processing is not available, the large sections may be further segmented, maintaining the anatomic relationships and submitted in multiple sections (e.g., anterior, central, and posterior portions of the segments). The most important sections are those that show the tumor in a relationship with margins and the globe. Large whole mount preparations allow the entire specimen to be submitted in only three large cassettes.

An alternative technique may be useful if the indication for exenteration is not orbital involvement by the transscleral spread of an intraocular tumor. The orbital tissues are dissected away from the eyeball by making horizontal sections through orbital tissue until the eye is encountered, and then dissociating the eye from the conjunctiva. The lab then processes the eye as described above. The margins of soft tissues of the superior and inferior aspects of the orbit can be painted with various colors of ink to identify the location within the orbit. This alternative technique generates more tissue sections than the conventional method described above and does not generate an aesthetic section containing the eyeball, eyelids, and orbital soft tissue. On the other hand, this technique eliminates the need for large cassettes and blocks, and provides for accurate mapping of the tissues involved. Both enucleation and exenteration specimens require experienced, specialized histotechnologists for adequate embedding and sectioning.

Orbital Excisional Biopsies

The excision of an orbital lesion serves as both a diagnostic and therapeutic maneuver. The indications are usually masses that cause proptosis or intraocular changes. In children, the most common lesions are benign dermoid cysts and hemangiomas. Rhabdomyosarcoma is the most common malignant primary orbital tumor in children. In adults, excisional biopsies are used for cavernous hemangiomas (usually retrobulbar, intraconal), solitary fibrous tumors, and secondary tumors including metastases. Optic nerve tumors (gliomas in children and meningiomas in adults) may be dissected from the eye and excised from the orbit.

Excisional biopsies in formalin should be measured and described, inked, and entirely submitted on cross sections for examination. In cases where there is suspicion of malignancy, especially in children, the specimen is submitted fresh for evaluation of the adequacy of sampling and tumor retrieval for genetic and molecular testing. The adequacy of sampling is evaluated by cryostat sections or by touch imprints (see below).

Incisional Orbital Biopsies

A biopsy performed for diagnostic purposes only (not therapeutic) is used in cases of large infiltrative lesions, extraocular muscle lesions, and lesions suspicious for lymphoma. In these cases, the biopsies are usually submitted for intraoperative evaluation to access adequacy of the sample, keeping the specimen fresh for molecular studies and flow cytometry. These biopsies are small and are better handled by touch imprints rather than cryostat sectioning because touch imprints preserve most of the tissue for permanent histopathology and testing (see below).

Occasionally, ophthalmologists obtain a biopsy from an optic nerve tumor. In these cases, the biopsy is usually small and obtained through the meninges, but occasionally, if the eye is already blind secondary to the tumor, the surgeon may select to obtain a segmental biopsy of the optic nerve with the lesion. In these samples, the margins (distal and proximal to the eye) should be submitted in different cassettes, the specimen inked and cross sections perpendicular to the axis of the optic nerve obtained to evaluate the tumor relationship with the meninges. This tissue is fixed in formalin.

Orbital Fine-Needle Aspiration Biopsy

Surgeons only rarely use FNAB to diagnose orbital tumors because the potential complications may outnumber the benefits. Excisional biopsies better treat many of the lesions that ophthalmologists might sample by FNAB. Nevertheless, some experienced orbital surgeons choose to use FNAB, and in these cases, the procedure is aided by imaging to avoid complications. Indications include either anteriorly located lesions (palpable) or large lesions with anterior access with a presumptive differential diagnosis of lymphoproliferative lesion or metastasis. Vascular or cystic lesions and lacrimal gland tumors are contraindications to this approach (8).

The disadvantages of orbital FNAB include poor cellular yield, a cytopathologic and not a histologic diagnosis, difficulty in interpreting the specimen, and low cellular yields requiring another biopsy procedure. The potential complications are retrobulbar hemorrhage, globe perforation, ptosis, extraocular motility dysfunction, and inadvertent entry intracranially. The samples are managed similarly to other FNAB specimens, either with rapid on-site evaluation (touch imprints, liquid-based cytology, RPMI, molecular analysis) or received prepared by the surgeon (smears and in fixative for further analysis).

Orbital Touch and Scrape Preparations

Cytologic evaluation by touch or scrape preparation is a constructive approach to handling small orbital biopsies. This technique allows for the processing and interpretation of small amounts of the sample, while allowing for preservation of most of the tissue for definite histopathologic diagnosis or other diagnostic tests, such as molecular pathology, flow cytometry, and cultures.

Clinical differential diagnoses that include lymphoma, rhabdomyosarcoma, and metastases to the orbit are rare indications for small incisional or excisional biopsies. If the specimen is an excisional biopsy, the margins are inked before bisecting the tissue. It is often helpful to obtain cells from the cut surface by picking up the tissue with forceps and gently placing the cut surface on a glass slide. This maneuver may be repeated on the same slide, adjacent to the previous touch specimen. It is customary to stain these touch preparations with a quick H&E stain, the same as used for frozen sections.

For larger specimens with firm, fibrous surfaces, it may be best to scrape the surface with a blade and to generate smears with the material obtained. If small pieces detach during this process, then another slide is placed on top of the first slide and pressed gently on the tissue as the second slide is drawn toward the end of the first slide. This "squash technique" generates a smear, and the slide against which the tissue is initially pressed usually contains most of the material that will be useful to generate a diagnosis.

In cases of lymphoproliferative lesions, material in RPMI is suitable for flow cytometry and the rest in formalin for histopathologic examination. In tumors in children, especially rhabdomyosarcoma, the tissue is divided into frozen samples for molecular and genetic testing and formalin for histopathologic examination (fig. 1-14).

LACRIMAL GLAND AND LACRIMAL SAC

Excisional Biopsies

Surgeons generally perform excisional biopsies of the lacrimal gland to avoid violating the capsule in benign mixed tumors, thereby minimizing the likelihood of seeding orbital tissue and promoting recurrence. These specimens are then measured, described, and inked before cross sectioning, to allow for the evaluation of margins and maintaining the integrity of the specimen. Occasionally, malignant tumors invade bone, and the specimen may necessitate decalcification.

Excisional biopsies of the lacrimal sac are rare. Epithelial tumors such as squamous papillomas or carcinomas usually fill the lumen of the sac. The specimen is measured and inked, and margins submitted in separate cassettes (nasal-distal and ductal-proximal margins). The remainder of the specimen is cross sectioned, and depending on the size of the lesion, may need to be further sectioned and embedded on edge to obtain proper orientation in the histologic sections.

Incisional Biopsies

Ophthalmologists frequently obtain biopsies from the palpebral lobe of the lacrimal gland to establish or rule out inflammatory lesions and lymphoma. Fresh tissue is sent for flow cytometry if touch imprints reveal adequate cellularity and suggest a diagnosis of lymphoma.

Surgeons often perform incisional biopsies of the lacrimal sac during a dacryocystorhinostomy procedure for lacrimal drainage obstruction.

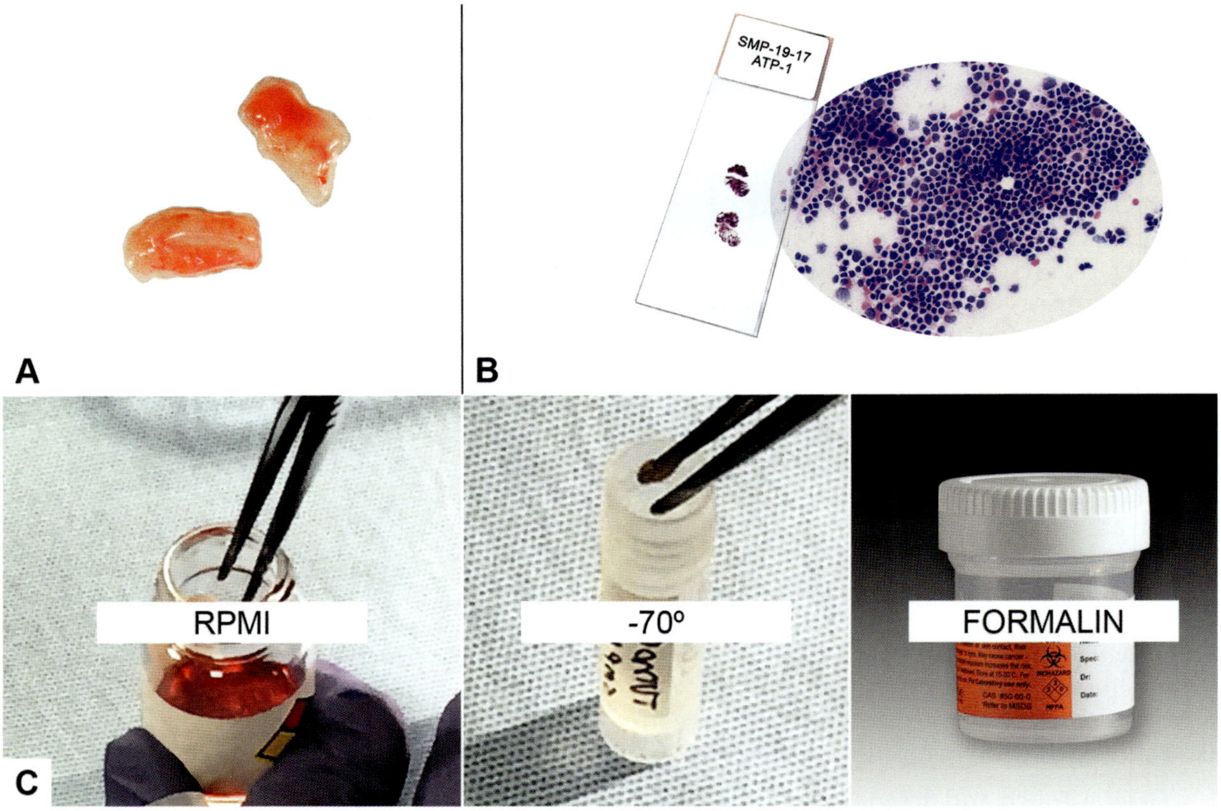

Figure 1-14

BIOPSY OF THE ORBIT

A: A small specimen is bisected fresh and the cut surface used to prepare a touch imprint.

B: The touch imprint shows the shape of the two pieces of the specimen. Cytologic preparation shows a cellular lymphocytic infiltrate.

C: A portion of the specimen is in RPMI solution for flow cytometry, and another piece is in formalin for histopathologic diagnosis. If needed, the specimen may be saved frozen (-70° centigrade) for possible molecular studies.

The pathology laboratory usually receives a few pieces of tissue from the lacrimal sac in formalin.

Fine-Needle Aspiration Biopsy

FNAB is seldom used in these areas because of the potential complications and the preference for excisional therapeutic and diagnostic biopsies.

PRETISSUE PROCESSING GEL EMBEDDING TECHNIQUE

Pretissue processing gel embedding is used in small specimens to secure the orientation of the specimen and to prevent tissue loss in processing (fig. 1-15). Histogel® or agar is used (9)

The gel is heated to achieve a liquid form, using a warm bath (with the tube in glassware containing hot water) or warmed by a hot plate or a microwave oven. Using a metal small embedding mold (for paraffin embedding), the liquid gel is poured into the well. The mold is at room temperature and in proximity to the stereoscopic microscope (preferably). Under microscopic observation (or using a loupe), the specimen is embedded into the gel in the desired position (usually on edge). The gel should be mildly firm while positioning the specimen so that it is held in place and not floating or falling on its side.

The metal mold containing the specimen embedded in the gel is gently moved on to a cold plate or ice pack with a flat surface to allow for cooling and solidification of the gel. A cassette with the case number is placed on top to identify the specimen at all times, especially if

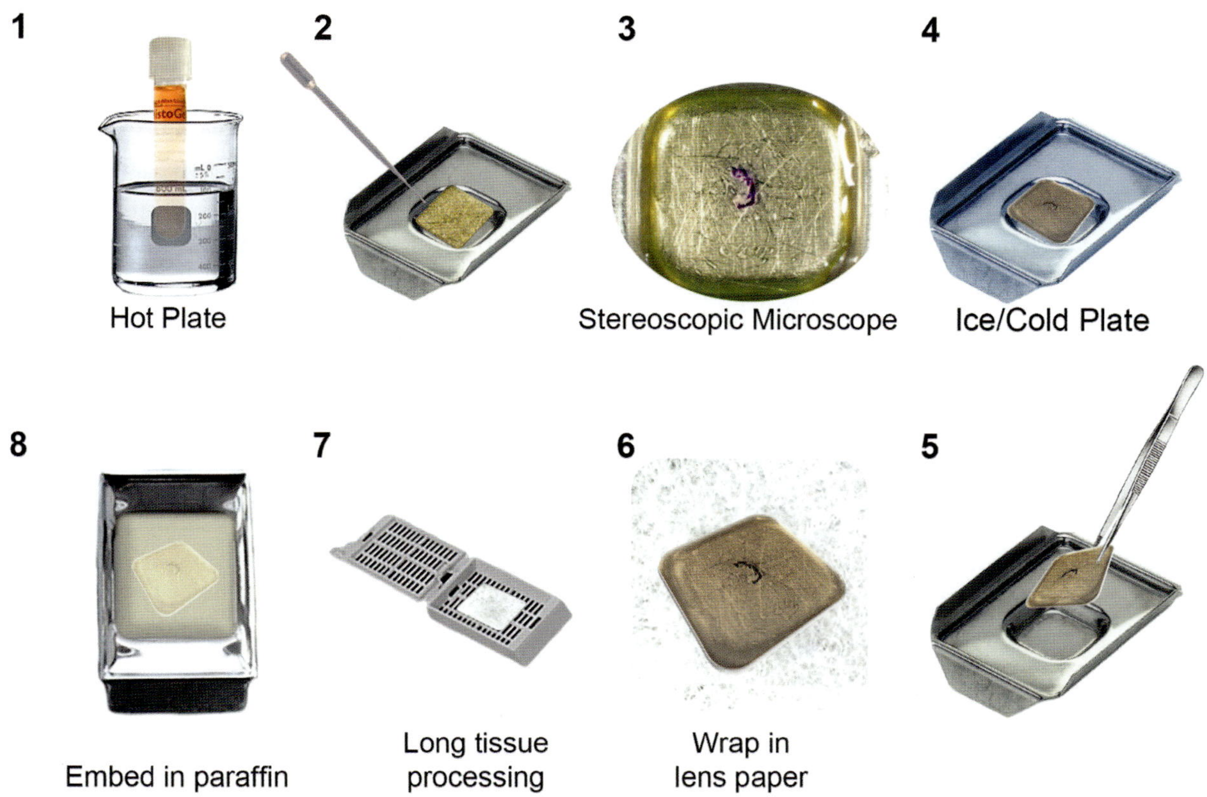

Figure 1-15

PRETISSUE PROCESSING GEL EMBEDDING

1. The gel (Histogel or agar) is melted in a hot bath using a hot plate or microwave oven. 2. Fluid gel fills the well of a small-diameter metal mold. 3. The tissue is in the gel with the desired orientation. 4. The embedded tissue in the metal mold is transferred to a cold surface to allow hardening of the gel. 5. The gel block is dislodged from the metal mold. 6. The gel block is wrapped in lens paper to prevent loss of tissue if the gel breaks or dissolves during processing. 7. The wrapped gel block is submitted in a cassette for a long tissue processing schedule to allow penetration of the reagents into the tissue. 8. The processed gel block is embedded in paraffin to obtain a paraffin block for sectioning.

the technique is repeated to embed other pieces of the specimen.

After the gel becomes solid, the gel block is gently dislodged and removed from the metal mold. The gel block is then wrapped in lens paper to prevent any loss of tissue should the gel fracture or melt during processing.

The wrapped gel block is then processed in a long processing schedule (not the short biopsy program) to allow for adequate infiltration of the reagents into the tissue. The processed gel block is positioned on the bottom of the well of the metal mold, with the flat surface down, to maintain the orientation of the tissue in the gel. The gel block then is embedded into paraffin to create the paraffin block that is sectioned to obtain the histologic slides.

REFERENCES

1. Eagle RC. Eye pathology: an atlas and text, 3rd ed. Philadelphia: Lippincott Williams & Wilkins; 2017.
2. Sastre X, Chantada GL, Doz F, et al. Proceedings of the consensus meetings from the International Retinoblastoma Staging Working Group on pathology guidelines for the examination of enucleated eyes and evaluation of prognostic risk factors in retinoblastoma. Arch Pathol Lab Med 2009;133:1199-202.
3. Folberg R, Verdick RE, Weingeist TA, Montague PR. Gross examination of eyes removed for ciliary body or choroidal melanoma. Ophthalmology 1986;93:1643-7.
4. Kim RS, Chevez-Barrios P, Bretana ME, Wong TP, Teh BS, Schefler AC. Histopathologic analysis of transvitreal fine needle aspiration biopsy needle tracts for uveal melanoma. Am J Ophthalmol 2017;174:9-16.
5. Kim RS, Chevez-Barrios P, Divatia M, Bretana ME, Teh B, Schefler AC. Yield, techniques, and complications of transvitreal and transscleral biopsies in small uveal melanoma. JAMA Ophthalmol 2018;136:482-8.
6. Chévez-Barrios P. Frozen section diagnosis and indications in ophthalmic pathology. Arch Pathol Lab Med 2005;129:1626-34.
7. Pfeiffer ML, Ozgur OK, Myers JN, et al. Sentinel lymph node biopsy for ocular adnexal melanoma. Acta Ophthalmol 2017;95:e323-e328.
8. Wiktorin AC, Dafgård Kopp EM, Tani E, Söderén B, Allen RC. Fine-needle aspiration biopsy in orbital lesions: a retrospective study of 225 cases. Am J Ophthalmol 2016;166:37-42.
9. LoRusso FJ, Font RL. Use of agar in ophthalmic pathology: a technique to improve the handling and diagnosis of temporal artery biopsies, subfoveal membranes, lens capsules, and other ocular tissues. Ophthalmology 1999;106:2106–218.

2 TUMORS OF THE CONJUNCTIVA AND CARUNCLE

ANATOMY AND HISTOLOGY

Conjunctiva

The conjunctiva is a mucous membrane that covers and connects the posterior surface of the eyelid to the anterior surface of the eyeball. Topographically, it is subdivided into the bulbar, forniceal, and palpebral (tarsal) regions (fig. 2-1) (1,2).

The bulbar conjunctival epithelium is continuous with the corneal epithelium at the region of the corneoscleral limbus; the tarsal conjunctival epithelium transitions into the epidermis in the eyelid margin (figs. 2-1, 2-2). The nonkeratinizing stratified cuboidal to columnar conjunctival epithelium is 2 to 5 cell layers in thickness and contains mucus-secreting goblet cells. There are more goblet cells in the fornix, the conjunctiva nasally, and the conjunctival fornix than elsewhere (1,2). Other cellular constituents of the conjunctival epithelium include melanocytes, Langerhans cells, and migratory intraepithelial lymphocytes (fig. 2-2) (1).

The conjunctival stroma, or substantia propria, is fibrovascular connective tissue composed of collagen fibrils, fibroblasts, blood vessels, lymphatic channels, peripheral nerves, and scattered acute and chronic inflammatory cells. The substantia propria in the bulbar and forniceal regions is loose, with haphazardly arranged stroma (fig. 2-2). The forniceal conjunctiva is redundant and folded to facilitate the eye movements (2). Aggregates of lymphocytes and plasma cells, with occasional reactive follicle formation, are found in the forniceal conjunctiva, similar to those observed in other mucosa-associated lymphoid tissues (MALT) (fig. 2-2) (2).

The palpebral substantia propria is more uniform and densely adheres to the underlying tarsal plate. Multiple tubular and cystic gland-like epithelial invaginations, called pseudoglands of Henle, extend into the palpebral substantia

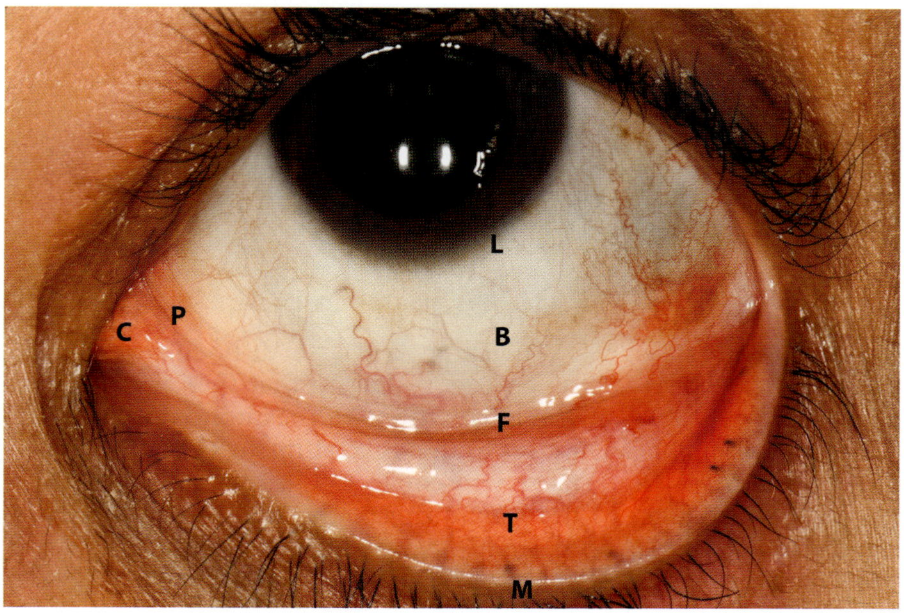

Figure 2-1

TOPOGRAPHY OF ANTERIOR SEGMENT

Topographic landmarks of the ocular segment: bulbar conjunctiva (B), forniceal conjunctiva (F), tarsal conjunctiva (T), eyelid margin (M), limbal conjunctiva (L), plica semilunaris (P), and caruncle (C).

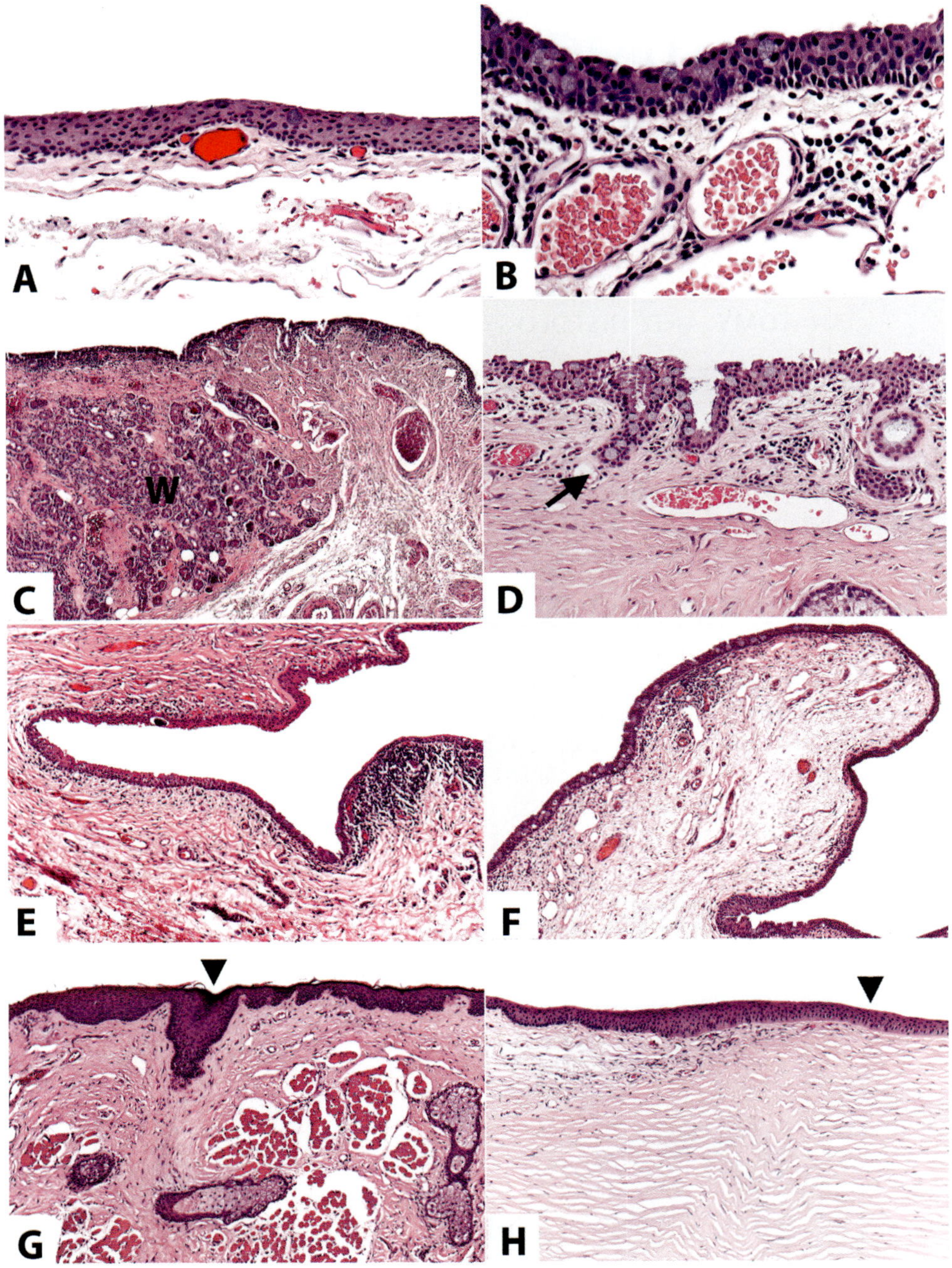

Figure 2-2

NORMAL CONJUNCTIVA

A: Temporal bulbar conjunctiva.

B: Nasal bulbar conjunctiva.

C: Tarsal conjunctiva with underlying tarsus and accessory lacrimal gland of Wolfring (W).

D: Tarsal conjunctiva with pseudoglands of Henle (arrow).

E: Forniceal conjunctiva with lymphoid aggregates.

F: Plica semilunaris.

G: Eyelid margin conjunctiva transitioning into epidermis at the region of Meibomian gland orifice (arrowhead).

H: Limbal conjunctiva transitioning into corneal epithelium at the corneoscleral junction (arrowhead) (All figures: Hematoxylin and eosin [H&E] stain).

propria (fig. 2-2). Acini and ducts of the accessory lacrimal glands of Wolfring and Krause appear in the palpebral and forniceal substantia propria, respectively (fig. 2-2) (1,2).

Nasally, bulbar conjunctiva forms a semilunar fold (plica semilunaris) (fig. 2-1), composed of goblet cell–rich epithelium and vascular stroma (fig. 2-2). Smooth muscle and cartilage occasionally appear in this location (1).

Caruncle

The caruncle is a small nodular region between the medial canthal skin and the plica semilunaris (fig. 2-1). The caruncular epithelium ranges from nonkeratinizing columnar epithelium with goblet cells, to stratified squamous nonkeratinizing epithelium, and in some cases, to keratinizing epithelium identical to skin epidermis. The caruncular stroma contains

Table 2-1

TYPES AND FREQUENCY OF CONJUNCTIVAL TUMORS[a]

	Number of Cases (%)			
Type of Tumor	Doheny 1970-2000	AFIP[b] 1984-1989	Wills Eye Hospital 1974-2016	SEER[b] Database 1973-2009
Choristomatous	29 (3%)	37 (3%)	101 (2%)	
Epithelial (total)	522 (48%)	509 (41%)	824 (17%)	
Epithelial (benign)			95 (2%)	
Epithelial (premalignant, malignant)			729 (15%)	311 (32%)
Melanocytic (total)	327 (30%)	542 (43%)	2590 (52%)	
Nevus			1163 (23%)	
Melanosis, melanocytosis, melanoma			1427 (29%)	246 (25%)
Hematolymphoid	113 (11%)	117 (9%)	482 (10%)	410 (42%)
Soft tissue (total)			247 (5%)	
Soft tissue (benign)			232 (5%)	
Soft tissue (malignant)			15 (<1%)	0
Other rare malignancies				12 (1%)
Secondary			66 (1%)	
Metastatic			6 (<1%)	
Adnexal glands	22 (2%)	24 (2%)	15 (<1%)	
Inflammatory, infectious, degenerative, and other non-neoplastic lesions			671 (13%)	
Total			5002	979

[a]Data from references 5, 7, and 8.
[b]AFIP = Armed Forces Institute of Pathology; SEER = Surveillance, Epidemiology, and End Results.

pilosebaceous units, sweat glands, accessory lacrimal glands of Popoff, smooth muscle fibers, and adipose tissue (fig. 2-2) (1,2).

CLASSIFICATION

Most conjunctival mass-forming lesions are degenerative or inflammatory, such as pinguecula, pterygium, exuberant granulation tissue, and granuloma (3,4). The most frequent conjunctival neoplasms include choristomas and epithelial, melanocytic, and lymphoid proliferations (Table 2-1) (5–8). Among the epithelial neoplasms, intraepithelial neoplasia and squamous cell carcinoma occur primarily in adults (5,7). Nevi are the most common conjunctival melanocytic lesions in all age groups, while atypical intraepithelial melanocytic proliferations and malignant melanoma are more frequent in adults (5,7). Choristomatous tumors (solid epibulbar dermoid, lipodermoid, complex choristomas, and others) occur in children (5,7). Lymphoproliferative neoplasms are identified with increasing frequency in the conjunctiva and ocular adnexal region (7). Many soft tissue tumors arise in the conjunctival substantia propria, analogous to their extraocular counterparts (5–8).

Secondary involvement of the conjunctiva occurs when neoplasms extend from adjacent sites, such as sebaceous carcinoma of the eyelid and caruncle, or when extraocular extension of uveal malignant melanoma occurs (6). Rarely, metastatic tumors involve the conjunctiva (6).

BENIGN EPITHELIAL NEOPLASMS AND TUMOR-LIKE LESIONS

Pinguecula and Pterygium

Definition. The literature considers *pinguecula* and *pterygium* as degenerative conjunctival lesions with multifactorial etiology, including ultraviolet (UV) irradiation, environmental factors (wind and dust leading to micro-trauma), chronic inflammation, dryness, and viral infection (9–11). Although recent studies suggest that pterygia may be neoplastic (11,12), the definitive data to support this hypothesis are lacking.

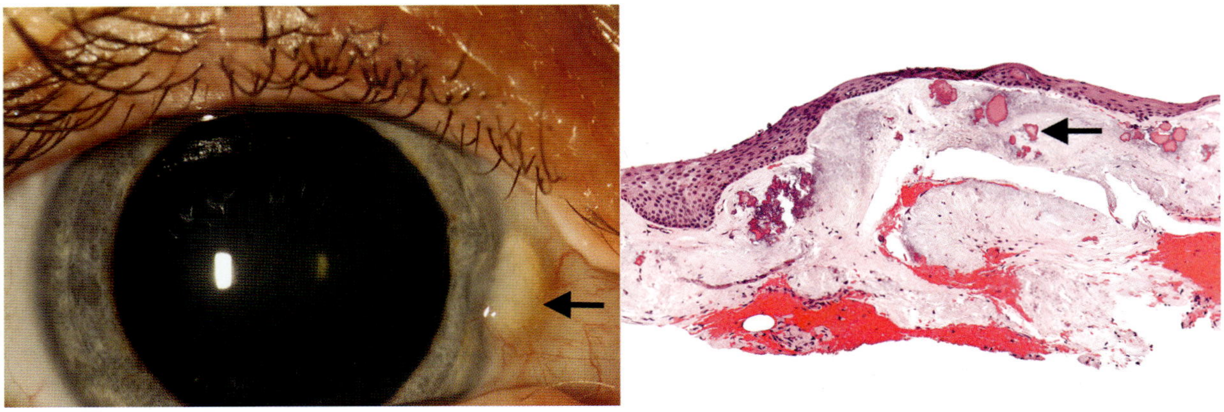

Figure 2-3

PINGUECULA

Left: Anterior segment photograph demonstrates a yellow-gray nodule (arrow) located nasally at the corneoscleral limbus of the right eye.

Right: Prominent actinic elastosis is present in the substantia propria with spheroidal bodies (arrow). The overlying limbal epithelium is variable in thickness, with foci of acanthosis and attenuation (H&E stain).

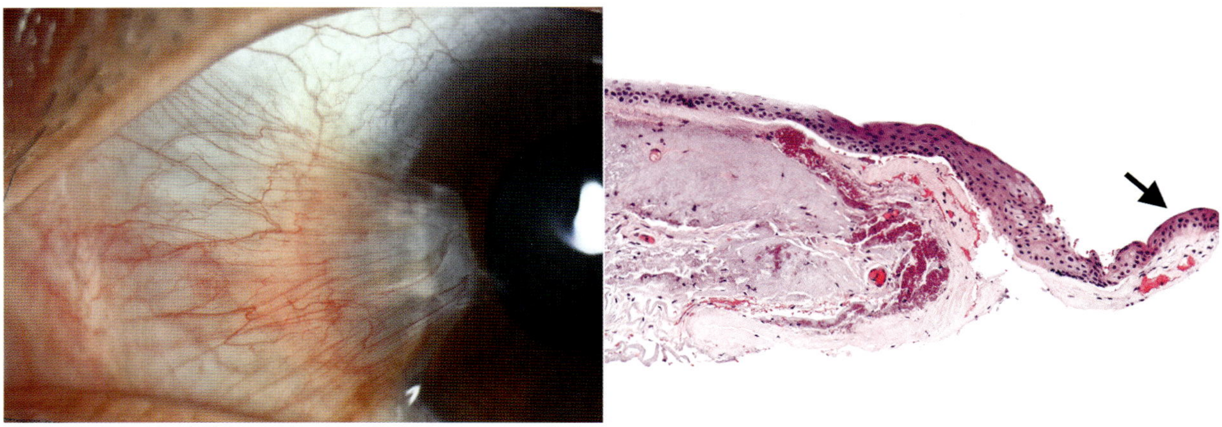

Figure 2-4

PTERYGIUM

Left: Vascularized wing-shaped conjunctival tissue extends onto the cornea.

Right: A segment of limbal conjunctiva has focally acanthotic epithelium. The underlying stroma contains prominent blood vessels and actinic elastosis. A corneal margin (arrow) is present (H&E stain).

Clinical Features and Epidemiology. A pinguecula is a small, yellowish nodule, frequently bilateral, situated at the corneoscleral limbus nasally and temporally (fig. 2-3) (9). A pterygium has a similar location, appearing as a triangular or wing-shaped fleshy and vascularized conjunctival membrane encroaching upon the superficial corneal layers (fig. 2-4) (10). Pinguecula and pterygia are more common in geographic regions with intense solar irradiation, known as the "pterygium belt," usually in older patients (10), and in some ethnic groups (13). Pterygia are associated with other benign and malignant squamous (14–17) and melanocytic (18) ocular surface lesions. These lesions can clinically masquerade as pterygia, and all pterygia should be submitted for pathologic evaluation (15–18).

Microscopic Findings. The identification of corneal tissue in pterygium is the definitive histopathologic feature distinguishing this lesion from pinguecula. Elastotic degeneration of collagen

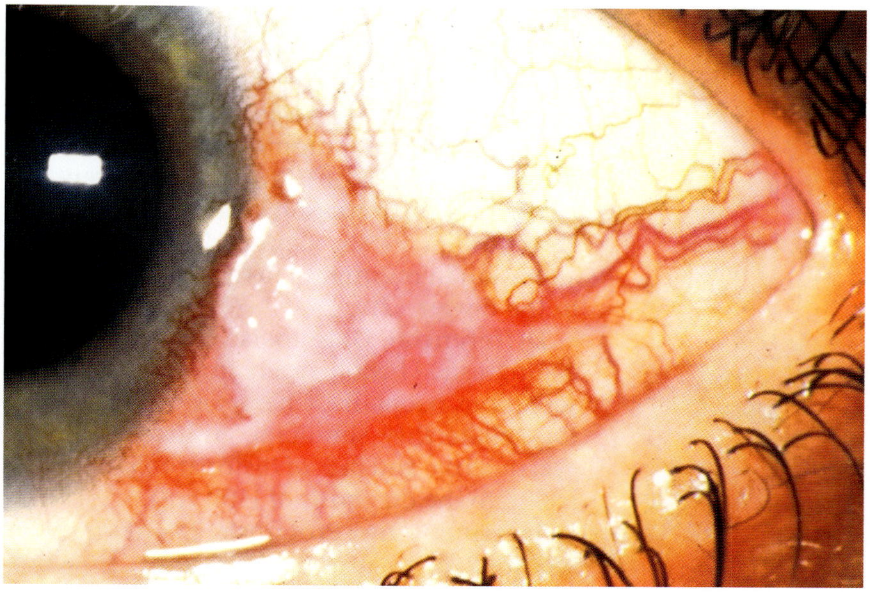

Figure 2-5

HEREDITARY BENIGN INTRAEPITHELIAL DYSKERATOSIS

Gray-white, inflamed, semilunar conjunctival lesion is in the interpalpebral limbal location. (Courtesy of Dr. R. C. Eagle, Jr., Philadelphia, PA.)

is identified, with occasional spheroidal bodies and variable chronic inflammatory infiltrates in both lesions. Fibrovascular stromal proliferation is prominent in a pterygium (19). The surface epithelium in both lesions may be either atrophic or hyperplastic, with loss of goblet cells, squamous metaplasia, and surface keratinization (8,19). Dysplasia is occasionally seen (15–17). The so-called recurrent pterygium may lack the histologic feature of elastotic degeneration, and it may be more appropriate to classify this lesion as exuberant hyperplastic scarring associated with increased vascularization (8).

Immunohistochemical and Molecular Genetic Findings. Pathologists are not required to use ancillary studies to evaluate most pinguecula and pterygia. The Ki-67 immunostain highlights dysplasia (20).

The literature makes reference to various genes and cellular pathways in the pathogenesis of pterygium, including the *p53* mutations (12), alterations in methylation in genes encoding metalloproteinase 2 (*MMP2*), hypermethylation of the *p16* gene promoter (21), and upregulation of proangiogenic cell-derived factor-1 (*SDF-1*) (22). Nevertheless, these studies are not used for diagnostic or prognostic purposes in pinguecula and pterygium.

Differential Diagnosis. The histologic differential diagnosis of pterygium and pinguecula includes ocular surface squamous neoplasia (15–17), conjunctival atypical intraepithelial melanocytic proliferation, and malignant melanoma (18). These tumors can co-exist with or simulate pinguecula and pterygium (15–18).

Treatment and Prognosis. Recurrence ranges from 3 to 42 percent, with lower rates following conjunctival auto-graft (23). Long-term follow-up following surgery is essential because of the risk of conjunctival carcinoma, which seems to be less likely when experienced surgeons excise pterygia (23,24).

Hereditary Benign Intraepithelial Dyskeratosis

Definition. *Hereditary benign intraepithelial dyskeratosis* (HBID) is a non-neoplastic hyperkeratotic and dyskeratotic epithelial disorder of ocular and oral mucosa (25,26).

Clinical Features and Epidemiology. HBID typically follows an autosomal dominant inheritance with incomplete penetrance (25,26), although sporadic cases have been described (27). This infrequently encountered condition is generally, but not exclusively, encountered in indigenous people of the Haliwa-Saponi tribe in North Carolina, United States, or their descendants in other geographic areas (25–27).

HBID manifests in childhood as bilateral gray-white, inflamed, semilunar conjunctival lesions in the interpalpebral limbus nasally and temporally. Similar lesions appear in the oral mucosa (fig. 2-5) (25,26).

Microscopic Findings. The lesions of HBID are sharply delineated from the surrounding

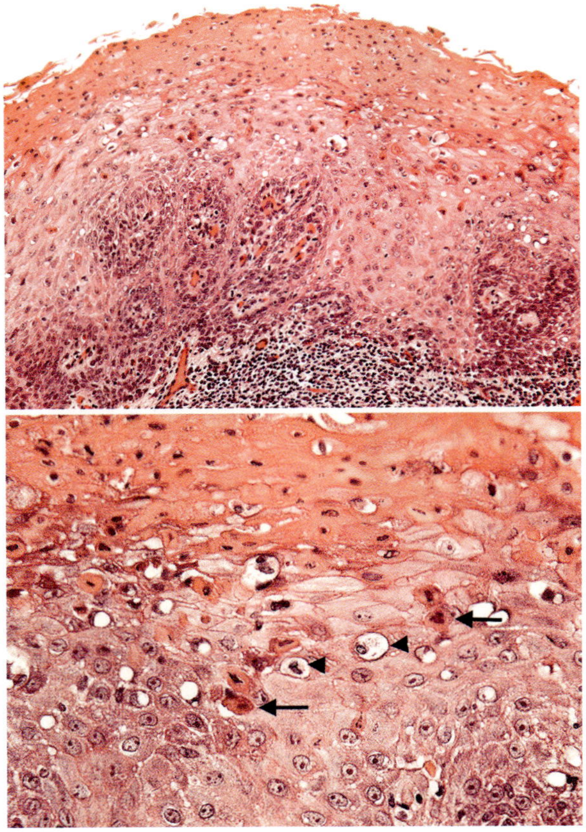

Figure 2-6

HEREDITARY BENIGN INTRAEPITHELIAL DYSKERATOSIS

Top: Epithelial acanthosis and pseudoepitheliomatous hyperplasia with an intense stromal chronic inflammatory infiltrate are seen.

Bottom: Higher magnification highlights prominent dyskeratosis (arrows) and nuclear pyknosis (arrowheads) in the middle and superficial epithelial layers, without appreciable nuclear atypia (H&E stain).

conjunctiva. Marked acanthosis with prominent dyskeratosis and nuclear pyknosis in the middle and superficial epithelial layers are seen. Nuclear atypia is absent. Chronic inflammatory infiltrates and capillary telangiectasia in the underlying substantia propria are characteristic features. The basement membrane remains intact (fig. 2-6) (25,26).

Cytologic Findings. Cytologic preparations from conjunctival brushings yield the typical cytologic features of HBID. These include rounded squamous epithelial cells with dense homogenous orange cytoplasm and hyperchromatic, pyknotic, or crenulated nuclei (28).

Molecular Genetic Findings. The *HBID* gene was localized to chromosome 4 (4q35), with duplication alleles potentially involved in causation (26,27). Molecular genetic studies for duplication in chromosome 4 (4q35) confirm the diagnosis of HBID (27,28). Recently, a novel missense mutation in the *NLRP1* gene on chromosome 17 was identified, suggesting that alteration in multiple genetic loci may give rise to the disease (29).

Differential Diagnosis. HBID should be distinguished histologically from ocular surface squamous neoplasia. Characteristic family history, bilaterality, simultaneous involvement of oral mucosa, and lack of histopathologic atypia facilitate the diagnosis of HBID (8).

Treatment and Prognosis. No treatment is required for asymptomatic lesions. In symptomatic patients, conjunctival lesions may be surgically excised. The use of limbal allograft transplantation and cyclosporine A may reduce recurrences (30). There is no progression to malignancy (25–30).

Papilloma

Definition. *Conjunctival (squamous cell) papilloma* is a benign lesion that is strongly associated with human papillomavirus (HPV) 6 and 11 (α-PV genus) infection (31,32), and less frequently with β-PV genus (31) and high-risk HPV infection (32).

Clinical Features and Epidemiology. Papillomas develop at any age but are more frequent in children and young adults (8,33). They are characterized clinically by multiple, translucent, finger-like projections of conjunctival epithelium with central hairpin-like vascular loops (2). The lesions in children and young adults are more common in caruncular, plica semilunaris, and forniceal locations, where papillomas are often multiple and typically pedunculated (fig. 2-7) (2,8,33). In older adults, they may originate from the limbus, where they typically present as a single lesion with sessile morphology (8,33).

Microscopic Findings. Squamous papillomas are exophytic, composed of fronds of variably acanthotic epithelium with goblet cells that enclose mildly chronically inflamed fibrovascular cores (fig. 2-8). Unlike viral skin lesions, it is rare to see HPV cytopathic sequelae, such as koilocytosis, hypergranulosis, and parakeratosis

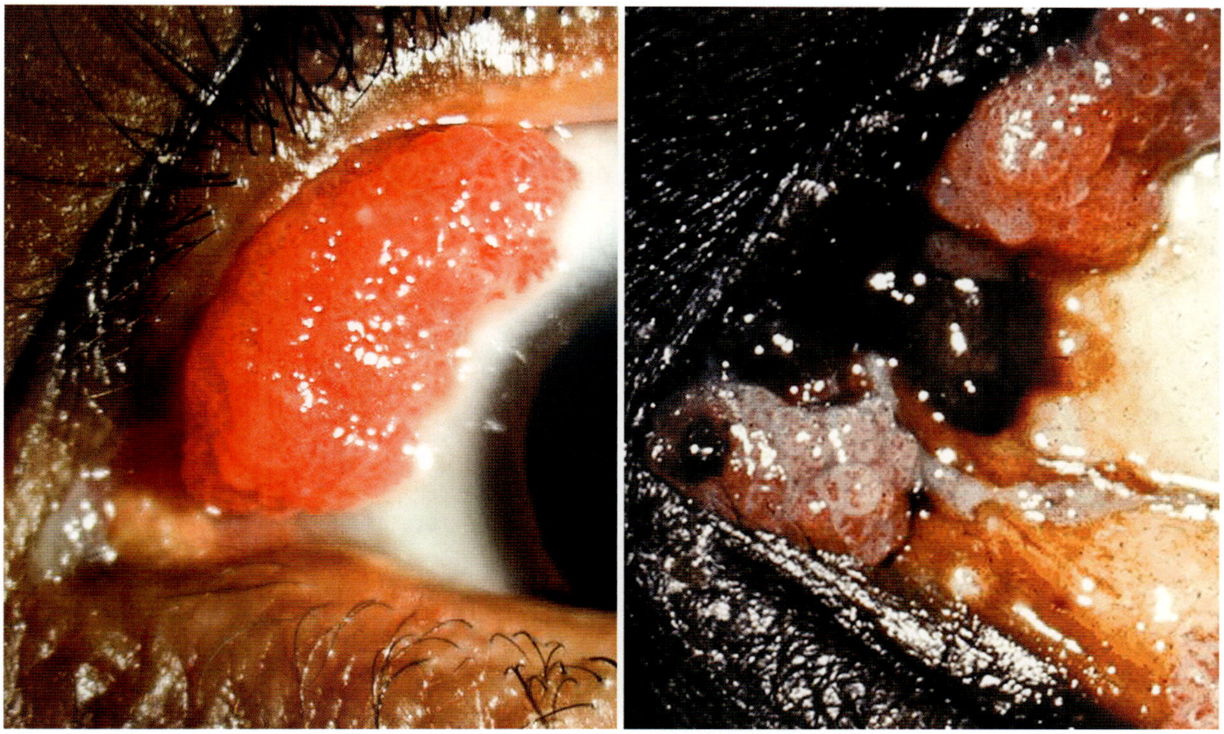

Figure 2-7

PEDUNCULATED CONJUNCTIVAL PAPILLOMA

Left: A pedunculated papilloma, composed of translucent epithelial fronds with central vascular hairpin-like loops, involves the forniceal conjunctiva.

Right: Multiple pedunculated papillomas with focal pigmentation arise from the fornix and caruncle (pigmented papilloma).

(2). Acute inflammatory cells are frequently seen within the surface epithelium (8).

Benign papillomas can harbor foci of epithelial dysplasia (6 to 20 percent) (32). Marked acanthosis, frequent absence of goblet cells, cytomorphologic atypia, and mitotic figures in the superficial epithelial layers (fig. 2-9) are characteristic (2). *Inverted papilloma*, a rare endophytic papilloma variant characterized by invaginated lobules of nonkeratinized squamous epithelium with a variable number of goblet cells, has also been described in the conjunctiva (34,35).

Immunohistochemical Findings. Although in situ hybridization and polymerase chain reaction (PCR)-based studies have been employed to identify and subtype HPV in conjunctival papillomas (31–33), no ancillary diagnostic techniques are employed routinely.

Differential Diagnosis. The differential diagnosis of conjunctival papilloma, particularly in the limbal location, includes papillary squamous cell carcinoma in situ (2). Pathologists must distinguish between inverted papilloma of the conjunctiva and low-grade infiltrating mucoepidermoid carcinoma (8). Cytomorphologic atypia, loss of polarity, and increased mitotic activity in the superficial epithelial layers should suggest a malignancy (2). Architectural disorganization and stromal desmoplasia are additional clues to the diagnosis of mucoepidermoid carcinoma (35).

Treatment and Prognosis. Treatment usually includes surgical excision and cryotherapy. The lesions may recur after excision. Oral cimetidine (36) and subconjunctival interferon alpha-2b are some of the adjuvant and neoadjuvant therapies successfully employed to reduce the risk of recurrence (37). The risk of progression to squamous cell carcinoma is low (38).

Amyloidosis

Definition. Abnormally folded proteins, which share the distinctive staining properties and fibrillary ultrastructural appearance of

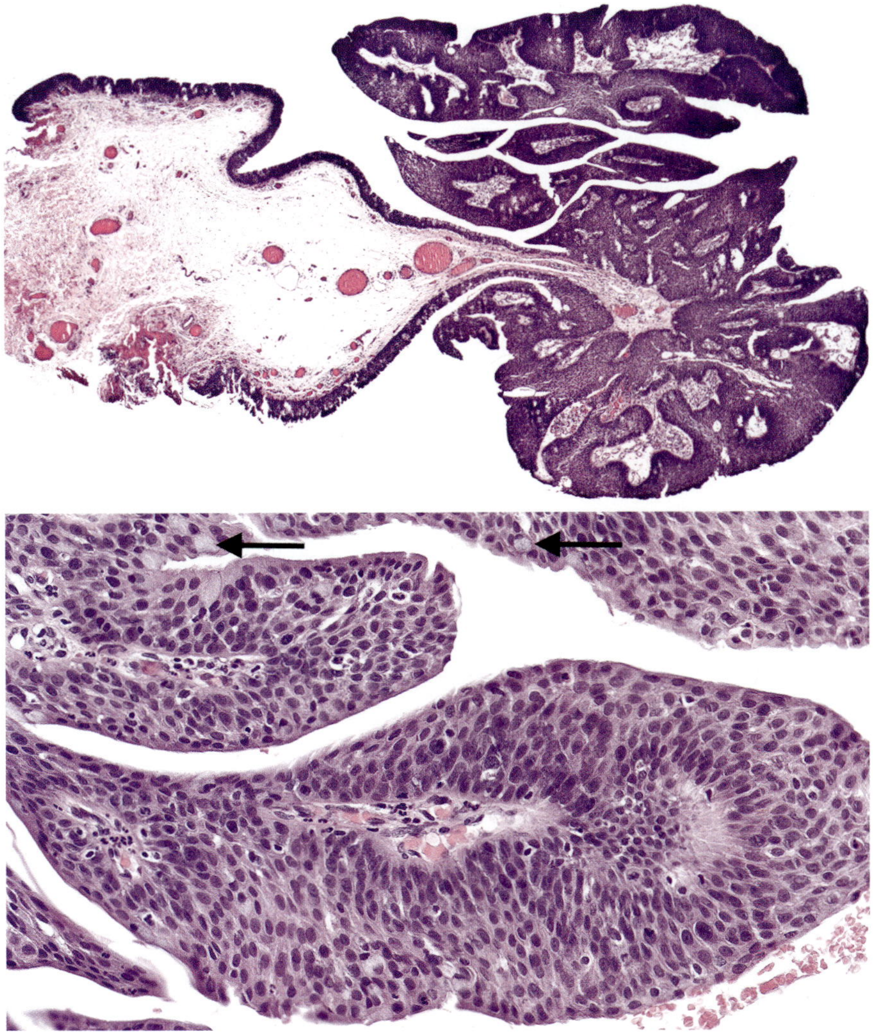

Figure 2-8

PEDUNCULATED SQUAMOUS PAPILLOMA

Top: A pedunculated lesion has a central stalk and radiating acanthotic epithelial fronds have central fibrovascular cores.

Bottom: The acanthotic epithelium focally retains the goblet cells (arrows) and demonstrates reactive changes. Mild acute and chronic inflammation is present (H&E stain).

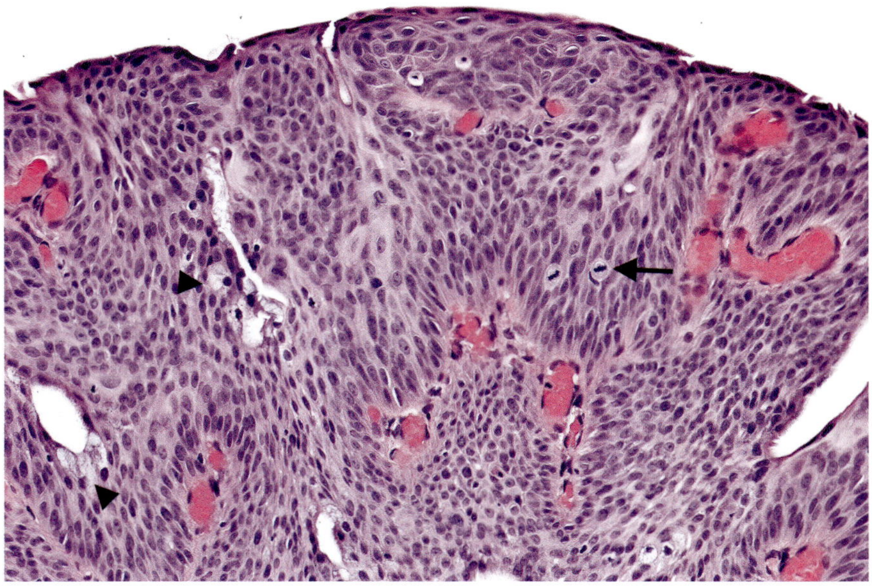

Figure 2-9

SQUAMOUS PAPILLOMA WITH DYSPLASIA

The tumor is characterized by focal acanthosis, loss of polarity, nuclear atypia, and superficial mitotic figures (arrow). Occasional goblet cells are retained (arrowheads) (H&E stain).

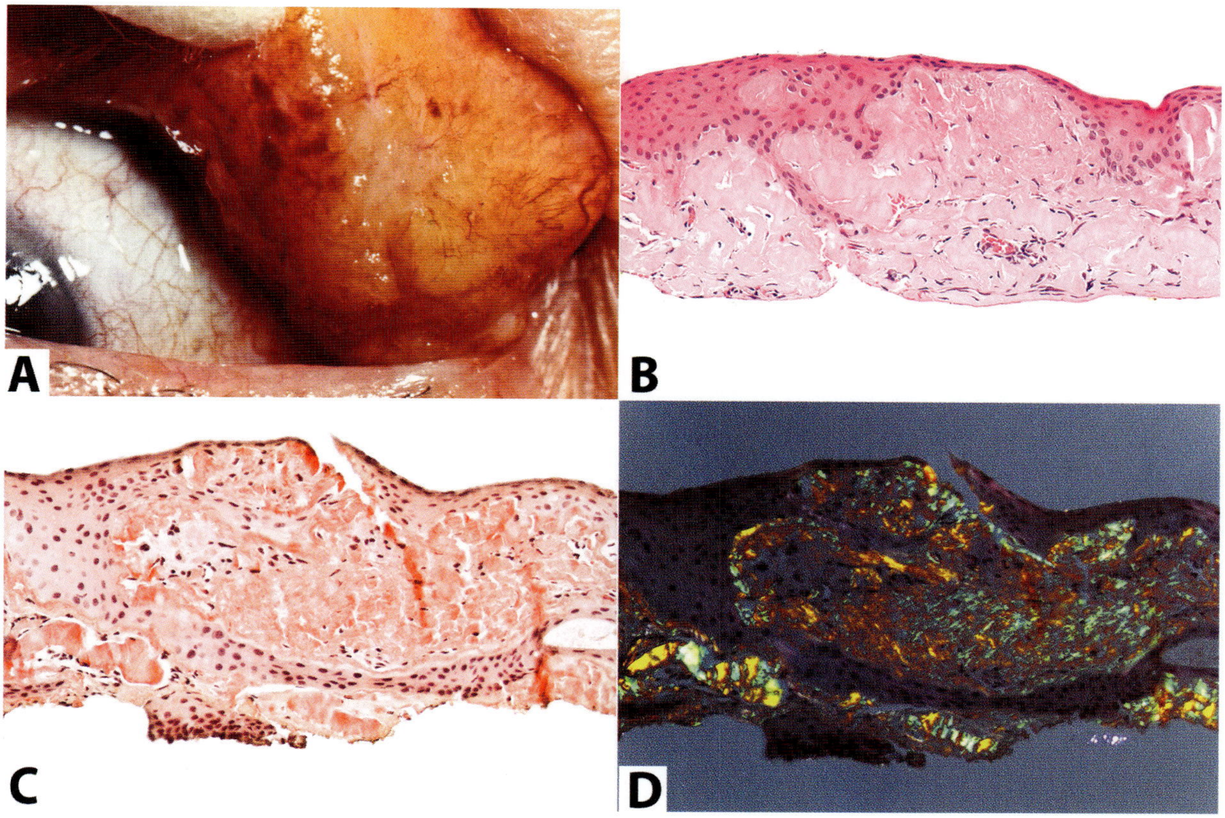

Figure 2-10

CONJUNCTIVAL AMYLOIDOSIS

A: A waxy-yellow nodule with punctate hemorrhages involves the palpebral conjunctiva (H&E stain). (Courtesy of Dr. R. C. Eagle, Jr., Philadelphia, PA.)

B: Amorphous, homogeneously eosinophilic deposits in the substantia propria are associated with a sparse chronic inflammatory infiltrate.

C,D: The deposits stain with Congo red (C) and demonstrate apple-green birefringence and dichroism with plane-polarized light (D).

amyloid, appear in the conjunctival substantia propria (39).

Clinical Features and Etiology. *Conjunctival amyloidosis* is usually seen in middle-aged adults. It presents as confluent waxy or yellow fusiform lesions or polypoidal papules, frequently associated with hemorrhage (fig. 2-10). Although the palpebral conjunctival location is most common, other conjunctival regions may be involved (40).

Amyloid deposition in the conjunctiva is most commonly an idiopathic primary localized process. The deposits are typically composed of a monoclonal light chain immunoglobulin (AL amyloid), secreted locally by plasma cells (40,41). Less frequently, conjunctival amyloidosis is induced by longstanding local inflammation, such as with trachoma (i.e., secondary localized amyloidosis, AA amyloid) (40). Occasionally, conjunctival amyloid deposition occurs in the context of primary conjunctival lymphoma, plasmacytoma, or systemic amyloidosis (40,42); the literature reports an association with systemic amyloidosis in 6 percent of patients with conjunctival amyloidosis (40).

Microscopic Findings. Amyloid appears as an amorphous, homogeneously eosinophilic deposit on a hematoxylin and eosin (H&E) stain. With the Congo red stain, amyloid is characterized by apple-green birefringence and dichroism with plane-polarized light (fig. 2-10) (40).

Immunohistochemical Findings. Immunohistochemical methods, sequencing, and mass spectrometry-based proteomic analysis are some of the techniques used in amyloid subtyping (39) but are not employed routinely.

Differential Diagnosis. Eosinophilic acellular proteinaceous deposits may mimic the histologic appearance of amyloid in the conjunctival stroma in two conditions: ligneous conjunctivitis (43) and lipoid proteinosis (44). The Congo red stain with polarization microscopy helps to distinguish amyloid from these entities.

Treatment and Prognosis. The management of conjunctival amyloidosis depends on the extent of local involvement and the systemic status of the patient. Local therapeutic options include conservative management with topical lubricants and excision. Recurrence following excision has reported in about 20 percent of patients (40).

DYSPLASTIC EPITHELIAL LESIONS AND MALIGNANT EPITHELIAL NEOPLASMS

Conjunctival Intraepithelial Neoplasia, Carcinoma in Situ, and Squamous Cell Carcinoma

Definition. The term *conjunctival intraepithelial neoplasia* (CIN) describes a wide range of neoplastic intraepithelial changes, ranging from varying grades of dysplasia to full-thickness intraepithelial neoplasia or *carcinoma in situ* (CIS) (8). *Squamous cell carcinoma* (SCC) of the conjunctiva is an invasive malignant conjunctival neoplasm with squamous differentiation, which can be keratinizing or nonkeratinizing (2). CIN, CIS, and SCC represent the clinical continuum of *ocular surface squamous neoplasia* (OSSN) (45,46).

Clinical Features and Epidemiology. The most common conjunctival neoplasm in the United States and the third most common lesion among ocular malignancies in an adult population (45) is OSSN, which includes the spectrum of CIN, CIS, and SCC. Worldwide, the mean age-standardized incidence is 0.18/year/100,000 for males and 0.08/year/100,000 for females. In higher risk populations, the incidence is as high as 3.4/year/100,000 for males and 3.0/year/100,000 for females (47).

Two broad epidemiologic patterns of disease are recognized. In the immunocompetent population, the incidence of OSSN increases with older age, male sex, sunlight exposure, and fair skin (45,46). Additional associations include exposure to toxic chemicals, smoking, and xeroderma pigmentosum (8). In the immunocompromised population, OSSN affects younger males and females equally, and is strongly associated with human immunodeficiency virus (HIV) and HPV 16 and 18 infections (47).

Intraepithelial and invasive squamous conjunctival lesions tend to develop in the sun-exposed interpalpebral fissure, especially at the limbus where the stem cells reside (2). The characteristic clinical appearance is a sharply demarcated, gelatinous or leukoplakic, placoid, papilliform, or nodular epithelial-based lesion, with corkscrew intrinsic vessels and feeder vessels. Involvement of the adjacent corneal epithelium is common (fig. 2-11) (33). Although invasive tumors tend to be larger and more elevated (33), the distinction between CIN and invasive SCC can be challenging clinically (46).

Microscopic Findings. Microscopic examination typically shows an abrupt transition between the normal conjunctival epithelium and the dysplastic epithelium. Intraepithelial dysplastic changes originate in the basal layers and extend toward the surface (2,8). These lesions should be classified according to the extent of intraepithelial atypia: CIN is classified as mild (CIN-I) if the lower one third of the epithelium is involved; moderate (CIN-II) if half to two thirds of the epithelium is involved; severe (CIN-III) if the abnormality extends into the superficial third of the epithelium; or carcinoma in situ (CIS) if there is full-thickness epithelial involvement (figs. 2-12–2-14).

The dysplastic epithelium is remarkable for increased cellularity, loss of polarity, nuclear atypia, increased mitotic activity in the superficial epithelial layers, and atypical mitotic figures (8,45). Dysplastic squamous cells may be highly differentiated, with abundant eosinophilic cytoplasm, or may lack appreciable keratinization, appearing as small atypical epithelial cells with a high nuclear to cytoplasm ratio and rounded or spindled morphology. Dyskeratosis and parakeratosis are common. Koilocytosis is occasionally identified in HPV-associated lesions (fig. 2-14) (2,8).

When intraepithelial squamous (or melanocytic) neoplasia extends onto the cornea, it remains superficial to the epithelial basement membrane and the underlying Bowman layer.

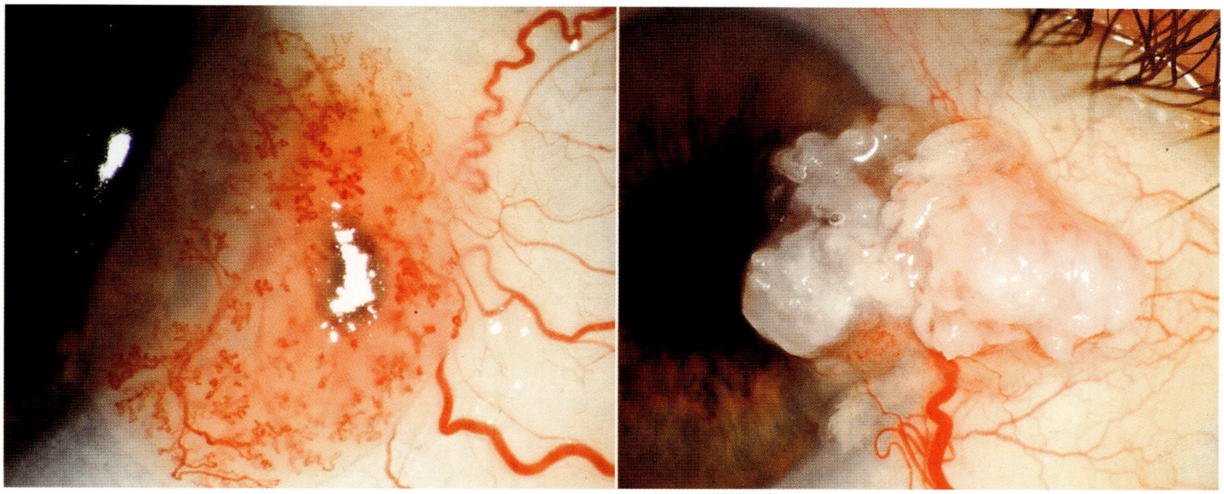

Figure 2-11

OCULAR SURFACE SQUAMOUS NEOPLASIA (OSSN)

Left: This is a gelatinous, papillomatous limbal mass with coarse corkscrew intrinsic blood vessels, prominent feeding vessels, and corneal invasion.

Right: A white "leukoplakic" limbal mass invades the cornea with dilated feeding vessels. In both cases, the cornea involvement remains superficial to Bowman layer and the corneal stroma. (Courtesy of Drs. J. Shields and C. Shields, Philadelphia, PA.)

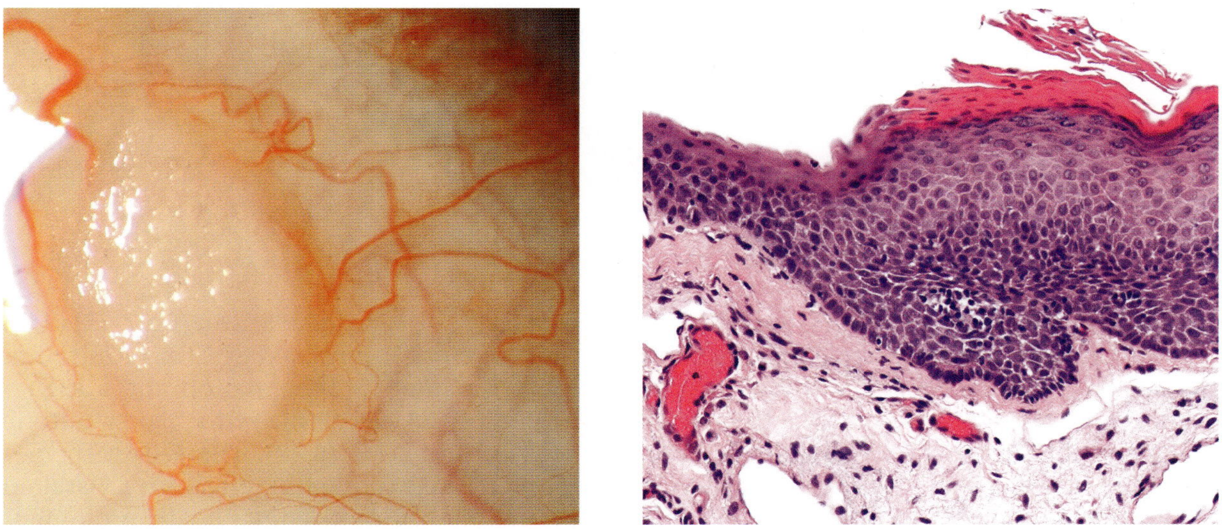

Figure 2-12

CONJUNCTIVAL INTRAEPITHELIAL NEOPLASIA WITH MILD DYSPLASIA (CIN-I)

Left: Slit-lamp photograph demonstrates an ovoid leukoplakic plaque associated with a dilated feeding vessel in the bulbar conjunctiva. (Courtesy of Drs. J. Shields and C. Shields, Philadelphia, PA.)

Right: Microscopically, the conjunctival epithelium is acanthotic, with surface keratinization and mild atypia in its inferior layers (H&E stain).

When ophthalmologists excise a pterygium, the superficial layers of the cornea are typically excised in a procedure known as a superficial lamellar keratectomy (a superficial shave biopsy of the cornea). However, Bowman layer is a natural barrier to the penetration of intraepithelial malignancy into the corneal stroma, and when intraepithelial squamous neoplasia

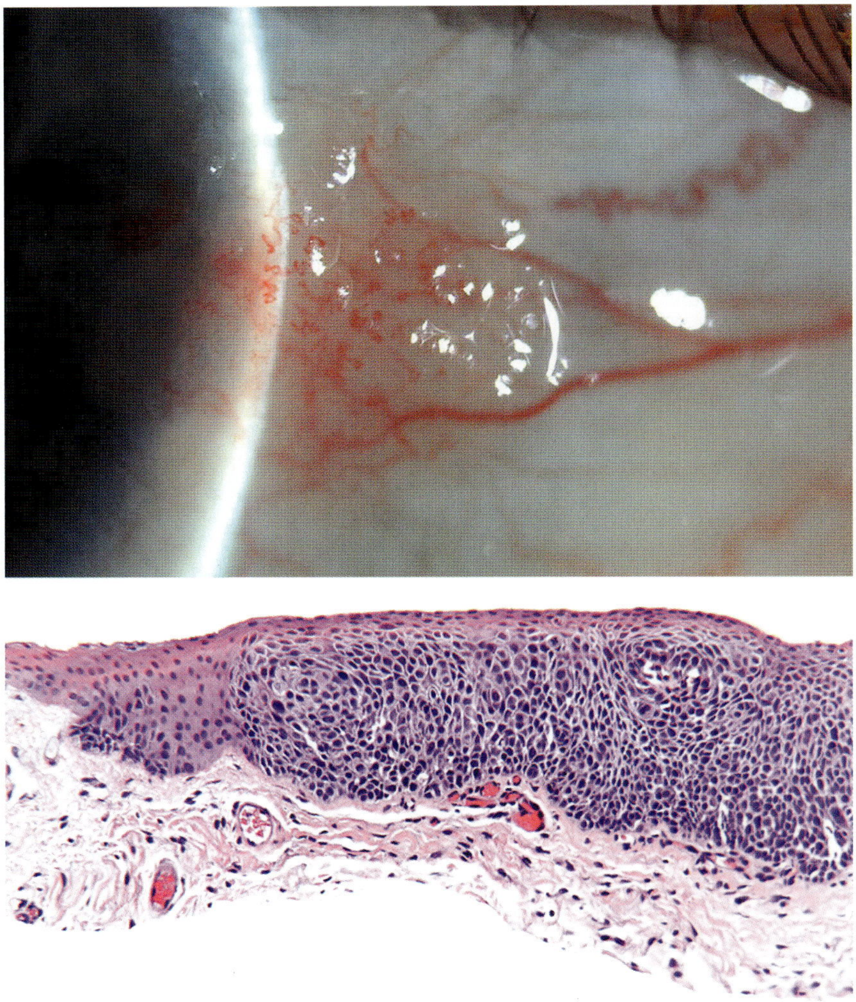

Figure 2-13

CONJUNCTIVAL INTRAEPITHELIAL NEOPLASIA WITH SEVERE DYSPLASIA (CIN-III)

Top: Slit-lamp photograph shows a papillomatous gelatinous vascular mass at the corneoscleral limbus with corneal involvement.

Bottom: The normal limbal conjunctival epithelium on the left undergoes an abrupt transition into an acanthotic and dysplastic epithelium, with dysplasia involving nearly the entire epithelial thickness. The underlying substantia propria has increased vascularity, actinic elastosis, and chronic inflammation (H&E stain).

(or intraepithelial melanocytic neoplasia) involves the corneal epithelium, surgery should remove only the epithelium, sparing Bowman layer. The surgeon may achieve this objective by debridement facilitated under topical anesthesia by wiping the epithelium off the surface with a cotton-tipped applicator soaked in absolute alcohol. The epithelium can be placed into filter paper, fixed in formalin, and embedded in paraffin. The corneal epithelium then becomes a biopsy specimen.

In SCC, there is stromal invasion by infiltrating lobules, nests, and cords of malignant cells, with variable degrees of squamous differentiation. Well-differentiated SCC is composed of cells closely resembling normal squamous epithelium, with eosinophilic cytoplasm, intercellular "bridges," and commonly, a concentric arrangement around a central focus of dyskeratosis or acellular keratin (fig. 2-15) (2,8). Moderately differentiated SCC has less prominent squamous differentiation and a higher degree of nuclear atypia with higher mitotic activity (fig. 2-15) (8,48). The cells in poorly differentiated tumors may feature a basaloid or spindle cell morphology, with prominent nuclear pleomorphism, brisk mitotic activity, and necrosis (fig. 2-16) (48).

Cytologic Findings. Impression cytology and exfoliation cytology have been used to diagnose OSSN. Cytologic characteristics are described as dysplasia, CIS, or invasive SCC (fig. 2-17). CIN is diagnosed when a mixture of dysplastic cells and intact, well-preserved malignant cells are seen in the specimen (45). Although cytologic diagnostic methods for this ocular surface disease have the advantages of low cost, noninvasive sampling, and the expedi-

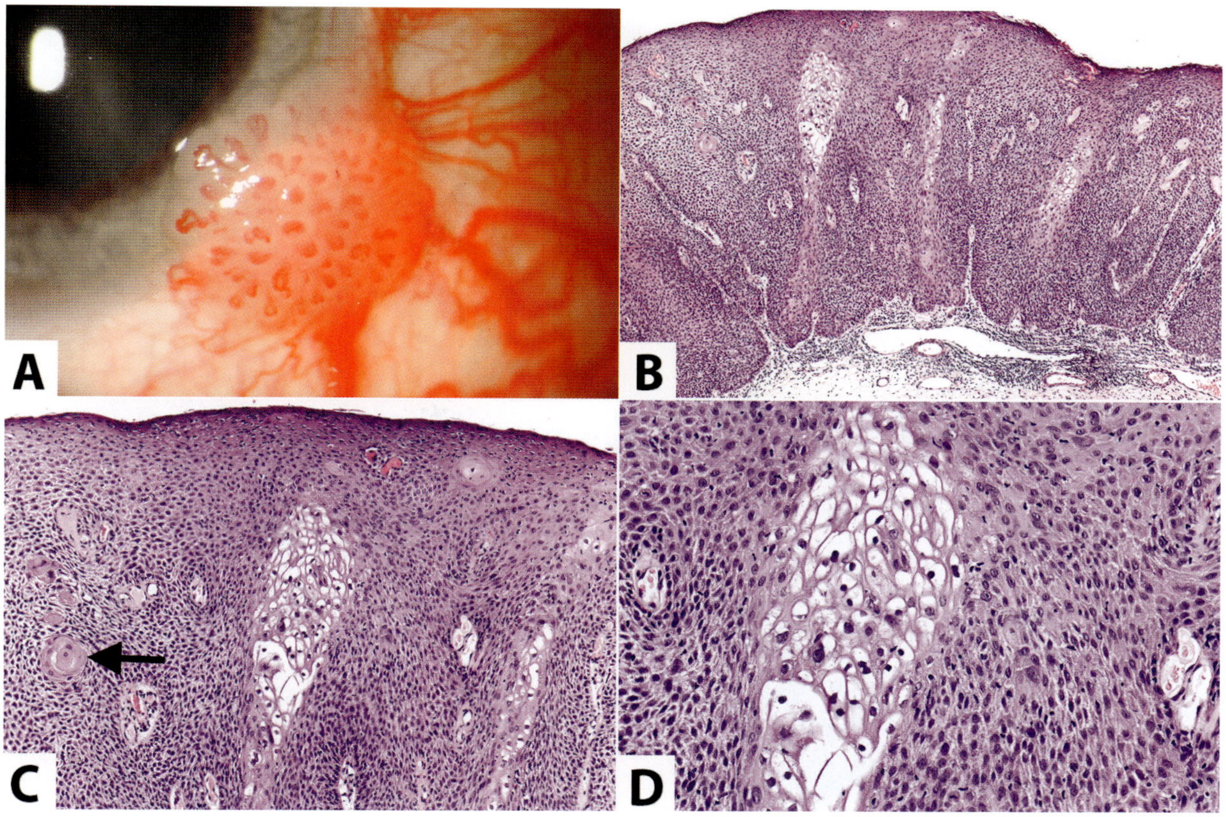

Figure 2-14

CONJUNCTIVAL SQUAMOUS CELL CARCINOMA IN SITU

A: A gelatinous papillomatous limbal mass has hairpin-like loops and prominent feeding vessels. (Courtesy of Drs. J. Shields and C. Shields, Philadelphia, PA.)

B: Microscopically, this nonkeratinizing carcinoma in situ is characterized by complete replacement of the epithelium by small atypical cells with a high nuclear to cytoplasmic ratio and overall papillomatous morphology with inflamed fibrovascular cores.

C: Rare foci of more prominent squamous differentiation with squamous eddy formation are present (arrow).

D: A central aggregate of cells with koilocytosis is also seen (H&E stain).

ency offered by the office setting, ocular surface cytology cannot reliably distinguish between the degrees of dysplasia, in situ carcinoma, and invasive carcinoma (49).

Immunohistochemical and Molecular Genetic Findings. Ancillary diagnostic studies are generally not routinely used for the diagnosis of CIN, CIS, or conjunctival SCC. Aberrant expression of p16, p53, p63 (45,50), and CK17, with loss of CK7 (51) may accompany CIN and SCC (fig. 2-18). The increased Ki-67 proliferative index may delineate the extent and degree of dysplasia (18). Telomerase reverse transcriptase promoter (*TERT*) mutations, supporting UV light mutagenesis, are found in almost half of OSSN cases (52).

Differential Diagnosis. The differential diagnosis of CIN includes benign conjunctival epithelial lesions, such as pinguecula and pterygium, pseudoepitheliomatous hyperplasia, and hereditary benign intraepithelial dyskeratosis (HBID, see above) (8). Intraepithelial involvement by malignant neoplasms, such as sebaceous carcinoma and melanoma, also needs to be considered (8). Invasive SCC is distinguished from other carcinoma variants (mucoepidermoid carcinoma, adenocarcinoma, sebaceous carcinoma) and conjunctival melanoma (8). Attention to the overall tumor morphology, including the abrupt transition from normal to a dysplastic epithelium, evidence of squamous differentiation with an absence of

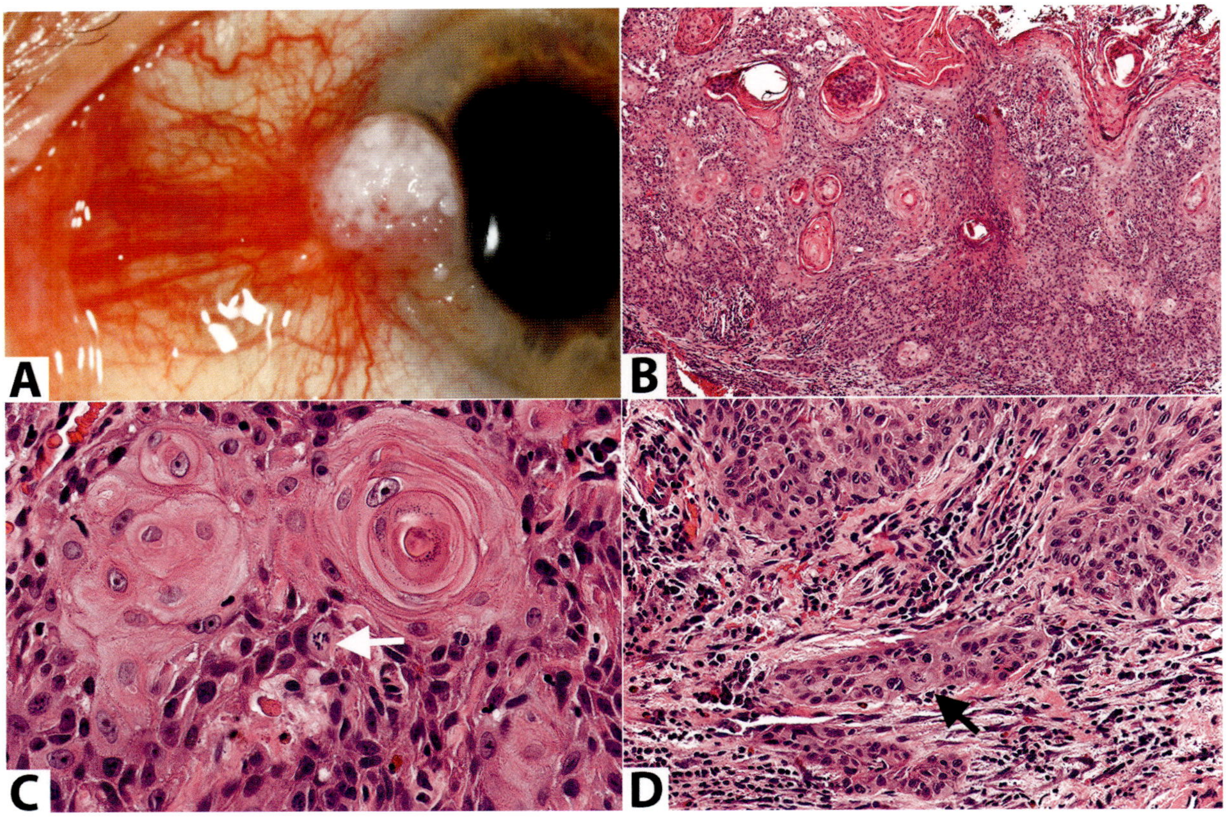

Figure 2-15

WELL- TO MODERATELY DIFFERENTIATED KERATINIZING SQUAMOUS CELL CARCINOMA AT LIMBUS

A: A leukoplakic papillomatous limbal mass with prominent feeding vessels is seen. (Courtesy of Drs. J. Shields and C. Shields, Philadelphia, PA.)

B: Microscopically, the neoplasm is characterized by full-thickness epithelial atypia, with epithelial acanthosis and prominent surface parakeratosis.

C: In situ component of the tumor with a well-differentiated morphology, prominent intercellular "bridges," squamous eddies, and keratin pearls. A rare mitotic figure (arrow) is present.

D: Invasive component of the neoplasm is characterized by moderately differentiated bands of infiltrating, mitotically active (arrow) carcinoma with less prominent squamous differentiation, in a background of desmoplastic and chronically inflamed stroma (H&E stain).

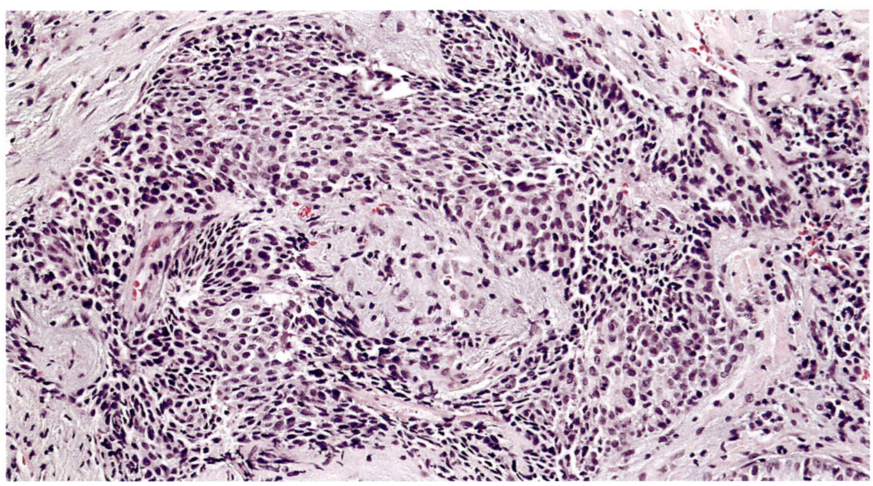

Figure 2-16

INVASIVE CONJUNCTIVAL SQUAMOUS CELL CARCINOMA, POORLY DIFFERENTIATED

Invasive neoplastic cells with minimal squamous differentiation, prominent nuclear pleomorphism, and brisk mitotic activity (H&E stain).

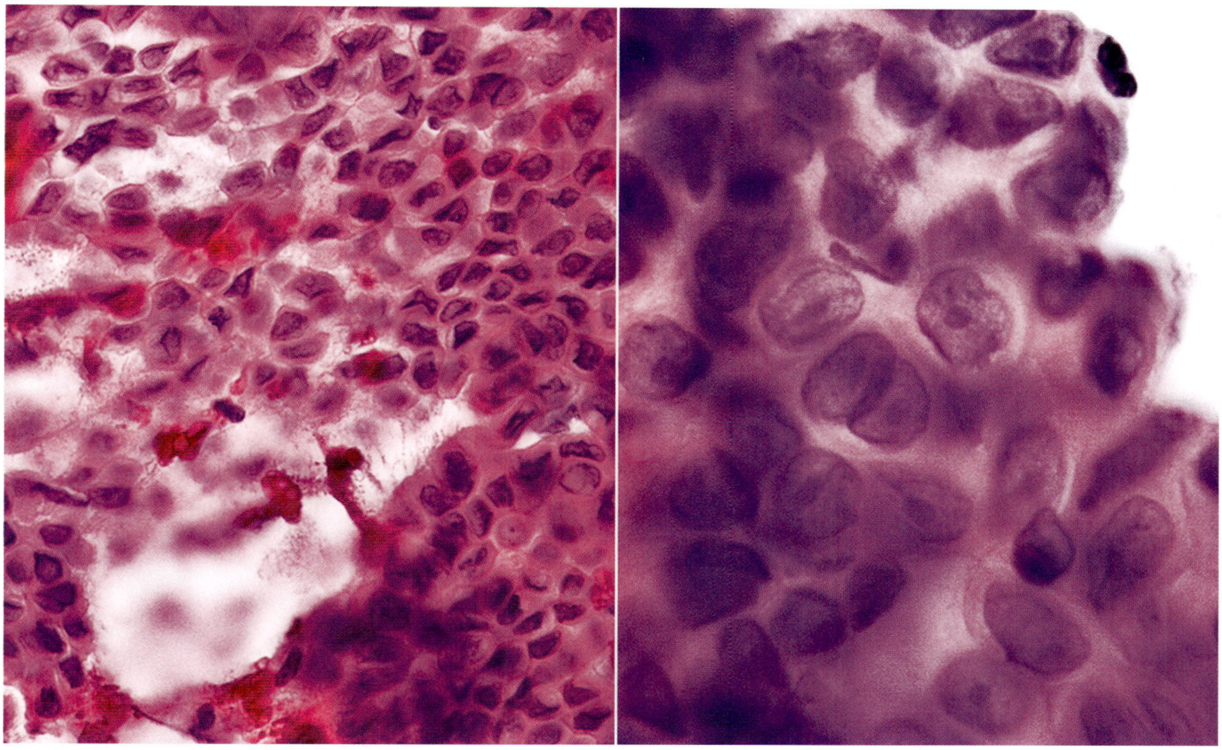

Figure 2-17

OCULAR SURFACE SQUAMOUS NEOPLASIA: EXFOLIATIVE CYTOLOGY

Sheets of rounded to polygonal squamous cells with irregular nuclei, prominent nucleoli, high nuclear to cytoplasmic ratio, and scant to moderate amounts of eosinophilic cytoplasm are seen (H&E stain).

intraepithelial mucin, and absence of sebaceous differentiation favor the diagnosis of CIN/SCC. In challenging cases, special stains for mucin and immunohistochemical studies supporting the presence of squamous differentiation (high molecular weight cytokeratins) (50,51), and excluding melanocytic differentiation (S-100 protein, SOX-10, Melan-A) (2,8) and sebaceous differentiation (androgen receptors, Ber-EP4, adipophilin) (53) are useful.

Treatment and Prognosis. Surgical excision with a 2- to 3-mm margin and adjuvant cryotherapy is the most accepted treatment of CIN and SCC. In recent years, multiple topical chemotherapeutic agents have been used as adjuvant and neoadjuvant therapeutic modalities and as monotherapy. These agents include mitomycin-C (MMC), 5-fluorouracil (5FU), and interferon alpha 2b (INFa2b) (45). CIN is a low-grade lesion, with a tendency for local recurrence and infrequent progression (3 percent) to invasive SCC (45).

SCC can be locally aggressive, invading the eye and orbit in 13 percent and 10 percent of cases, respectively (fig. 2-19) (8). Recurrence and locally aggressive behavior may accompany incomplete excision, higher Ki-67 proliferative index (54), and an immunocompromised state (48,55). Mortality from direct extension into the cranium and metastases are uncommon, with figures of 0 to 8 percent quoted (54,56). Tumor-related mortality has been shown to be higher in locally advanced disease (57).

CONJUNCTIVAL SQUAMOUS CELL CARCINOMA VARIANTS

Mucoepidermoid Carcinoma

Conjunctival mucoepidermoid carcinoma is a rare mucin-producing variant of squamous cell carcinoma, accounting for only 0.3 percent of all dysplastic and malignant squamous lesions (58). Similar to conventional SCC, mucoepidermoid carcinoma develops in the limbal and

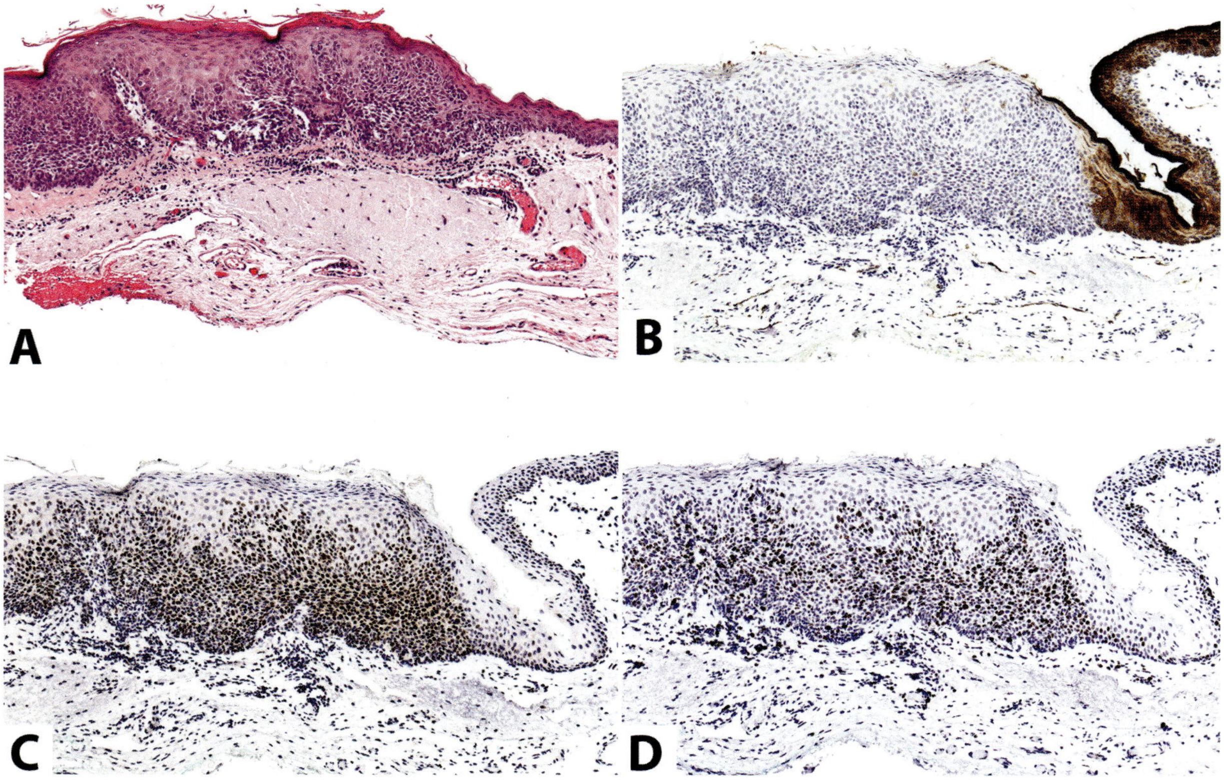

Figure 2-18

CONJUNCTIVAL INTRAEPITHELIAL NEOPLASIA

A: This conjunctival intraepithelial neoplasia is characterized by prominent squamous differentiation with surface parakeratosis. The dysplasia ranges from mild to severe. Severe actinic elastosis and chronic inflammation are present in the underlying substantia propria.

B: There is an abrupt transition from normal conjunctival epithelium with retained CK7 immunoreactivity to a dysplastic epithelium, which is negative for CK7.

C: The expression of p53 is increased in the region of dysplasia.

D: The Ki-67 proliferative index is increased in the dysplastic epithelium (H&E stains).

bulbar conjunctiva as a white-gray or fleshy mass (58). Patients are of an average age of 67 years (fig. 2-20) (8).

This neoplasm features a mixture of epidermoid and mucin-producing cells in various proportions. Frequently, the presence of intracellular mucin is subtle and requires ancillary stains, such as mucicarmine, Alcian blue, and colloidal iron for identification (fig. 2-20) (8,55). When compared to its salivary gland counterpart, conjunctival mucoepidermoid carcinoma lacks prominent intermediate cells and large pools of extracellular mucin surrounded by the neoplastic epithelium (58). The presence of squamous pearls, commonly seen in conventional conjunctival SCC, is the exception rather than the rule (8). Mucoepidermoid carcinoma expresses CK7, typically lost in conventional conjunctival SCC (58).

Mucoepidermoid carcinoma is more aggressive than conventional SCC, demonstrating a high rate of local recurrence and locally aggressive behavior, with invasion into the eye ocular or orbit (see figs. 2-24, 2-25) (8,55,58). There are reports of uncommon regional and distal metastases (8,55). Thus, management of this tumor requires precise local control by an experienced surgeon (58). Because the clinical presentation of mucoepidermoid carcinoma is typically similar to conventional SCC, histopathologic evaluation plays a critical role in guiding treatment (55,58).

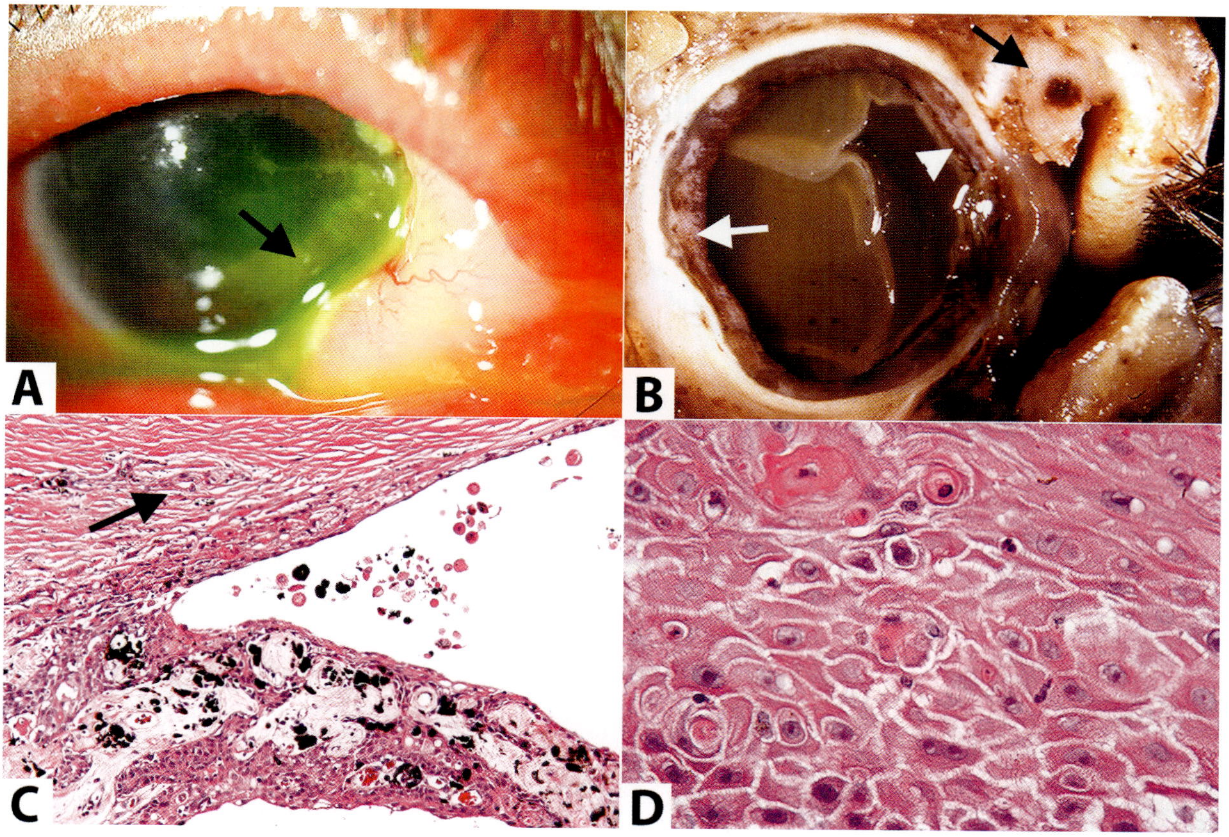

Figure 2-19

INVASIVE CONJUNCTIVAL CARCINOMA WITH INTRAOCULAR INVOLVEMENT

A: A white limbal mass is associated with corneal edema and retrocorneal gray-white material (arrow), suggestive of intraocular invasion by the tumor. The surface of the eye was stained with fluorescein dye to highlight surface corneal pathology, lending the yellow-green tinge to the ocular surface.

B: Gross evaluation of orbital exenteration shows a bulbar and anterior orbital white-gray, focally hemorrhagic and necrotic mass. There is patchy involvement of the anterior chamber structures (arrowhead) and posterior uveal tract (white and black arrows) by the gray infiltrating tumor.

C: Microscopic evaluation of the anterior segment shows invasive squamous cell carcinoma, which penetrates through the limbal emissary canals (arrow) to invade the anterior chamber angle structures and iris. Dyskeratotic tumor cells and macrophages are floating in the anterior chamber.

D: The tumor is well-differentiated with prominent keratinization. Intraocular invasion by conjunctival squamous cell carcinoma is a rare event (H&E stain).

Spindle Cell Carcinoma

Spindle cell carcinoma (sarcomatoid carcinoma) of the conjunctiva is an uncommon variant of SCC, with only a few cases reported in the literature (8,55). This neoplasm appears typically as a localized, elevated, smooth-surfaced, reddish limbal or bulbar conjunctival mass in elderly patients (8).

Microscopically, conjunctival spindle cell carcinoma features malignant spindle cells with vesicular nuclei, large basophilic or eosinophilic nucleoli, and variable mitotic figures, arranged in fascicles and associated with stromal desmoplasia, reminiscent of sarcoma (fig. 2-21) (8,55). There may be subtle continuity between the invasive spindle cell population and the overlying epithelium, which features a variable degree of dysplasia (8).

The neoplastic cells are characteristically positive for vimentin and may also express actin and desmin. A broad panel of cytokeratins and p63 or p40 may be needed to demonstrate the presence of epithelial differentiation and avoid misdiagnosis as atypical fibroxanthoma,

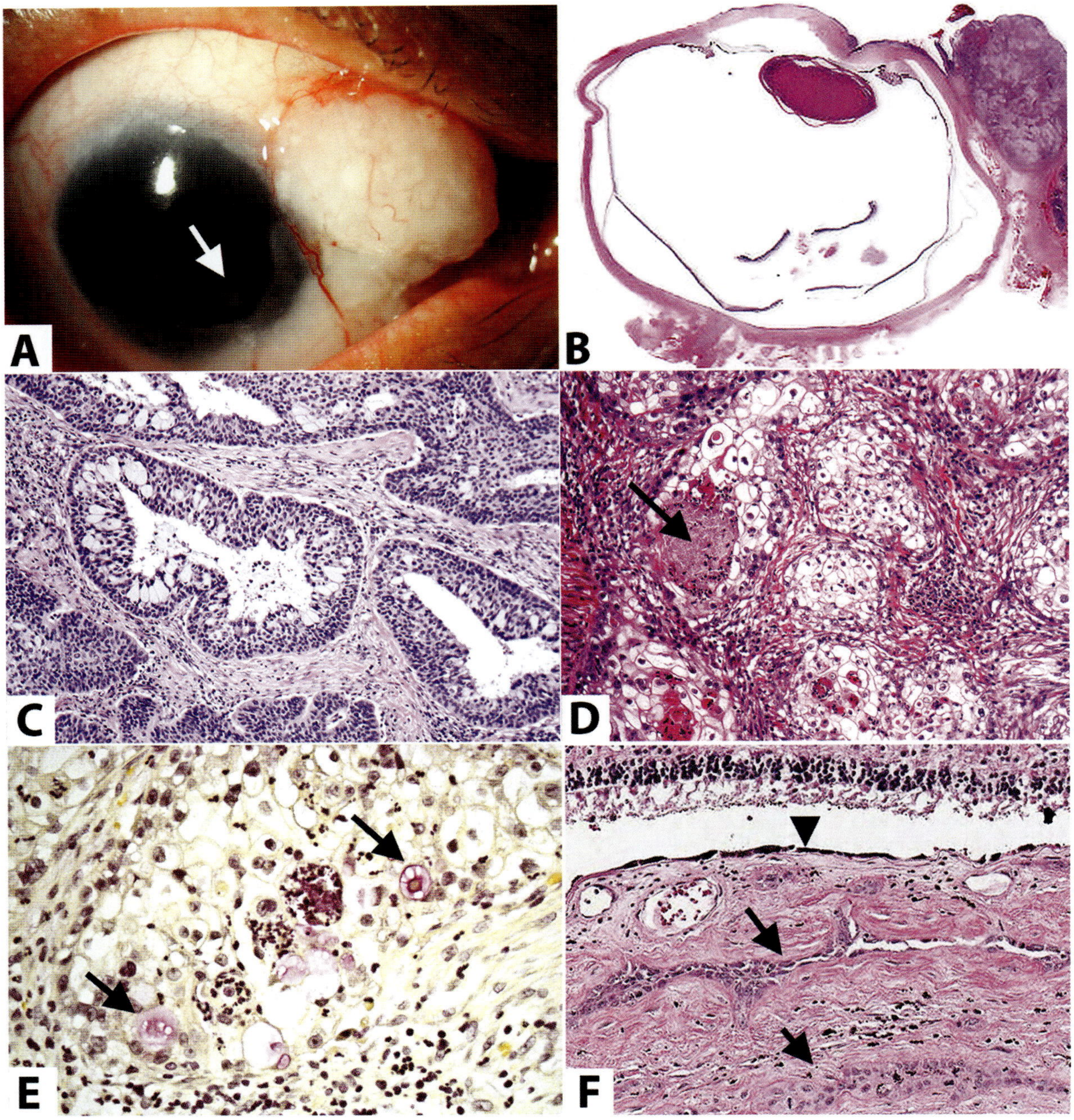

Figure 2-20

MUCOEPIDERMOID CARCINOMA OF THE CONJUNCTIVA WITH INTRAOCULAR AND ORBITAL INVASION

A: A heterogeneous, white-gray, dome-shaped mass involves the bulbar and limbal conjunctiva and the anterior chamber (arrow).

B: Low-power view of partial orbital exenteration highlights the large epibulbar and anterior orbital mass (B,C,D,F: H&E stain).

C: The orbital component of the low-grade neoplasm features cystic islands of invasive epithelial cells with goblet cells and squamoid foci.

D: In other areas, the neoplasm is composed predominantly of nests of clear cells with comedonecrosis (arrow).

E: Mucicarmine stain in the region of clear cell differentiation highlights rare foci of intracytoplasmic mucin (arrows) (E,G: mucicarmine stain).

F: The intraocular component of the neoplasm is characterized infiltrating bands and islands of carcinoma (arrows) in the choroidal stroma, beneath the retinal pigment epithelium (arrowhead) and the overlying retina.

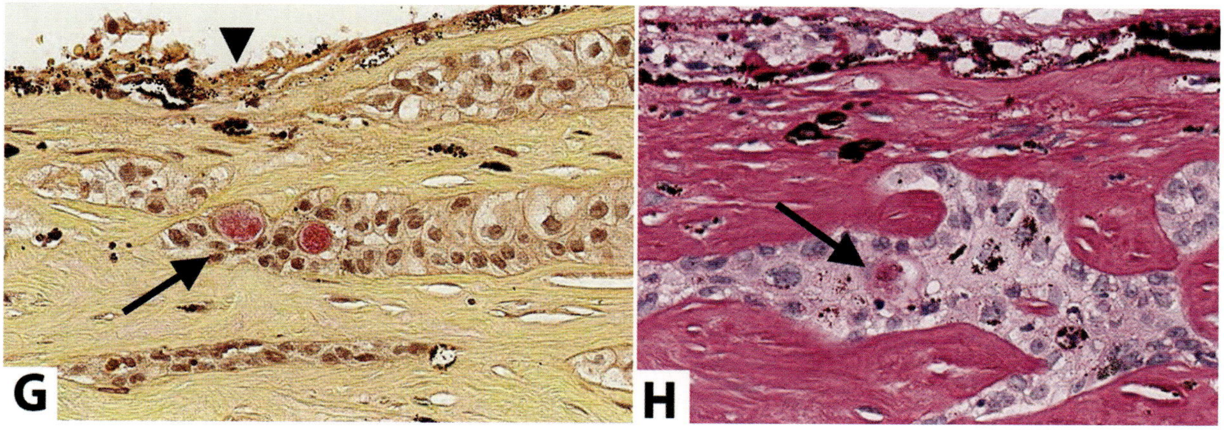

Figure 2-20, continued

G: Mucicarmine stain highlights a focus of intracytoplasmic mucin in the invasive intraocular carcinoma (arrow). Retinal pigment epithelium is denoted with an arrowhead.

H: Periodic acid–Schiff (PAS) post-diastase stain similarly highlights the mucin (arrow).

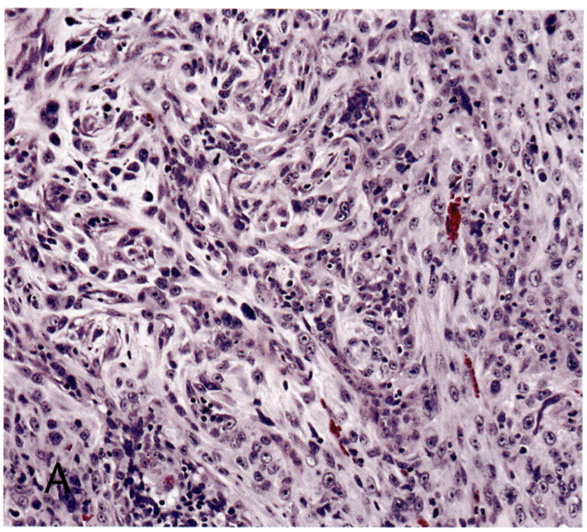

Figure 2-21

SPINDLE CELL CARCINOMA OF CONJUNCTIVA

A: An invasive neoplasm is composed of anaplastic spindle cells, arranged in fascicles and associated with stromal desmoplasia (H&E stain).

B: The neoplastic cells express vimentin.

C: Some tumor cells also express pancytokeratin.

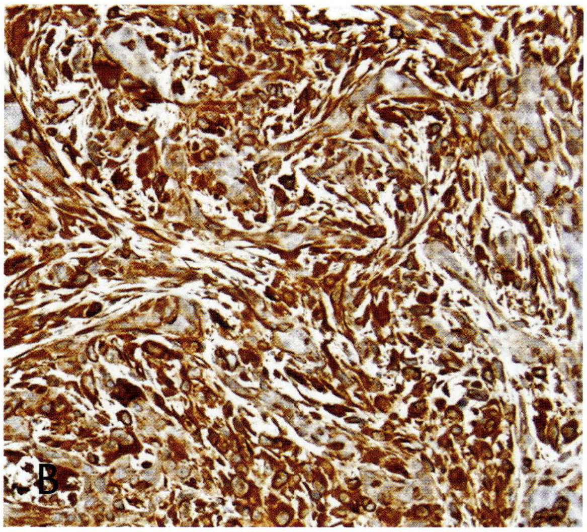

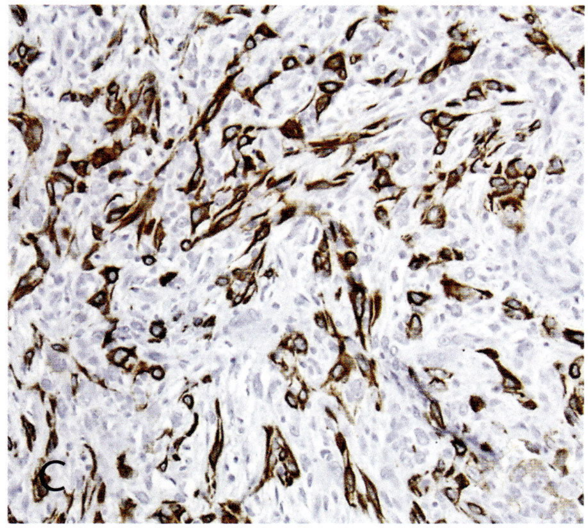

fibrosarcoma, leiomyosarcoma, amelanotic melanoma, or exuberant granulation tissue (55).

Spindle cell carcinoma of the conjunctiva is treated by complete surgical excision. Recurrence with locally aggressive behavior, including intraocular spread, has been reported (8).

Other Conjunctival Carcinomas

Other carcinomas can primarily involve the conjunctiva but are rare. *Basal cell carcinoma* arises from the conjunctiva and caruncle without any adjacent eyelid tumor. The neoplasm is indistinguishable from basal cell carcinoma in other locations (8,59).

Primary conjunctival intraepithelial *sebaceous carcinoma* without underlying sebaceous gland tumor is infrequent. The histopathology is identical to the intraepithelial spread of sebaceous carcinoma arising in sebaceous glands (8,60).

Lymphoepithelioma-like carcinoma is another rare tumor described in the conjunctiva. It is morphologically similar to and has a strong association with Epstein-Barr virus, analogous to its nasopharyngeal counterpart (8). See chapter 3 for additional discussion on periocular basal cell carcinoma and sebaceous carcinoma, and the Nonmelanocytic Tumors of the Skin Fascicle in this series for additional information.

PATHOLOGIST REPORT FOR EPITHELIAL TUMORS

Recommendations from the Task Force for Staging of Cancer of the Eye of the American Joint Committee on Cancer (AJCC), 8th edition, appears in Appendix 1. These recommendations apply only to carcinoma of the conjunctiva, consisting predominantly of SCC and corneal squamous intraepithelial neoplasia, along with other histologic subtypes (mucoepidermoid carcinoma, sebaceous gland carcinoma, adenosquamous carcinoma, spindle cell carcinoma, and basal cell carcinoma) (61). Nonepithelial tumors of the conjunctiva are not staged using these criteria. Tumor staging depends upon the review of a biopsy. The 8th edition staging system is different from prior editions by more precisely defined disease extent for early invasive lesions (T1 and T2). This change has implications for management, as emerging studies support adjuvant treatment in some of these patients (61).

Clinical assessment should document inspection, measurement, and documentation, with photography of all affected conjunctival surfaces, palpation of regional lymph nodes, and examination of ipsilateral sinuses. Anterior segment ultrasound imaging and ultrasonography are useful for measuring tumor thickness and for identifying invasion of adjacent structures. A radiologic evaluation may be necessary to stage locally invasive and nodal disease (61).

Complete resection of the primary site, if possible, is required for pathologic staining. The specimen should be submitted on a piece of filter paper, to ensure a flat orientation. Histopathologic evaluation should assess the depth of invasion by the tumor and peripheral and deep margins. Sentinel lymph node biopsy is investigational. Beyond the factors used to assign T, N, or M categories, the system does not require additional prognostic factors for stage groupings (61).

BENIGN MELANOCYTIC TUMOR-LIKE LESIONS AND NEOPLASMS

Melanocytic Nevi

Various types of melanocytic nevi have been recognized in the conjunctiva, analogous to their cutaneous counterparts. Common acquired melanocytic (nevocellular) nevi are the most frequent, followed by the blue and combined melanocytic nevi (62–64), conditions described in detail below. Other rare nevus variants described in the conjunctiva include spindle and epithelioid (Spitz) nevus (8), granular cell nevus, balloon cell nevus, inverted (clonal) cell nevus (65), and melanocytoma (8). See *Melanocytic Tumors of the Skin*, Fascicle 10 of this series for additional information on these lesions.

Definition. Melanocytic nevi are benign conjunctival melanocytic lesions composed of modified melanocytes, called nevus cells, derived embryologically from the neural crest (2).

Clinical Features and Epidemiology. Conjunctival nevocellular nevi first appear in childhood as small, circumscribed, amelanotic or pigmented lesions that move freely over the scleral surface, and are associated with feeder and intrinsic vessels. These tumors are most frequent in the bulbar (commonly juxtalimbal) conjunctiva, caruncle, and plica semilunaris, and are rare in the fornix and palpebral

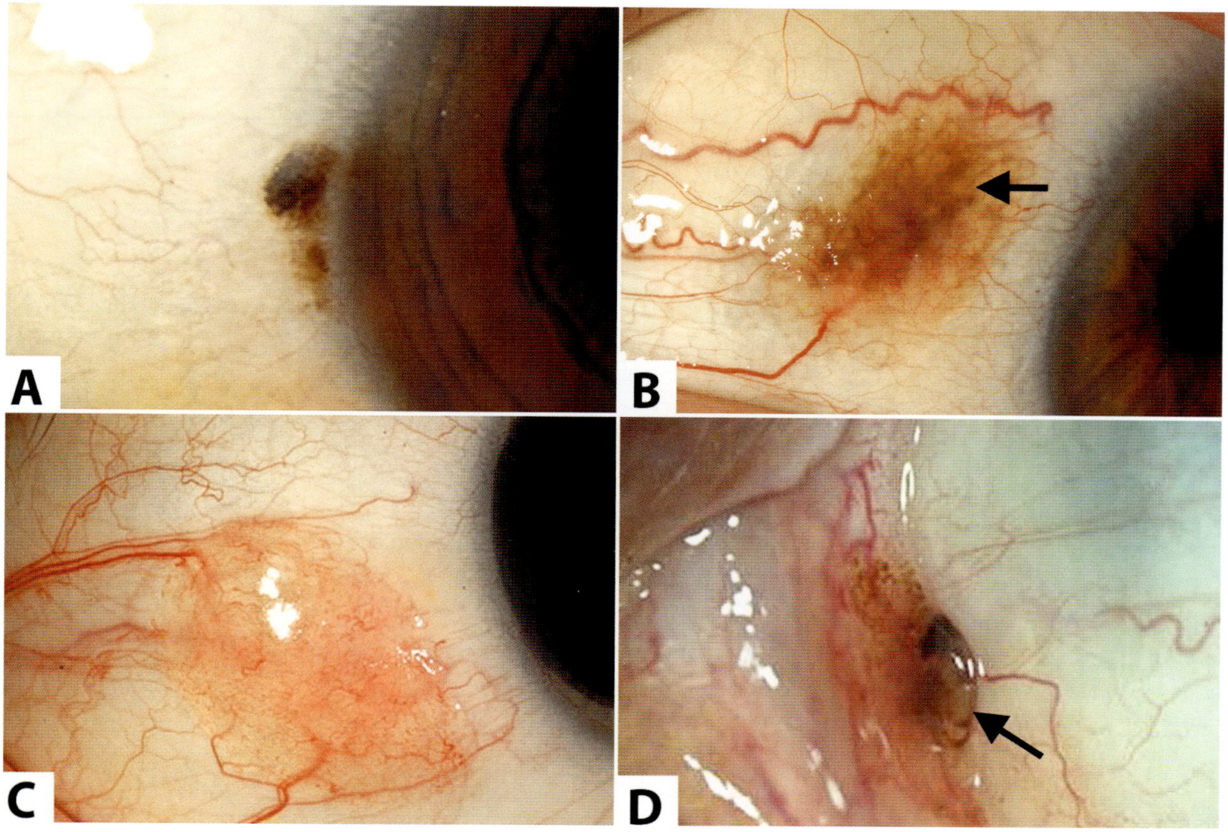

Figure 2-22

CONJUNCTIVAL MELANOCYTIC NEVI

There is a clinical spectrum of conjunctival nevi: flat pigmented macule on bulbar conjunctival surface (A), elevated, circumscribed pigmented nodule with cysts (arrow) and feeder vessels (B), amelanotic vascular nodule with feeder vessels (C), and variably pigmented nodule in plica semilunaris with cysts (arrow)(D). (All figures courtesy of Drs. J. Shields and C. Shields, Philadelphia, PA.)

conjunctiva (fig. 2-22) (62,63). Conjunctival nevocellular nevi occur equally in males and females, and are more common in whites (63).

Junctional nevi are typically diagnosed in childhood or adolescence and are overall rare in biopsy specimens (63). *Compound nevi*, the most common type of nevi biopsied, account for 70 to 78 percent of conjunctival nevi and are diagnosed at a median age of 21 to 23 years (range, 5 to 74 years) (63). *Subepithelial nevi*, regarded as the last stage of conjunctival nevocellular nevus ontogeny, are diagnosed, at the mean age of 46 years (range, 12 to 87 years), and comprise 15 to 24 percent of conjunctival nevi biopsied. Both compound and subepithelial nevi frequently appear elevated and contain epithelial cysts (63).

Blue nevi account for fewer than 4 percent of conjunctival and caruncular nevi, and are distributed with even frequency in the bulbar, forniceal, and palpebral conjunctiva and in the caruncle (63,64). Most blue nevi are diagnosed between the third and fifth decades of life, when they present as dark brown to black, sharply demarcated, unifocal or multifocal, slightly elevated conjunctival masses, with a thickness of approximately 2 mm (64). Unlike the common acquired nevocellular conjunctival nevi, blue nevi lack epithelial cysts and rarely contain intrinsic or feeder vessels (63,64).

Although 4 to 8 percent of conjunctival melanocytic nevi become increasingly pigmented and enlarge during puberty and pregnancy, most lesions are stable in size and color over time and do not extend into the cornea (62,63). Table 2-2 summarizes the critical distinguishing clinical characteristics between conjunctival

Table 2-2

CLINICAL CHARACTERISTICS OF CONJUNCTIVAL MELANOCYTIC LESIONS[a]

Lesion Name	Clinical Characteristics
Nevus	Most lesions apparent at a young age Mostly stable, occasional change with hormonal surges Flat or nodular Mobile over sclera Variable pigmentation Cysts frequent Seldom encroach onto the cornea and rarely seen in the palpebral (tarsal) conjunctiva
Melanoma	Most lesions present at an older age (around 60 years) Usually in lightly pigmented individuals Nodular Immobile Vascular Variable pigmentation
Primary acquired melanosis	Flat Most lesions present at an older age (around 60 years) Usually in lightly pigmented individuals Unilateral Waxing and waning Variable pigmentation with intraepithelial "dusting"
Conjunctival Skin Tone Associated Melanosis (C-STAM)	Darkly pigmented individuals Bilateral, asymmetric Can enlarge with age Often densest around the limbus Variable pigmentation with intraepithelial "dusting"

[a]Data from references 63 and 66.

nevi and other melanocytic lesions (63,66). Because the clinical and histopathologic differential diagnosis of conjunctival nevi primarily includes malignant melanoma, the importance of careful documentation of clinical history and exam findings, and communication of these findings to the pathologist, cannot be overemphasized.

Microscopic Findings. Junctional nevi are composed of nests of round to spindle-shaped melanocytes with oval nuclei and occasional small nucleoli, arranged along the epithelial-subepithelial junction (fig. 2-23) (8,62). Compound nevi contain nevus cells both within the epithelium and the substantia propria, with maturation toward the deeper layers. Plump cells (type "A") are seen in the junctional nests and superficial substantia propria, and lymphocytoid (type "B") nevus cells in the deeper layers of the subepithelial compartment. In older patients, the deepest cells may display spindle-like neuroid differentiation (type "C") (62). Solid and cystic epithelial rests are common. The cysts are lined by conjunctival epithelium and goblet cells (figs. 2-24, 2-25). Polyhedral balloon cells and spindle cells are also found.

The presence of multinucleated cells and pleomorphism does not imply malignancy. Mitotic activity, if present, appears in the superficial layer. The nevi may be associated with a prominent stromal and intraepithelial, predominantly lymphocytic, inflammatory infiltrate, often with eosinophils. These lesions tend to appear in late childhood or early adolescence and are termed *inflamed conjunctival nevi* (62).

Subepithelial nevi lack junctional components and are intrastromal, with epithelial cysts seen in approximately half of the lesions (fig. 2-26). Compared to junctional and compound nevi, subepithelial nevi are more likely to be amelanotic or minimally pigmented, and are rarely associated with a lymphocytic infiltrate (62,67).

Blue nevi are characterized by a proliferation of dendritic and/or fusiform melanocytes in the subepithelial connective tissues (fig. 2-27) (62,64). Except for combined or collision

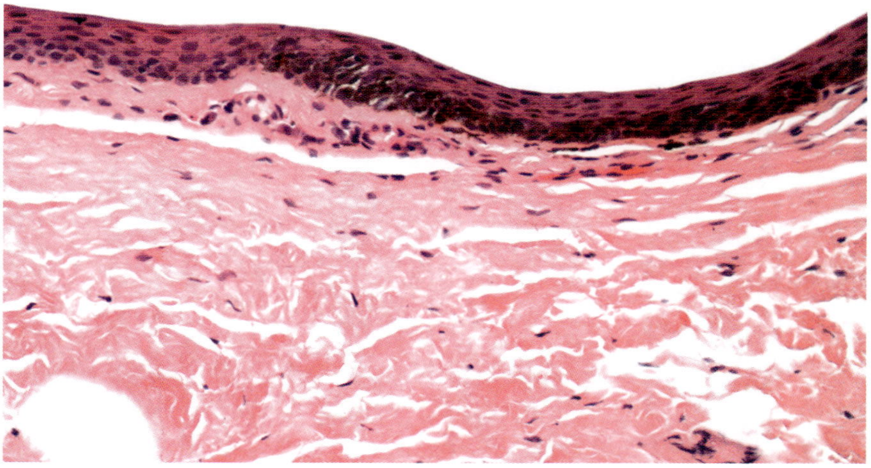

Figure 2-23

JUNCTIONAL MELANOCYTIC NEVUS IN A CHILD

Confluent nests of round to spindle-shaped melanocytes with oval nuclei and inconspicuous nucleoli are arranged along the epithelial-subepithelial junction (H&E stain). (Courtesy of Dr. R. C. Eagle, Jr., Philadelphia, PA.)

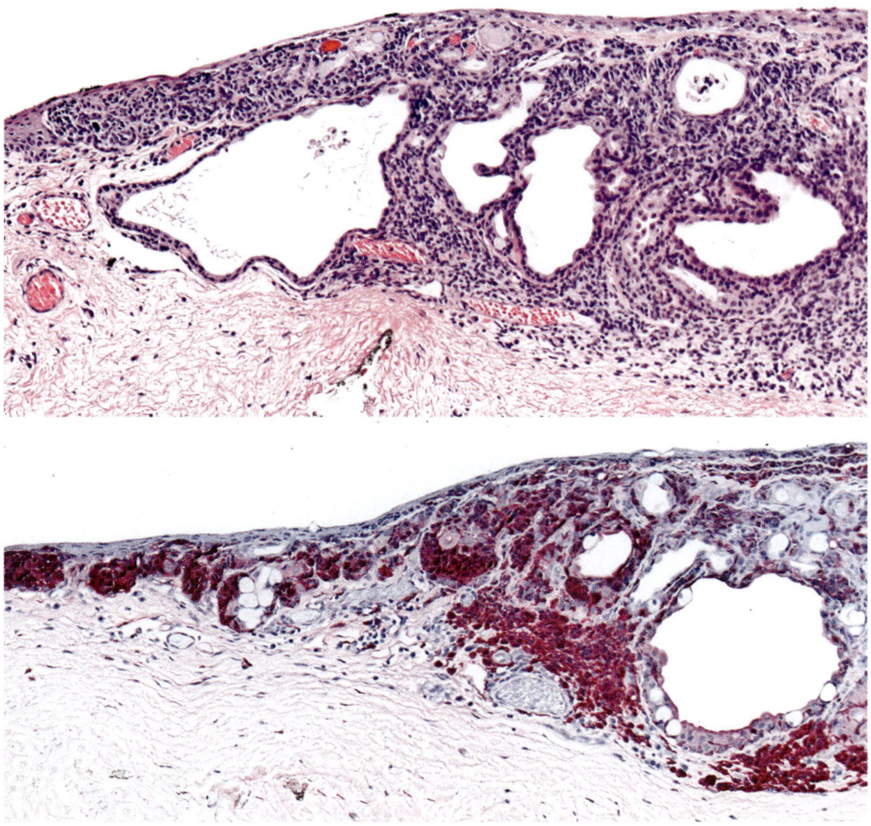

Figure 2-24

COMPOUND CYSTIC MELANOCYTIC NEVUS: PERIPHERAL ASPECT

Top: Nests of nevus cells are within the epithelium and the substantia propria. Solid and cystic conjunctival epithelial rests are present in the substantia propria, surrounded by the nevus cells, a frequent feature in a conjunctival melanocytic nevus (H&E stain).

Bottom: Both the junctional and subepithelial components of the lesion express S-100 protein.

lesions, such as a *combined nevus*, which incorporates features of both nevocellular and blue nevi (fig. 2-28), there is no associated junctional melanocytic proliferation or entrapped epithelial cysts (62,64). *Common blue nevi* are characterized by the presence of variably pigmented, spindle-shaped dendritic melanocytes, which reveal slender, branching networks of dendritic processes in the substantia propria. Melanophages and stromal fibrosis are frequent (fig. 2-27) (62,64). *Cellular blue nevi* have dense aggregates of cytologically bland spindle or epithelioid amelanotic melanocytes, frequently set in a background of common blue nevi (fig. 2-29) (62). In benign blue nevi variants, the mitotic figures are characteristically absent (62).

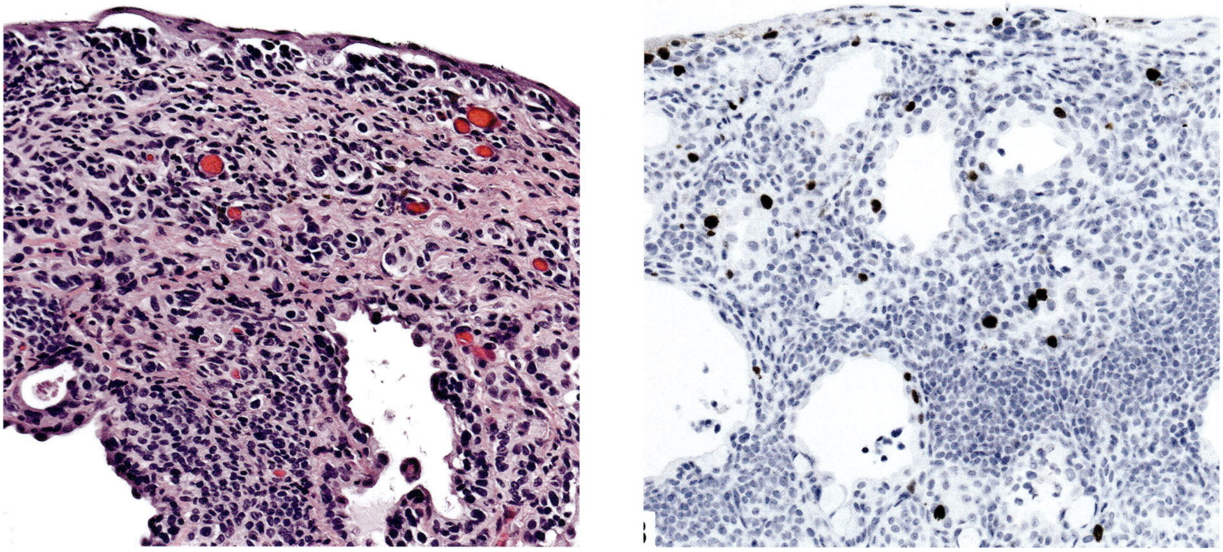

Figure 2-25

COMPOUND CYSTIC MELANOCYTIC NEVUS: CENTRAL ASPECT

Left: The junctional and subepithelial components of the lesion are seen, with maturation toward the deeper layers. Similar to the periphery of the lesion (fig. 2-24), nevocellular nevus cells in the substantia propria surround cystic rests of conjunctival epithelium (H&E stain).

Right: Ki-67 immunostain is mostly positive in the epithelium and labels rare nevus cell nuclei.

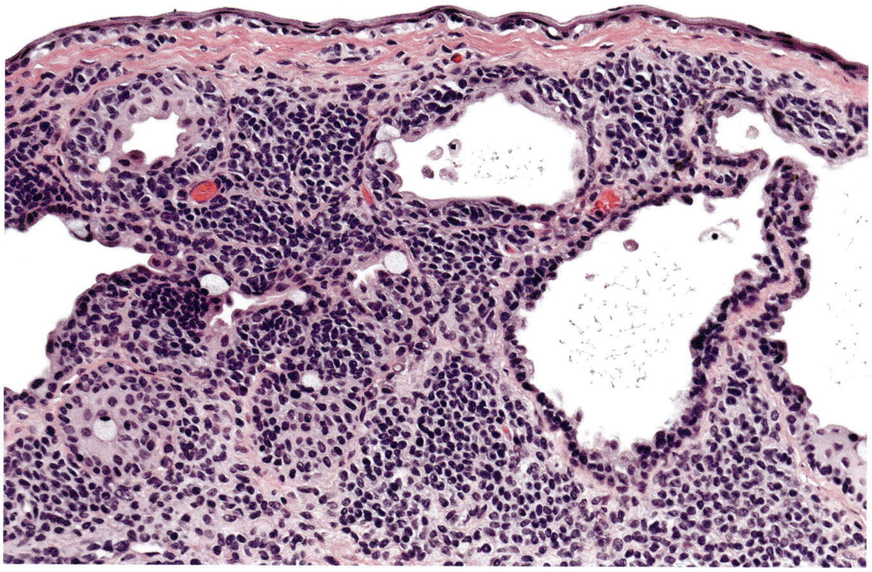

Figure 2-26

SUBEPITHELIAL CYSTIC MELANOCYTIC NEVUS

Nests of nevus cells are within the epithelium and the substantia propria, with maturation toward the deeper layers. Solid and cystic epithelial rests are present in the substantia propria (H&E stain).

Immunohistochemically, all nevocellular nevi are positive for S-100 protein, SOX-10, and MART-1 and frequently weakly positive for HMB-45 within the superficial junctional zone (65,68). Ki-67 immunoreactivity is generally confined to the junctional zone and rare superficial stromal melanocytes, where the mean proliferative index is approximately 1 to 2 percent (figs. 2-24, 2-25) (65,67,68). Blue nevi can be diffusely positive for HMB-45 (69).

Molecular Genetic Findings. *BRAF* V600E mutations are identified in half of the acquired nevocellular conjunctival nevi (70). The genetic profile of conjunctival blue nevi remains obscure. Cutaneous blue nevi harbor mutations in *GNA11* and *GNAQ* (71).

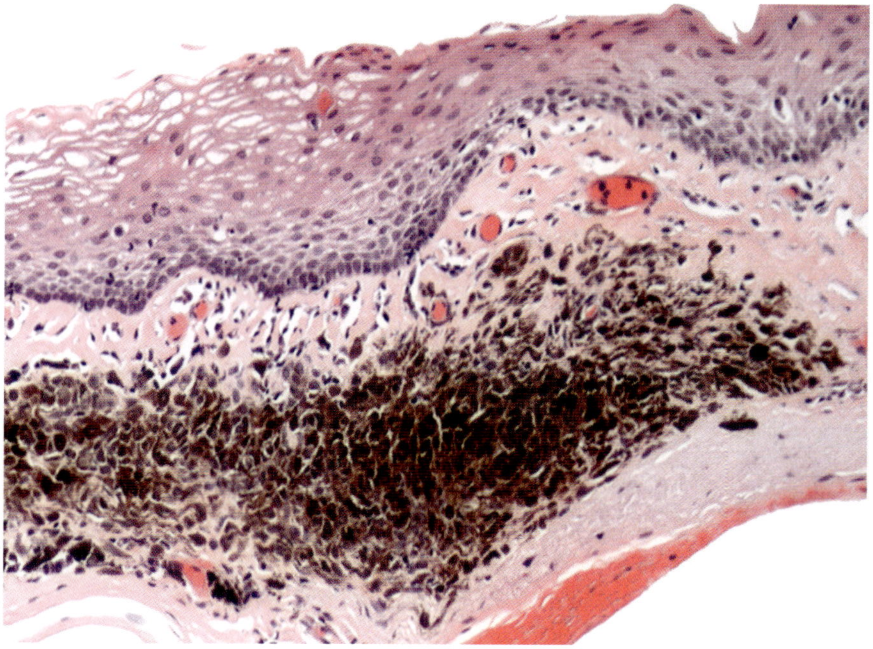

Figure 2-27

COMMON BLUE NEVUS OF THE BULBAR CONJUNCTIVA

Microscopic evaluation demonstrates intensely pigmented, ovoid and spindle-shaped dendritic melanocytes in the substantia propria, associated with stromal fibrosis and melanophagic infiltrate (H&E stain). (Courtesy of Dr. R. C. Eagle, Jr., Philadelphia, PA.)

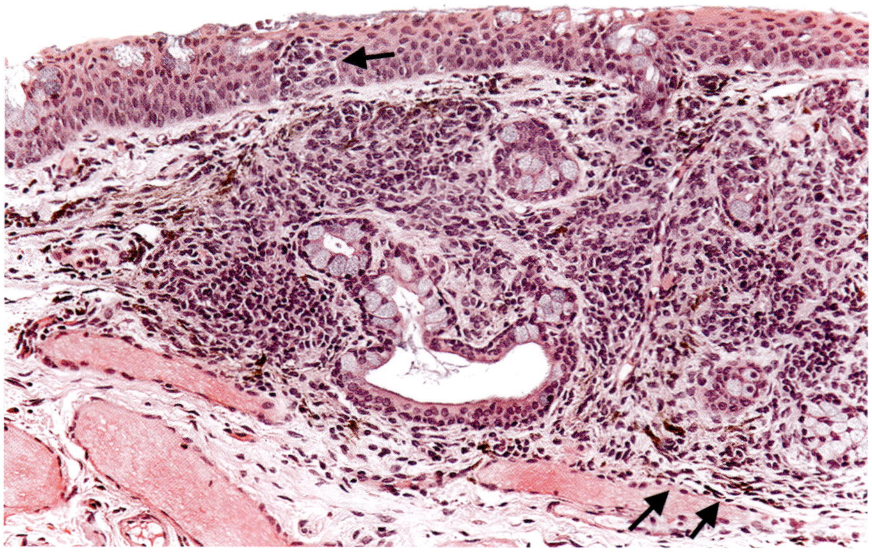

Figure 2-28

COMPOUND COMBINED CYSTIC MELANOCYTIC NEVUS OF THE CONJUNCTIVA

The lesion combines the features of compound cystic melanocytic nevus with common blue nevus. A junctional nevus nest is highlighted with a single arrow. The deep stromal pigmented spindle melanocytes reminiscent of blue nevus are labeled with two arrows (H&E stain).

Differential Diagnosis. Histopathologically, junctional melanocytic nevus may resemble primary acquired melanosis (see below). Junctional nevus in an adult should be considered primary acquired melanosis with atypia until proven otherwise (2). Invasive conjunctival melanoma is a chief diagnostic consideration for other nevus variants. Invasive melanoma frequently lacks symmetry, maturation, and epithelial cysts, and tends to demonstrate higher-grade nuclear atypia and mitotic activity (fig. 2-30). Additionally, most invasive conjunctival melanomas are associated with an atypical intraepithelial melanocytic proliferation (62).

Immunohistochemically, invasive conjunctival melanoma is more likely to be diffusely positive for HMB-45, with a mean Ki-67 proliferative index of 16 to 17 percent (fig. 2-30) (67,68). Despite careful histopathologic examination, a group of indeterminate melanocytic lesions exists that cannot be unequivocally classified as benign or malignant (72).

Extraocular extension of uveal melanoma, with secondary conjunctival involvement, is a

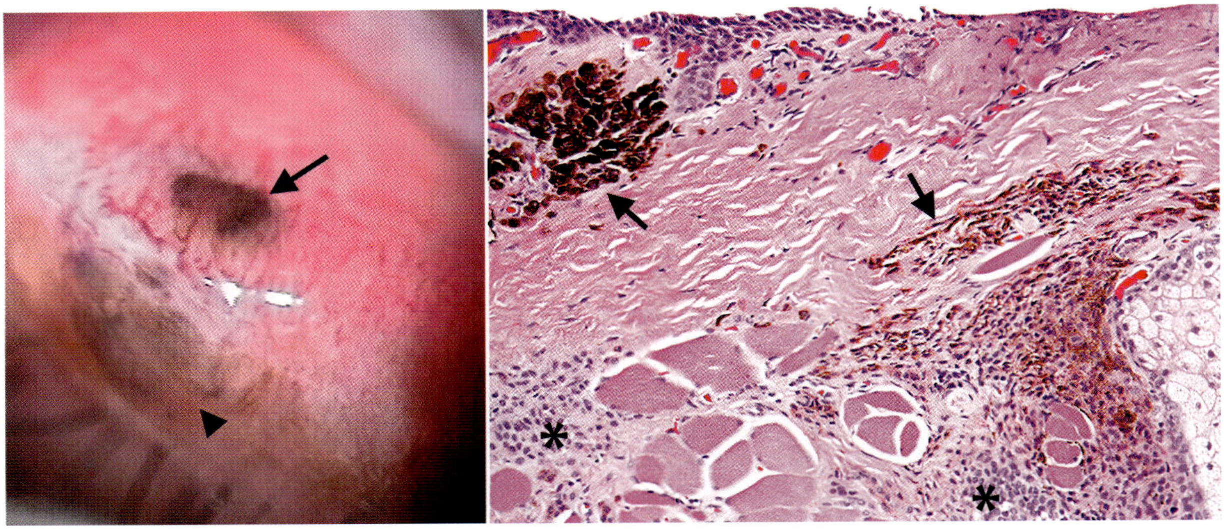

Figure 2-29

CELLULAR BLUE NEVUS OF EYELID MARGIN AND TARSAL CONJUNCTIVA IN A CHILD

Left: Brown to bluish, minimally elevated lesion involves the eyelid margin (arrowhead) and palpebral conjunctiva (arrow).
Right: Microscopically, the lesion is composed of densely pigmented ovoid, spindle, and dendritic common blue nevus cells (arrows) within fibrotic stroma, and densely cellular amelanotic ovoid melanocytes (asterisks) (H&E stain).

diagnostic consideration in subepithelial and blue nevi. Attention to the overall architecture of the lesion, degree of nuclear atypia, and mitotic activity is helpful in establishing the diagnosis (62). Molecular genetic studies may be of potential benefit (70,73).

Treatment and Prognosis. Serial observation with slit lamp photography is the preferred management for conjunctival nevi. Any change that is suggestive of malignant transformation should be treated like melanoma (63,66). The risk of malignant transformation is less than 1 percent (66).

NON-NEVOID INTRAEPITHELIAL MELANOCYTIC PROLIFERATIONS

General Features

Comprehending the complex and sometimes controversial histopathologic classification of conjunctival pigmented lesions requires an understanding of the clinical challenges faced by ophthalmologists who are responsible for managing patients with these lesions. Congenital pigmentation of the ocular surface, a condition that is known as *congenital melanosis oculi,* results from hyperpigmentation in the underlying choroid. In this condition, the sclera may be blue, and the pigment is beneath, not within, the conjunctiva.

Some forms of conjunctival pigmentation are secondary to the use of topical medications that deposit pigment in the conjunctiva. Formerly, silver nitrate was used topically as a treatment for conjunctivitis and epinephrine was used topically as a treatment for open-angle glaucoma; both of these medications used chronically deposit pigment in the conjunctiva. Pigmentation used as eyeliner in some Middle Eastern cultures, kohl, can also be deposited in the conjunctiva. Conjunctival pigmentation may appear in systemic conditions such as Addison disease and Peutz-Jeghers syndrome. Conjunctival pigmentation and intraepithelial melanocytic hyperplasia also occur in response to local inflammation or an unrelated conjunctival neoplasm (fig. 2-31). All of these conditions are secondary, not primary, and they are known as *secondary acquired melanosis* [in a clinical context] (2).

Patients with dark skin may also have bilateral, flat brown conjunctival pigmentation. This pigmentation is typically more intense at the limbus and fades into the fornix (fig. 2-32). The literature uses a variety of names for this normal finding (2). One term that frequently appears in articles and books is racial melanosis,

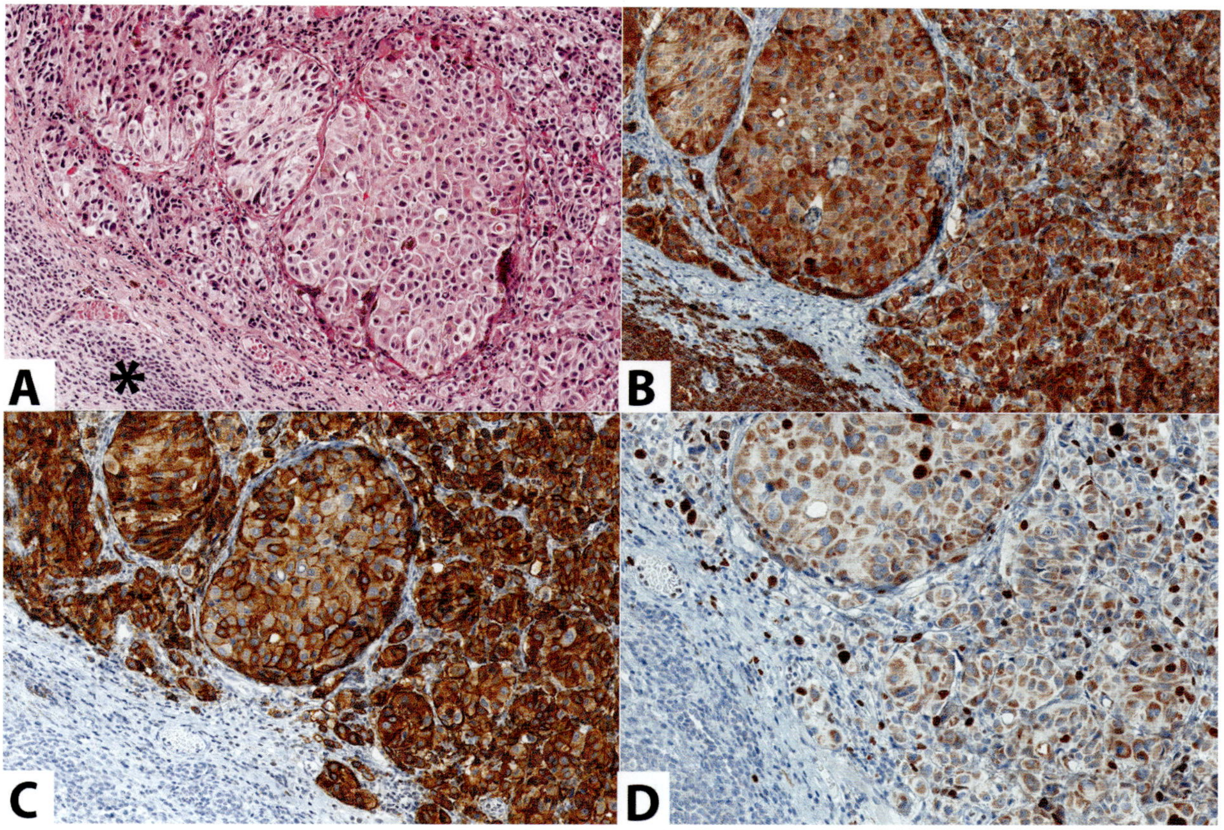

Figure 2-30

CONJUNCTIVAL MELANOMA IN ASSOCIATION WITH A NEVUS

A: Bland lymphocytoid and spindle melanocytes of conjunctival nevus lack appreciable nuclear pleomorphism and hyperchromasia (asterisk). In contrast, the nests of invasive malignant melanoma are characterized by markedly pleomorphic ("bizarre") epithelioid cells with enlarged hyperchromatic nuclei (H&E stain).

B: The nevus and melanoma components of the lesion diffusely express Melan-A.

C: HMB-45 is positive in melanoma cells only and negative in the nevus.

D: Ki-67 highlights the frequent immunoreactive cells in the melanoma, while the nevus cells are essentially negative.

a name that is medically incorrect and socially offensive. The appearance of brown conjunctival pigmentation associates with a dark skin tone, independent of "race." True, this clinical finding is common in Blacks and is normal and of no concern. And yet, this finding may also be present in the conjunctiva of people who identify as "white" with "swarthy" skin tones.

We propose a new term to describe flat, brown conjunctival pigmentation in dark-skinned individuals. *Conjunctival skin tone-associated melanosis (C-STAM)* is appropriate clinically and for use by pathologists. Some pathologists may cringe at the use of "melanosis." The pigmentation is melanin and not one of many other brown or black pigments that deposit in the conjunctiva. In C-STAM, pathologists should not detect melanocytic hyperplasia or atypia. This lesion is not a precursor of conjunctival melanoma.

Pathologists should be aware of the clinical setting before rendering diagnoses based on histopathology alone. The same findings characteristic of C-STAM in the context of unilateral flat, brown conjunctival pigmentation in a fair-skinned individual leads to a different diagnosis.

The majority of conjunctival pigmented lesions are not congenital or secondary. These are lesions that are acquired, generally unilateral, and tend to develop in fair-complexioned individuals (fig. 2-33). These lesions are classified clinically as *primary acquired melanosis* (PAM) (2,74–78). When observed, PAM may be histologically nothing

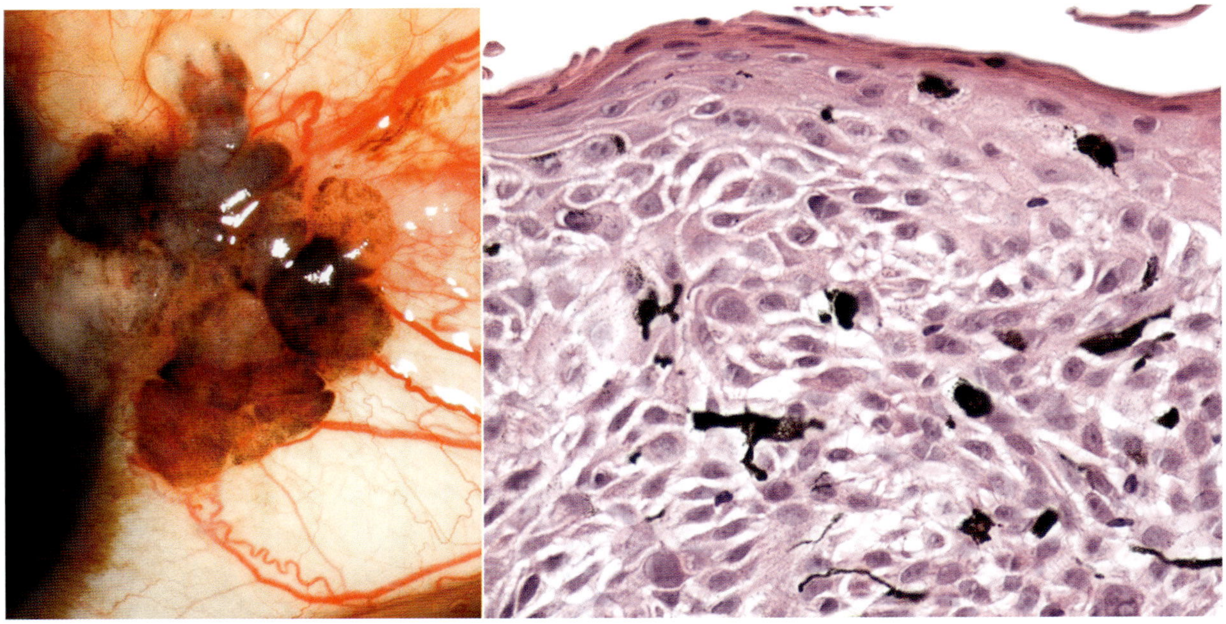

Figure 2-31

SECONDARY CONJUNCTIVAL MELANOSIS (CONJUNCTIVAL MELANOCYTIC INTRAEPITHELIAL NEOPLASIA WITHOUT ATYPIA, INTRAEPITHELIAL MELANOCYTIC PROLIFERATION WITHOUT ATYPIA)

Left: A pigmented nodule is in the interpalpebral limbal region.

Right: An increased number of cytologically normal-appearing melanocytes with focal intraepithelial migration are present in conjunctival squamous cell carcinoma excised from a darkly complexioned patient (H&E stain). (Courtesy of Dr. R. C. Eagle, Jr., Philadelphia, PA.)

more than the deposit of melanin pigment in the epithelium without melanocytic hyperplasia or atypia. On the other hand, the lesion may be a proliferation of atypical melanocytes in the conjunctival epithelium that, if untreated, may progress to conjunctival melanoma. The mortality rate of conjunctival melanoma is approximately 25 percent (79), and the best treatment is the prevention of progression from an intraepithelial precursor lesion to a potentially life-threatening invasive malignancy.

This is a formidable clinical challenge: there are no criteria available to permit the clinical discrimination between acquired unilateral pigmented lesions that pose a high risk of progression to melanoma from those lesions at low risk. For this reason, ophthalmologists are counseled to obtain a biopsy, excisional or incisional, of pigmented conjunctival lesions. It is then the responsibility of the pathologist to assign levels of risk to the clinical lesion designated as primary acquired melanosis based on the histologic criteria discussed below. The decision point is binary: either the lesion poses no risk of progression to melanoma (the source of clinical pigmentation is the deposit of conjunctival pigmentation in the absence of melanocytic hyperplasia or melanocytic hyperplasia without atypia), or the lesion histologically is a proliferation of atypical melanocytes within the conjunctival epithelium.

Various histologic nomenclatures have been proposed to describe the pathology of lesions that are referred to as primary acquired melanosis. A simple binary classification that was favored by the World Health Organization (WHO) is *primary acquired melanosis without atypia* for those lesions without risk of progression to melanoma and *primary acquired melanosis with atypia* for those lesions with histologic features associated with a risk of progression to melanoma (75,80).

Experts have proposed other classification schemes (77,78). One subdivides conjunctival melanosis into *hypermelanosis, conjunctival melanocytic intraepithelial neoplasia (C-MIN) with and without atypia,* and *melanoma in situ* (77). Another subdivides conjunctival melanosis into *intraepithelial nonproliferative melanocytic pigmentation* and *intraepithelial melanocytic proliferation*

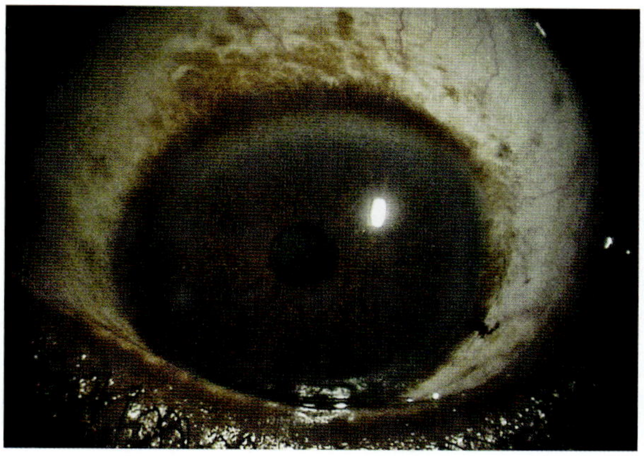

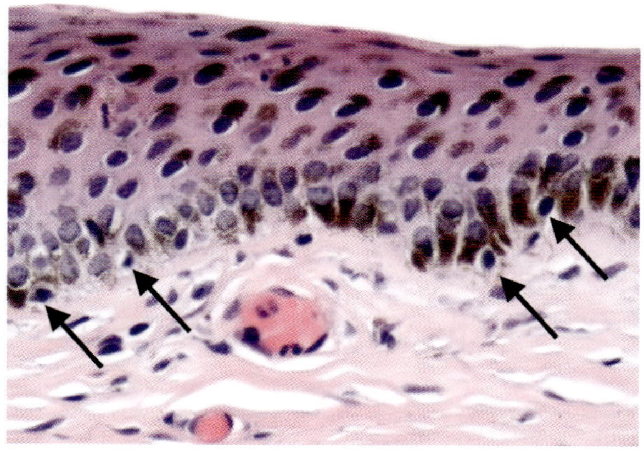

Figure 2-32

CONJUNCTIVAL SKIN TONE ASSOCIATED MELANOSIS (C-STAM)

Left: The conjunctival pigmentation is densest at the limbus in a darkly complexioned individual.

Right: Microscopic evaluation demonstrates a segment of bulbar conjunctiva with increased production of melanin by melanocytes (arrows) that are normal in size, location, and number. The melanin is present in the supranuclear location in the basal and superficial epithelial cells (H&E stain). (Courtesy of Dr. R. C. Eagle, Jr., Philadelphia, PA.)

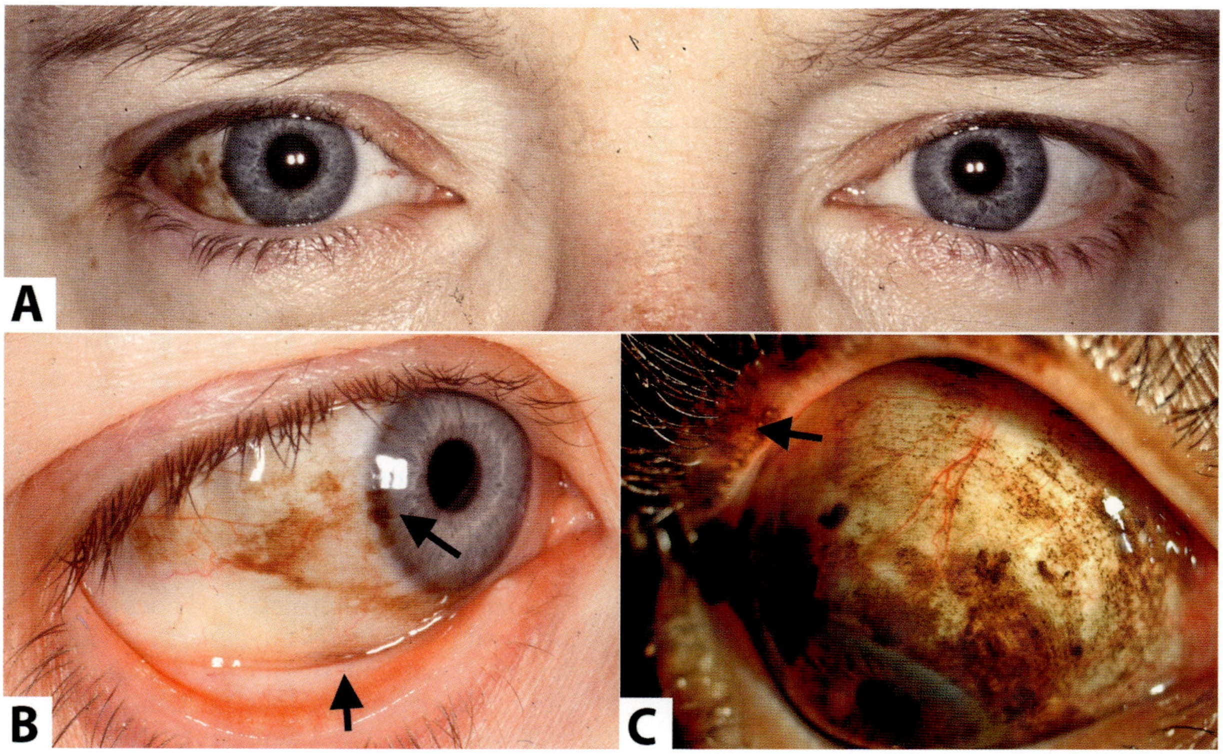

Figure 2-33

PRIMARY ACQUIRED MELANOSIS

A: Increased unilateral conjunctival pigmentation of the right eye in a lightly complexioned individual. (A,B: courtesy of Drs. J. Shields and C. Shields, Philadelphia, PA.)

B: Higher magnification highlights the involvement of the cornea and forniceal conjunctiva (arrows).

C: In a darker complexioned patient, there is diffuse pigmentation of the bulbar and forniceal, conjunctiva, and eyelid margin (arrow).

Table 2-3

CONSENSUS PROPOSED HISTOLOGIC CRITERIA FOR INTEGRATED CLASSIFICATION OF CONJUNCTIVAL MELANOCYTIC INTRAEPITHELIAL LESIONS[a]

WHO[b] Grade	Definition	Histologic Description
Low Grade	Benign melanosis	Conjunctival hypermelanosis (increased pigment in epithelial cells without melanocytic hyperplasia or atypia) Melanocytic hyperplasia without atypia (melanocytes with condensed round nuclei, smaller than basal epithelial cell, inconspicuous nucleoli, and inconspicuous cytoplasm)
	Low-grade CMIL[b]	Predominantly basilar melanocytic hyperplasia with mild atypia (dendritic or small polyhedral, non-epithelioid melanocytes with irregular nuclear contours, inconspicuous nucleoli, and inconspicuous or scant cytoplasm)
High-Grade	High-grade CMIL	Significant non-basilar hyperplasia of atypical melanocytes involving <50% of epithelium without nests and/or melanocytes with epithelioid morphology
	Melanoma in situ	Atypical melanocytes with nesting and/or >50% full-thickness intraepithelial migration

[a]Data from reference 80a.
[b]WHO = World Health Organization; CMIL = conjunctival melanocytic intraepithelial lesion.

(IMP) with and without atypia (78). Recently, a panel of international experts compared the reproducibility of three major classification systems, PAM, C-MIN, and the new four-tiered WHO classification. The authors concluded that these three systems had comparable strengths and weaknesses and proposed a revision to the new WHO classification that is, like the PAM system, two-tiered (80a). The revised classification will be studied further. The proposed simplification of the new WHO classification appears in Table 2-3. Additionally, to maximize the diagnostic yield, surgeons need to be familiar with the guidelines for submission of conjunctival biopsies on melanocytic lesions, summarized in the American Joint Committee on Cancer (AJCC) manual (61).

Clinical Features

PAM usually presents in white adults around 60 years of age, but can occur in any race and at any age (66). The characteristic clinical appearance is unilateral, mottled, or "dusted" conjunctival pigmentation, which waxes and wanes over time and can cover broad aspects of the conjunctiva (66,81). Although it is impossible to clinically distinguish PAM with atypia from PAM without atypia (75), clinical predictors of aggressive behavior include a larger size of the lesion and involvement of nonbulbar conjunctiva and cornea (fig. 2-33) (81).

Conjunctival melanoma represents 5 percent of all ocular melanomas. The annual incidence ranges from 0.1 to 0.7 per one million persons and is higher in whites and people over 50 years (82,83). Most conjunctival melanomas appear as solitary ovoid to irregularly shaped, placoid or nodular vascular lesions in the sun-exposed interpalpebral bulbar conjunctiva, often contiguous with the limbus (fig. 2-34) (79,84). Although 60 percent of these tumors are pigmented, 20 percent are amelanotic and 20 percent are mixed (85). Deeply invasive tumors are adherent to the underlying sclera, a feature that helps distinguish conjunctival melanoma from a nevus (66,85). Flat conjunctival tissue displaying variable pigmentation frequently surrounds the thickened portion of the tumor, reflecting the underlying intraepithelial disease (79,84). There may be contiguous incursion into the eyelid margin (fig. 2-34) (79,84).

Histologic Findings

Normal Conjunctival Melanocytes. Conjunctival melanocytes are generally small and dendritic, without significant nuclear hyperchromasia or conspicuous nucleoli. They appear widely scattered along the basal layer of the epithelium (fig. 2-35) (77,78). The melanin transferred to the conjunctival epithelial cells is typically not visible to the naked eye (77).

Conjunctival Skin Tone-Associated Melanosis (C-STAM). Microscopically, there is increased production of melanin by melanocytes that are normal in size, location, and number

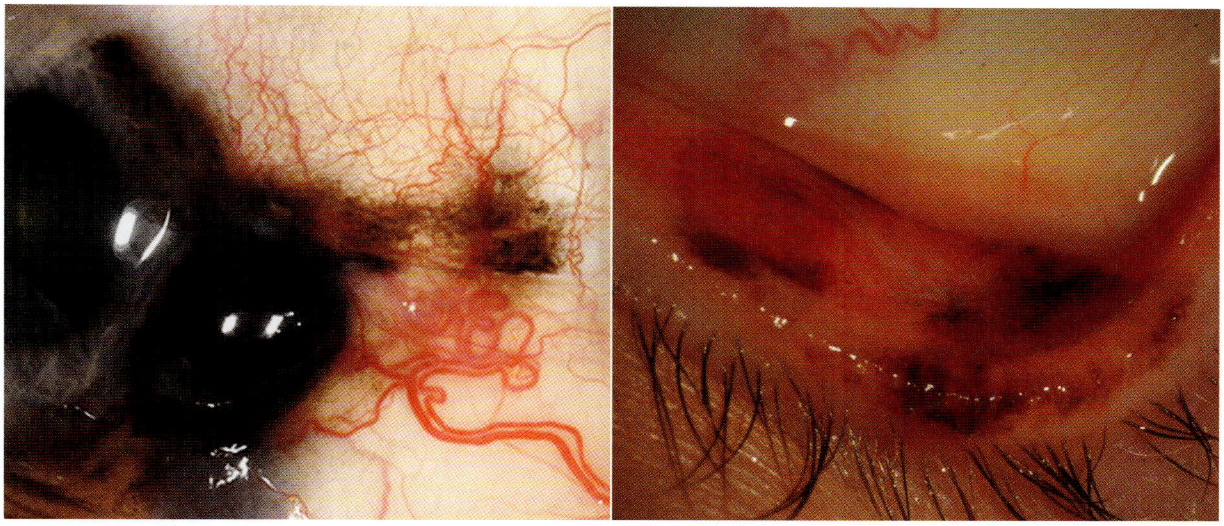

Figure 2-34

CONJUNCTIVAL MELANOMA

Left: Slit-lamp photograph shows a solitary ovoid to irregularly shaped, vascular nodule in the sun-exposed interpalpebral bulbar and limbal conjunctiva in a background of patchy flat diffuse pigmentation (primary acquired melanosis). (A,B: courtesy of Drs. J. Shields and C. Shields, Philadelphia, PA.)

Right: A poorly circumscribed, variably pigmented patch in the palpebral conjunctiva and eyelid margin, with minimal foci of nodularity.

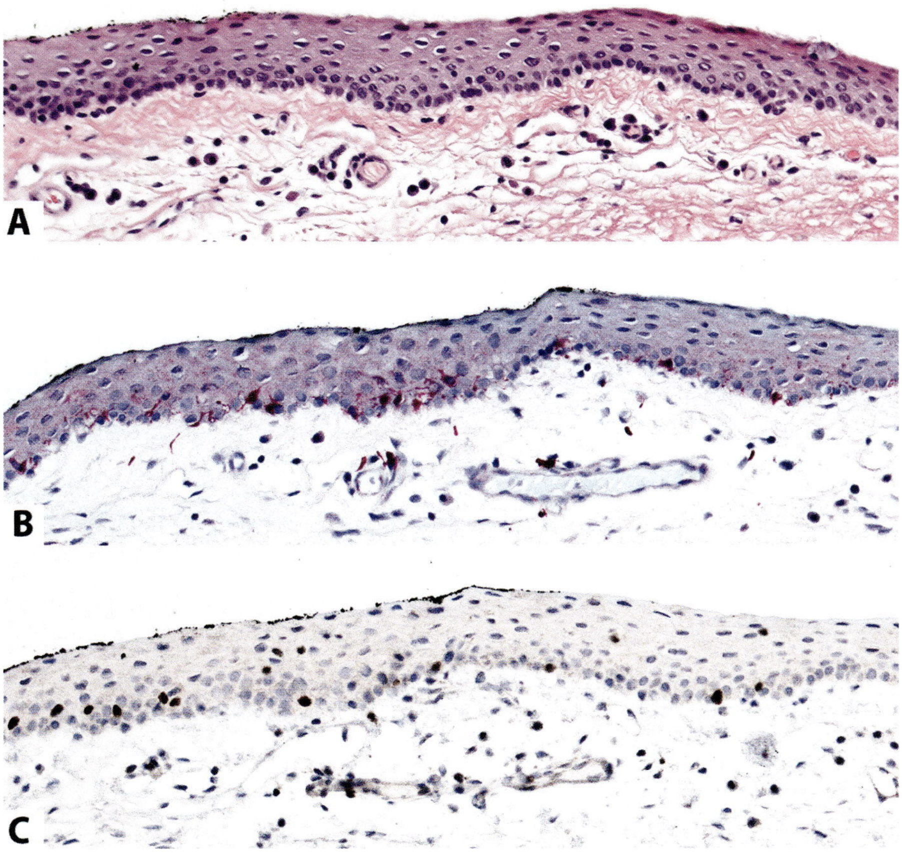

Figure 2-35

NORMAL CONJUNCTIVAL MELANOCYTES

H&E-stained preparation (A) and the corresponding SOX-10- (B) and S-100 protein- (C) stained preparations demonstrate widely scattered small dendritic melanocytes, distributed along the basal epithelial layer.

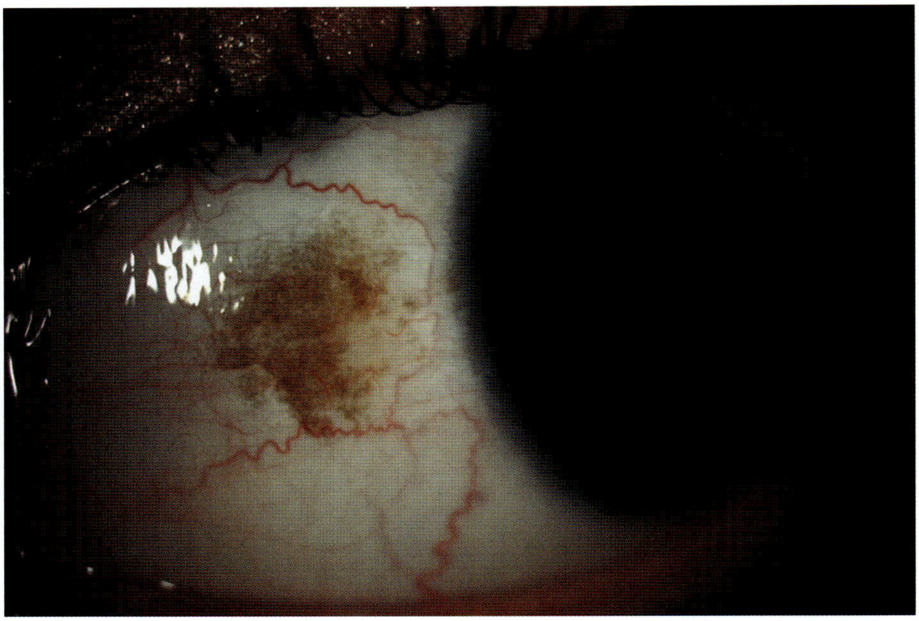

Figure 2-36

EPHELIS (FRECKLE)

A small melanotic macule is seen in the interpalpebral conjunctiva. Clinically, this lesion is indistinguishable from primary acquired melanosis, and biopsy is required to determine the absence or presence of atypia. Some authorities do not recognize the lesion "ephelis" in the conjunctiva for this reason.

(77). Occasionally, there is a slightly increased number of morphologically normal melanocytes (66). The melanin pigment is most dense in the basal epithelial cells, where it is typically supranuclear in location. A similar supranuclear pigment is present, to a lesser extent, in the superficial keratinocytes (fig. 2-32) (77).

The definitive diagnosis of conjunctival skin tone-associated melanosis (C-STAM) requires clinical-pathologic correlation. In the absence of clinical context, similar microscopic findings are seen with "ephelis" and primary and secondary acquired melanosis without atypia (2,77,78).

There are no reported cases of progression by complexion-associated melanosis to melanoma (66). However, isolated unilateral or highly asymmetric pigment, particularly in a nonbulbar conjunctival location, should raise a high index of suspicion even in darkly pigmented individuals. In such cases, excisional biopsy is prudent (66).

Ephelis (Primary Conjunctival Hypermelanosis, Epithelial Nonproliferative Melanocytic Pigmentation). Some authorities recognize the existence of ephelides, or freckles, in the conjunctiva while others regard these lesions as examples of primary acquired melanosis (fig. 2-36). Microscopically, there is increased production of melanin by melanocytes that are normal in size, location, and number. The melanin pigment is most dense in the basal epithelial cells, where it is typically supranuclear in location (77). The definitive diagnosis of an ephelis requires clinical-pathologic correlation, mindful of the reality that similar microscopic findings occur in complexion-associated melanosis and primary and secondary acquired melanosis without atypia (2,77,78).

Primary Acquired Melanosis Without Atypia (Conjunctival Melanocytic Intraepithelial Neoplasia Without Atypia, Intraepithelial Melanocytic Proliferation Without Atypia). PAM without atypia microscopically manifests as increased melanin production by cytologically normal-appearing melanocytes that are either normal or minimally increased in number (fig. 2-37) (2,8), and that lack features typical of melanocytic atypia. PAM without atypia has virtually no risk of progression to melanoma. This description corresponds to the new simplified WHO classification as WHO low grade-benign melanosis (see Table 2-3).

PREMALIGNANT AND MALIGNANT MELANOCYTIC LESIONS

Primary Acquired Melanosis with Atypia (Conjunctival Melanocytic Intraepithelial Neoplasia with Atypia, Intraepithelial Melanocytic Proliferation with Atypia)

Microscopic Findings. There is a spectrum of epithelial involvement by atypical melanocytes. Melanocytic atypia has been

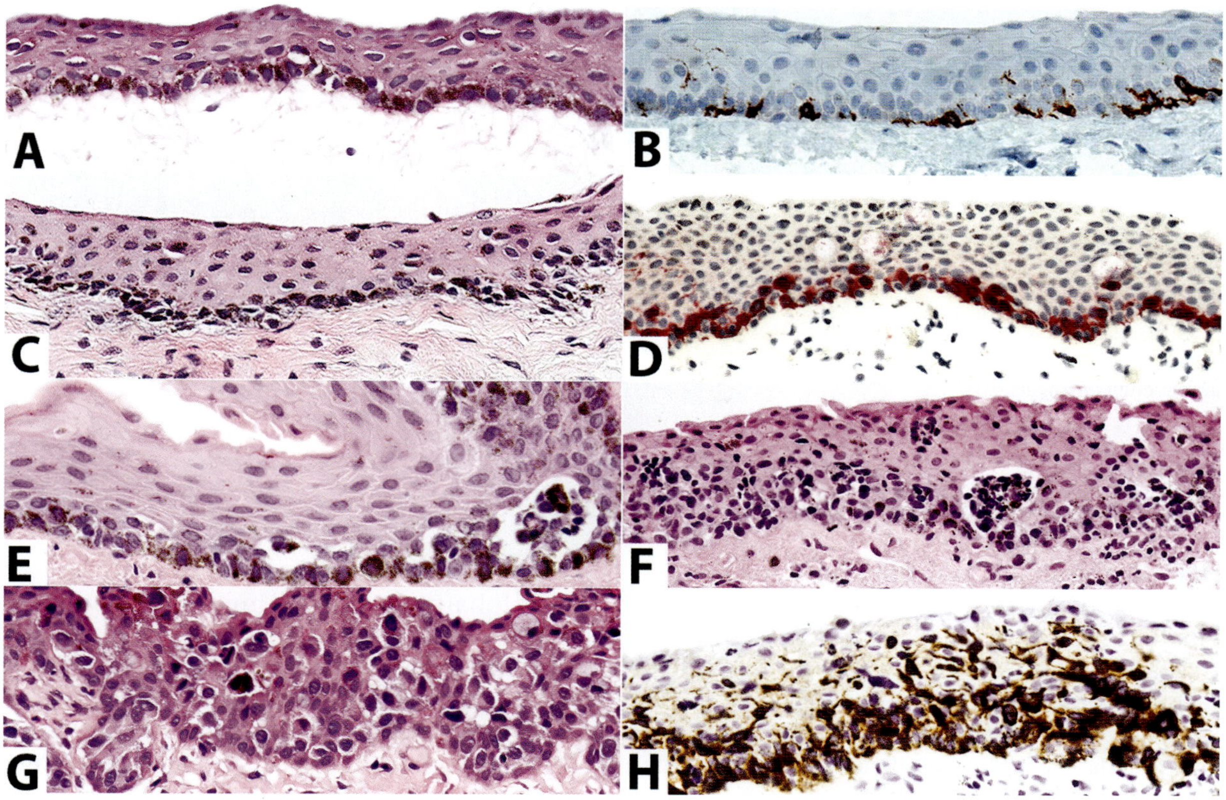

Figure 2-37

PRIMARY ACQUIRED MELANOSIS

A,B: Primary acquired melanosis without atypia (conjunctival epithelial hypermelanosis, nonproliferative melanocytic pigmentation) is characterized by increased melanin in the conjunctival epithelial cells, with cytologically normal-appearing melanocytes that are normal in number, as confirmed with corresponding Melan-A immunostain (B) (A,C,E,F,G: H&E stain).

C: Primary acquired melanosis with mild atypia (conjunctival melanocytic intraepithelial neoplasia with mild atypia, intraepithelial melanocytic proliferation with mild atypia) is characterized by increased melanin in the conjunctival cells with minimally enlarged atypical melanocytes that lack their normal lymphocytoid morphology and are mildly increased in number (basilar hyperplasia pattern).

D: Corresponding Melan-A immunostain of primary acquired melanosis with mild atypia highlights the mild melanocytic hyperplasia (B,D,H: Melan-A immunostain).

E: Primary acquired melanosis with mild to moderate atypia (conjunctival melanocytic intraepithelial neoplasia with mild to moderate atypia, intraepithelial melanocytic proliferation with mild to moderate atypia) is characterized by increased numbers of minimally atypical melanocytes with inconspicuous nucleoli and scant cytoplasm, confined to the basal layer of the epithelium in a single-cell lentiginous pattern and with focal nesting.

F: Primary acquired melanosis with severe atypia (conjunctival melanocytic intraepithelial neoplasia with severe atypia, intraepithelial melanocytic proliferation with severe atypia) is characterized by proliferation of atypical melanocytes with hyperchromatic nuclei and frequent intraepithelial migration in the form of nests and pagetoid scatter.

G,H: Primary acquired melanosis with severe atypia/melanoma in situ is characterized by markedly atypical intraepithelial melanocytes with enlarged nuclei, prominent nucleoli and epithelioid morphology with prominent intraepithelial migration, and nearly complete epithelial replacement, as highlighted with Melan-A immunostain (H).

traditionally classified as mild, moderate, and severe, although with limited interobserver and intraobserver reproducibility (77). In *PAM with mild atypia*, the atypical melanocytes are small with polyhedral, spindle, or dendritiform morphology, inconspicuous nucleoli, and scant cytoplasm. They typically reside in the basal layer of the epithelium in a single-cell lentiginous pattern (fig. 2-37) (2,75,76). This corresponds to the proposed simplified new WHO classification as low-grade conjunctival melanocytic intraepithelial lesion (CMIL, Table 2-3).

PAM with severe atypia shows markedly atypical intraepithelial melanocytes featuring enlarged

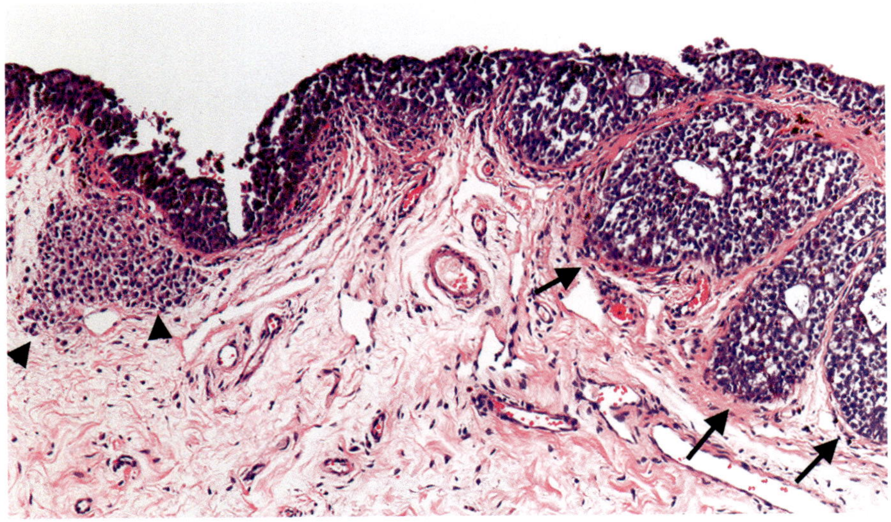

Figure 2-38

PRIMARY ACQUIRED MELANOSIS WITH SEVERE ATYPIA/MELANOMA IN SITU

The lesion replaces the pseudoglands of Henle (arrows), mimicking stromal invasion. In contrast, the true invasive melanoma focus is characterized by a permeative architecture (arrowheads) (H&E stain).

nuclei, prominent nucleoli, and epithelioid morphology. There may be intraepithelial migration in the form of nests and pagetoid scatter (fig. 2-37) (2,75,76). *PAM with moderate atypia* features intermediate findings (fig. 2-37) (75). This description corresponds to the proposed simplified new WHO classification as high grade CMIL (Table 2-3).

The terminology of *melanoma in situ* is controversial, with some experts suggesting that the term is acceptable when there is nearly complete replacement of the epithelium by atypical melanocytes, while others use the term for lesions containing epithelioid melanocytes (fig. 2-37) (77). Others avoid using the term altogether. Melanoma in situ is recognized as a component of high-grade CMIL (Table 2-3).

Immunohistochemical Findings. Some advocate the use of immunohistochemical stains as melanocytic markers. The expression of Melan-A, SOX-10, S-100 protein, MITF, or HMB-45 may be helpful in assessing the extent of the intraepithelial melanocytic proliferation. These markers, preferably performed with a red chromagen, may also help identify microinvasion (fig. 2-37) (2). Other experts, mindful of the trend to avoid the overuse of immunohistochemistry, do not order these markers routinely and instead restrict their use to the rare problematic case.

Cytologic Findings. Although the literature references the use of cytology in the diagnosis of these lesions, the authors discourage its use in routine practice because a biopsy yields information that is more relevant to the clinical management than either impression cytology or exfoliative cytology. Impression cytology demonstrates atypical melanocytes in the superficial epithelium, diagnostic of either PAM with atypia or melanoma, with a sensitivity of approximately 70 percent (86). Exfoliative cytology shows increased intraepithelial melanocytes with varying degrees of atypia, with sensitivity and specificity of 85 and 78 percent, respectively, for higher grades of atypia (87).

Molecular Genetic Findings. The genetic profile of PAM with atypia is obscure. *BRAF* mutation has appeared in 2 of 8 (20 percent) cases unassociated with nevi (88). There is a report of *TERT* promoter mutations in 2 of 25 (8 percent) cases of PAM with atypia (89). These techniques are not employed routinely in the evaluation of these cases.

Differential Diagnosis. The distinction of PAM with mild atypia from PAM without atypia or secondary melanosis can be challenging, and is frequently observer dependent. Clinical correlation in such cases may be helpful. Similarly, differentiation of a junctional nevus from PAM in the absence of clinical data may be difficult. Junctional nevus in an adult is PAM until proven otherwise (2,8). Observation of cytologic atypia, a mixture of nests and single cell pagetoid scatter, and overall irregular involvement of the epithelium support the diagnosis of PAM with atypia (8). Palpebral conjunctival PAM may involve pseudoglands of Henle and may be misinterpreted as invasive melanoma (fig. 2-38) (8).

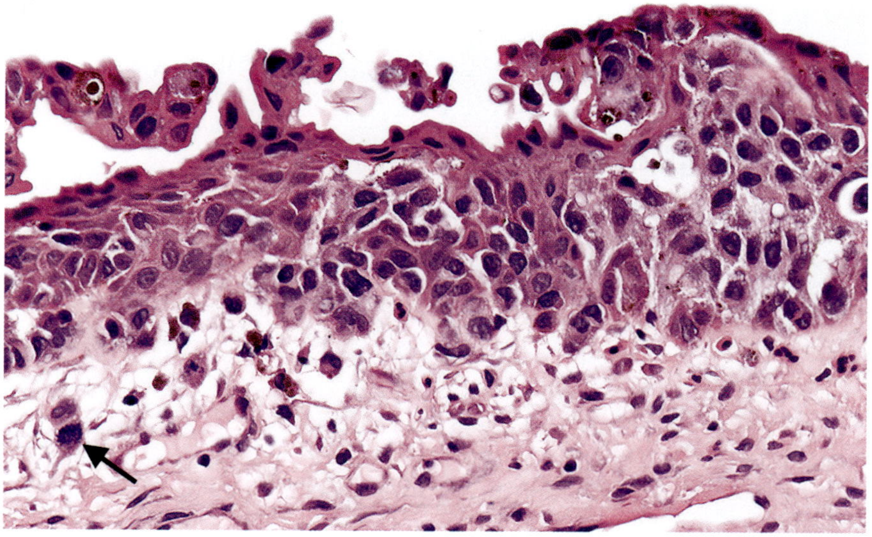

Figure 2-39

CONJUNCTIVAL MICROINVASIVE MELANOMA

Intraepithelial proliferation of markedly atypical epithelioid melanocytes with focal invasion of the substantia propria by mitotically active malignant melanoma cells (arrow) (H&E stain).

Treatment and Prognosis. PAM with atypia carries a high risk for recurrence and can progress to conjunctival malignant melanoma. The high-risk microscopic features consistently associated with progression to invasive melanoma are epithelioid morphology and nonbasilar growth pattern (75,81). The rate of progression to invasive malignant melanoma ranges from 13 (81) to 90 percent (90) for PAM with severe atypia and 0 (81) to 20 percent (75) for PAM with mild to moderate atypia.

Map biopsies (1- to 2-mm sampling biopsies of all quadrants of the bulbar and palpebral conjunctiva) are recommended to assess the extent of disease, followed by excision of the lesion with cryotherapy. Neoadjuvant chemotherapy with mitomycin-C or interferon-alpha-2b may be useful in a subset of patients with diffuse disease. Recurrence after biopsy or cryotherapy of PAM with atypia is 30 percent (66).

Melanoma

Definition. The invasion of atypical melanocytes into the subepithelial connective tissue, no matter how superficial, qualifies the lesion as *melanoma* (fig. 2-39) (8).

Microscopic Findings. The surface epithelium of melanoma is usually attenuated and may be ulcerated. It contains atypical melanocytes. The neoplastic cells are frequently epithelioid, with large vesicular nuclei, prominent central nucleoli, and eosinophilic cytoplasm. A mixture of other cell types, including small polyhedral cells, spindle cells, and balloon cells, may be present (fig. 2-40) (8). Lymphocytes are often associated with the base and infiltrating edges of the lesion. The degree of pigmentation and necrosis are variable (8). Examination of the conjunctiva adjacent to the melanoma may reveal an atypical intraepithelial melanocytic proliferation, a nevus, or both (see fig. 2-30) (91).

Cytologic Findings. Although impression and exfoliative cytology have been described as sensitive and specific methods for detection of conjunctival melanoma (86,87), full-thickness biopsy remains the gold standard for diagnosis and staging of this neoplasm.

Molecular Genetic Findings. Between 30 and 50 percent of conjunctival melanomas harbor *BRAF* mutations (73,85,88), which are associated with younger patient age, male gender, sun-exposed location, and nevus precursor lesion (85,88). Although the prognostic significance of *BRAF* mutation in conjunctival melanoma is currently controversial (85,88), it offers a potential avenue for targeted therapies (92).

Other common molecular genetic alterations include *NRAS* mutations (93) and *TERT* promoter mutations (94) in approximately 20 to 30 percent of tumors. The mutation signature in the affected genes appears to be consistent with UV-induced DNA damage (73).

Differential Diagnosis. Early conjunctival melanoma needs to be distinguished from melanocytic nevus. Clinical correlation, with attention to the patient's age and tumor

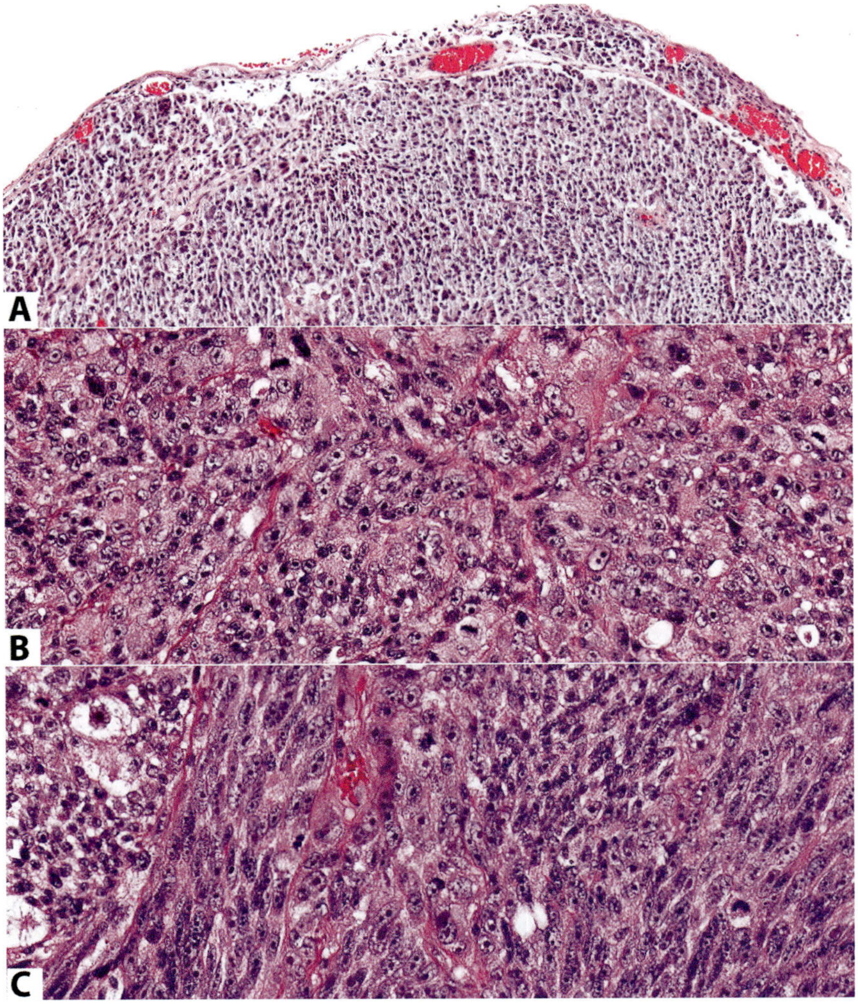

Figure 2-40

CONJUNCTIVAL MELANOMA

A: Low-power photomicrograph demonstrates an ulcerated conjunctival nodule with intraepithelial and invasive melanoma.

B,C: The neoplastic cells are epithelioid (B) and spindled (C) in morphology, with prominent nuclear pleomorphism, central large nucleoli, and brisk mitotic activity (H&E stain).

characteristics, is important (66,79). Histopathologically, the presence of significant nuclear atypia, mitotic activity, necrosis, atypical intraepithelial melanocytic proliferation lateral to the subepithelial component of the main tumor (76), diffuse staining for HMB-45, and increased Ki-67 proliferative index (greater than 5 percent) (67,68,76) help establish the diagnosis of conjunctival melanoma. Despite the careful histopathologic examination, a group of indeterminate melanocytic lesions exists that cannot be unequivocally classified as benign or malignant (72).

The histologic differential diagnosis may include extraocular extension of uveal melanoma and metastatic melanoma. In both scenarios, the overlying conjunctival epithelium is typically uninvolved by tumor. Contiguity with the underlying iris or ciliary body melanoma can be demonstrated clinically (8). Nonpigmented spindle cell melanoma of the conjunctiva may resemble the spindle cell variant of conjunctival squamous cell carcinoma and other soft tissue spindle cell tumors. Immunohistochemical studies confirming the melanocytic nature of the tumor are useful in such cases (2,8).

Treatment and Prognosis. Conjunctival melanoma has high potential for recurrence and regional (fig. 2-41) and distant metastases, with a 10-year melanoma-related mortality rate of 20 to 30 percent (79,95). Tumor thickness of greater than 2 mm, ulceration, epithelioid cell type, increased mitotic activity, and lymphovascular invasion are associated with higher rates of metastases and tumor-related death (95). Additionally, de novo origin, nonbulbar conjunctival location,

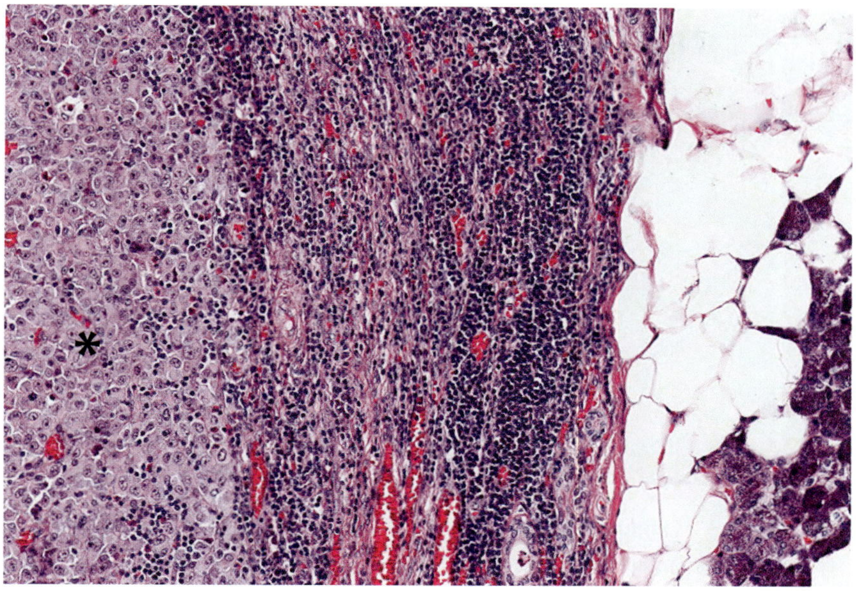

Figure 2-41

CONJUNCTIVAL MALIGNANT MELANOMA, METASTATIC TO REGIONAL (INTRAPAROTID) LYMPH NODE

Amelanotic epithelioid melanoma (asterisk) is surrounded by a narrow rim of lymphoid tissue and parotid gland parenchyma (H&E stain).

and nodular and multicentric growth patterns are also associated with poor prognosis (84,85).

Most conjunctival melanomas are managed by excision and adjuvant cryotherapy. Neoadjuvant chemotherapy with mitomycin-C or interferon-alpha-2b may be useful in a subset of patients with diffuse disease. Radiotherapy and enucleation are reserved for larger, complicated tumors. Sentinel lymph node biopsy is not used routinely in the management of conjunctival melanoma, and is controversial (66).

PATHOLOGIST REPORT FOR MELANOCYTIC TUMORS

The system presented in Appendix 2, from the Taskforce for Staging of Cancer of the Eye of the American Joint Committee on Cancer (AJCC), 8th edition, is used to stage melanomas arising from the bulbar and palpebral conjunctiva and the caruncle (61). The staging system applies only to malignant melanoma of the conjunctiva, and does not include primary cutaneous melanoma with secondary invasion of the conjunctiva. The 8th edition staging system is different from prior editions by 1) modifying T categories to describe circumferential extent of disease, 2) specifying in N category whether a biopsy was performed, and 3) removal of histologic grading.

Clinical staging should include inspection of the entire conjunctiva, caruncle, cornea, and eyelid margins, with documentation of circumferential extent (in clock hours) of the affected regions. Supplemental drawings and photography are recommended to capture the extent of the entire tumor. Clinical exam should assess for local invasion and palpable regional lymphadenopathy. Imaging for locally advanced and metastatic disease is also performed (61).

Pathologic examination should provide a precise histologic diagnosis of pathologic process(es), describing the relationship to anatomic structures and surgical margins. For invasive conjunctival melanoma, the following histomorphologic features should be detailed in the report: the status of all tumor margins including the presence or absence of PAM with atypia at the lateral margins, thickness (measured from the surface of the conjunctival epithelium), and presence or absence of surface ulceration. Additional features, when present, should be noted on the report: lymphatic or vascular invasion, perineural invasion, microsatellites, an immune response suggestive of regression. Nevus, when present, should be noted in the report. Some authorities include cell typing (spindle cell melanoma, epithelioid melanoma, and anaplastic melanoma) in their reports and mitotic counts per square millimeter according to the recommendations of the AJCC (61). Beyond the factors used to assign T, N, and M categories, no additional prognostic factors are required for stage groupings (61).

BRAF mutational status of the tumor cells should be considered for potential targeted

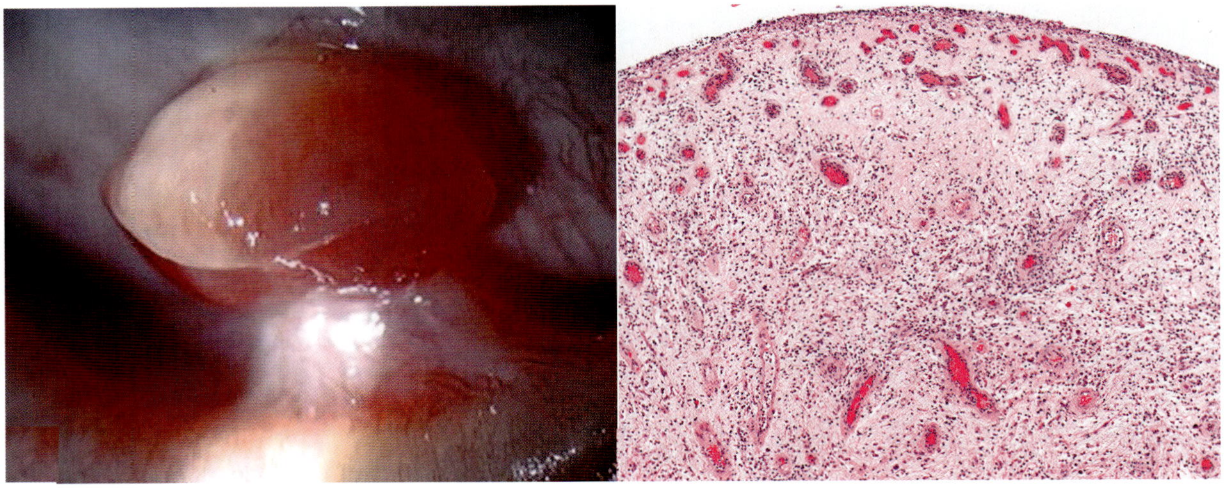

Figure 2-42

EXUBERANT GRANULATION TISSUE (PYOGENIC GRANULOMA)

Left: A lobulated, vascular, fleshy mass involves the tarsal conjunctiva.

Right: Microscopically, the lesion is composed of exuberant granulation tissue, consisting of proliferating capillaries that radiate from the stalk of the lesion, in a background of acute and chronic inflammatory infiltrate, fibroblasts, and fibrin (H&E stain).

therapies in an eventuality of metastatic disease. *BRAF* mutational status may be determined using either molecular or immunohistochemical techniques (61).

MESENCHYMAL NEOPLASMS

Vascular and Lymphatic Neoplasms and Tumor-Like Lesions

Various benign and malignant vascular and lymphatic neoplasms and tumor-like lesions involve the conjunctival substantia propria, including *telangiectasis, lymphangiectasis, capillary* and *cavernous hemangioma, lymphangioma,* and *Kaposi sarcoma* (Table 2-1) (3–6). See *Nonmelanocytic Tumors of the Skin* Fascicle in this series for a detailed discussion of these entities.

Exuberant granulation tissue can involve conjunctiva as a reparative response to another lesion and trauma, presenting as a lobulated, fleshy, vascular mass (fig. 2-42). This lesion is known as *pyogenic granuloma*, but is unrelated to the capillary hemangioma, pyogenic granuloma type recognized by dermatopathologists (2). Microscopically, the mass of granulation tissue is dominated by proliferating capillaries that radiate from the stalk of the lesion, and an acute or chronic inflammatory infiltrate in a background of fibrin and fibrosis (fig. 2-42) (2).

Rhabdomyosarcoma. Rhabdomyosarcoma, botryoid type, can rarely involve the conjunctiva in a pattern similar to other mucosal surfaces (fig. 2-43). Children usually present with a fleshy, gelatinous, lobulated, rapidly growing mass, which in early stages is unassociated with inflammatory signs (96). See *Tumors of Soft Tissues* Fascicle in this series for a detailed discussion of this entity.

Other Mesenchymal Conjunctival Tumors. Other rare conjunctival stromal tumors include *myxoma, peripheral nerve sheath tumors*, and *submucosal neuroma*, which have the microscopic appearance and systemic associations similar to nonocular counterparts. The literature refers to the existence of fibrocytic neoplasms *(fibrous histiocytoma, atypical fibroxanthoma,* and *undifferentiated pleomorphic sarcoma)* in the conjunctiva (2,4,33). See *Nonmelanocytic Tumors of the Skin* and *Tumors of the Soft Tissues* Fascicles in this series for additional information on these entities.

TUMOR-LIKE CONGENITAL LESIONS

Benign congenital lesions composed of mature tissue elements not generally found in the conjunctival location are classified as *choristomas* (Table 2-1) (2,5,33). *Solid epibulbar dermoids* are circumscribed tumors most frequently located in the inferotemporal limbus

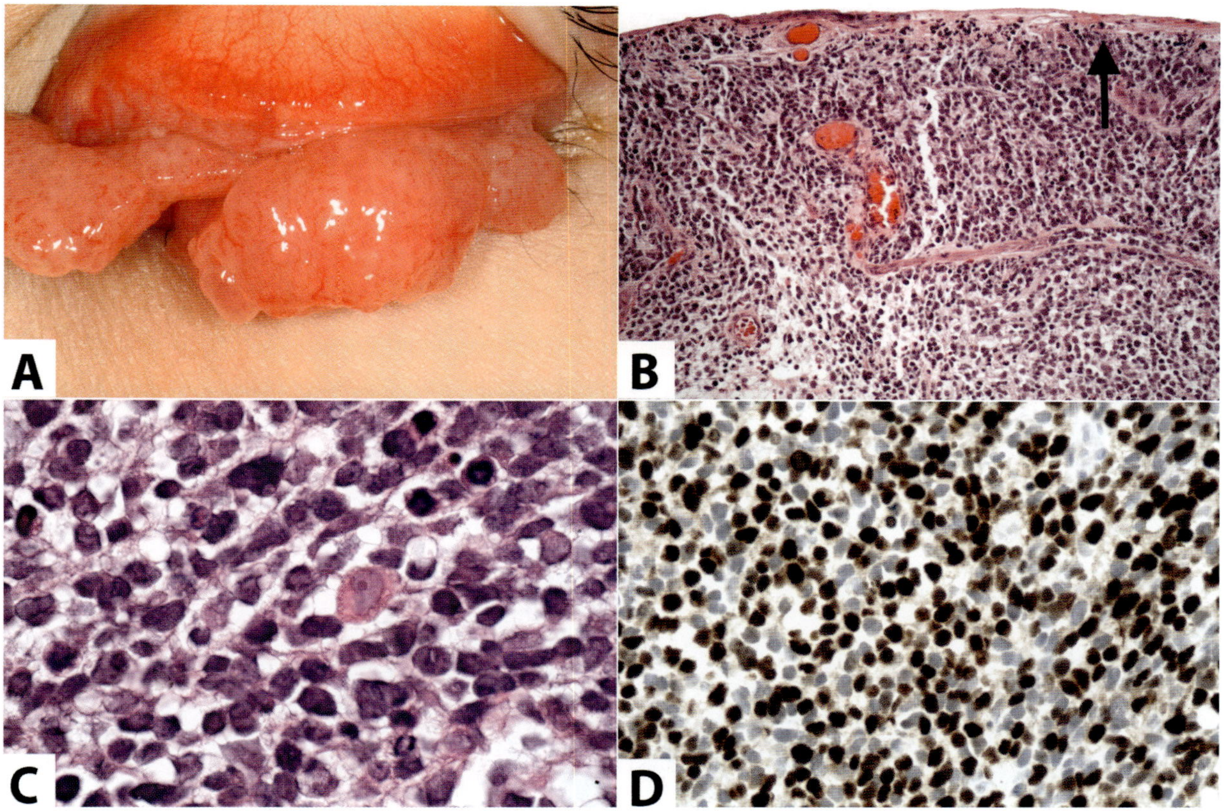

Figure 2-43

BOTRYOID RHABDOMYOSARCOMA

A: A fleshy, lobulated, vascular mass involves the forniceal conjunctiva.

B: Microscopic evaluation demonstrates an infiltrate of pleomorphic, round to ovoid, dyscohesive cells in the substantia propria. The condensation of neoplastic cells beneath the attenuated conjunctival epithelium (arrow) is analogous to a cambium layer seen in extraocular botryoid rhabdomyosarcoma.

C,D: Higher magnification highlights the primitive mesenchymal neoplastic cells (C) that express myogenin (D) (H&E stains). (A: courtesy of Drs. J. Shields and C. Shields, Philadelphia, PA; B–D, courtesy of Dr. R. C. Eagle, Jr., Philadelphia, PA.)

(fig. 2-44). Microscopically, a nonkeratinized or keratinized stratified squamous epithelium, resembling epidermis, covers the lesion. There is a dense paucicellular collagenous stroma that contains pilosebaceous units and sweat glands. Lobules of adipose tissue are found at the base of the lesion (2-44) (2,33). When cartilage or lacrimal gland tissue (or a combination of these elements) contribute to the lesion, the term *complex choristoma* is used (fig. 2-45) (2,33).

Dermolipoma or *lipodermoid* usually involves the bulbar conjunctiva temporally or superotemporally, and is primarily composed of mature adipose tissue (fig. 2-46) (2,33). Bilateral epibulbar dermoids and dermolipomas occur in two thirds of patients with oculo-auriculo-vertebral (Goldenhar) syndrome. There is an association between complex choristomas and epidermal nevus syndromes (2,33).

Other choristomatous lesions of the ocular surface include *ectopic lacrimal gland, episcleral osseous choristoma* (33), and *smooth muscle hamartoma* (8). Ophthalmologists tend to manage symptomatic epibulbar choristomatous lesions by surgical resection. There are no reports of malignant transformation.

TUMORS OF THE CARUNCLE

The caruncular region can spawn a large variety of tumors and tumor-like lesions because it contains cutaneous, mucous membrane, and accessory lacrimal gland elements (97,98). Tumors of caruncle are rare, however, comprising about 4 percent of all conjunctival tumors (97). The

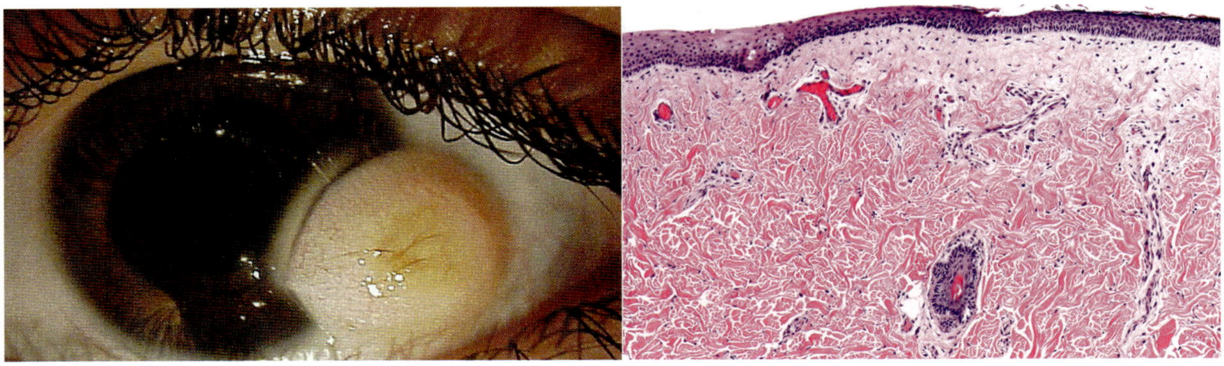

Figure 2-44

SOLID EPIBULBAR LIMBAL DERMOID

Left: A gray nodule at the limbus inferotemporally is associated with hair and an arc of intracorneal lipid.
Right: Microscopically, the lesion is surfaced focally by nonkeratinized and focally by keratinized epithelium, with coarse collagen bundles in the stroma. Pilosebaceous units and a few lobules of adipose tissue are also present (H&E stain).

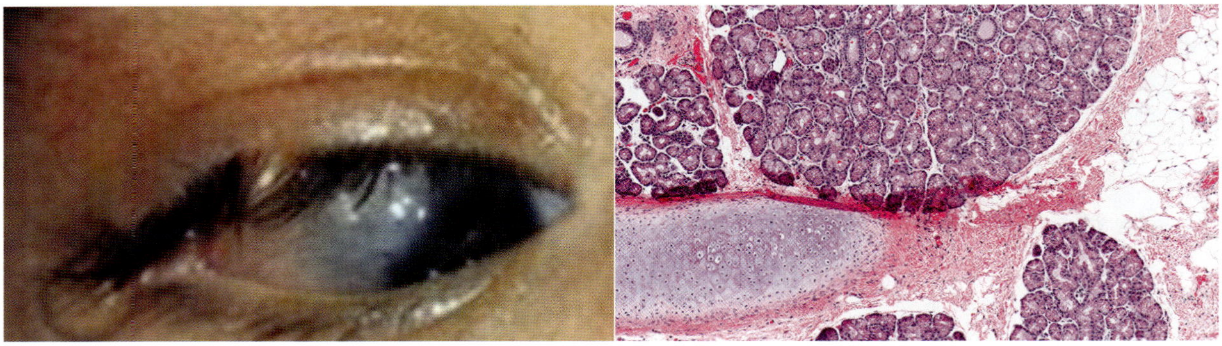

Figure 2-45

COMPLEX CHORISTOMA IN A PATIENT WITH AN EPIDERMAL NEVUS SYNDROME VARIANT

Left: A fleshy mass involves the bulbar conjunctiva and an adjacent cutaneous fibroma.
Right: Microscopically, the lesion is composed of mature cartilage, smooth muscle, and adipose tissue in a background of coarse collagen (H&E stain).

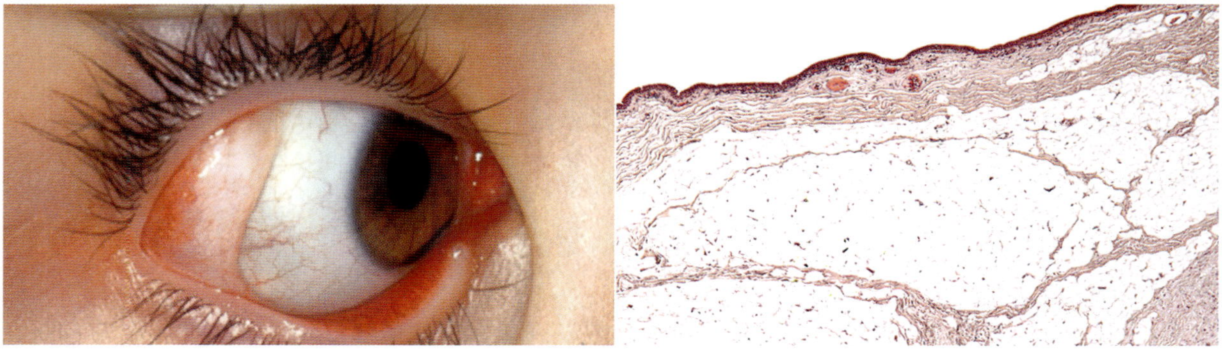

Figure 2-46

DERMOLIPOMA (LIPODERMOID)

Left: A yellow, fatty-appearing mass is seen in the bulbar conjunctiva superotemporally.
Right: Lobules of adipose tissue are in a background of coarse collagen surfaced by nonkeratinizing stratified squamous epithelium (H&E stain).

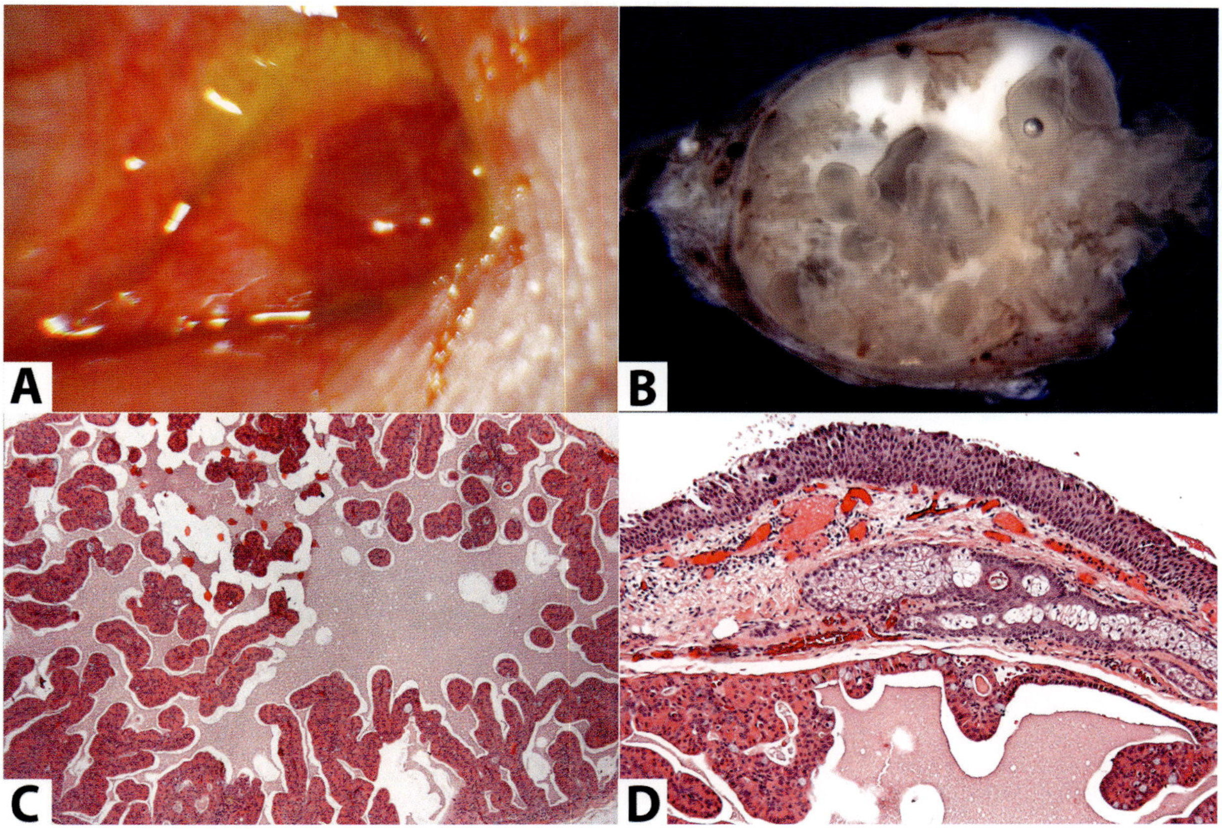

Figure 2-47

ONCOCYTOMA

A: A circumscribed reddish mass involves the caruncle.

B: Gross photograph of a circumscribed mass with heterogeneous solid and cystic components.

C,D: Microscopically, the mass is composed of a cystadenomatous proliferation of tall columnar cells with eosinophilic cytoplasm, central round nuclei, and granular eosinophilic luminal material (H&E stain).

most common biopsied caruncular lesions are *squamous papillomas, nevi,* and *sebaceous neoplasms* (97,98). Squamous papilloma and melanocytic nevus have been described earlier in this chapter. Periocular sebaceous neoplasms are discussed in chapter 3. Additional information on melanocytic nevi and sebaceous neoplasms appears in the Fascicles *Melanocytic Tumors of the Skin* and *Nonmelanocytic Tumors of the Skin.*

Oncocytoma

Oncocytoma, or *oncocytic adenoma,* can rarely arise in the accessory lacrimal glands of caruncle (97–99). This benign neoplasm presents as a circumscribed orange, brown, or bluish caruncular mass, mimicking hemangioma and melanocytic tumor (fig. 2-47) (97–99).

Caruncular oncocytomas are composed of a circumscribed cystadenomatous proliferation of tall, columnar epithelial cells with small round nuclei and intensely eosinophilic cytoplasm. Eosinophilic granular material is present in the lumen (fig. 2-47) (98,99).

Although the literature records the appearance of oncocytic adenocarcinoma in the caruncle, there are no well-documented reports of malignant transformation of caruncular oncocytoma (100). Observation and local resection are acceptable management options (33).

SECONDARY AND METASTATIC TUMORS

Extension of tumor into the conjunctiva from adjacent sites (secondary tumors) and metastases from other primary sites are rare, accounting for 1 percent of all conjunctival tumors in an oncology referral center (Table 2-1) (5). The most common neoplasms to involve conjunctiva secondarily are eyelid carcinomas and extraocular extension of uveal melanoma (5). Attention to the overall architecture of the lesion, absence of an in situ component, and correlation with clinical findings are important in establishing the diagnosis.

Metastases to conjunctiva typically occur in a setting of widespread metastatic disease in a patient with a known primary tumor (101). The most common primary sites are breast, lung, and skin (cutaneous melanoma) (5,101). Conjunctival metastatic lesions are typically solitary, with either bulbar or palpebral involvement (101). Satellite foci of metastatic melanoma must be differentiated from primary conjunctival melanoma (8).

REFERENCES

1. Last RJ. The appendages of the eye. In: Wolff E, Last RJ. Anatomy of the eye and orbit, including the central connections, development, and comparative anatomy of the visual apparatus, 6th ed. London: Lewis; 1968;182-240.
2. Eagle RC Jr. Conjunctiva. In: Eagle RC Jr. Eye pathology: an atlas and text, 3rd ed. Philadelphia: Wolters Kluwer; 2017;55-79.
3. Akinsola FB, Mbadugha CA, Onakoya AO, Adefule-Ositelu AO, Aribaba OT, Rotimi-Samuel A. Pattern of conjunctival masses seen at Guinness Eye Centre Luth Idi-Araba. Nig Q J Hosp Med 2012;22:39-43.
4. Grossniklaus HE, Green WR, Luckenbach M, Chan CC. Conjunctival lesions in adults. A clinical and histopathologic review. Cornea 1987;6:78-116.
5. Shields CL, Alset AE, Boal NS, et al. Conjunctival tumors in 5002 Cases. Comparative analysis of benign versus malignant counterparts. The 2016 James D. Allen Lecture. Am J Ophthalmol 2017;173:106-33.
6. Shields CL, Chien JL, Surakiatchanukul T, Sioufi K, Lally SE, Shields JA. Conjunctival tumors: review of clinical features, risks, biomarkers, and outcomes—the 2017 J. Donald M. Gass Lecture. Asia Pac J Ophthalmol (Phila) 2017;6:109-20.
7. Hassan WM, Bakry MS, Hassan HM, Alfaar AS. Incidence of orbital, conjunctival and lacrimal gland malignant tumors in USA from Surveillance, Epidemiology and End Results, 1973-2009. Int J Ophthalmol 2016;9:1808-13.
8. Font RL, Croxatto JO, Rao NA. Tumors of the conjunctiva and caruncle. In: Tumors of the eye and ocular adnexa. AFIP Atlas of Tumor Pathology, 4th series, Fascicle 5. Washington DC: American Registry of Pathology; 2006:1-40.
9. Asokan R, Venkatasubbu RS, Velumuri L, Lingam V, George R. Prevalence and associated factors for pterygium and pinguecula in a South Indian population. Ophthalmic Physiol Opt 2012;32:39-44.
10. Zhou WP, Zhu YF, Zhang B, Qiu WY, Yao YF. The role of ultraviolet radiation in the pathogenesis of pterygia. Mol Med Rep 2016;14:3-15.
11. Liu L, Wu J, Geng J, Yuan Z, Huang D. Geographical prevalence and risk factors for pterygium: a systematic review and meta-analysis. BMJ Open 2013;3:e003787.
12. Liu T, Liu Y, Xie L, He X, Bai J. Progress in the pathogenesis of pterygium. Curr Eye Res. 2013;38:1191-7.
13. Zhong H, Chen Q, Li J, Shen W, et al. Ethnic variations in pterygium in a rural population in southwestern China: the Yunnan Minority Eye Studies. Ophthalmic Epidemiol 2016;23:116-21.
14. Detorakis ET, Kymionis G, Tsatsos M, Spandidos DA. Pterygium concomitant with other ocular surface lesions: clinical implications and pathogenetic links. Exp Ther Med 2016;11:69-72.
15. Hirst LW, Axelsen RA, Schwab I. Pterygium and associated ocular surface squamous neoplasia. Arch Ophthalmol 2009;127:31-2.
16. Endo H, Kase S, Suzuki Y, Kase M. Coincidence of inflamed conjunctival carcinoma in situ and primary pterygium. Case Rep Ophthalmol 2016;7:208-12.
17. Artornsombudh P, Sanpavat A, Tinnungwattana U, Tongkhomsai V, Sansopha L, Tulvatana W. Prevalence and clinicopathologic findings of conjunctival epithelial neoplasia in pterygia. Ophthalmology 2013;120:1337-40.
18. Perra MT, Colombari R, Maxia C, et al. Finding of conjunctival melanocytic pigmented lesions within pterygium. Histopathology 2006;48:387-93.
19. Jaros PA, DeLuise VP. Pingueculae and pterygia. Surv Ophthalmol 1988;33:41-9.

20. Ohara M, Sotozono C, Tsuchihashi Y, Kinoshita S. Ki-67 labeling index as a marker of malignancy in ocular surface neoplasms. Jpn J Ophthalmol 2004;48:524-9.
21. Li W, Liu J, Galvin JA. Epigenetics and common ophthalmic diseases. Yale J Biol Med 2016;23; 89:597-600.
22. Bamdad S, Khademi B, Chenari N, Taseh A, Razmkhah M. Stromal cell derived factor-1, CXCR4 and CXCR7 gene transcripts in pterygia. J Curr Ophthalmol 2016;29:28-32.
23. Clearfield E, Hawkins BS, Kuo IC. Conjunctival autograft versus amniotic membrane transplantation for treatment of pterygium: findings from a cochrane systematic review. Am J Ophthalmol 2017;182:8-17.
24. Oellers P, Karp CL, Sheth A, et al. Prevalence, treatment, and outcomes of coexistent ocular surface squamous neoplasia and pterygium. Ophthalmology 2013;120:445-50.
25. Yanoff M. Hereditary benign intraepithelial dyskeratosis. Arch Ophthalmol1968;79:291-3.
26. Allingham RR, Seo B, Rampersaud E, et al. A duplication in chromosome 4q35 is associated with hereditary benign intraepithelial dyskeratosis. Am J Hum Genet 2001;68:491-4.
27. Bui T, Young JW, Frausto RF, Markello TC, Glasgow BJ, Aldave AJ. Hereditary benign intraepithelial dyskeratosis: report of a case and re-examination of the evidence for locus heterogeneity. Ophthalmic Genet 2016;37:76-80.
28. Cummings TJ1, Dodd LG, Eedes CR, Klintworth GK. Hereditary benign intraepithelial dyskeratosis: an evaluation of diagnostic cytology. Arch Pathol Lab Med 2008;132:1325-8.
29. Soler VJ, Tran-Viet KN, Galiacy SD, et al. Whole exome sequencing identifies a mutation for a novel form of corneal intraepithelial dyskeratosis. J Med Genet 2013;50:246-54.
30. Cai R, Zhang C, Chen R, Bi Y, Le Q. Clinicopathological features of a suspected case of hereditary benign intraepithelial dyskeratosis with bilateral corneas involved: a case report and mini review. Cornea 2011;30:1481-4.
31. Mlakar J, Kocjan BJ, Hošnjak L, et al. Morphological characteristics of conjunctival squamous papillomas in relation to human papillomavirus infection. Br J Ophthalmol 2015;99:431-6.
32. Verma V, Shen D, Sieving PC, Chan CC. The role of infectious agents in the etiology of ocular adnexal neoplasia. Surv Ophthalmol 2008;53:312-31.
33. Shields CL1, Shields JA. Tumors of the conjunctiva and cornea. Surv Ophthalmol 2004;49:3-24.
34. Stagner AM, Jakobiec FA, Chi A, Bradshaw SH, Mendoza SD. Conjunctival inverted squamous papilloma: a case report with immunohistochemical analysis and review of the literature. Surv Ophthalmol 2015;60:263-8.
35. Jakobiec FA, Harrison W, Aronian D. Inverted mucoepidermoid papillomas of the epibulbar conjunctiva. Ophthalmology 1987;94:283-7.
36. Kaliki S, Arepalli S, Shields CL, et al. Conjunctival papilloma: features and outcomes based on age at initial examination. JAMA Ophthalmol 2013;131:585-93.
37. Singh M, Gautam N, Gupta A, Kaur M. Interferon alfa-2b in the management of recurrent conjunctival papillomatosis. Indian J Ophthalmol 2016;64:778-80.
38. Bredow L, Martin G, Reinhard T, Mittelviefhaus H, Auw-Haedrich C. Recurrent conjunctival papilloma progressing into squamous cell carcinoma with change of HPV-finding during the course. Br J Ophthalmol 2009;93:1437, 1451.
39. Picken MM. Amyloidosis-where are we now and where are we heading? Arch Pathol Lab Med 2010;134:545-51.
40. Demirci H, Shields CL, Eagle RC Jr, Shields JA. Conjunctival amyloidosis: report of six cases and review of the literature. Surv Ophthalmol 2006:419-33.
41. Cheong-Leen, R. Primary localised conjunctival amyloidosis presenting as subconjunctival haemorrhage. Eye 2001;15:679-80.
42. Reynolds MM, Veverka KK, Gertz MA, et al. Ocular manifestations of systemic amyloidosis. Retina 2018;38:1371-6.
43. Schuster V, Seregard S. Ligneous conjunctivitis. Surv Ophthalmol 2003;48:369-88.
44. Kamath SJ, Marthala H, Manapragada B. Ocular manifestations in lipoid proteinosis: a rare clinical entity. Indian J Ophthalmol 2015;63:793-5.
45. Hamam R, Bhat P, Foster CS. Conjunctival/corneal intraepithelial neoplasia. Int Ophthalmol Clin 2009;49:63-70.
46. Honavar SG. Ocular surface squamous neoplasia: are we calling a spade a spade? Indian J Ophthalmol 2017;65:907-9.
47. Gichuhi S, Sagoo MS, Weiss HA, Burton MJ. Epidemiology of ocular surface squamous neoplasia in Africa. Trop Med Int Health 2013;18:1424-43.
48. Chisi SK, Kollmann MK, Karimurio J. Conjunctival squamous cell carcinoma in patients with human immunodeficiency virus infection seen at two hospitals in Kenya. East Afr Med J 2006;83:267-70.
49. Semenova EA, Milman T, Finger PT, et al. The diagnostic value of exfoliative cytology vs histopathology for ocular surface squamous neoplasia. Am J Ophthalmol 2009;148:772-8.
50. Jung SM, Lin HC, Chu PH, et al. Expression of cell cycle-regulatory proteins, MIB-1, p16, p53, and p63, in squamous cell carcinoma of conjunctiva: not associated with human papillomavirus infection. Virchows Arch 2006;448:301-5.

51. Jakobiec FA, Mendoza PR, Colby KA. Clinicopathologic and immunohistochemical studies of conjunctival large cell acanthoma, epidermoid dysplasia, and squamous papilloma. Am J Ophthalmol 2013;156:830-46.
52. Mudhar HS. Update on conjunctival pathology. Indian J Ophthalmol 2017;65:797-807.
53. Eagle RC Jr. Eyelid. In: Eagle RC Jr. Eye pathology: an atlas and text, 3rd ed. Philadelphia: Wolters Kluwer; 2017;269-96.
54. McKelvie PA, Daniell M, McNab A, Loughnan M, Santamaria JD. Squamous cell carcinoma of the conjunctiva: a series of 26 cases. Br J Ophthalmol 2002;86:168-73.
55. Mittal R, Rath S, Vemuganti GK. Ocular surface squamous neoplasia—review of etio-pathogenesis and an update on clinico-pathological diagnosis. Saudi J Ophthalmol 2013;27:177-86.
56. Yousef YA, Finger PT. Squamous carcinoma and dysplasia of the conjunctiva and cornea: an analysis of 101 cases. Ophthalmology 2012;119:233-40.
57. Miller CV, Wolf A, Klingenstein A, et al. Clinical outcome of advanced squamous cell carcinoma of the conjunctiva. Eye (Lond) 2014;28:962-7.
58. Rankin JK, Jakobiec FA, Zakka FR, Foster CS. An improved approach to diagnosing and treating conjunctival mucoepidermoid carcinoma. Surv Ophthalmol 2012;57:337-46.
59. Østergaard J, Boberg-Ans J, Prause JU, Heegaard S. Primary basal cell carcinoma of the caruncle with seeding to the conjunctiva. Graefes Arch Clin Exp Ophthalmol 2005;243:615-8.
60. Margo CE, Grossniklaus HE. Intraepithelial sebaceous neoplasia without underlying invasive carcinoma. Surv Ophthalmol 1995;39:293-301.
61. Amin MB, Edge SB, Greene FL, et al., eds. AJCC cancer staging manual, 8th ed. Switzerland: Springer; 2017.
62. Folberg R, Jakobiec FA, Bernardino VB, Iwamoto T. Benign conjunctival melanocytic lesions. Clinicopathologic features. Ophthalmology 1989;96:436-61.
63. Shields CL, Fasiuddin AF, Mashayekhi A, Shields JA. Conjunctival nevi: clinical features and natural course in 410 consecutive patients. Arch Ophthalmol 2004;122:167-75.
64. Berman EL, Shields CL, Sagoo MS, Eagle RC Jr, Shields JA. Multifocal blue nevus of the conjunctiva. Surv Ophthalmol 2008;53:41-9.
65. Jakobiec FA, Colby K, Bajart AM, Saragas SJ, Moulin A. Immunohistochemical studies of atypical conjunctival melanocytic nevi. Arch Ophthalmol 2009;127:970-80.
66. Oellers P, Karp CL. Management of pigmented conjunctival lesions. Ocul Surf 2012;10:251-63.
67. Jakobiec FA, Bhat P, Colby KA. Immunohistochemical studies of conjunctival nevi and melanomas. Arch Ophthalmol 2010;128:174-83.
68. Reddy HS, Keene CD, Chang SH, Jian-Amadi A, Cimino PJ. Immunohistochemical profiling including beta-catenin in conjunctival melanocytic lesions. Exp Mol Pathol 2017;102:198-202.
69. Kim HJ, McCormick SA, Nath S, Moskowitz B, Milman T. Melanocytic nevi of the tarsal conjunctiva: clinicopathologic case series with review of literature. Ophthal Plast Reconstr Surg 2010;26:438-42.
70. Goldenberg-Cohen N, Cohen Y, Rosenbaum E, et al. T1799A BRAF mutations in conjunctival melanocytic lesions. Invest Ophthalmol Vis Sci 2005;46:3027-30.
71. Roh MR, Eliades P, Gupta S, Tsao H. Genetics of melanocytic nevi. Pigment Cell Melanoma Res 2015;28:661-72.
72. Grossniklaus HE, Margo CE, Solomon AR. Indeterminate melanocytic proliferations of the conjunctiva. Trans Am Ophthalmol Soc 1999;97:157-68; discussion 168-70.
73. Swaminathan SS, Field MG, Sant D, et al. Molecular characteristics of conjunctival melanoma using whole-exome sequencing. JAMA Ophthalmol 2017;135:1434-7.
74. Zembowicz A, Mandal RV, Choopong P. Melanocytic lesions of the conjunctiva. Arch Pathol Lab Med 2010;134:1785-92.
75. Folberg R, McLean IW, Zimmerman LE. Primary acquired melanosis of the conjunctiva. Hum Pathol 1985;16:129-35.
76. Jakobiec FA, Folberg R, Iwamoto T. Clinicopathologic characteristics of premalignant and malignant melanocytic lesions of the conjunctiva. Ophthalmology 1989;96:147-66.
77. Damato B, Coupland SE. Conjunctival melanoma and melanosis: a reappraisal of terminology, classification and staging. Clin Exp Ophthalmol 2008;36:786-95.
78. Jakobiec FA. Conjunctival primary acquired melanosis: is it time for a new terminology? Am J Ophthalmol 2016;162:3-19.
79. Shields CL, Markowitz JS, Belinsky I, et al. Conjunctival melanoma: outcomes based on tumor origin in 382 consecutive cases. Ophthalmology 2011;118:389-95.
80. Eberhart CG, Coonland SE, Folberg R, Margo C, Rao N. Conjunctival melanocytic intraepithelial neoplasia. In: Grossniklaus HE, Eberhart CG, Kivelä TT, eds. WHO classification of tumours of the eye, 4th ed. Lyon: IARC Press; 2018.
80a. Milman T, Eiger-Moscovich M, Henry RK, et al. Validation of the newly proposed WHO classification system for conjunctival melanocytic intraepithelial lesions: a comparison with the 'C-MIN' and 'PAM" classification schemes. Am J Ophthalmol 2020. [in press]

81. Shields JA, Shields CL, Mashayekhi A, et al. Primary acquired melanosis of the conjunctiva: risks for progression to melanoma in 311 eyes. The 2006 Lorenz E. Zimmerman lecture. Ophthalmology 2008;115:511-9.
82. Yu GP, Hu DN, McCormick S, Finger PT. Conjunctival melanoma: is it increasing in the United States? Am J Ophthalmol 2003;135:800-6.
83. Hu DN, Yu G, McCormick SA, Finger PT. Population-based incidence of conjunctival melanoma in various races and ethnic groups and comparison with other melanomas. Am J Ophthalmol 2008;145:418-23.
84. Damato B, Coupland SE. An audit of conjunctival melanoma treatment in Liverpool. Eye (Lond) 2009;23:801-9.
85. Larsen AC. Conjunctival malignant melanoma in Denmark: epidemiology, treatment and prognosis with special emphasis on tumorigenesis and genetic profile. Acta Ophthalmol 2016;94:1-27.
86. Paridaens AD, McCartney AC, Curling OM, Lyons CJ, Hungerford JL. Impression cytology of conjunctival melanosis and melanoma. Br J Ophthalmol 1992;76:198-201.
87. Keijser S, van Luijk CM, Missotten GS, Veselic-Charvat M, de Wolff-Rouendaal D, de Keizer RJ. Predictive value of exfoliative cytology in pigmented conjunctival lesions. Acta Ophthalmol Scand 2006;84:188-91.
88. Larsen AC, Dahl C, Dahmcke CM, et al. BRAF mutations in conjunctival melanoma: investigation of incidence, clinicopathological features, prognosis and paired premalignant lesions. Acta Ophthalmol. 2016;94:463-70.
89. Koopmans AE, Ober K, Dubbink HJ, et al. Prevalence and implications of TERT promoter mutation in uveal and conjunctival melanoma and in benign and premalignant conjunctival melanocytic lesions. Invest Ophthalmol Vis Sci 2014;55:6024-30.
90. Sugiura M, Colby KA, Mihm MC Jr, Zembowicz A. Low-risk and high-risk histologic features in conjunctival primary acquired melanosis with atypia: clinicopathologic analysis of 29 cases. Am J Surg Pathol 2007;31:185-92.
91. Folberg R, McLean IW, Zimmerman LE. Malignant melanoma of the conjunctiva. Hum Pathol 1985;16:136-43.
92. Mor JM, Heindl LM. Systemic BRAF/MEK inhibitors as a potential treatment option in metastatic conjunctival melanoma. Ocul Oncol Pathol 2017;3:133-41.
93. Griewank KG, Murali R, Schilling B, et al. TERT promoter mutations in ocular melanoma distinguish between conjunctival and uveal tumours. Br J Cancer 201323;109:497-501.
94. Griewank KG, Westekemper H, Schilling B, et al. Conjunctival melanomas harbor BRAF and NRAS mutations—response. Clin Cancer Res 2013;19:6331-2.
95. Esmaeli B, Roberts D, Ross M, Fellman M, Cruz H, Kim SK, Prieto VG. Histologic features of conjunctival melanoma predictive of metastasis and death (an American ophthalmological thesis). Trans Am Ophthalmol Soc 2012;110:64-73.
96. Taylor SF, Yen KG, Zhou H, Patel BC. Primary conjunctival rhabdomyosarcoma. Arch Ophthalmol 2002;120:668-9.
97. Shields CL, Shields JA. Tumors of the caruncle. Int Ophthalmol Clin 1993;33:31-6.
98. Levy J, Ilsar M, Deckel Y, Maly A, Pe'er J. Lesions of the caruncle: a description of 42 cases and a review of the literature. Eye (Lond) 2009;23:1004-18.
99. Say EA, Shields CL, Bianciotto C, Eagle RC Jr, Shields JA. Oncocytic lesions (oncocytoma) of the ocular adnexa: report of 15 cases and review of literature. Ophthal Plast Reconstr Surg 2012;28:14-21.
100. Harris GJ, Paul S, Hunt BC. Oncocytic adenocarcinoma of the orbit. Ophthal Plast Reconstr Surg. 2017;33:e18-20.
101. Kiratli H, Shields CL, Shields JA, DePotter P. Metastatic tumours to the conjunctiva: report of 10 cases. Br J Ophthalmol 1996;80:5-8.

3 TUMORS OF THE EYELIDS

ANATOMY AND HISTOLOGY

The eyelids are folds of skin with highly modified epidermal appendages, associated soft tissues, and conjunctival lining that cover and protect the eye (1,2). Each eyelid is composed of six layers, which from anterior to posterior are: the epidermis, dermis, orbicularis oculi muscle, tarsus, and palpebral (tarsal) conjunctiva (1).

The eyelid epidermis is one of the most delicate areas of skin in the body. Only five or six cells in thickness, the epidermis lacks prominent rete ridges and is covered by a thin layer of surface keratin (2). A variety of epidermal appendages are present in the eyelids: eccrine glands and modified hair (eyelashes, also known as cilia) with associated sebaceous glands of Zeis and apocrine glands of Moll.

The eyelid dermis is composed of papillary and reticular layers, and contains blood vessels, lymphatic channels, and peripheral nerves. The subcutaneous fat is absent in the palpebral region, and the eyelid dermis rests directly on the skeletal muscle, the orbicularis oculi.

The tarsi are flat, semilunar plates, which contribute to the rigidity of the eyelids. They are composed of collagenous tissue and elastic fibers, and contain a vertically arranged row of Meibomian glands (25 in the upper eyelid and 20 in the lower eyelid). Meibomian gland duct orifices open into the eyelid margin at the mucocutaneous junction.

The palpebral conjunctiva is composed of stratified columnar epithelium with goblet cells and underlying substantia propria (stroma), which is firmly adherent to the tarsus. The conjunctival epithelium transitions into stratified squamous in the eyelid margin near the mucocutaneous junction (1–3).

Accessory lacrimal glands have histologic features identical to those seen in the main lacrimal gland, and are found in the fornix of conjunctival substantia propria (glands of Krause) and at the edge of the tarsal plate distal from the eyelid margin (glands of Wolfring or Ciaccio) (1–3). Lacrimal puncta and canaliculi, lined by stratified squamous nonkeratinizing epithelium, are located medially in the eyelid, surrounded by the orbicularis oculi muscle fibers (which in that location is called Horner muscle).

In the upper eyelid, the striated fibers of the palpebral portion of the levator palpebrae superioris transition into a dense fibrous membrane, the levator aponeurosis, which inserts like an apron into the deeper aspect of the orbicularis muscle anterior to the tarsal plate. The smooth muscle of Müller in the upper eyelid originates posterior to the levator muscle and inserts into the superior margin of the tarsal plate. In the lower eyelid, the capsulopalpebral fascia and inferior tarsal muscle are analogous anatomically and histologically to the levator aponeurosis and Müller muscle (1,3). The orbital septum is a thin membrane of fibrous and elastic connective tissue, which has multiple attachments to the orbicularis muscle, the levator palpebrae superioris, and the tarsal plates. All pathologic processes located behind the orbital septum are considered to be intraorbital (3).

The absence of fat in the dermis of the eyelid overlying the tarsus is significant. If the pathologist identifies tumor admixed with adipose connective tissue, then the neoplasm may have violated the orbital septum and the surgeon should be alerted to the possibility of orbital invasion in the pathology report.

The anatomic structures of the eyelids and their relation to the anterior orbit and the globe are shown in figure 3-1. Figure 3-2 illustrates the histology of the eyelids.

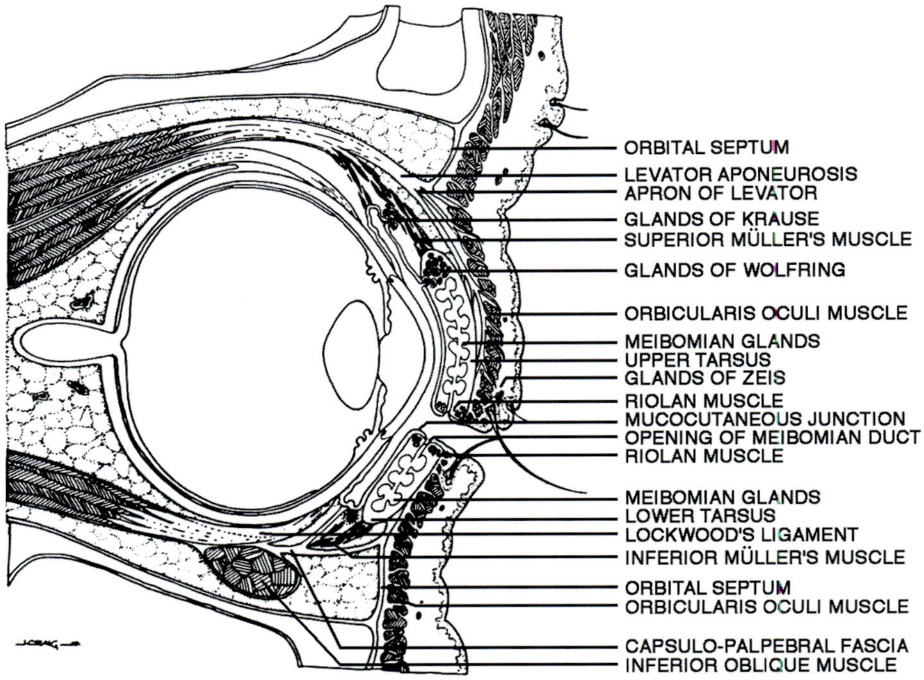

Figure 3-1

EYELID STRUCTURES

Schematic drawing outlining the important eyelid structures.

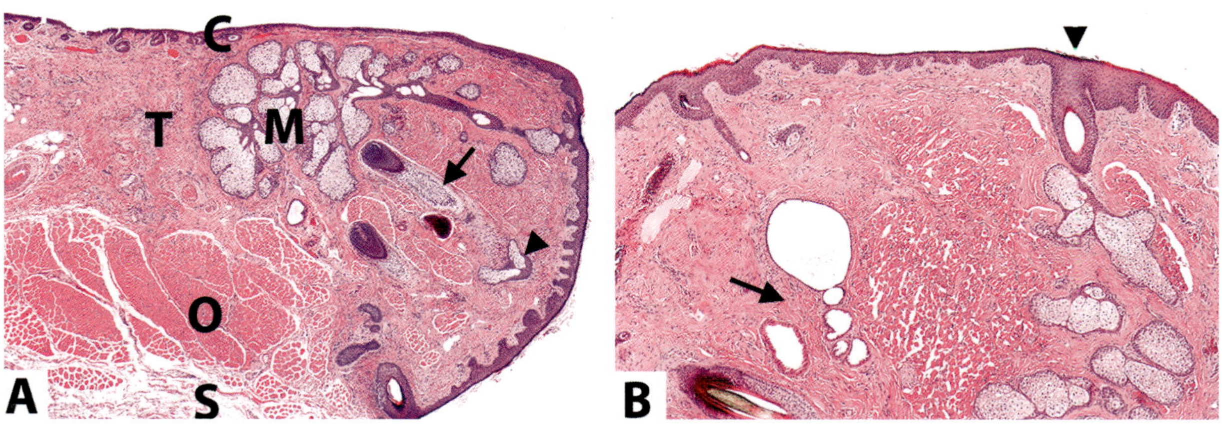

Figure 3-2

EYELID HISTOLOGY

A: Full-thickness section of the lower eyelid with the following regions from anterior to posterior: skin (S), orbicularis oculi muscle (O), tarsus (T) with meibomian glands (M), and tarsal (palpebral) conjunctiva (C). The cilia (eyelashes) and associated sebaceous glands of Zeis are designated with an arrow and arrowhead, respectively.

B: Eyelid margin tissue with mucocutaneous junction (arrowhead) at the site of Meibomian gland duct. Apocrine glands (arrow) of Moll are also present.

CLASSIFICATION

Eyelid tumors are the most common group of neoplasms in daily ophthalmology practice (4). Because the eyelid contains many tissue types, a wide variety of benign and malignant tumors develop. These include benign, premalignant, and malignant neoplasms with epidermal, adnexal, and melanocytic differentiation. There is a broad spectrum of benign and malignant soft tissue tumors, neoplasms of accessory lacrimal gland origin, and hematolymphoid malignancies (Table 3-1) (3–7). Although most tumors involving the eyelid tissues can also appear in nonperiocular locations, some neoplasms have a particular predilection for the eyelid, such

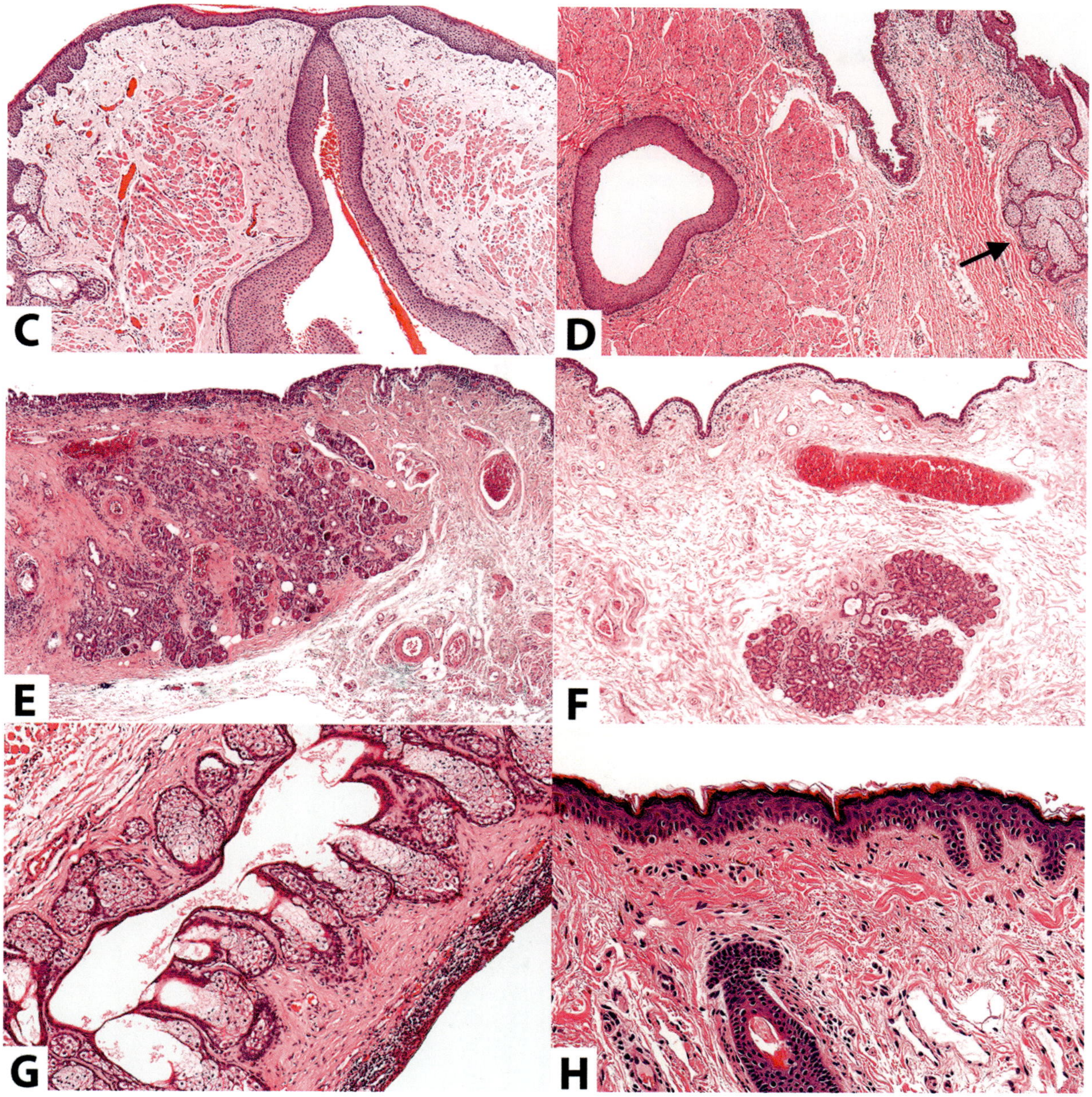

Figure 3-2, continued

C: Lacrimal punctum with surrounding Horner muscle.
D: Lacrimal canaliculus with surrounding Horner muscle and an adjacent caruncle with sebaceous glands of Popoff (arrow).
E: Gland of Wolfring at the edge of the tarsal plate.
F: Gland of Krause in the fornix.
G: Meibomian gland with its central duct and holocrine secretion is vertically oriented within the tarsal plate.
H: Delicate eyelid epidermis and dermis (Hematoxylin and eosin [H&E] stain).

as sebaceous gland carcinoma and endocrine mucin-producing sweat gland carcinoma (8,9).

The relative frequencies of benign and malignant eyelid tumors are available from only a limited number of biopsy series, influenced by multiple factors, including the geographic location, the genetic background of the population, access to medical care, and surgical setting (3–7). Despite this variability, benign tumors largely predominate over malignant ones, with the

Table 3-1

TYPES AND FREQUENCY OF EYELID TUMORS[a]

	Number of Cases (%)				
Type of Tumor	United States (3) 1970-2000	Switzerland (4) 1989-2007	Turkey (5) 2008-2012	Japan (6) 1990-2004	India (7) 1957-1991
Tumors of the Epidermis (total)	**862 (60%)**	**3704 (67%)**	**908 (59%)**	**44 (50%)**	
Benign					
Squamous papilloma	237 (17%)	1063 (19%)	336 (22%)	6 (7%)	16 (8%)
Seborrheic keratosis	120 (8%)	876 (16%)	266 (17%)	9 (10%)	
Others	36 (3%)	678 (12%)	156 (10%)	7 (7%)	20 (10%)
Premalignant/low malignant potential					
Actinic keratosis	31 (2%)	198 (4%)	90 (6%)		
Bowen disease		15 (<1%)	4 (<1%)		
Keratoacanthoma		35 (<1%)	34 (2%)		
Malignant					
Basal cell carcinoma	410 (29%)	772 (14%)	21 (1%)	9 (10%)	19 (9%)
Squamous cell carcinoma	28 (2%)	67 (1%)	1 (<1%)	2 (2%)	9 (4%)
Adnexal Tumors (total)	**328 (23%)**	**579 (11%)**	**202 (13%)**		
Sebaceous gland tumors					
Sebaceous gland hyperplasia	7 (<1%)	14 (<1%)	8 (<1%)		
Sebaceous adenoma	4 (<1%)	5 (<1%)	18 (1%)		
Sebaceous carcinoma	82 (6%)	29 (<1%)	0	9 (10%)	11 (5%)
Tumors of sweat glands					
Hidrocystoma	10 (<1%)	326 (6%)	164 (11%)		
Syringoma	8 (<1%)	45 (1%)	5 (<1%)		
Mucinous sweat gland carcinoma	14 (1%)				
Others	144 (10%)	10 (<1%)			
Tumors of hair follicles					
Trichilemmoma	1 (<1%)	47 (1%)	3 (<1%)		
Trichoepithelioma	4 (<1%)	29 (<1%)	2 (<1%)		
Pilomatrixoma	14 (1%)	30 (<1%)	0		
Others	40 (3%)	10 (<1%)	2 (<1%)		
Melanocytic Tumors (total)	**172 (12%)**	**846 (15%)**	**279 (18%)**		
Nevocellular nevi	159 (11%)	810 (15%)	272 (18%)	14 (16%)	15 (7%)
Melanocytic nevi variants	4 (<1%)	29 (<1%)	7 (<1%)		
Malignant melanoma	9 (<1%)	7 (<1%)	0		1 (<1%)
Vascular Tumors (total)	**26 (2%)**	**83 (2%)**	**18 (1%)**		**26 (13%)**
Capillary hemangioma	11 (<1%)	32 (<1%)	14 (1%)		
Arteriovenous hemangioma		28 (<1%)	0		
Cavernous hemangioma	10 (<1%)	18 (<1%)	4 (<1%)		
Others	5 (<1%)	5 (<1%)	0		
Tumors of Neural Origin (total)	**10 (<1%)**	**17 (<1%)**	**10 (<1%)**		**22 (11%)**
Neurofibroma	10 (<1%)	4 (<1%)	5 (<1%)		
Solitary circumscribed neuroma		10 (<1%)	3 (<1%)		
Traumatic neuroma		3 (<1%)	2 (<1%)		
Hematolymphoid Tumors	**23 (2%)**	**0**	**0**	**4 (5%)**	
Miscellaneous Tumors/Lesions	**29 (2%)**	**268 (5%)**	**124 (8%)**		
Merkel cell carcinoma	4 (<1%)	4 (<1%)	0		
Dermatofibroma	3 (<1%)	9 (<1%)	1 (<1%)		
Xanthelasma	5 (<1%)	246 (4%)	116 (8%)		
Xanthogranuloma	2 (<1%)	6 (<1%)	2 (<1%)		
Others	15 (1%)	3 (<1%)	4 (<1%)	3 (3%)	
Secondary and Metastatic Tumors	**7 (<1%)**	**4 (<1%)**	**1 (<1%)**	**2 (2%)**	
Total	1,436	5,504	1,541	87	207

[a]Data from references 3 through 7.

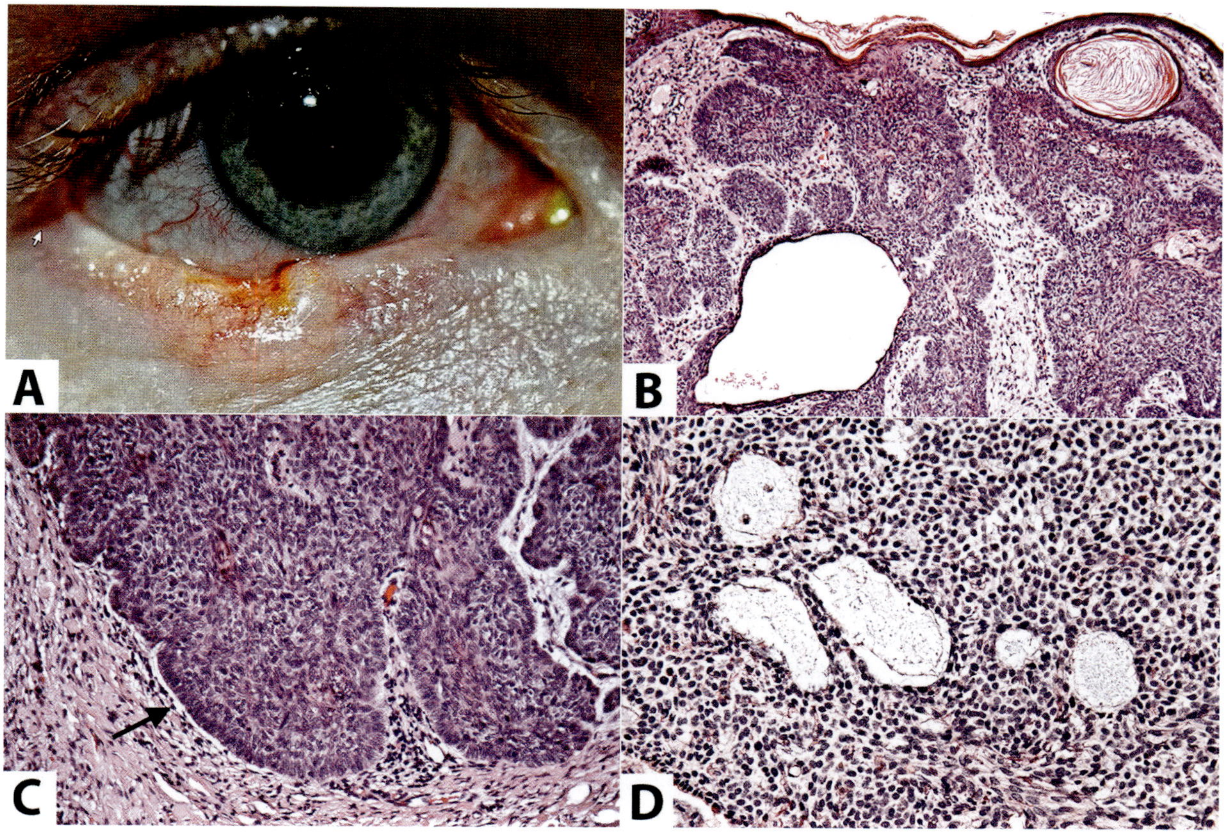

Figure 3-3

NODULAR (SOLID) BASAL CELL CARCINOMA

A: A pearly, ulcerated nodule with telangiectasia is identified along the lower eyelid margin.

B: Nests of basaloid cells with infundibular differentiation extend from the epidermis into the dermis.

C: The peripheral cells in the neoplastic nests are arranged in a palisading fashion perpendicular to the retracted desmoplastic stroma (arrow).

D: Focus of pseudoglandular adenoid differentiation within the nodule (B–D: H&E stain).

most frequent subtypes including squamous papilloma and seborrheic keratosis, followed by hidrocystoma and melanocytic nevus. In most geographic locations and ethnic groups, basal cell carcinoma is the most frequent malignant tumor, followed by squamous carcinoma and sebaceous carcinoma (3–7).

Because of proximity to the eyeball, sinonasal sinuses, and cranium, the oncologic staging and management of eyelid neoplasms are tailored to the periocular region (10,11).

BASAL CELL CARCINOMA

Definition. *Basal cell carcinoma* (BCC) is a neoplasm derived from, or closely related to, basal keratinocytes and follicular or epidermal stem cells (12).

Clinical Features. BCC is by far the most common eyelid malignancy, accounting for 86 to 96 percent of all malignancies (10). BCC is an indolently growing, pearly nodule with central ulceration and telangiectasia (*noduloulcerative BCC*) (fig. 3-3). These lesions tend to develop in the sun-exposed lower eyelid and medial canthal region, although they also grow in a diffuse infiltrative fashion, appearing as an indurated, fibrotic plaque with poorly defined margins (*morpheaform*, or *fibrosing/sclerosing BCC*) (fig. 3-4) (12).

Risk factors include fair skin type, sun exposure, ionizing radiation, advanced age, immunosuppression, and personal history of nonmelanoma skin cancer. BCC is increased in patients with albinism, xeroderma pigmentosum, and

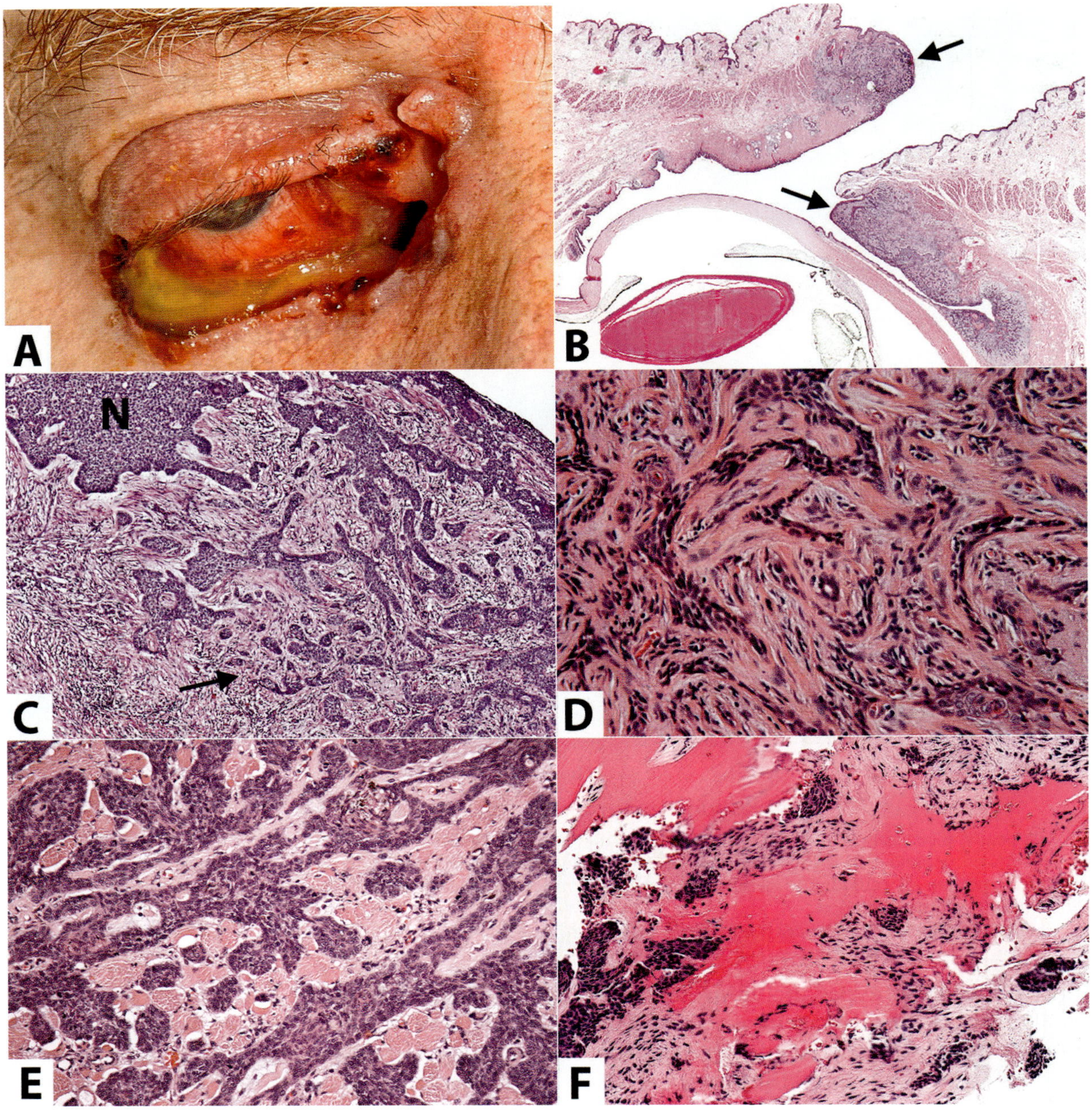

Figure 3-4

INFILTRATING (SCLEROSING OR MORPHEAFORM) BASAL CELL CARCINOMA

A: Diffuse and ill-defined thickening of the periocular tissues is associated with ulceration in the lower eyelid, medial canthus and medial component of the upper eyelid. This clinical appearance is described as a "rodent ulcer."

B: Orbital exenteration specimen at scanning magnification: diffuse involvement of the upper and lower eyelids by basal cell carcinoma (arrows).

C: The neoplasm features focally a mixed histology, with superficial nodular growth pattern (N) and a deeper infiltrative morphology (arrow).

D: Typical appearance of morpheaform basal cell carcinoma as bands of basaloid cells infiltrating into the markedly fibrotic stroma.

E: Cords or bands of neoplastic cells infiltrate into the orbicularis muscle.

F: Bone invasion is present (B–F: H&E stain).

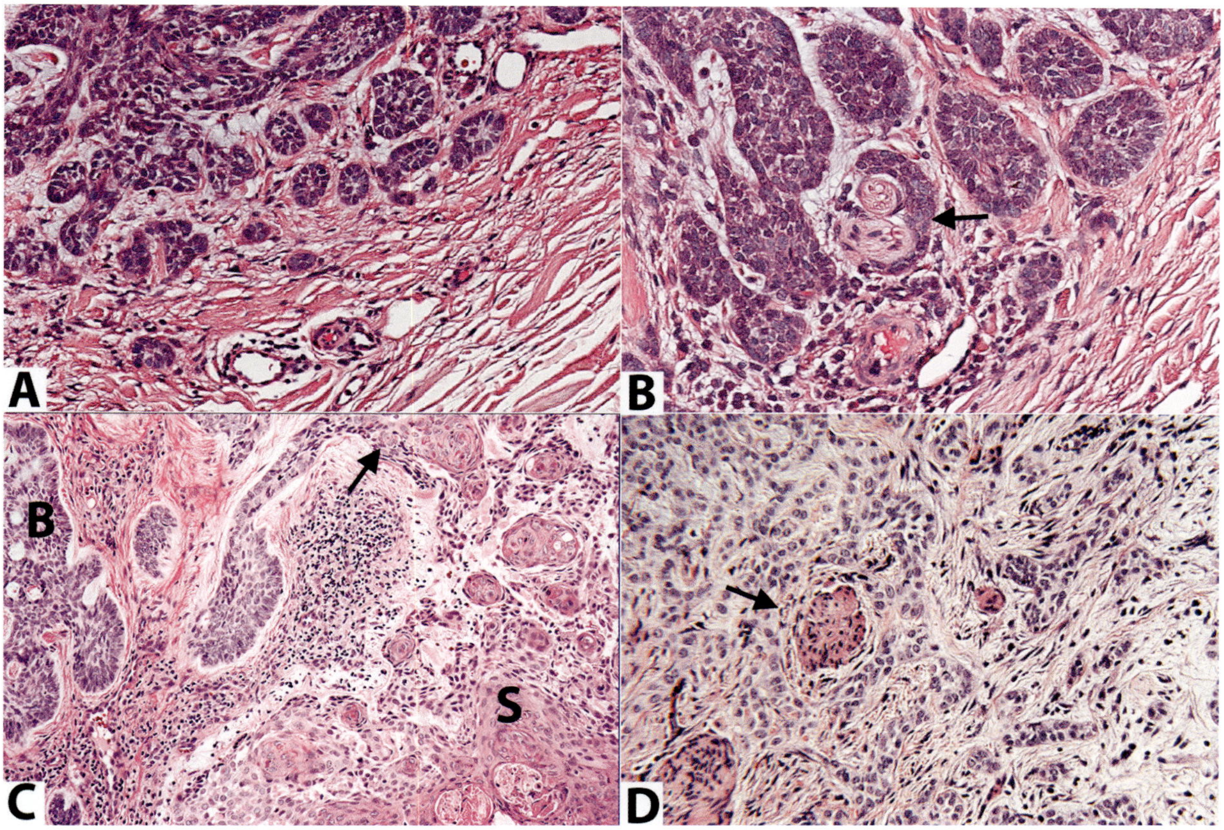

Figure 3-5

BASAL CELL CARCINOMA: HIGH-RISK HISTOLOGIC SUBTYPES

A: Basal cell carcinoma with micronodular growth pattern. Small basaloid cell nests infiltrate the stroma.

B: Micronodular basal cell carcinoma with perineural invasion (arrow).

C: Basosquamous carcinoma shows dimorphic differentiation with a typical basal cell carcinoma (B) and squamous cell carcinoma (S) morphology and a region of transition between the two types (arrow).

D: Metatypical basal cell carcinoma is characterized by infiltration of the stroma by bands of neoplastic cells with morphology intermediate between the basal and squamous cell carcinoma. Perineural invasion is present (arrow) (all figures H&E stain).

nevoid basal cell carcinoma (Gorlin-Goltz) syndrome (10,12).

Microscopic Findings. The histologic subtypes of nonperiocular BCC also develop in the eyelid skin (13,14). *Nodular (solid) BCC* is the most common histologic subtype, accounting for two thirds of periocular BCCs (13,14). This subtype corresponds to the noduloulcerative clinical BCC variant. It is characterized histopathologically by nodular masses of basaloid cells with peripheral palisading that extend into the desmoplastic dermis, with a unique separation artifact between the two. The neoplastic cells resemble basal epidermal cells, but have more elongated nuclei with a higher nuclear to cytoplasmic ratio, lack intercellular "bridges," and have more readily identifiable mitotic figures and apoptotic bodies. The terms keratotic, cystic, or adenoid types apply to differentiated solid BCC with adnexal structures (fig. 3-3) (15).

The aggressive *infiltrating (sclerosing/morpheaform)* BCC is the second most common histologic subtype, accounting for approximately 17 percent of all periocular BCCs (13,14). This subtype has a predilection for facial (and periocular) skin, and is characterized by elongated strands of basaloid cells embedded in a dense fibrous stroma. Infiltrating BCC permeates deeply into the adjacent dermis, and in more advanced stages, into the orbital structures and paranasal sinuses, and rarely, into the eye itself (fig. 3-4) (15).

Table 3-2

TYPICAL IMMUNOHISTOCHEMICAL PATTERNS FOR COMMON PERIOCULAR TUMORS[a]

	Malignant Melanoma	Squamous Cell Carcinoma	Merkel Cell Carcinoma	Sebaceous Carcinoma	Basal Cell Carcinoma
Screening	S-100 protein	p63, EMA	CK20	EMA, AR, adipophilin	Ber–EP4
EMA[b]	– –/+[c]	+	+	+	– –/+
Androgen receptors	–	– – /+	–	+	– /+
Ber-EP4	–	– – /+	+/ –	– /+	+
CEA	N/A	– /+	–	– /+	– /+
Adipophilin	N/A	– – /+	–	+	–
CK7	–	– /+	+/ –	+/ –	variable
CK20	–	– – /+	+	–	–
CAM5.2	– – /+	– – /+	+	+/ –	variable
p63	–	+	– /+	+	+
S-100A	+	–	– /+	N/A	–
MART–1	+	–	–	–	–
HMB45	+	–	–	–	–

[a]Data from reference 16.
[b]EMA = epithelial membrane antigen; CK = cytokeratin; AR = androgen receptors; CEA = carcinembryonic antigen.
[c]0% (–), <10% (– – /+), up to 50% (– /+), 50-80% (+/ –), ->80% (+), not available (N/A).

Superficial and *micronodular BCC* subtypes are less common. They account for approximately 10 percent and 5 percent of periocular BCCs, respectively (fig. 3-5) (13).

Up to 39 percent of periocular BCCs have mixed histology, with the most common combinations being nodular with infiltrative (fig. 3-4) and nodular with superficial BCC. Squamous differentiation appears in approximately 5 percent of tumors. Perineural invasion has been reported in approximately 1 percent of eyelid BCCs (fig. 3-5) (13), but is as high as 19 percent in BCCs with orbital invasion (10). The literature reports aggressive histology (infiltrative, micronodular, and basosquamous/metatypical BCC) in about 30 percent of periocular BCCs, which is more frequent than in nonfacial skin (fig. 3-5) (14).

Molecular Genetic Findings. Ancillary studies are generally not required for the diagnosis of BCC (16). BCC in patients with nevoid basal cell carcinoma syndrome harbors a germline mutation in a tumor suppressor *PTCH1* gene, a regulator of sonic hedgehog signaling pathway. Ultraviolet light (UV) signature mutations in *PTCH1* and other mediators of sonic hedgehog signaling appear in up to 80 percent of sporadic BCCs. While this insight into the molecular genetics of BCC is not helpful at this time for diagnostic or prognostic purposes, this information has led to the emergence of targeted therapies (see below) (12).

Differential Diagnosis. BCC is easily recognized histologically. In rare tumors that lack normal basaloid differentiation, histopathologic considerations primarily include squamous cell carcinoma and sebaceous carcinoma. In challenging cases, a panel of immunohistochemical stains help elucidate the tumor origin (Table 3-2).

Treatment and Prognosis. The morpheaform subtype and tumors in the medial canthus location are more likely to be incompletely resected. These tumors tend to recur and may eventually involve the orbit. Following complete resection of BCC with negative margins, the rate of local recurrence is less than 1 percent at 5-year follow-up. In contrast, the reported recurrence rate can be as high as 38 percent at 5 years for incompletely excised tumors. The recurrence rate of morpheaform BCC has been recorded as high as 3.8 percent at 5-year follow-up, even after resection with negative surgical margins. The rate of metastasis of BCC is low (0.03 percent), with regional lymph nodes being the most common metastatic site. BCCs can rarely cause mortality by direct invasion into the cranium (10).

The mainstay of treatment for orbital and periocular BCCs is surgical excision, sometimes

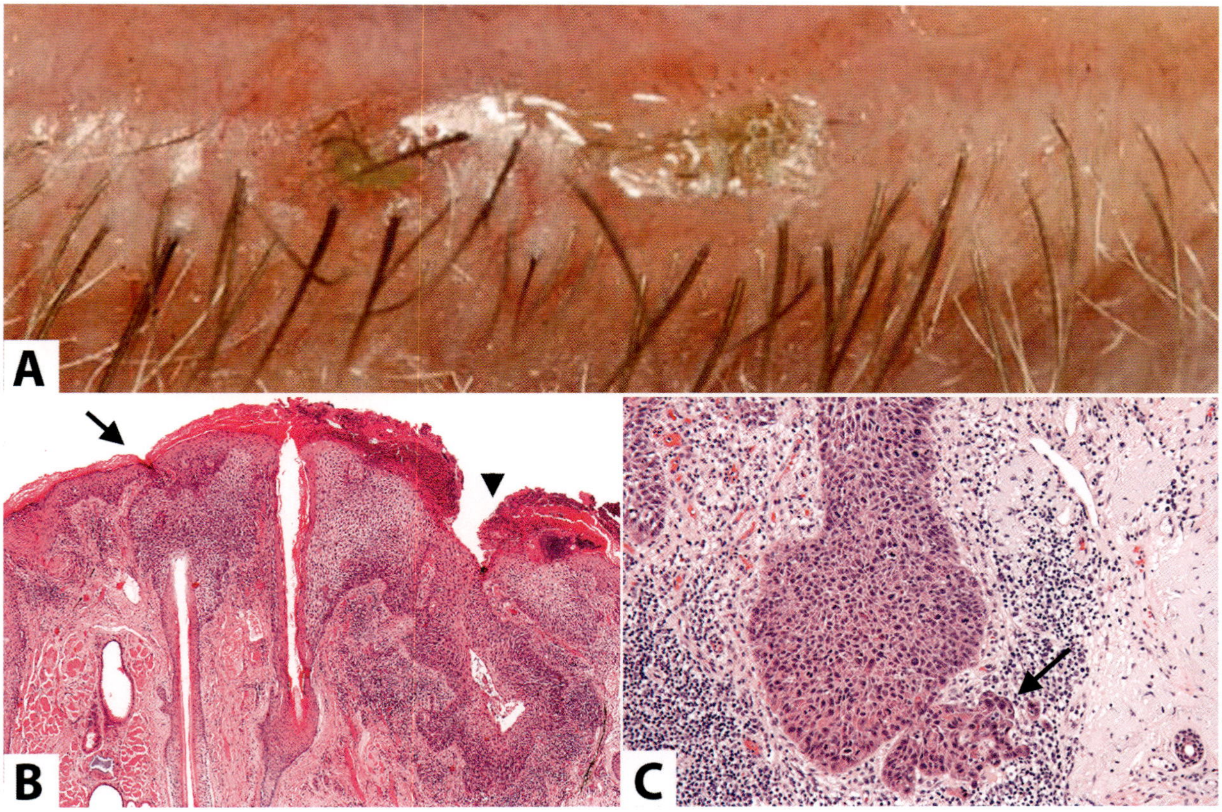

Figure 3-6

MICROINVASIVE SQUAMOUS CELL CARCINOMA ARISING IN ACTINIC KERATOSIS

A: A hyperkeratotic, focally ulcerated and minimally thickened plaque involves the anterior eyelid margin.

B: The epidermis in the region of eyelashes is acanthotic and parakeratotic, with variable degrees of dysplasia (arrow) and focal carcinoma in situ (arrowhead).

C: A focus of microinvasion (arrow) into the actinic-damaged and chronically inflamed dermis is identified (B,C: H&E stain).

supplemented by adjuvant radiation therapy. In tumors with orbital invasion or in a setting of diffuse and multifocal disease, orbital exenteration (removal of the globe, eyelids, and orbital soft tissues) may be required (10,17). The literature recommends frozen section control of surgical margins or Mohs micrographic surgery for aggressive BCC variants (3).

In some patients with locally advanced periocular BCC, surgery may not be desirable. Considerations against surgical intervention include loss of visual function that can accompany radical resection, unresectable tumors that invade cranium, metastatic disease, or advanced tumors in patients with multiple co-morbidities who cannot tolerate radical surgery. Such patients may be candidates for targeted therapies directed against hedgehog signaling pathway mediators; these been employed successfully in the management of select cases of periocular BCC (17,18). FDA-approved agents include vismodegib and sonidegib (17).

SQUAMOUS CELL CARCINOMA AND ITS PRECURSORS

Definition. Eyelid *squamous cell carcinoma* (SCC) is an invasive tumor, believed to be derived from suprabasal epidermal keratinocytes (12,15).

Clinical Features. Cutaneous SCC accounts for approximately 5 percent of all eyelid malignancies (12,15). The tumor presents as a hyperkeratotic ulcerated nodule or plaque in the sun-exposed lower eyelid region in patients in the sixth or seventh decade of life (fig. 3-6). Men are affected two to three times as commonly as women (10).

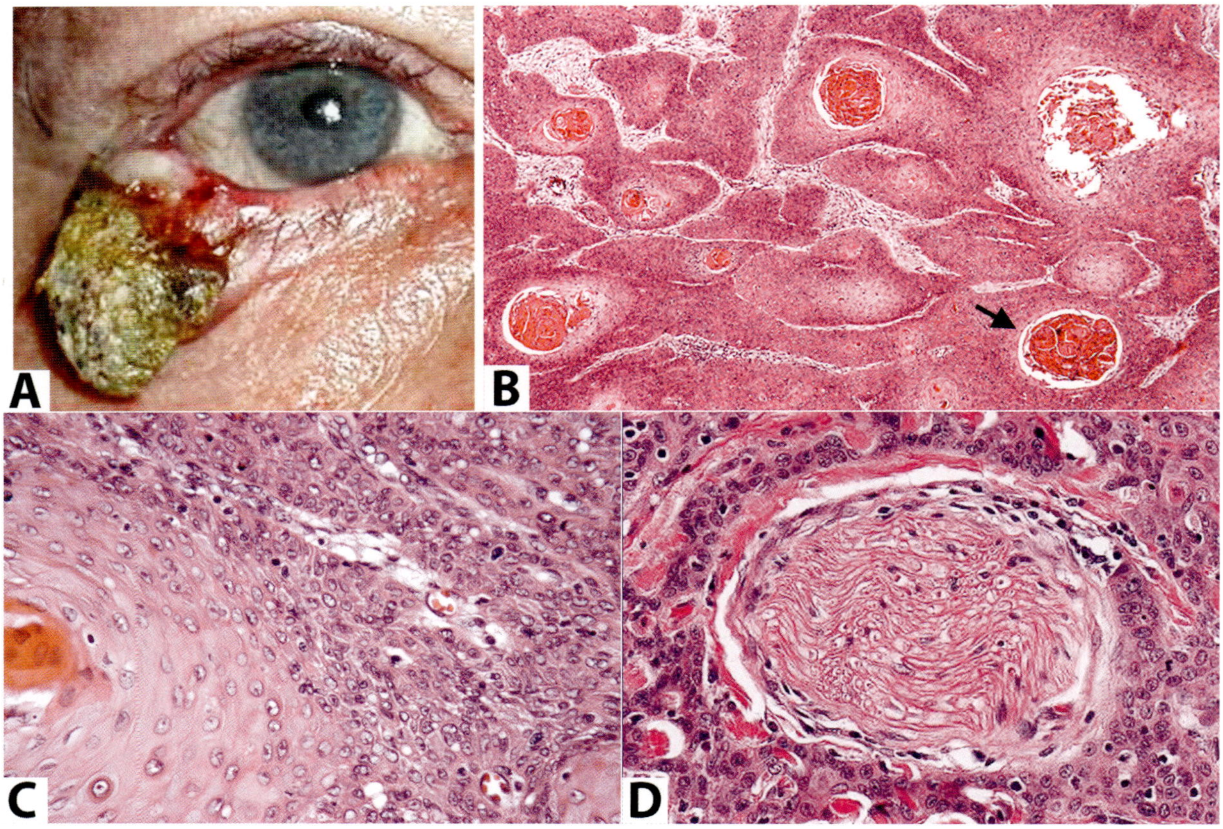

Figure 3-7

INVASIVE SQUAMOUS CELL CARCINOMA

A: A hyperkeratotic and ulcerated mass in the lower lateral eyelid margin and the eyelid.
B: Nests of keratinizing squamous cells with numerous keratin pearls (arrow).
C: Most of the tumor is well to moderately differentiated.
D: Poorly differentiated foci with perineural invasion at the leading edge of the tumor (B–D: H&E stain).

UV radiation is believed to be the leading environmental risk factor for the development of SCC (12). Other risk factors include immunosuppression, exposure to ionizing radiation and chemicals (such as hydrocarbons and arsenic), a high-fat diet, and infection with human papillomavirus (HPV) (10). The inherited genetic factors implicated in the pathogenesis of SCC include light skin pigmentation and defective DNA repair mechanisms (12).

Microscopic Findings. Eyelid SCC progresses from an in situ intraepidermal lesion (typically actinic keratosis and less frequently, Bowen disease) to an invasive tumor, composed of nests of malignant cells with variable degrees of squamous differentiation and keratinization (fig. 3-6) (12). In rare instances in which no precursor lesion is identified, de novo origin has been postulated (3,15).

Most eyelid SCCs are well- to moderately differentiated tumors (10). Well-differentiated SCC closely resembles the surface epidermis, and is composed of polygonal cells with abundant acidophilic cytoplasm, easily identifiable intercellular "bridges" (desmosomes), and foci of keratinization that manifest as keratin pearls (fig. 3-7) (15). Poorly differentiated SCC is characterized by prominent cellular pleomorphism and atypia (anaplasia), with nuclear hyperchromasia, absence of intercellular "bridges," keratinization of individual cells, and presence of atypical mitotic figures (fig. 3-7) (3,15).

Microscopic perineural invasion occurs in 8 to 14 percent of facial and periorbital SCCs (fig.

3-7). It may be heralded by the clinical signs of trigeminal distribution sensory deficit, ophthalmoplegia, orbital pain, or facial palsy (10). Spindle cell SCC and adenoid SCC are some of the SCC variants recognized in the eyelid skin (3).

Molecular Genetic Findings. Epidermal growth factor receptor (EGFR) signaling activation and overexpression have been noted in cutaneous SCC of the head and neck, and are associated with the risk of metastases (18). The key genes mutated in sporadic SCC include *TP53*, *RAS*, and *p16/CDKN2A* (12). Molecular studies and other ancillary diagnostic methods are generally not employed for diagnosis and prognostication of eyelid SCC.

Differential Diagnosis. The histologic diagnosis can be challenging in poorly differentiated tumors, raising the possibility of other eyelid carcinomas (BCC, sebaceous carcinoma, and sweat gland carcinoma). In these instances, ancillary immunohistochemical stains are of value (Table 3-2). Keratoacanthoma (discussed in greater detail below) should be distinguished from conventional SCC. For keratoacanthoma, the clinical history (rapid growth with potential early involution) and overall tumor morphology are important (15). Metastatic SCC to eyelid is an infrequent occurrence, but must be considered in tumors that lack evidence of origin from the epidermis.

Treatment and Prognosis. Eyelid SCC can behave aggressively, with a significant risk of local tissue destruction, invasion into the orbit, perineural invasion, and lymphatic or hematogenous metastasis. Reported rates of regional lymph node metastasis vary from 1.3 to 24.0 percent (19), with most regional metastases occurring in the parotid, preauricular, and submandibular nodes (12). Distant metastases are much less common, reported in 6.2 percent of cases (10), and disease-related mortality has been estimated to be low (less than 2 percent) (19). Locally advanced disease (tumors 10 to 20 mm in thickness or larger and those involving the full-thickness of the eyelid), poorly differentiated histology, perineural invasion, recurrent tumors, and immunosuppression are some of the risk factors associated with poor prognosis (19).

The mainstay of periocular SCC treatment is surgical excision, either wide excision with or without frozen section control of margins, or Mohs micrographic surgery (19,20). Adjuvant radiotherapy may be considered in cases with perineural invasion (10). Recently, the EGFR inhibitor, erlotinib, was used experimentally in the management of locally advanced and metastatic cutaneous SCC, including periocular SCC (18).

SEBACEOUS LESIONS

Compared to the extraocular skin, sebaceous gland tumors are frequent in a periocular location due to the abundance of sebaceous glands. Sebaceous tumors in the periocular region typically originate in the Meibomian glands and glands of Zeis, and less frequently, in the caruncle and skin of the eyebrow. The spectrum of sebaceous lesions ranges from sebaceous gland hyperplasia to sebaceous carcinoma (2).

Sebaceous Gland Hyperplasia

Common in the periocular tissues of older people, *sebaceous gland hyperplasia* presents as a yellowish, elevated, soft, frequently umbilicated nodule (fig. 3-8). Hypertrophic and hyperplastic, fully mature sebaceous gland lobules in a symmetric arrangement are clustered around a central dilated sebaceous duct (fig. 3-8) (2,15). These benign lesions do not progress to sebaceous carcinoma. Solitary sebaceous gland hyperplasia is generally not associated with Muir-Torre syndrome (see below) (21).

Sebaceous Adenoma and Sebaceous Epithelioma

Sebaceous adenomas and *epitheliomas* present as tan, yellow, or reddish, occasionally ulcerated nodules or polypoid lesions on the eyelids of older people (fig. 3-9) (15). Histologically, there is a disorganized proliferation of sebaceous gland lobules that vary in size and shape, and frequently involve the surface epidermis (21,22). Expansion of germinative basaloid cells beyond the typical one or two layers is characteristic of the sebaceous lobules in sebaceous adenomas. Mature sebocytes, however, still predominate over germinative cells (fig. 3-9). Sebaceous epithelioma (sebaceoma) is distinguished from other sebaceous lesions by a predominance of immature cells. Occasionally, it may be infiltrative (fig. 3-10). Holocrine secretion can be prominent in sebaceous adenoma and sebaceoma, with resultant cystic-appearing morphology (fig. 3-9). Keratoacanthoma-like epidermal changes appear in some lesions.

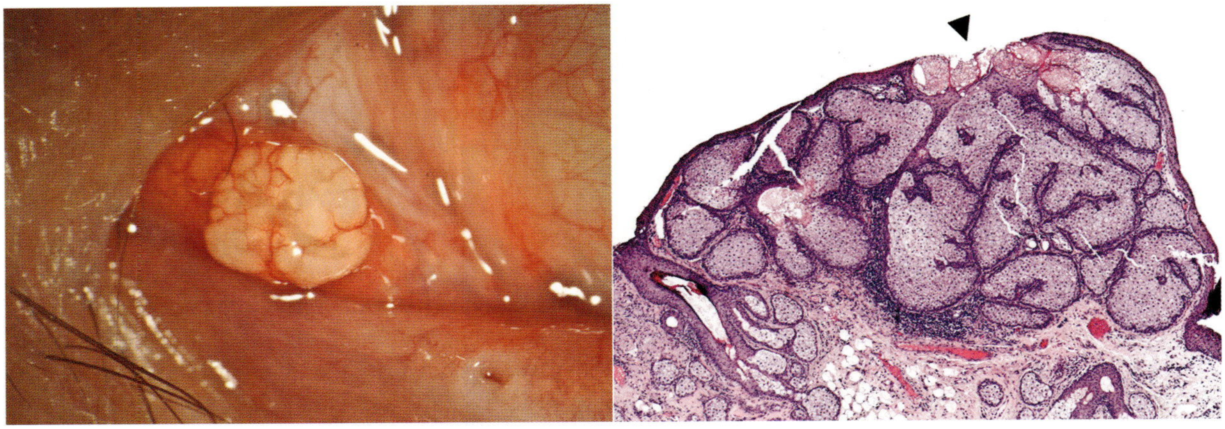

Figure 3-8

SEBACEOUS GLAND HYPERPLASIA: CARUNCLE

Left: A yellow, lobulated, well-circumscribed caruncular mass. (Courtesy of Drs. J. Shields and C. Shields, Philadelphia, PA.)

Right: Hypertrophic and hyperplastic sebaceous gland lobules are arranged around a central dilated duct (arrowhead) (H&E stain).

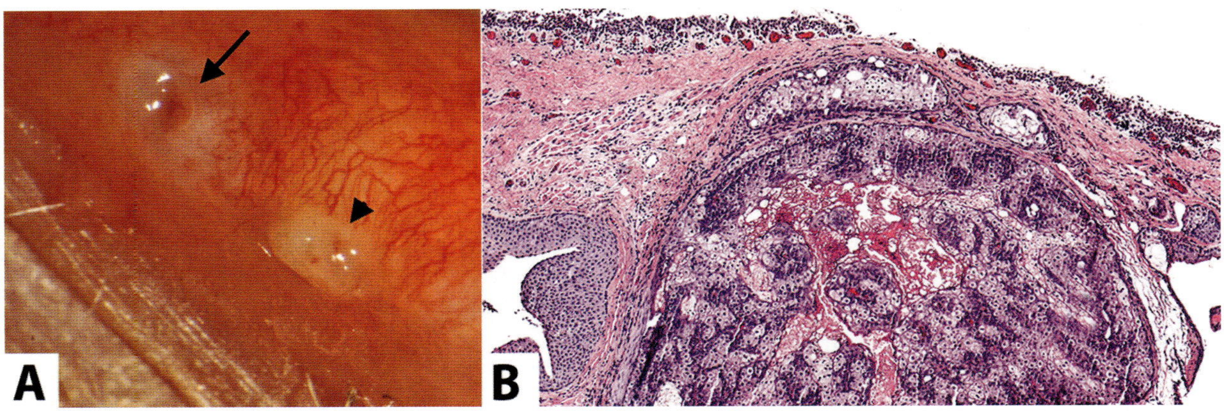

Figure 3-9

SEBACEOUS ADENOMA IN A PATIENT WITH MUIR-TORRE SYNDROME

A: A yellow vaguely papillomatous, nonulcerated nodule involves the posterior lamella of the eyelid margin (arrowhead), adjacent to the lacrimal punctum (arrow).

B: The nodule is composed of a well-circumscribed but haphazardly arranged proliferation of sebaceous gland lobules with central comedonecrosis, lending the lesion a somewhat cystic appearance clinically.

There is a controversy regarding the appropriate nomenclature for sebaceous epithelioma, with the term *well-differentiated sebaceous carcinoma* applied by some experts to these lesions (22). It is likely that sebaceous adenoma, epithelioma, and well-differentiated sebaceous carcinoma represent a morphologic continuum.

Sebaceous adenoma and epithelioma have been considered specific markers of *Muir-Torre syndrome* (MTS), with reported association ranging from 25 to 85 percent (12). MTS, a variant of Lynch syndrome, has autosomal dominant transmission and is characterized by the association of cutaneous lesions (sebaceous neoplasms and keratoacanthomas) with a visceral malignancy, characteristically colorectal or genitourinary carcinoma. Cutaneous manifestations of MTS can occur before (22 percent), concurrent with (6 percent), or after (56 percent) the diagnosis of a visceral malignancy (12). The diagnosis of sebaceous adenoma and epithelioma in the periocular region, therefore, warrants investigation for MTS.

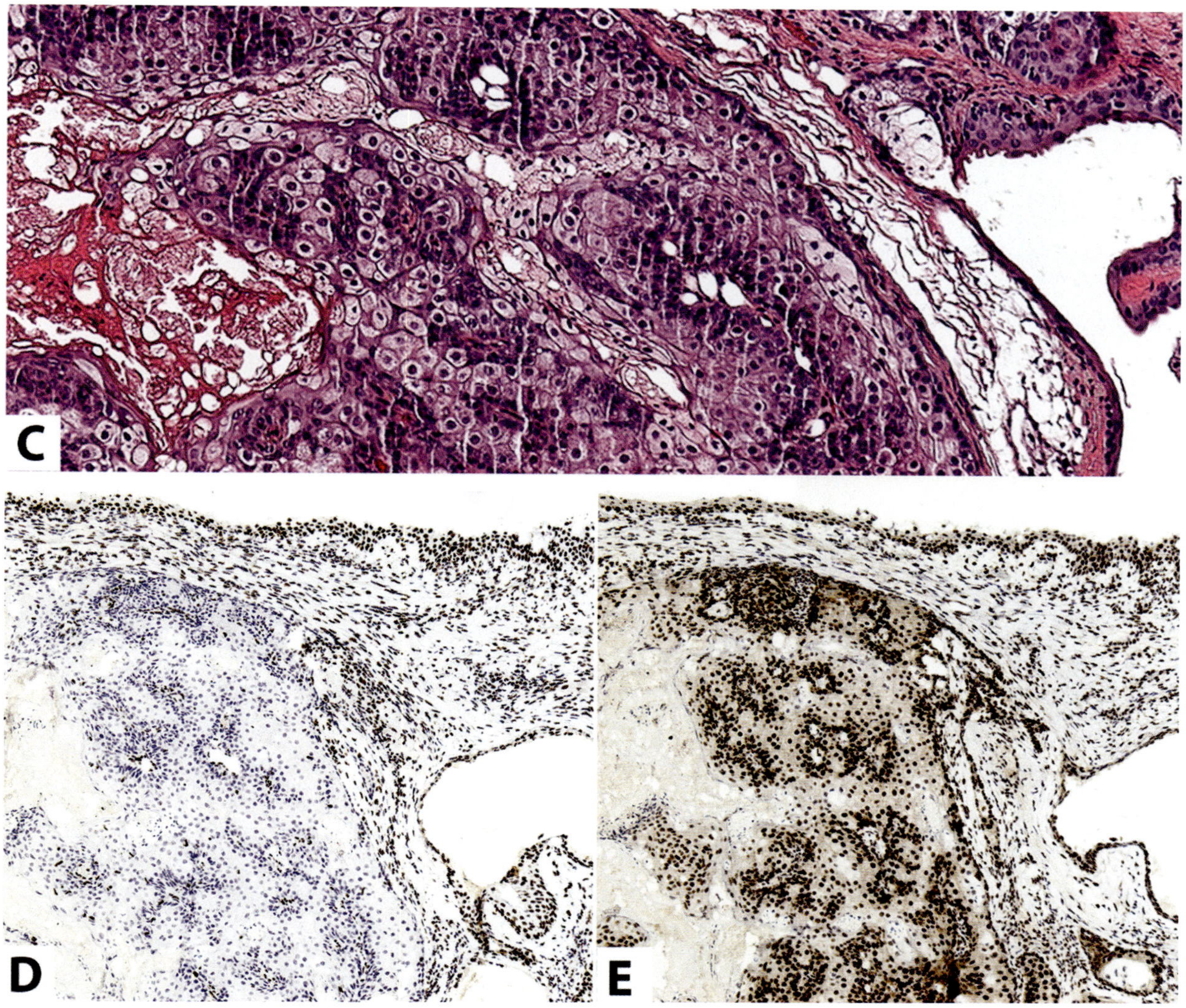

Figure 3-9, continued

C: At higher magnification, the hamartomatous arrangement of sebaceous gland lobules with greater than normal number of germinative cells is evident. Mature sebocytes predominate (B,C: H&E stain).

D: There is loss of expression of MSH-2 in the lesion with continued expression in the adjacent stromal cells and lymphocytes.

E: The neoplastic cells retain expression of MLH-1.

A germline mutation frequently causes a germline MTS in one or more of the DNA mismatch repair (*MMR*) genes. The four DNA mismatch repair genes implicated in MTS are *MSH2* (over 90 percent of cases), *MLH1* (less than 10 percent of cases), and less frequently, *MSH6* and *PMS2* (12). Thus, testing for *MMR* gene status has become an integral part of the workup of sebaceous adenoma and epithelioma.

Immunohistochemical studies to assess MMR protein expression, using a panel of antibodies, help to assess MMR status in these tumors. In the head and neck region, approximately half of sebaceous adenomas and epitheliomas are MMR deficient (fig. 3-9) (23). Studies have shown that with a panel of four antibodies, immunohistochemistry is as sensitive as microsatellite instability (MSI) analysis for predicting germline *MMR* mutation (12) (positive predictive value of 37 to 50 percent) (24). This predictive data have been obtained on all sebaceous lesions, irrespective of site (24), and may not apply to the periocular sebaceous adenoma and epithelioma. It is vital to understand that abnormal immunohistochemistry, including the absence of MSH2, is not diagnostic of MTS and should

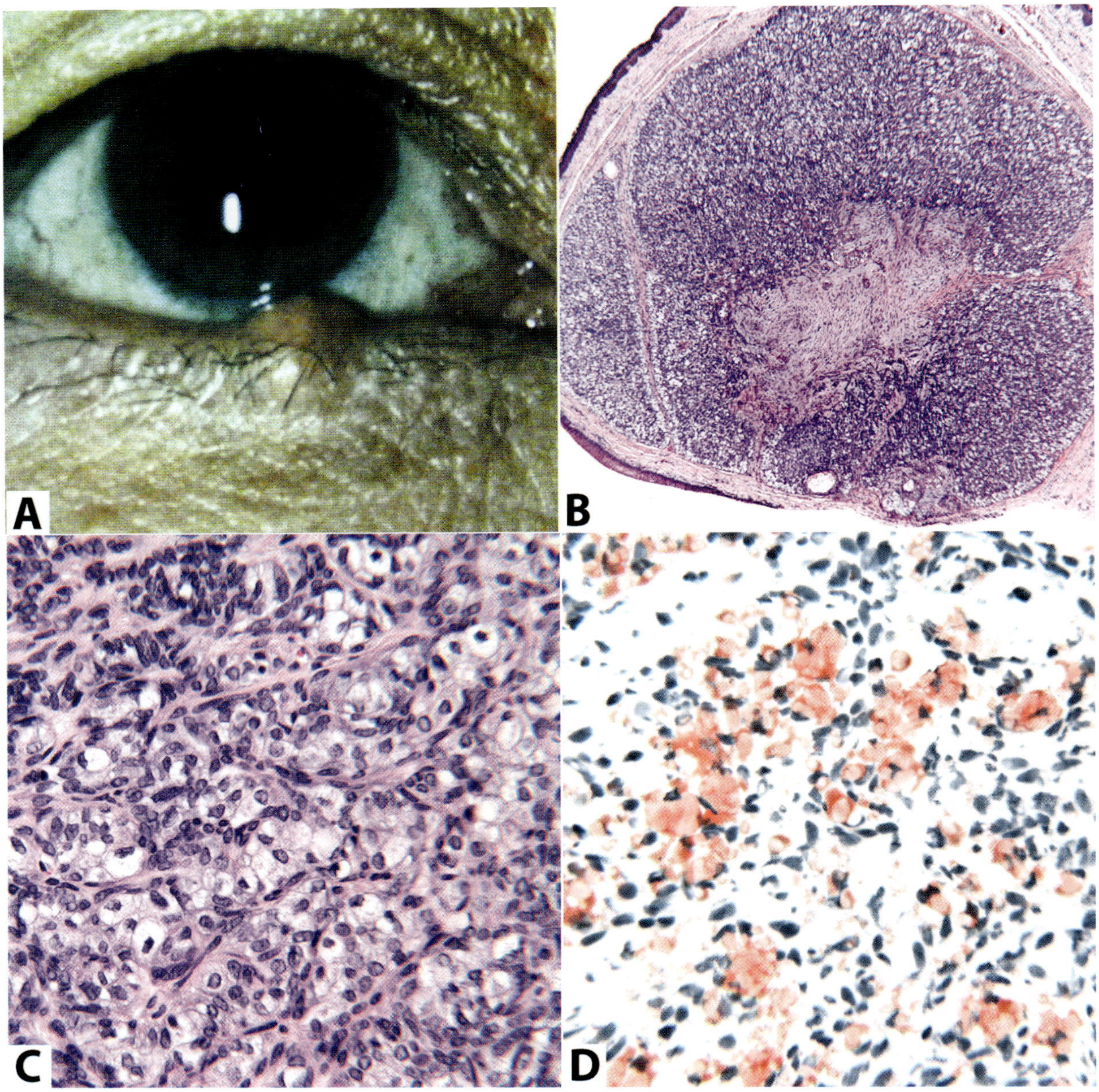

Figure 3-10

SEBACEOUS EPITHELIOMA (SEBACEOMA)

A: A yellow to orange nodule without ulceration or madarosis (eyelash loss) is at the eyelid margin.
B: The well-circumscribed arrangement of basophilic lobules is associated with the ducts (B,C: H&E stain).
C: Higher power assists in identifying the predominance of germinative cells over mature sebocytes.
D: Oil red O stain confirms the presence of lipid within the neoplastic cells.

be interpreted cautiously in conjunction with family history and germline genetic testing (24).

Sebaceous adenoma and epithelioma may be locally aggressive, but almost never metastasize. Surgical excision is thus appropriate treatment for these lesions (22).

Sebaceous Carcinoma

Definition. *Sebaceous (gland) carcinoma* is a malignant neoplasm arising from various sebaceous glands of ocular adnexa (15).

Clinical Features. Sebaceous carcinoma is more frequent in the head and neck region, with

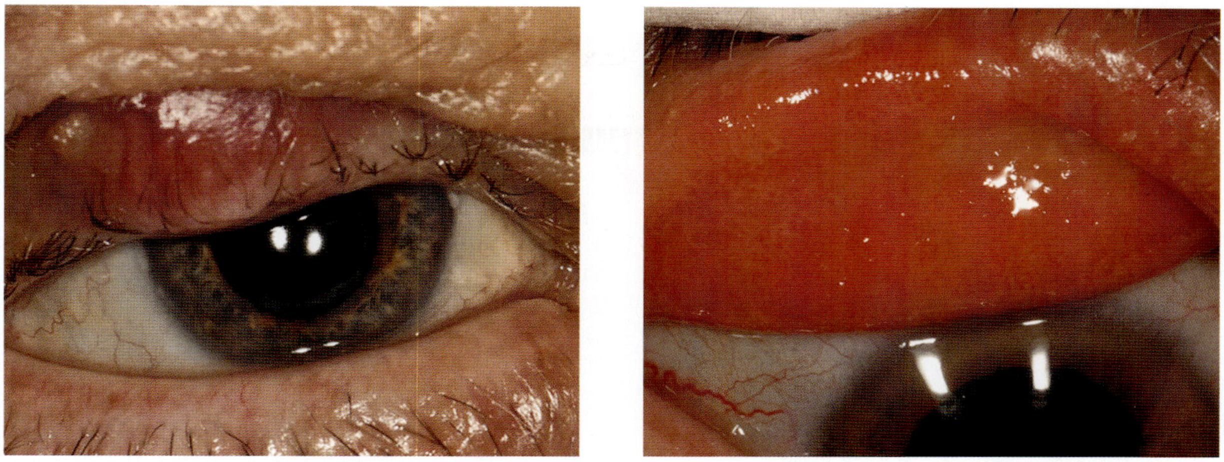

Figure 3-11

SEBACEOUS CARCINOMA

Left: A red-yellow, nonulcerated nodule involves the upper eyelid and eyelid margin, and is associated with loss of eyelashes (madarosis). This presentation can mimic a chalazion.

Right: Diffuse thickening of the upper eyelid margin in this patient with sebaceous carcinoma. The eyelid is everted, permitting visualization of the conjunctival lining of the tarsus, the palpebral conjunctiva. The papillary conjunctival reaction masquerades clinically as blepharoconjunctivitis. (Courtesy of Drs. J. Shields and C. Shields, Philadelphia, PA.)

the eyelid being the most common site, accounting for 40 percent of all cases. The frequency of sebaceous carcinoma varies greatly by geographic location and ethnicity (25). In the United States, sebaceous carcinoma frequency is comparable to that of SCC (approximately 5 percent of malignant eyelid tumors) (2). In the Asian-Indian population, sebaceous carcinoma is a common eyelid tumor, accounting for 28 to 60 percent of all eyelid malignancies (25). This neoplasm is more common in women and the elderly (2). Prior irradiation is an important risk factor (9).

Periocular sebaceous carcinoma presents as a solitary tan, yellow, or erythematous eyelid nodule, associated with distortion of normal eyelid margin architecture and loss of eyelashes (madarosis) (fig. 3-11) (9). Upper eyelid location is seen in two thirds of tumors, reflecting the greater mass of Meibomian gland tissue in the upper eyelid (2). The second most frequent presentation is diffuse unilateral thickening of the eyelid that resembles blepharitis, associated with conjunctival changes that resemble conjunctivitis (fig. 3-11). This presentation reflects diffuse intraepithelial tumor spread (9).

Microscopic Findings. Sebaceous carcinoma frequently grows in lobules reminiscent of normal sebaceous glands. Central comedonecrosis, which reflects an aberrant attempt at holocrine secretion by malignant cells, is common. The neoplastic nuclei are large and ovoid, with vesicular chromatin and multiple nucleoli. Mitotic figures are frequent. Well-differentiated tumors have a predominance of cells with abundant vacuolated cytoplasm, reflecting the presence of intracytoplasmic lipid. Poorly differentiated tumors lack appreciable cytoplasmic lipidization (fig. 3-12) (2). Most sebaceous carcinomas in the periocular region tend to be moderately differentiated.

A characteristic and diagnostically helpful feature of periocular sebaceous carcinoma is its ability to invade and replace eyelid epidermis and conjunctival epithelium in a pagetoid fashion (fig. 3-13) (2). Recognition of this pattern has led to recommendation of sampling of the conjunctival cul-de-sac by map biopsies (1- to 2-mm biopsies in multiple conjunctival locations) to assess the intraepithelial tumor burden (9). Conjunctival epithelium replaced by sebaceous carcinoma has a tendency to shed. This denuded appearance, while not diagnostic of sebaceous carcinoma, should suggest its existence and prompt careful evaluation of tissue (levels on the index tissue and/or review of additional tissue samples) (fig. 3-13) (26). As a corollary, a map biopsy of the conjunctiva in which no epithelium is identified cannot, with certainty, be declared negative. Instead, a

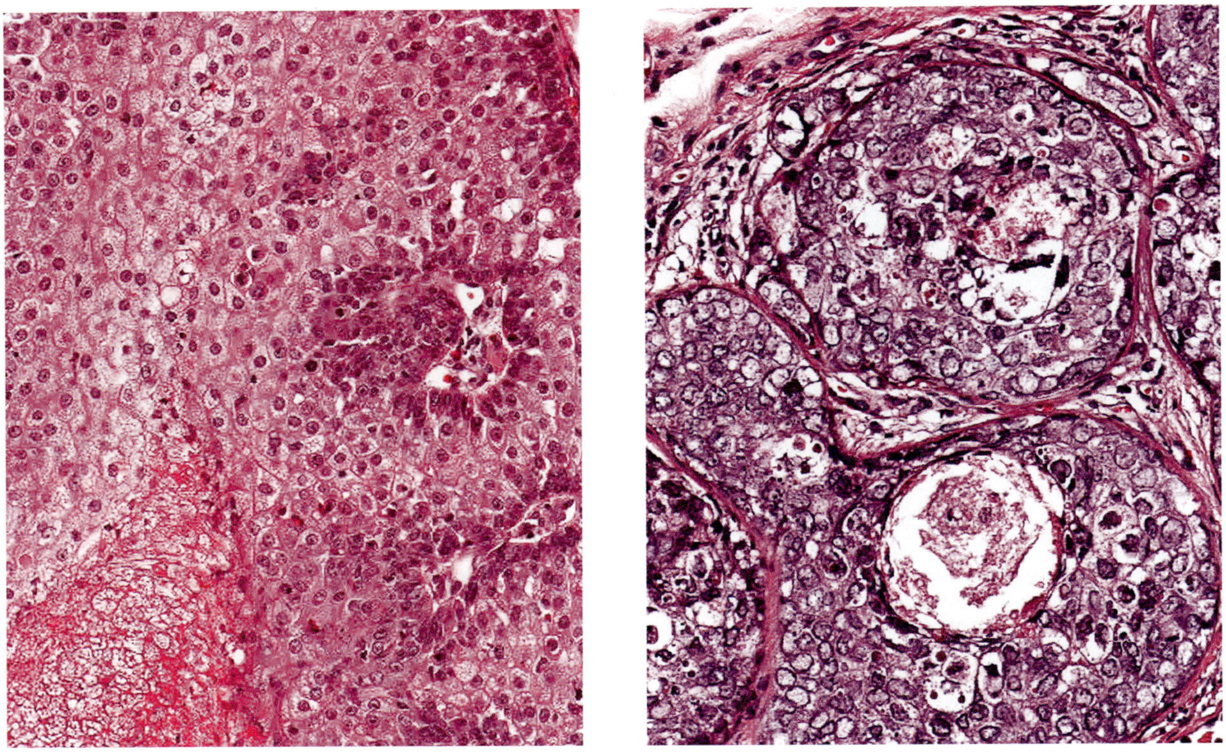

Figure 3-12

SEBACEOUS CARCINOMA

Left: Well-differentiated sebaceous carcinoma is characterized by a morphology reminiscent of normal sebaceous gland lobules, with predominance of cells that resemble mature sebocytes. Comedonecrosis is present.

Right: Poorly differentiated sebaceous carcinoma replaces the Meibomian glands. The neoplastic cells feature ovoid nuclei with vesicular chromatin, scant cytoplasm with minimal vacuolation, and brisk mitotic activity with apoptosis and comedonecrosis (H&E stain).

comment should state that intraepithelial sebaceous carcinoma cannot be excluded when the entire epithelium has sloughed away.

Immunohistochemical and Molecular Genetic Findings. The diagnosis of sebaceous carcinoma is usually based on morphologic findings alone, without the aid of ancillary diagnostic techniques, mainly when an experienced ocular pathologist reviews the biopsies (2). In challenging cases, ancillary diagnostic studies are performed. Traditionally, special histochemical stains for lipid, such as oil red O (fig. 3-10) and Sudan black IV, have been used to detect lipid in sebaceous gland neoplasms. These methods, however, have a sensitivity of approximately 40 percent, require fresh-frozen tissue, and the stain intensity fades over time (27).

Although there is no perfect immunohistochemical stain for sebaceous carcinoma, cytoplasmic expression of adipophilin in a characteristic vesicular pattern has been shown to be sensitive and specific for periocular sebaceous carcinoma (fig. 3-14) (27). Other useful immunohistochemical markers include androgen receptors (AR) (fig. 3-14) and epithelial membrane antigen (EMA). The immunohistochemical panels currently employed to support the diagnosis of sebaceous carcinoma are imperfect, but are used to evaluate tumors in cases where the histogenesis is in question (Table 3-2) (16).

When patients present with a history of multiple sebaceous neoplasms, the possibility of MTS (see above) requires consideration. However, unlike sebaceous carcinomas in non-head and neck locations, periocular sporadic sebaceous carcinomas infrequently harbor *MMR* defects (23,28). In a recent next-generation sequencing analysis of periocular and extraocular sebaceous carcinomas, the authors found that mutations in *TP53* and *RB1* were the most

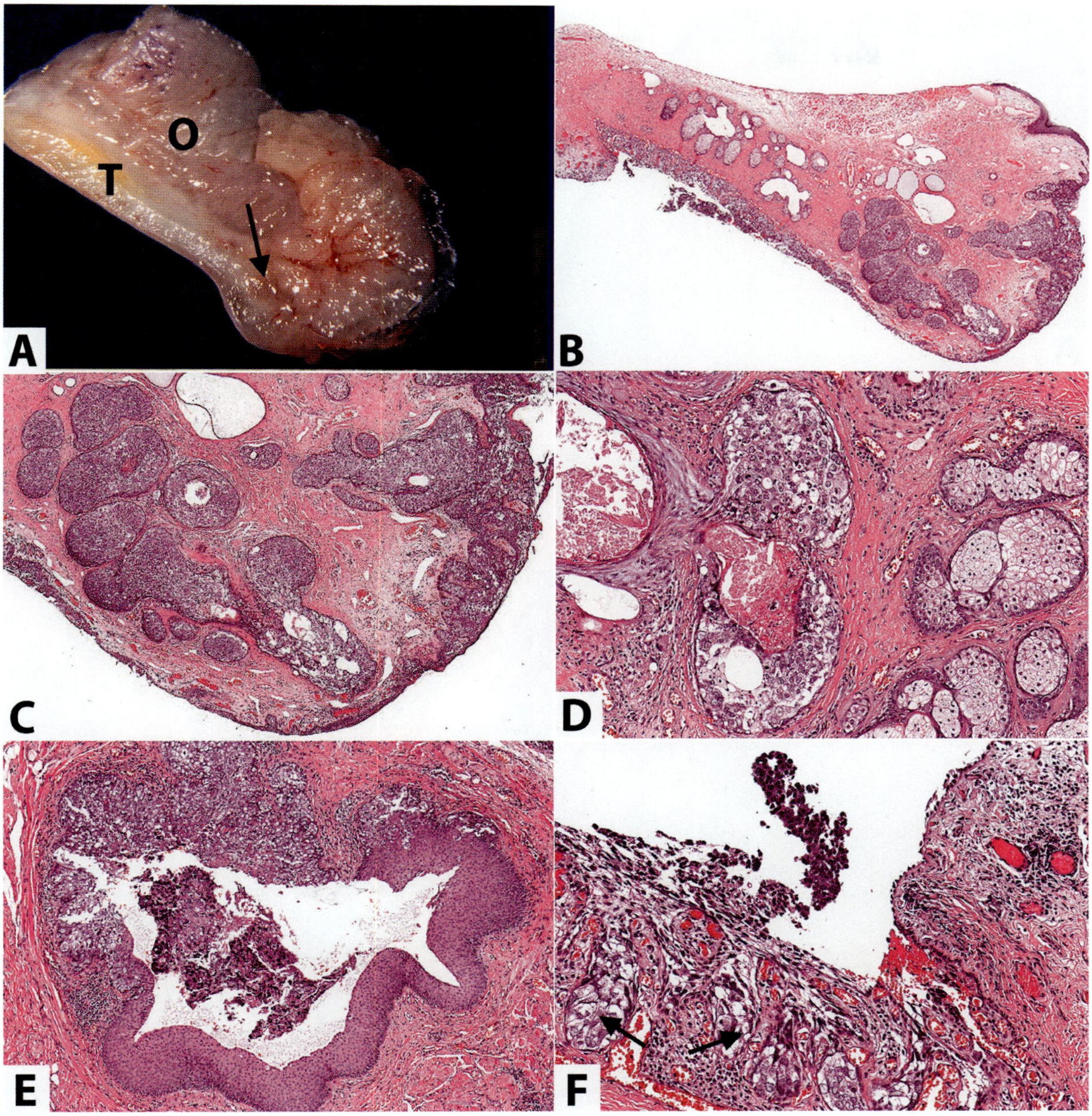

Figure 3-13

SEBACEOUS CARCINOMA: CHARACTERISTIC PATTERNS

A: A skin-sparing resection of the upper eyelid includes the conjunctiva with tarsus (T) and the overlying orbicularis oculi muscle (O). A yellow-tan, finely lobulated mass arises from the tarsus (arrow) to involve the eyelid margin.

B: Corresponding scanning magnification view of the specimen illustrated in A. The cellular mass replaces the Meibomian gland at the eyelid margin (B–F: H&E stain).

C: Higher magnification of the area illustrated in B. The lobulated tumor in the Meibomian gland region extends into the adjacent anterior lamella of the eyelid. The tumor invades the conjunctival epithelium and the eyelid margin epidermis.

D: Side by side comparison of the normal Meibomian gland lobules on the right with the in situ moderately differentiated sebaceous carcinoma on the left. Comedonecrosis is present.

E: Focal intraepithelial involvement of the lacrimal canaliculus. The malignant cells are shed into the lumen.

F: Malignant cells replace the epithelium of the conjunctival pseudoglands of Henle (arrows). The surface epithelium, replaced by the malignant cells, is shed from the underlying stroma (refer to the text).

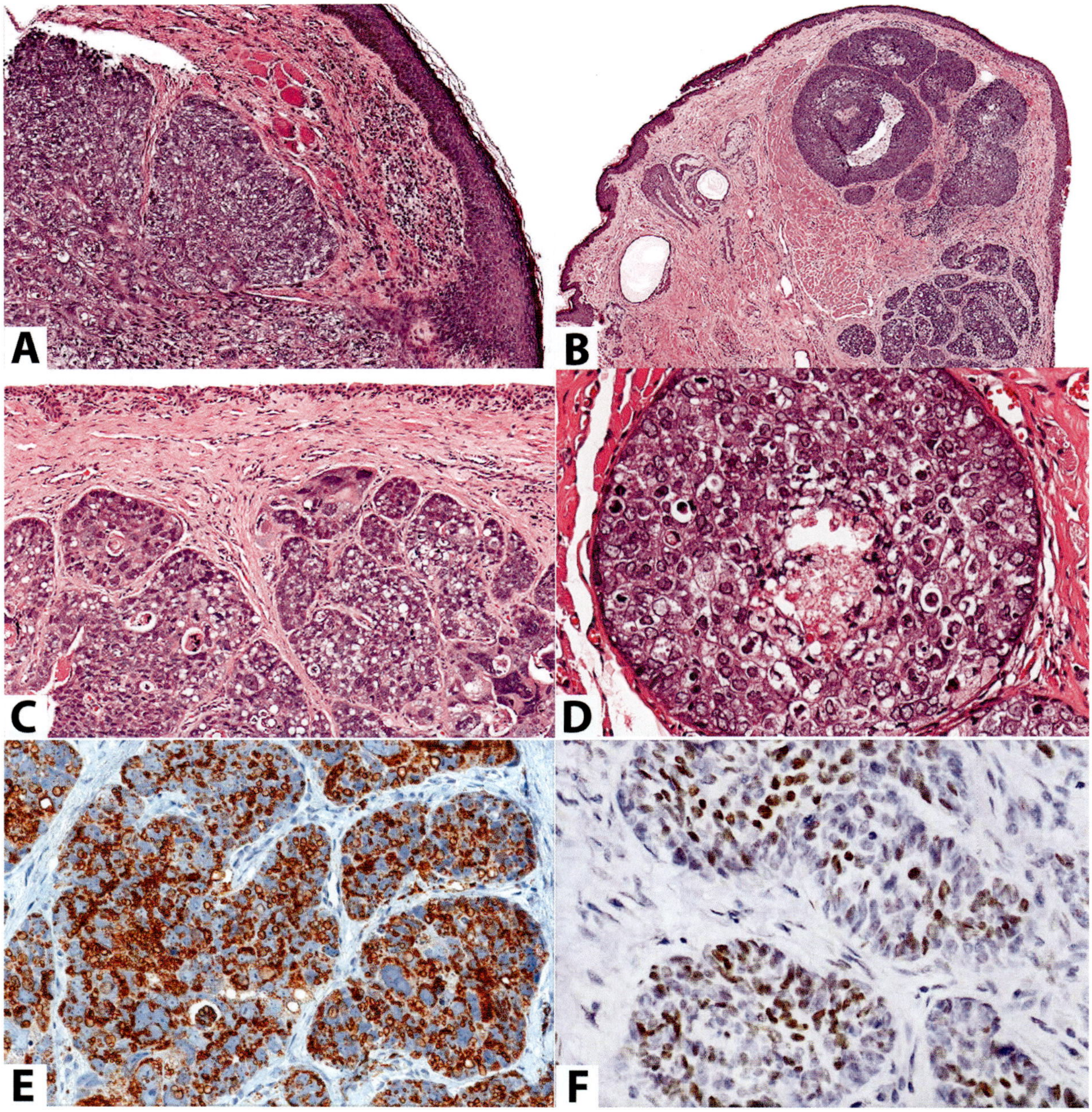

Figure 3-14

SEBACEOUS CARCINOMA MIMICKING BASAL CELL CARCINOMA (BCC)

A: Superficial biopsy with nodules of basaloid cells that extend from the epidermis. Even though a rent in the tissue superiorly resembles the retraction artifact of BCC, the peripheral palisading and stromal desmoplasia that characterize BCC are not present (A–D: H&E stain).

B: Full-thickness eyelid resection of the same tumor. A very cellular mass is centered in the tarsus. The mass arises from a Meibomian gland, with secondary extension into the tissue of the eyelid margin.

C: The Meibomian gland is replaced by a neoplasm that is poorly differentiated, with only focal intracytoplasmic lipid vacuoles.

D: The nuclear features of this lesion are characteristic of a high-grade tumor; comedonecrosis is present.

E: Adipophilin immunostain shows characteristic cytoplasmic staining in large vacuoles.

F: The neoplastic cells express androgen receptors (AR).

common and unique to ocular sebaceous carcinomas, and corresponded with modified TP53 and RB1 protein expression. Only extraocular sebaceous carcinomas harbored mutations explicitly affecting DNA repair and chromatin remodeling pathways, with defects in MMR protein expression, and a microsatellite instability (MSI)-high phenotype (29). Although no specific guidelines for ancillary *MMR* gene status testing are currently available for periocular sebaceous carcinoma, in practice, this investigation is generally limited to tumors in patients with clinical concern for MTS and sebaceous carcinoma with well-differentiated histology.

Differential Diagnosis. Sebaceous carcinoma can confound both ophthalmologists and pathologists. Lesions arising deep in the eyelid, such as Meibomian gland carcinoma, typically cause no abnormalities in the overlying skin until late in the course of the disease. Meibomian gland carcinoma may be confused with chalazion. It is recommended, therefore, that recurrent or atypical chalazia should be submitted for pathologic evaluation. Chronic unilateral blepharoconjunctivitis is another clinically challenging presentation of sebaceous carcinoma. Thus, an elderly patient with unilateral chronic keratoconjunctivitis unresponsive to medical therapy should undergo diagnostic biopsy (2).

Eyelid sebaceous carcinoma is commonly misdiagnosed (35 percent of cases are initially misdiagnosed on pathologic evaluation of biopsy or surgical resection specimens) (10). Histopathologically, poorly differentiated sebaceous carcinoma can mimic BCC, SCC, mucoepidermoid carcinoma, sweat gland carcinoma, and metastases (27). Careful attention to tumor morphology may help distinguish these lesions. Although frequently basophilic, sebaceous carcinoma has higher degree of nuclear atypia and mitotic activity than most BCCs. Additionally, sebaceous carcinoma nests lack the peripheral palisading and retraction artifact that characterize BCC (fig. 3-14) (2).

Some sebaceous carcinomas have eosinophilic cytoplasm and distinct cell borders reminiscent of SCC (10). Careful inspection for sebaceous differentiation helps establish the diagnosis of sebaceous carcinoma. Additionally, intraepithelial pagetoid spread is not typical of BCC and conventional SCC, and may serve as a helpful morphologic feature to diagnose sebaceous carcinoma (fig. 3-15) (2). As mentioned previously, immunohistochemical panels may also be helpful in distinguishing between these tumors (16).

Treatment and Prognosis. Sebaceous carcinoma can spread by direct extension to involve vital structures and can metastasize to regional lymph nodes and distal sites, such as lung, liver, brain, and skull (2). Reported metastatic rates are 8 to 18 percent for nodal metastasis and 3 to 8 percent for distant metastasis (10). Factors associated with poor prognosis include origin in the upper eyelid, tumor diameter greater than 10 mm, origin from Meibomian glands, duration of symptoms before diagnosis for more than 6 months, infiltrative growth pattern, poorly differentiated histology, extensive pagetoid spread, orbital invasion, and lymphovascular invasion (2,3).

Surgical resection is the mainstay of treatment of sebaceous carcinoma. Because most tumors are localized to the deep eyelid tissues, skin sparing resection of the eyelid is preferable to facilitate the reconstruction (fig. 3-13).

The practice of frozen section control of margins in the management of sebaceous carcinoma is controversial. Most ophthalmic pathologists do not recommend performing frozen sections on sebaceous carcinoma because of its propensity for pagetoid spread and because of the inherent challenges that frozen section artifact introduces into the interpretation of sebaceous lesions.

There is a general belief that irradiation is not highly effective in the management of sebaceous carcinoma. However, adjuvant radiotherapy has been used in select cases of recurrent locally aggressive tumors. Topical chemotherapy with mitomycin C has been used in recent years for the treatment of intraepithelial sebaceous carcinoma, supplanting the traditional orbital exenteration. Currently, orbital exenteration is reserved for tumors with extensive orbital invasion or widespread eyelid involvement that make local resection impossible (9).

ADNEXAL TUMORS

Endocrine Mucin-Producing Sweat Gland Adenocarcinoma

Definition. *Endocrine mucin-producing sweat gland carcinoma* (EMPSGC), a recently recognized variant of cutaneous mucinous carcinoma, is defined as a neoplasm with neuroendocrine

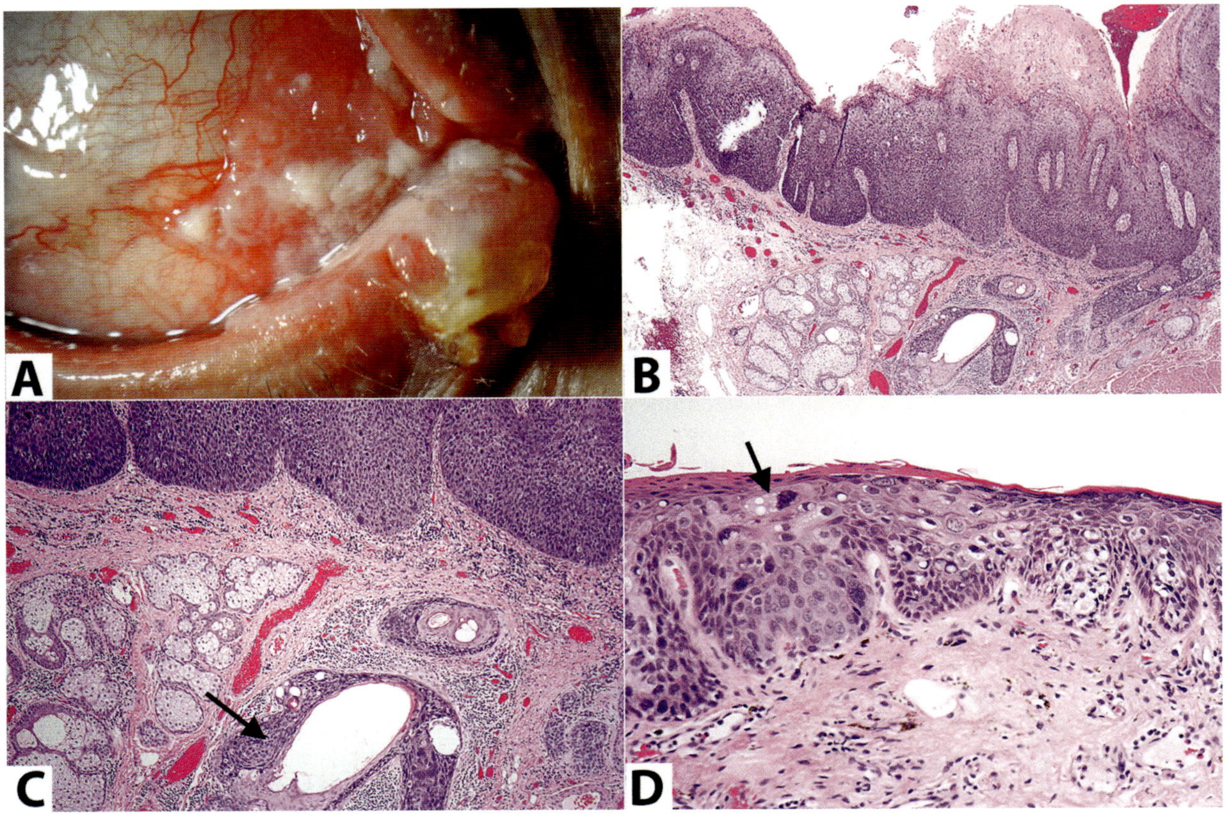

Figure 3-15

SEBACEOUS CARCINOMA MIMICKING SQUAMOUS CELL CARCINOMA

A: A fleshy mass in the caruncle with focal leukoplakia.

B: The epithelium is massively acanthotic and parakeratotic with cellular atypia and extension into an adnexal appendage, reminiscent at scanning magnification of squamous cell carcinoma in situ.

C: With higher magnification, involvement of the sebaceous gland duct (arrow) and adjacent sebaceous lobules by the malignant cells is seen.

D: Pagetoid involvement of the eyelid margin epithelium by sebaceous carcinoma cells (arrow) with hyperchromatic nuclei and vacuolated cytoplasm (B–D: H&E stain).

differentiation and a solid, papillary, and cystic growth pattern, analogous to endocrine ductal carcinoma in situ/solid papillary carcinoma of the breast (30).

Clinical Features. EMPSGC is nearly exclusively localized to the periocular skin of older patients, with an average patient age of 70 years (range, 48 to 87 years) (8,31). The data on sex distribution are controversial, with some studies indicating female predominance (31), while others report male predilection (8).

EMPSGC usually presents as a slow-growing, solitary, skin-colored or erythematous, placoid, nodular, pedunculated or cystic, nonulcerated growth in the eyelid margin and the adjacent periocular skin. It is equally distributed between the upper and lower eyelids (fig. 3-16) (8,30,31). A rare presentation as a multifocal tumor has been reported (31).

Microscopic Findings. EMPSGC is a multinodular, mixed cystic and solid dermal-based neoplasm that commonly shows areas of papillary and cribriform architecture, analogous to intraductal carcinoma/solid papillary carcinoma of the breast. Focal epidermal connection has been noted in some cases. The neoplastic cells are oval to polygonal, with eosinophilic cytoplasm and occasional basophilic intracytoplasmic mucin, which can be highlighted with histochemical stains (periodic acid–Schiff [PAS]), mucicarmine, Alcian blue, and colloidal iron). There is mild to moderate nuclear pleomorphism and mitotic

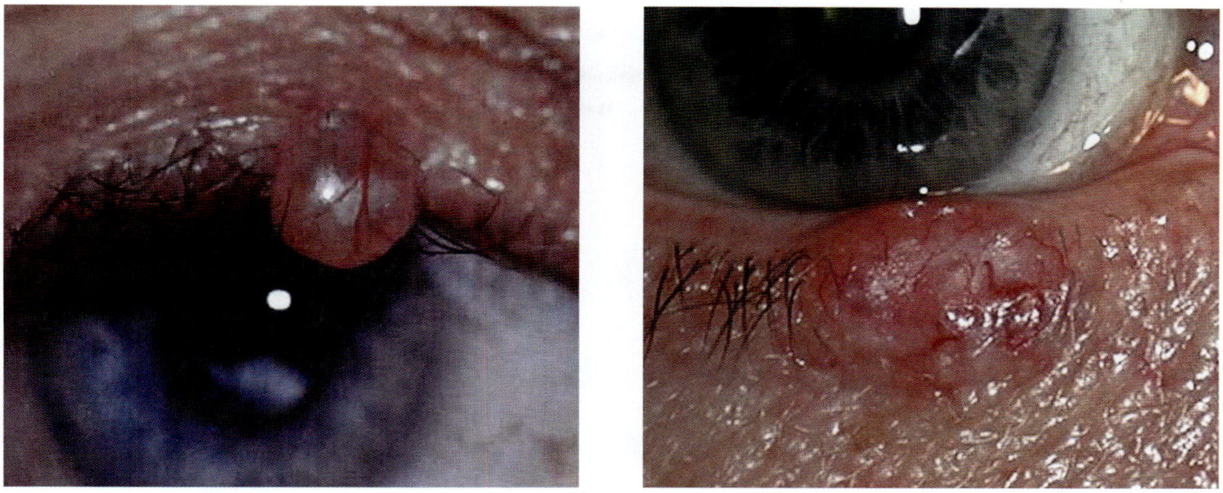

Figure 3-16

ENDOCRINE MUCIN-PRODUCING SWEAT GLAND CARCINOMA (EMPSGC)

Left: EMPSGC can present as a cystic bluish translucent nodule with focal vascularity in the eyelid margin, mimicking hidrocystoma.

Right: This EMPSGC presents as mostly solid-appearing mass with telangiectatic vessels, associated with eyelash loss, in the anterior lamella of the eyelid margin, mimicking BCC.

figures are not abundant (8,30,31). The "salt and pepper" stippled chromatin pattern, suggestive of neuroendocrine differentiation, has been reported by some experts (30).

Characteristically, there is no necrosis, perineural invasion, or lymphovascular invasion. Invasive mucinous carcinoma, characterized by islands of neoplastic cells floating in the abundant pools of mucin, occasionally coexist with EMPSGC, suggesting that EMPSGC is either a precursor to a variant of cutaneous mucinous carcinoma or a point on the continuum of these lesions to cutaneous mucinous carcinoma (fig. 3-17) (8,30).

Immunohistochemical Findings. Although the diagnosis of EMPSGC can be suggested by tumor morphology, immunohistochemical studies are useful in confirming endocrine differentiation and distinguishing between the in situ and invasive tumor components (8,30,31). EMPSGC has an immunohistochemical pattern similar to that of endocrine carcinoma of the breast. The neoplastic cells typically express CK7 and low molecular weight cytokeratin (CAM5.2), while they are negative for CK20. Expression of BRST-2, estrogen and progesterone receptors (31,32), and, recently, GATA-3 (32) is also characteristic.

Most published studies of EMPSGC report positivity with at least one or more neuroendocrine markers. These include chromogranin, synaptophysin, neuron-specific enolase, CD57, and CD56. Chromogranin and synaptophysin appear to be the most useful neuroendocrine markers (fig. 3-17). Immunohistochemical evidence of neuroendocrine differentiation can be focal. Thus, a negative result in a small biopsy does not rule out an EMPSGC (31). Positive staining with neuroendocrine markers, however, is also seen in primary cutaneous mucinous carcinoma (8,31). In one study, similar immunohistochemical expression patterns were observed in both EMPSGC and presumed primary mucinous sweat gland carcinoma of periocular skin (8).

p63, smooth muscle actin, and calponin highlight the myoepithelial cells rimming the cysts and nodules of in situ tumor (fig. 3-17). Absence of expression of these markers supports the diagnosis of invasive carcinoma (8,30,31).

Molecular Genetic Findings. The molecular genetic data on EMPSGC are almost nonexistent. Next-generation sequencing analysis was performed on one EMPSGC, which failed to demonstrate the mutational signatures of papillary carcinoma of the breast and mucinous carcinomas from noncutaneous sites, raising the

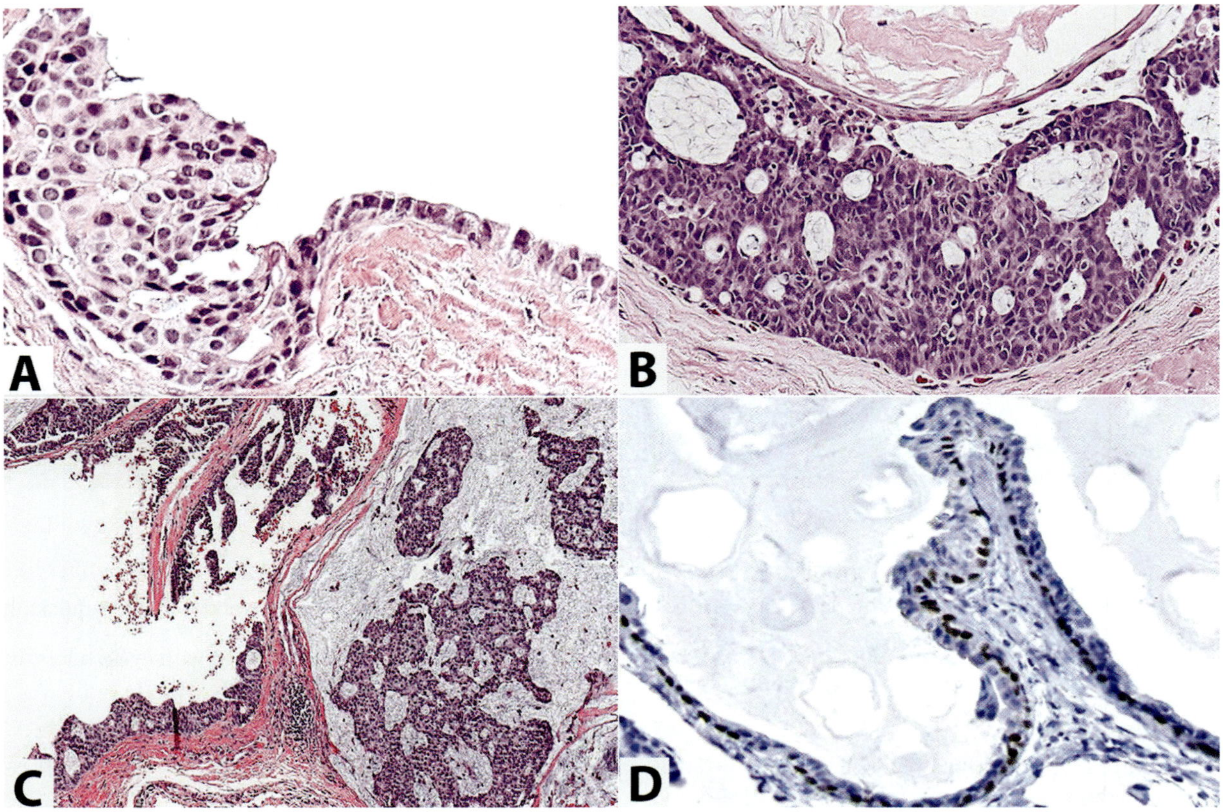

Figure 3-17

ENDOCRINE MUCIN-PRODUCING SWEAT GLAND CARCINOMA

A: Cystic appearance of EMPSG, with focal sweat duct-type bilayer and focal cribriform multilayered architecture. The neoplastic cells are characterized by mild nuclear pleomorphism without brisk mitotic activity or necrosis (A–C: H&E stain).

B: Cribriform lining in EMPSGC is reminiscent of a ductal carcinoma in situ.

C: Typical appearance of EMPSGC on the left, with papillary and cribriform proliferation of the epithelial lining, and an adjacent focus of invasive mucinous carcinoma, characterized by nests of neoplastic cells floating in an abundant pool of mucin.

D: The myoepithelial cells in the neoplastic lining are highlighted with p63 immunostain.

possibility that EMPSGC may harbor molecular profiles distinct from these tumors (33).

Differential Diagnosis. The histopathologic differential diagnosis of EMPSGC includes a range of primary skin neoplasms and cutaneous metastases (31). Cases with a mixed solid-cystic growth pattern can resemble hidradenoma, hidradenocarcinoma, and apocrine adenoma. Hidradenomas typically exhibit a mixture of polygonal cells with eosinophilic cytoplasm and with glycogen-rich clear cytoplasm. Duct-like structures, areas of squamoid differentiation, and stromal hyalinization are common, and mitoses are generally absent. A cystic variant of a well-differentiated hidradenocarcinoma can resemble EMPSGC, but is characterized by more prominent nuclear atypia, mitotic activity, tumor necrosis, and clear cell change.

Apocrine adenomas share many of the architectural features of EMPSGC, presenting as intradermal solid cystic nodules often with papillary and cribriform structures. The presence of copious eosinophilic cytoplasm with "decapitation" secretion, lack of intracytoplasmic mucin, and large epithelioid nuclei with prominent nucleoli favor apocrine adenoma.

A partial biopsy limited to sampling of the cystic component of an EMPSGC can be mistaken for a hidrocystoma. The presence of cytologic atypia, often associated with a clinging growth pattern and intracytoplasmic mucin, should point toward the diagnosis of EMPSGC. Positive staining with neuroendocrine markers can conclusively distinguish EMPSGC from any of the entities listed above (31).

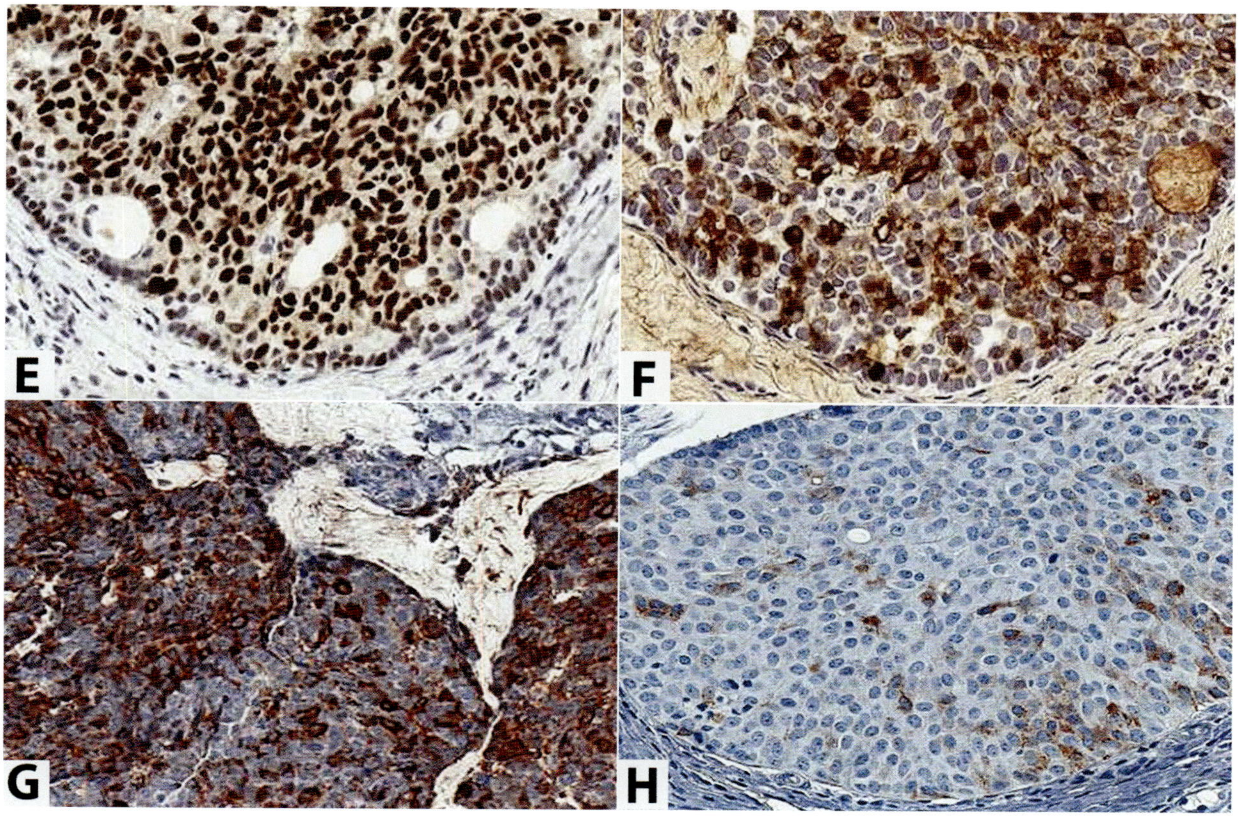

Figure 3-17, continued

E: The nuclei of neoplastic cells express estrogen receptors diffusely.
F: Strong expression of BRST-2 (GCDFP-15) in this tumor.
G: The neoplastic cells express chromogranin.
H: Rare neoplastic cells express synaptophysin.

EMPSGC may be confused with an adenoid variant of nodular BCC. Notably, rare cases of BCC have been reported to show immunohistochemical or ultrastructural features of neuroendocrine differentiation. More striking peripheral palisading, retraction from the surrounding myxoid stroma, higher nuclear to cytoplasmic ratio of the cells that show prominent apoptosis, and occasionally, areas of necrosis are typically seen in BCC (31).

Separating EMPSGC from invasive mucinous carcinoma is largely based on architectural features because of the cytologic and immunophenotypic similarity of these tumors. As mentioned, EMPSGC can coexist with invasive mucinous carcinoma, suggesting that these tumors in a periocular location may represent a continuum (8).

When metastases are considered, breast carcinoma is the leading consideration based on tumor morphology and immunophenotype. Documentation of an in situ component helps exclude metastases. Although data on the mutational profile of EMPSGC are scant (see above), the absence of the mutations seen in breast carcinoma may be diagnostically useful. In routine practice, correlation with clinical data (presence of breast primary and evidence of metastatic disease) is a more practical approach than molecular profiling (31).

Treatment and Prognosis. Follow-up information from published cases suggests that the overall prognosis is good, even for tumors associated with invasive mucinous carcinoma. Although a local recurrence rate of 30 percent has been reported, distant metastases have not been documented (8,31). This observation has led some experts to suggest that EMPSGC may be a precursor of cutaneous mucinous carcinoma variant with endocrine differentiation and

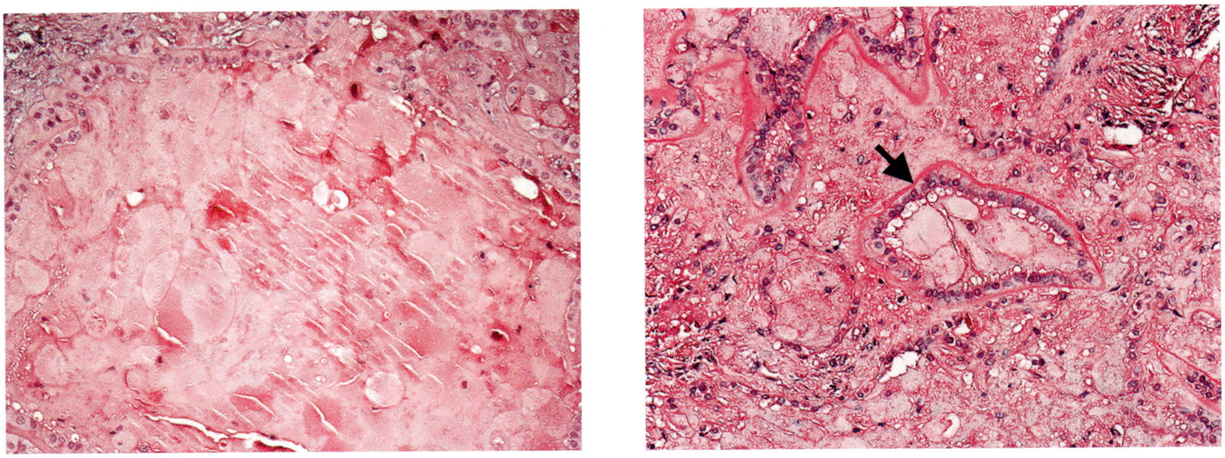

Figure 3-18

PHAKOMATOUS CHORISTOMA

Left: Ribbons of cuboidal cells with eosinophilic cytoplasm and centrally placed nuclei, reminiscent of lens epithelium, surround a central aggregate of cells with abundant eosinophilic cytoplasm and condensed small nuclei, reminiscent of bladder cells of Wedl. A densely fibrotic stroma is present in the periphery.

Right: Periodic acid–Schiff (PAS)-stained preparation highlights the thick basement membrane material associated with the cuboidal cells (arrow), corresponding to a lens capsule.

a favorable prognosis (30). Others, remarking on similarities in immunophenotype and biologic behavior of the isolated EMPSGC, the EMPSGC with coexistent mucinous carcinoma component, and the isolated mucinous sweat gland carcinoma, suggest that these lesions in a periocular location represent a morphologic continuum of a single entity (8).

Treatment of EMPSGC is primarily surgical, with local excision to ensure clearance of margins. Given the high predilection for cosmetically sensitive areas, Mohs micrographic surgery may be an appropriate line of management (31).

CONGENITAL LESIONS

Phakomatous Choristoma

Definition. *Phakomatous choristoma (Zimmerman tumor)* is a rare sporadic congenital lesion of presumed lenticular anlage origin. According to one theory, phakomatous choristoma occurs when some epithelial cells that differentiate into lens tissue are trapped in the subcutaneous tissue of the lower eyelid during fetal development (34,35).

Clinical Features. Phakomatous choristoma characteristically presents as a small, firm, rubbery, slowly growing nodule in the medial aspect of the lower eyelid (34). Although typically an isolated finding, rare lesions coexist with other ocular anomalies (35). Intraoperative findings include adherence to adjacent structures, such as the inferior tarsus, orbital septum, lacrimal sac, and soft tissues of the eyelid. The specimen may fragment secondary to difficulties encountered during surgical excision (34).

Gross Findings. Gross examination demonstrates a tan-pink, firm, rubbery mass, with variable degrees of encapsulation and a cut surface that is homogeneously white. There are reports of a faint reticulated pattern of white bands identified grossly (34).

Microscopic Findings. Histologically, islands of epithelial cells grow irregularly in a dense fibrous tissue matrix in phakomatous choristoma. An epithelial component consists of cuboidal cells with abundant eosinophilic cytoplasm and centrally located nuclei resembling lens epithelial cells, and large cells with abundant eosinophilic cytoplasm, resembling bladder cells of Wedl (fig. 3-18). Thick PAS-positive basement membrane material, analogous to lens capsule, surrounds the epithelial islands (34).

Immunohistochemical Findings. The cuboidal epithelial cells generally express S-100 protein, crystallins, and vimentin, supporting the lenticular origin of phakomatous choristoma (34).

Differential Diagnosis. Although the clinical differential diagnosis can be broad due to the

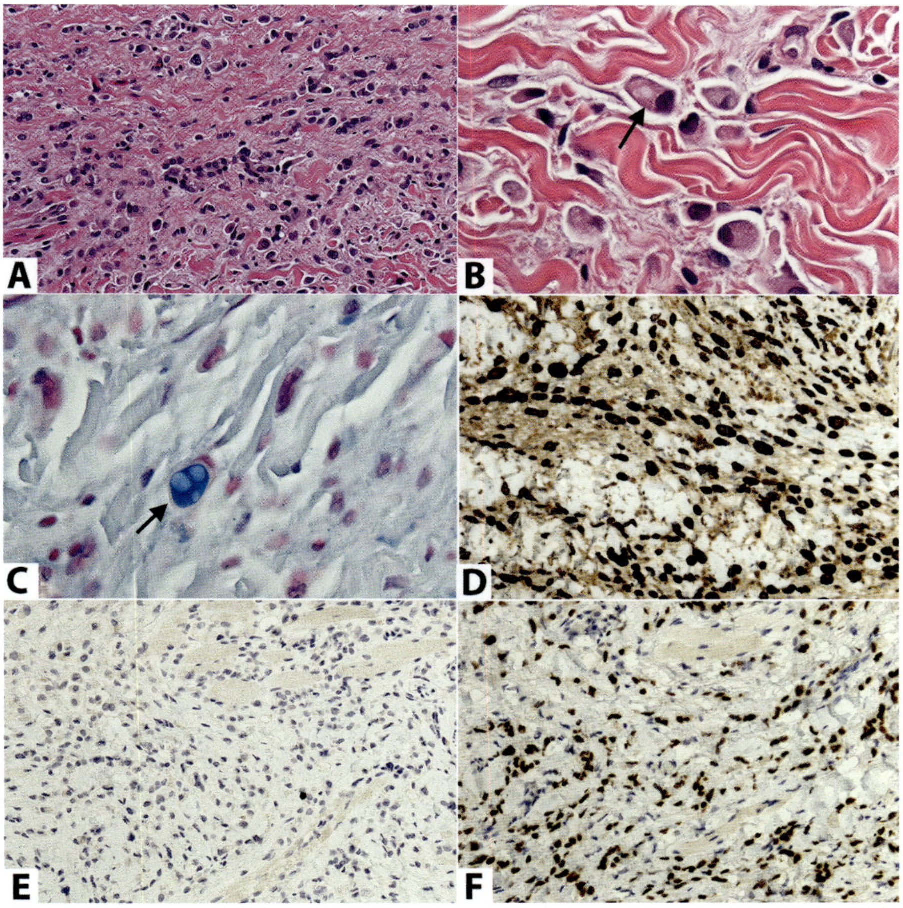

Figure 3-19

BREAST CARCINOMA METASTATIC TO THE EYELID

A: Atypical cells with pleomorphic nuclei infiltrate the fibrotic stroma in a single file arrangement.

B: Higher magnification assists in identifying histiocytoid and signet ring morphologies with intracytoplasmic mucin vacuoles (arrow) (A,B: H&E stain).

C: Alcian blue stain highlights the intracytoplasmic mucin that indents the malignant cell nucleus (arrow).

D: The neoplastic cells diffusely express CK7.

E: There is no expression of CK20.

F: GATA-3 is expressed in the malignant cell nuclei.

rarity of this lesion, the microscopic findings of phakomatous choristoma are characteristic (34).

Treatment and Prognosis. Simple surgical excision of this benign lesion is curative (34).

SECONDARY AND METASTATIC TUMORS

Metastasis to the eyelids is rare, representing less than 1 percent of all malignant eyelid tumors (Table 3-1) (3–7,36). More than 50 percent of eyelid metastases are from a breast primary and frequently occur in a setting of widespread metastatic disease (fig. 3-19). Other common primary tumors that metastasize to eyelids are prostate and lung carcinomas, cutaneous melanoma, and renal cell carcinoma. One fifth of patients have no history of primary cancer (36).

The development of a rapidly progressive subcutaneous nodule in the eyelid is characteristic of metastases to this area (15). Microscopically, metastatic tumors frequently retain the morphology and immunophenotype of the tumor of origin. A distinctive involvement of the eyelid stroma in a single file arrangement occurs in metastatic carcinomas (fig. 3-19) (36).

The distinction between primary eyelid carcinomas and metastases with overlapping histologic features is made when an in situ component of the primary tumor is recognized. Immunohistochemical panels documenting the immunophenotype of the tumor of origin are helpful. In the absence of an in situ component, the distinction of metastatic ductal carcinoma of the breast from EMPSGC and invasive mucinous sweat gland carcinoma can be challenging (see discussion above) (31). Similarly, the signet ring and histiocytoid variants of sweat gland carcinoma can look morphologically and immunophenotypically identical to metastatic lobular carcinoma of the breast (fig. 3-19) (37). In these challenging cases, correlation with clinical and oncologic data is essential (31,37). Management of the metastatic disease focuses on systemic oncologic control of the tumor.

LESIONS OCCURRING IN EYELID AND SKIN

The discussion of lesions mentioned in this section is limited to site-specific findings. Most of these tumors are identical to counterparts in extraocular locations regarding clinical features, histopathology, and prognostic attributes. For additional information on these lesions, please see the Fourth Series Fascicles, *Melanocytic Tumors of the Skin* (37a) and *Nonmelanocytic Tumors of the Skin* (37b).

Benign Epidermal Tumors

Benign epidermal tumors are the most common neoplasms identified in the eyelid skin (Table 3-1) (3–7). Of these, *fibroepithelial polyp (squamous papilloma)* and *seborrheic keratosis* are the most common (3–7). These lesions share clinical and morphologic features with nonperiocular counterparts (15).

Epidermal Tumors of Low Malignant Potential

Keratoacanthoma. Keratoacanthoma is a cryptic tumor that has generated longstanding debate regarding its etiology (38). A lesion with predilection for sun-exposed facial skin, keratoacanthoma occasionally involves the eyelid of middle-aged to older individuals as a dome-shaped tumor with rolled borders and a central keratin-filled crater (fig. 3-20) (15). Two striking features of keratoacanthoma are its clinical behavior with spontaneous regression after rapid growth and its nosological position on the border between benignity and malignancy. Although traditionally regarded as a reactive process, or a form of pseudoepitheliomatous hyperplasia, keratoacanthoma is currently believed to be a tumor of hair follicle derivation because it expresses markers found in the follicular isthmus and infundibulum and has the triphasic nature (proliferative [early growth], stabilization [well-developed stage], and regression phases) that mimic the hair cycle (38).

The neoplastic nature of keratoacanthoma is suggested by observation of mutations in these tumors, including the molecular genetic alterations that characterize keratoacanthomas associated with several syndromes, such as MTS (see above). Additionally, rare keratoacanthomas have evolved into frank SCC or behave in a malignant fashion with regional lymph node metastases. These observations perpetuate the controversy on whether keratoacanthoma is a benign neoplasm or a variant of SCC, so-called SCC, keratoacanthoma type (38).

The distinction between conventional SCC and keratoacanthoma is important due to prognostic differences between the two lesions. The diagnosis of keratoacanthoma is based on the clinical presentation of a crater-like tumor, rapid (weeks to months) growth with a triphasic course, and histopathologic examination of a suitable biopsy specimen (38). In the eyelid, excision of the entire tumor is recommended to facilitate histopathologic diagnosis and ensure surgical clearance of margins (39).

Histopathologically, keratoacanthoma is a cup-shaped nodular elevation with thickened epidermis containing islands of well-differentiated squamous epithelium that may be infiltrated by neutrophils and surrounding a central mass of keratin. The lesion base is often well-demarcated from the adjacent dermis by an inflammatory reaction. The cellular characteristics of keratoacanthoma are similar to those of SCC; the architecture is a key feature for establishing the diagnosis (fig. 3-20) (15,38,39).

There continues to be a controversy regarding the wait and see approach in anticipation of eventual involution versus surgical excision of solitary keratoacanthoma (38). In the periocular region, early excision of keratoacanthoma has been advocated, not only to maintain normal eyelid function, but also to prevent further tissue destruction and invasion into deeper tissues. Complete excision is also advised based on the observation that a significant percentage of patients with clinically suspected periocular keratoacanthomas have been found to have invasive SCC on pathologic evaluation of excised tissue (39).

Adnexal Tumors

A variety of benign and malignant adnexal neoplasms with follicular, apocrine, and eccrine differentiation develop in the eyelid skin. Of these, eccrine and apocrine hidrocystomas are the most common (Table 3-1) (3–7).

Apocrine Hidrocystoma. This tumor frequently presents along the eyelid margin, at the medial and lateral canthi of an adult, as a translucent, often bluish cystic nodule that originates from the duct of the apocrine gland of Moll (fig. 3-21). The bluish coloration results

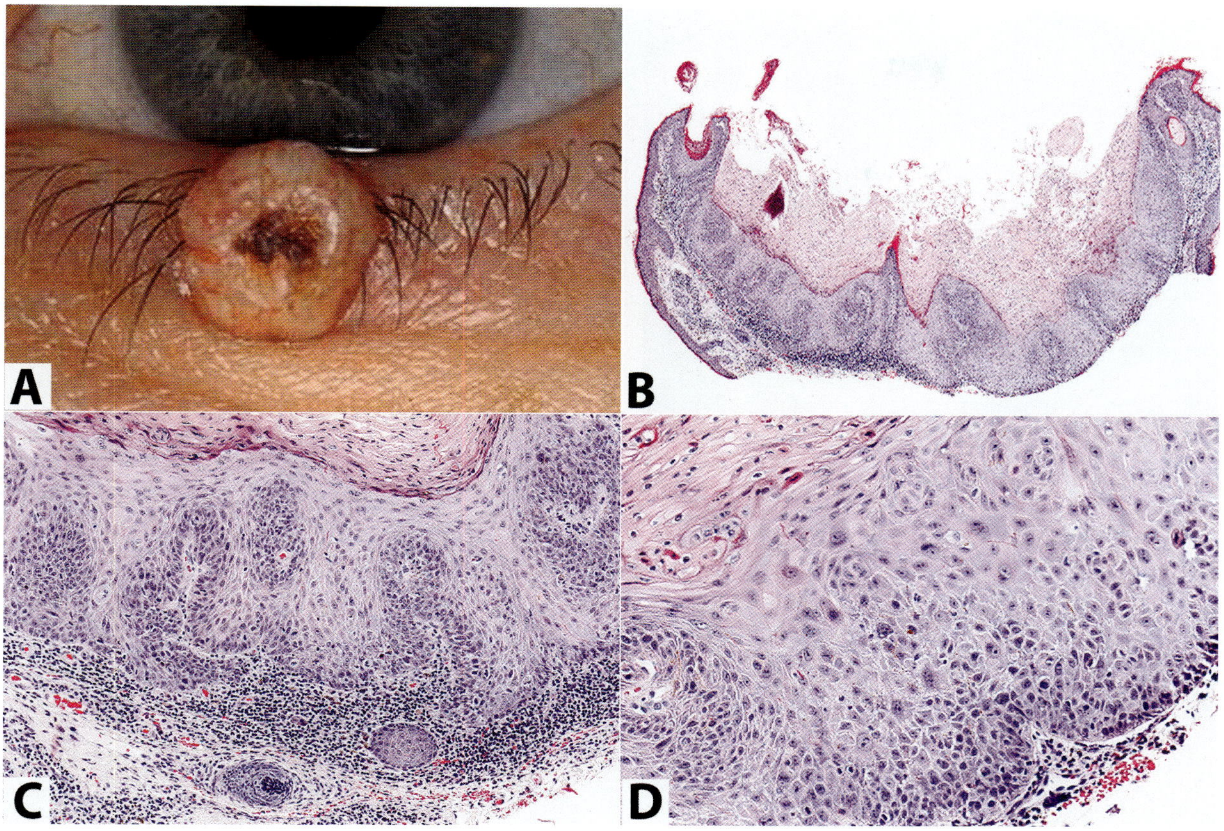

Figure 3-20

KERATOACANTHOMA

A: A dome-shaped nodule with rolled edges and a central keratin-filled crater involves the anterior surface of the lower eyelid margin.

B: Scanning magnification helps to identify the characteristic crateriform shape of keratoacanthoma, with overhanging walls, central parakeratin, and a blunt, pushing (noninfiltrative) margin, sharply demarcated from the underlying dermis (B–D: H&E stain).

C: Higher magnification assists in identifying the pushing margin of keratoacanthoma, with a prominent interface inflammatory infiltrate.

D: At a cellular level, keratoacanthoma is characterized by epidermal dysplasia and surface parakeratin. A small incisional biopsy can be misinterpreted as squamous cell carcinoma.

from the presence of iron, lipofuscin, or both, in the apocrine lining and secretions that lend a brown color to this dermal-based cyst, which then appears blue because of Tyndall effect.

Histopathologically, apocrine hidrocystoma is characterized by a multilocular or unilocular cystic appearance. The apocrine bilayer lining is composed of an inner cuboidal layer of cells with apical snouts and decapitation secretion and an outer layer of flattened myoepithelial cells (fig. 3-21) (2,15).

Eccrine Hidrocystoma. This is a ductal retention cyst of eccrine sweat glands, which appears clinically as a solitary translucent cystic lesion on the eyelid of an adult, but not in the eyelid margin. Histopathologically, eccrine hidrocystoma generally has a unilocular cystic architecture, lined by a cuboidal eccrine epithelial bilayer. Occasionally, only a single layer of flattened cells lines the cyst (fig. 3-22) (2,15).

Syringoma. This is another notable benign eccrine sweat gland tumor with a predilection for the periocular skin of young and middle-aged women. The lesions are frequently multiple and appear as a cluster of nodules, usually 1 to 3 mm in size (fig. 3-23) (15).

Histopathologically, syringoma appears as a well-circumscribed lesion in the upper dermis,

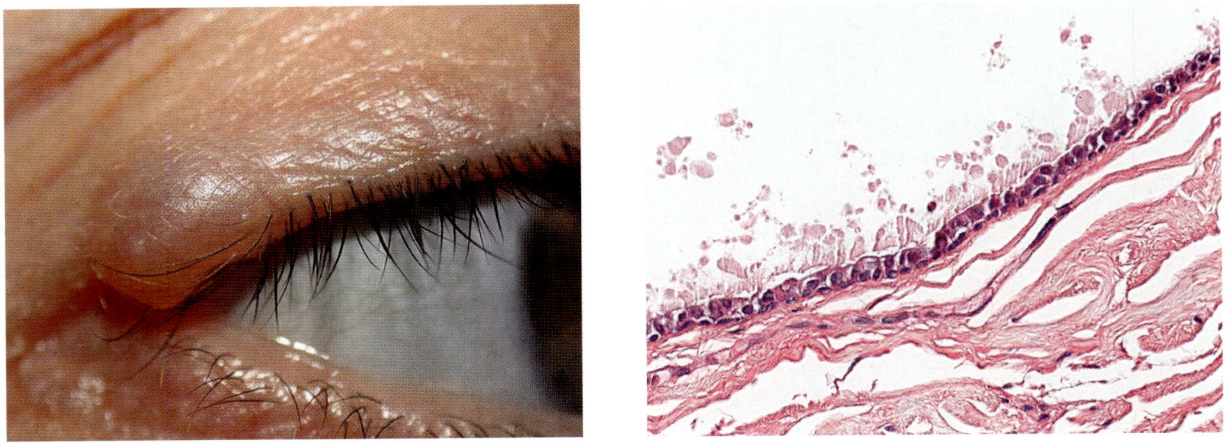

Figure 3-21

APOCRINE HIDROCYSTOMA

Left: A translucent, bluish cyst is present at the lateral canthus, adjacent to the eyelashes.
Right: The cyst lining is composed of an apocrine bilayer, with outer spindle myoepithelial cells and inner cuboidal cells with apocrine decapitation secretion (H&E stain).

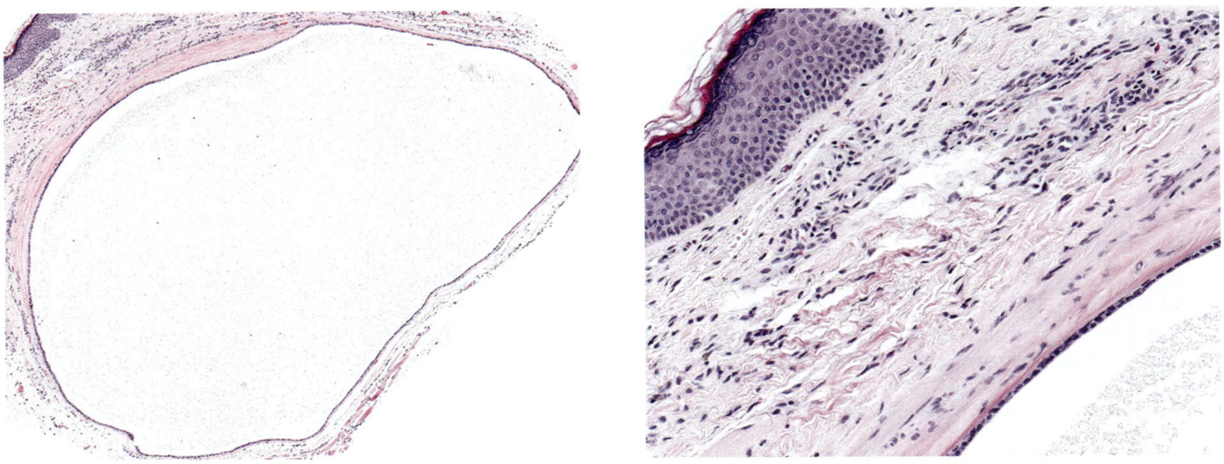

Figure 3-22

ECCRINE HIDROCYSTOMA

Left: A dermal-based unilocular cyst at scanning magnification.
Right: At higher magnification, the cuboidal and flattened epithelial bilayer lining of the cyst is identified (H&E stain).

composed of cuboidal epithelial cells. The cells are arranged in small islands. Ovoid and comma-shaped ductules are lined by an eosinophilic cuticle with amphophilic luminal material, embedded in the sclerotic stroma (fig. 3-23). Some lesions contain clear cells, so-called *clear cell syringoma* (2,15).

Microcystic Adnexal Carcinoma. Syringoma may be confused with microcystic adnexal carcinoma, a locally aggressive variant of ductal eccrine carcinoma, which is characterized by proliferation of sweat gland ductules in a desmoplastic stroma, reminiscent of syringoma. The key distinguishing features facilitating the diagnosis of microcystic adnexal carcinoma are poor circumscription with infiltration of deep dermis, the presence of horn cysts and abortive hair follicles, and perineural invasion. The distinction between syringoma and microcystic adnexal carcinoma may be impossible in a small

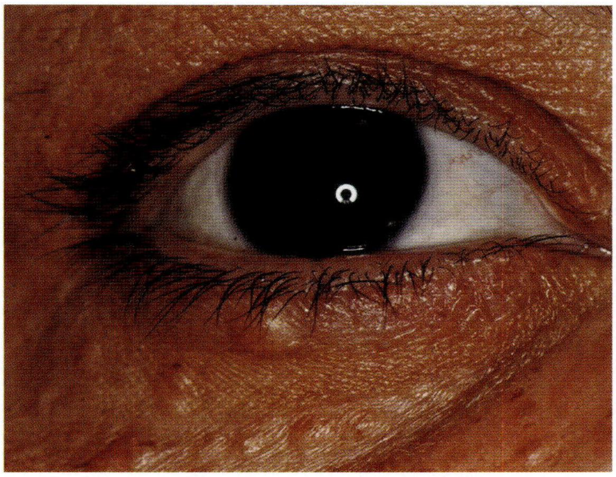

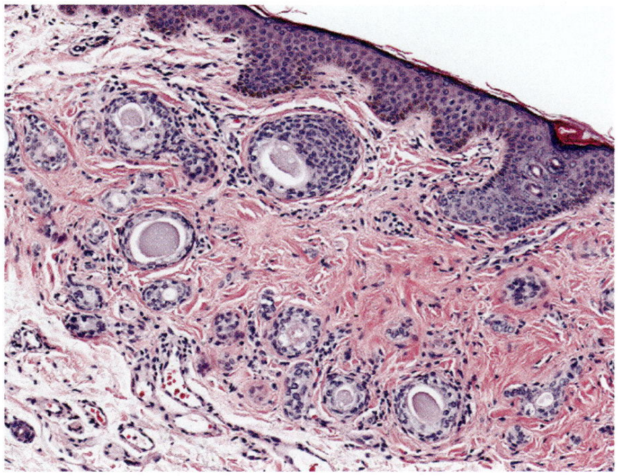

Figure 3-23

SYRINGOMA

Left: Multiple small nodules in the lower eyelid are characteristic of syringoma. In Blacks, these nodules may be more or less pigmented than the surrounding uninvolved skin, or the same color.

Right: There is a circumscribed proliferation of ovoid and comma-shaped sweat gland ductules within the superficial sclerotic dermis. The ductules are lined by several layers of cuboidal cells and contain luminal amphophilic acellular material (H&E stain).

superficial biopsy, and correlation with the clinical appearance of the lesion may be helpful. In clinically suspicious tumors, a deeper biopsy should be obtained (40).

Vascular Tumors and Tumor-Like Lesions

A variety of vascular lesions involve the periocular skin. These include benign neoplasms and tumor-like lesions, such as *juvenile capillary hemangioma* (fig. 3-24), *nevus flammeus, lobular capillary hemangioma (pyogenic granuloma), cavernous hemangioma,* and *arteriovenous malformations.* Malignant neoplasms of vascular derivation, such as *Kaposi sarcoma* (fig. 3-25) and *angiosarcoma,* can also involve the periocular tissues (15). These lesions have clinical and microscopic features similar to nonperiocular cutaneous counterparts (figs. 3-24, fig. 3-25).

Merkel Cell Carcinoma

Merkel cell carcinoma (MCC) is an aggressive cutaneous neoplasm with neuroendocrine differentiation (41). Although MCC has a predilection for the head and neck skin, only 5 to 10 percent of these tumors involve the periocular skin.

MCC usually presents in the upper eyelid, close to the eyelid margin, as a painless, solitary, multilobulated, violaceous nodule without ulceration (fig. 3-26) (41). Microscopic findings are similar to MCC in the nonperiocular skin (fig. 3-26).

MCC of the eyelid displays an aggressive clinical course with a high rate of local recurrence (14 percent), regional lymph node invasion (20 percent), and metastasis (5 percent) (42). MCC of the periocular region is staged according to general cancer protocols for this lesion.

Cutaneous T-Cell Lymphomas

Cutaneous eyelid lymphomas are extremely rare (Table 3-1) (3–7). When eyelid skin is involved, the most common lymphomas are cutaneous *T-cell lymphoma of mycosis fungoides type* (fig. 3-27) and *CD30-positive T-cell lymphoproliferative disorder/primary cutaneous anaplastic large cell lymphoma* (43,44). Eyelid involvement by mycosis fungoides generally presents in a setting of known cutaneous disease (43). Thus, biopsy of affected eyelid skin is generally not performed. In contrast, CD30-positive T-cell lymphoproliferative disorders, such as lymphomatoid papulosis, can present with an isolated eyelid involvement, which manifests as a strikingly erythematous, centrally ulcerated nodule. In this clinical setting, a diagnostic biopsy is typically done (44).

These hematolymphoid malignancies in the eyelid share patient demographic characteristics and clinical, microscopic, immunophenotypic, and molecular genetic features with similar processes in the nonperiocular skin (43,44).

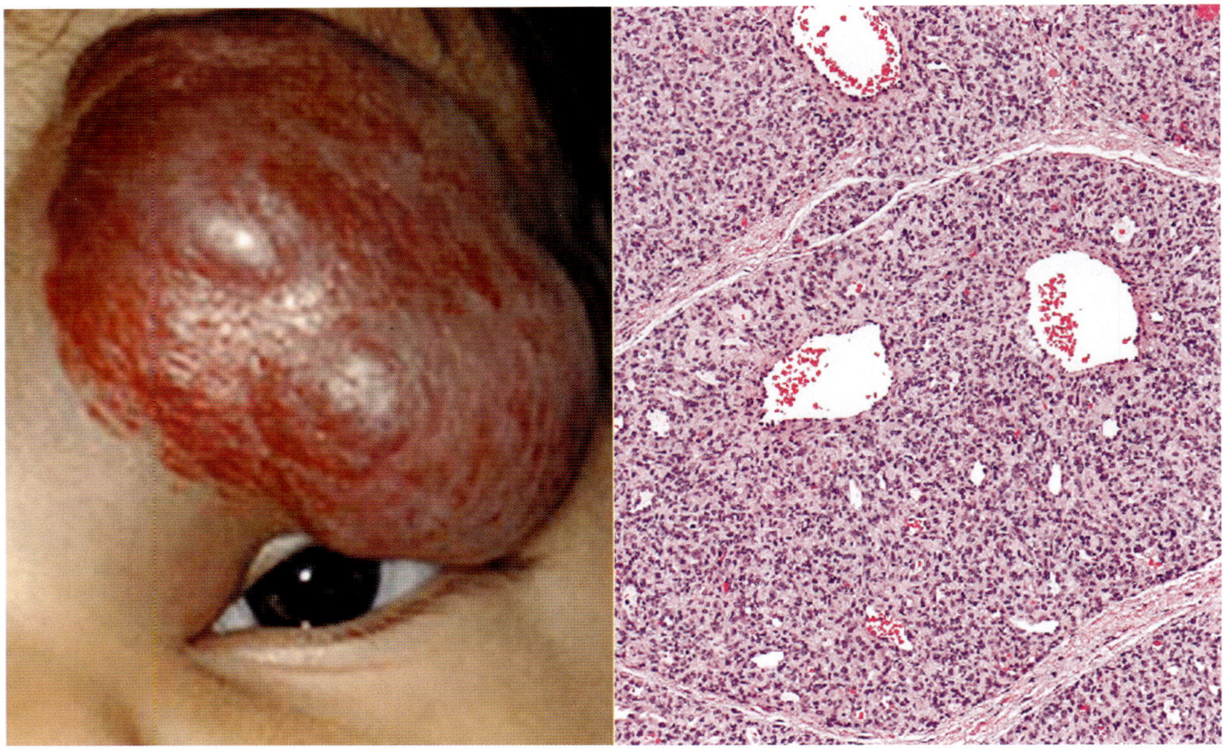

Figure 3-24

JUVENILE (INFANTILE) CAPILLARY HEMANGIOMA

Left: A strawberry-like vascular mass involves the upper eyelid of an infant.
Right: An unencapsulated and lobular proliferation of capillary sized vascular channels, lined by plump endothelial cells, is clustered around centrally positioned larger vessels (H&E stain).

Melanocytic Lesions and Neoplasms

Benign Melanocytic Proliferations. Periocular skin can give rise to a broad spectrum of benign melanocytic proliferations and neoplasms. These include *ephelides, lentigines, nevocellular nevi* (fig. 3-28), and nevus variants that are clinically and microscopically similar to the lesions observed in nonperiocular cutaneous sites (2,3).

Nevus of Ota is a lesion of particular interest, because of its localization in periocular skin and noncutaneous associations (45,46). Nevus of Ota typically presents as a congenital gray-blue unilateral periocular pigmentation in the distribution of the first two branches of the trigeminal nerve (fig. 3-29). The histopathologic features include the proliferation of intensely pigmented dendritic melanocytes in the deep dermis, widely separated by collagen bundles. In a full-blown syndrome of *oculodermal melanocytosis*, similar dendritic melanocytes involve the ocular tissues, including the episclera and uveal tract (fig. 3-29). The deep orbital tissues and the ipsilateral meninges may be affected (45).

Patients with nevus of Ota are at risk for deep cutaneous/orbital melanoma, uveal melanoma, and central nervous system (CNS) melanoma that can arise in the nevus. Recent studies have demonstrated *GNAQ* and *GNA11* mutations in all of these tumors, suggesting that these presumably initiating mutational events explain the link between these neoplasms (45,46). See chapter 8 for additional information on this topic.

Premalignant and Malignant Melanocytic Proliferations

Dysplastic melanocytic nevi and *malignant melanomas* are rare tumors in the eyelid skin. Malignant melanoma of the eyelid comprises less than 1 percent of all eyelid malignancies. Periocular cutaneous malignant melanoma tends to appear on the lower eyelid and characteristically presents as a longstanding

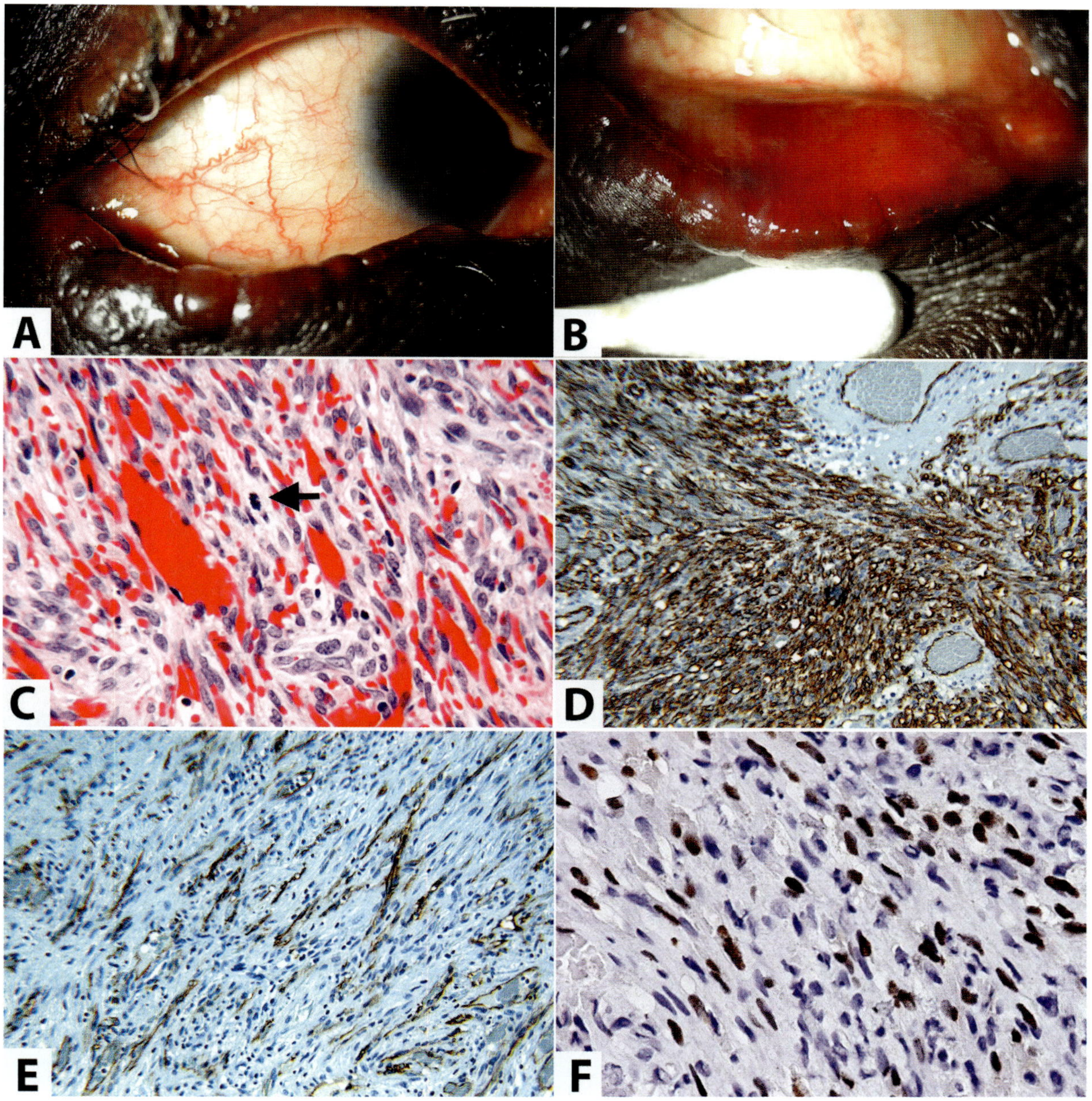

Figure 3-25

KAPOSI SARCOMA

A,B: A violaceous nonulcerated mass involves the lower eyelid tissues, including the skin, eyelid margin, and palpebral conjunctiva.

C: The neoplasm is composed of a cellular proliferation of mildly to moderately pleomorphic spindle cells, lining slit-like spaces, and elongated vascular channels. Occasional mitotic figures (arrow) are present (H&E stain).

D: The neoplastic cells diffusely express the vascular and lymphatic endothelial marker CD31.

E: Focal expression of lymphatic endothelial marker D2-40 is present.

F: Human herpes virus (HHV)-8 is expressed in the neoplastic nuclei.

pigmented lesion with gradual or recent growth, occasionally associated with ulceration. The lentigo maligna melanoma subtype has been most commonly reported in a periocular location, likely reflecting the role of chronic sun damage in the pathogenesis of these tumors (fig. 3-30).

A careful clinical examination is essential to distinguish primary cutaneous melanoma of the

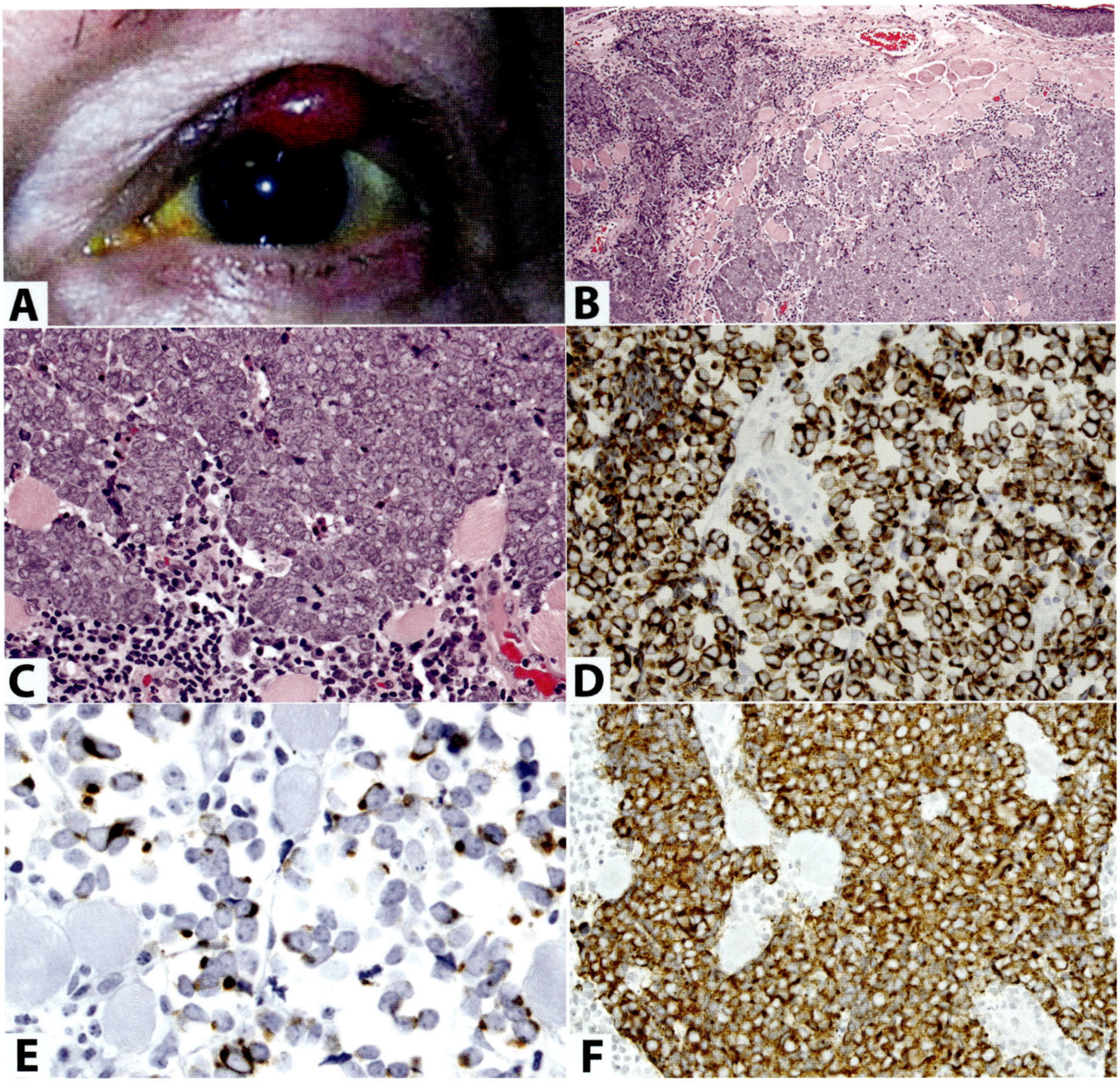

Figure 3-26

MERKEL CELL CARCINOMA

A: A violaceous, nonulcerated nodule in the upper eyelid is associated with eyelash loss (madarosis).

B: An infiltrative neoplasm is in the eyelid stroma, without involvement of the surface epidermis (B,C: H&E stain).

C: The neoplastic cells are distinctive for their round nuclei with finely granular chromatin, suggestive of neuroendocrine differentiation.

D: There is diffuse and strong cytoplasmic expression of CK20 in a dot-like paranuclear pattern.

E: Rare cells express CK7 in a paranuclear dot-like fashion.

F: Chromogranin is expressed diffusely.

eyelid from primary conjunctival melanoma with cutaneous extension (see chapter 2). Malignant melanoma of the eyelid should be evaluated with cancer staging protocols for cutaneous malignant melanoma.

PATHOLOGIST REPORT FOR PRIMARY CARCINOMA OF THE EYELID

The system presented in Appendix 3, from the Taskforce for Staging of Cancer of the Eye of the American Joint Committee on Cancer (AJCC),

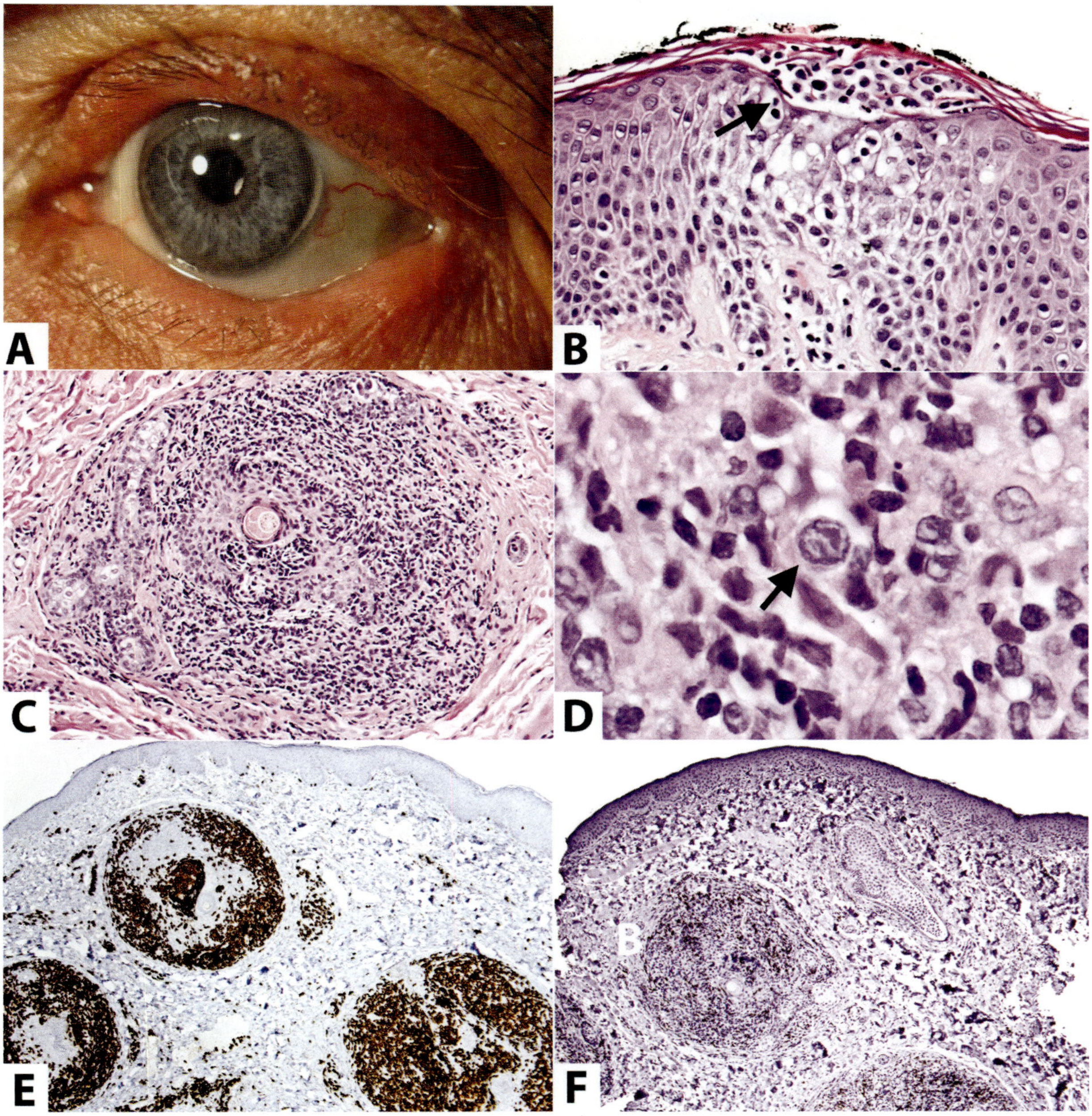

Figure 3-27

MYCOSIS FUNGOIDES

A: An erythematous nodule at the lower lateral eyelid margin.

B: Intraepidermal single cell lymphocyte exocytosis and the aggregate (arrow) of atypical lymphocytes (Pautrier microabscess) were identified in an earlier biopsy at an nonophthalmic cutaneous site.

C: A folliculotropic atypical lymphocytic infiltrate, compatible with folliculotropic form of mycosis fungoides in the eyelid biopsy illustrated in A.

D: High magnification enables identification of variably sized atypical lymphocytes with prominent nuclear folds (cerebriform nuclei, arrow) (B–D: H&E stain).

E,F: The neoplastic cells express CD3 diffusely (E), and aberrantly lack expression of CD7 (F).

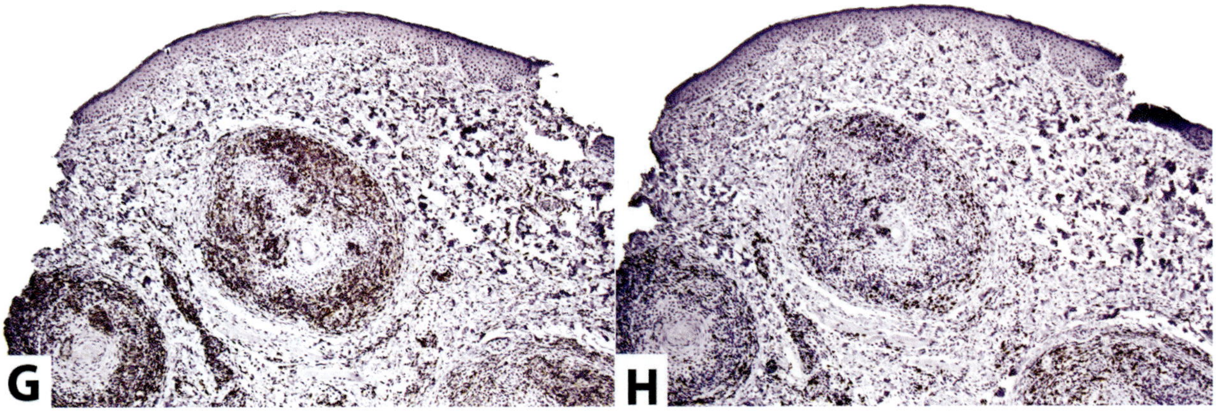

Figure 3-27, continued

G,H: There is overexpression of CD4 (G) compared to CD8 (H).

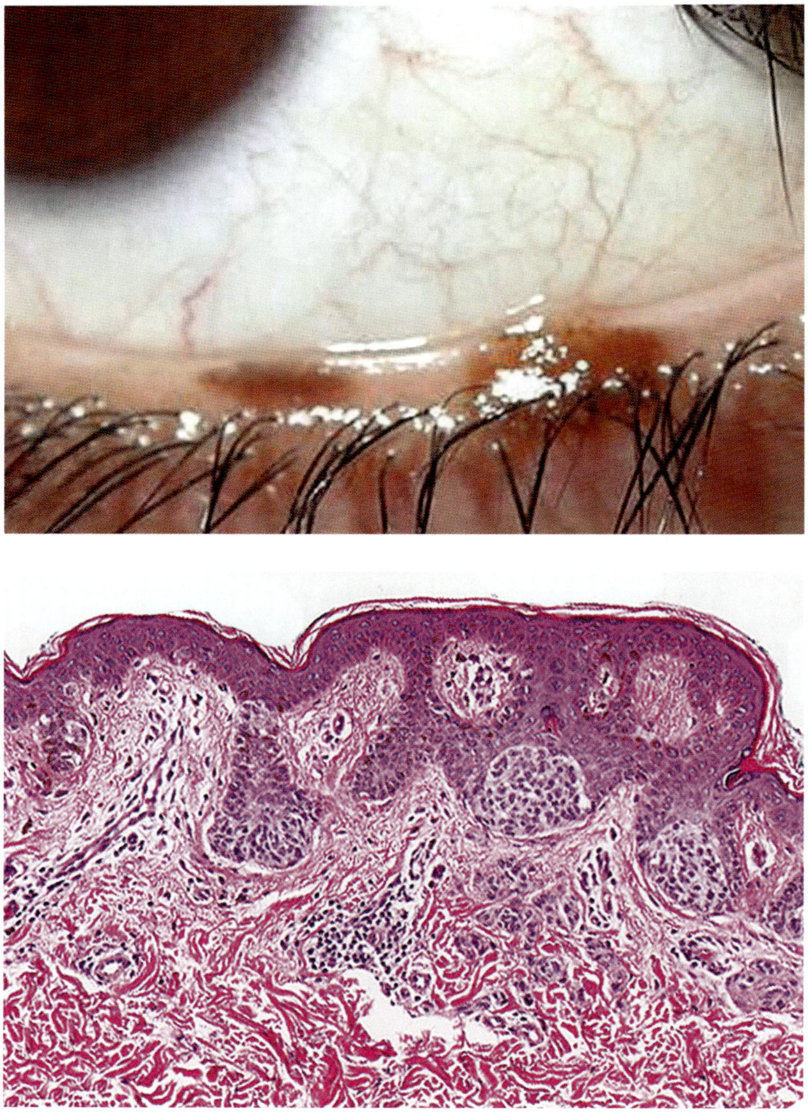

Figure 3-28

MELANOCYTIC NEVUS

Top: Clinical photograph demonstrates two pigmented, circumscribed, and elevated lesions, involving the eyelid margin.

Bottom: Microscopically, the lesion is characterized by nests of nevocellular nevus cells at the epidermal-dermal junction and in the underlying stroma, compatible with a compound nevocellular nevus (H&E stain).

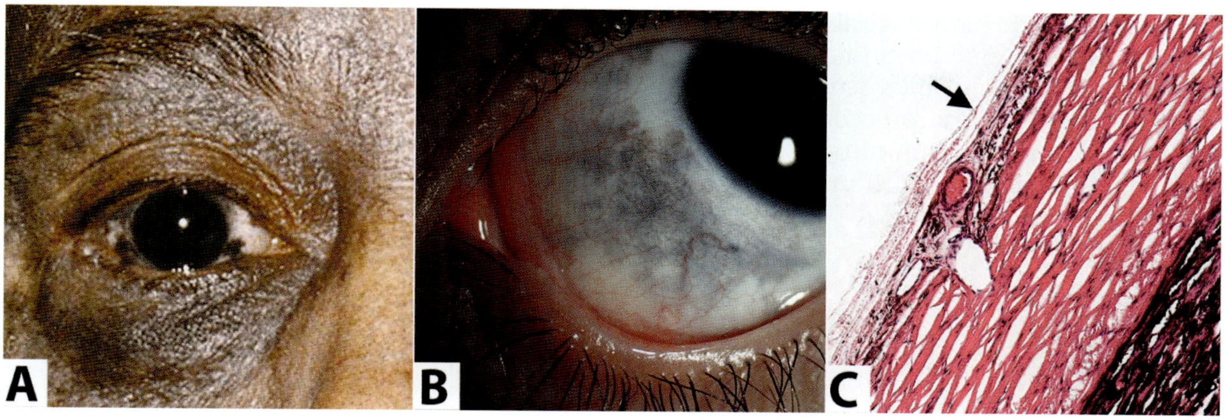

Figure 3-29

NEVUS OF OTA

A: This patient has oculodermal melanocytosis (nevus of Ota). The gray-blue pigmentation of the periocular skin and ocular tissues (episcleral and uvea) reflects a deeply situated and highly pigmented nevus.

B: Grayish discoloration of the ocular surface in a patient with ocular melanocytosis.

C: Intensely pigmented melanocytes in the episclera, accompanied in this case by choroidal thickening by highly pigmented and cytologically bland cells oriented uniformly in one plane (arrow) (H&E stain).

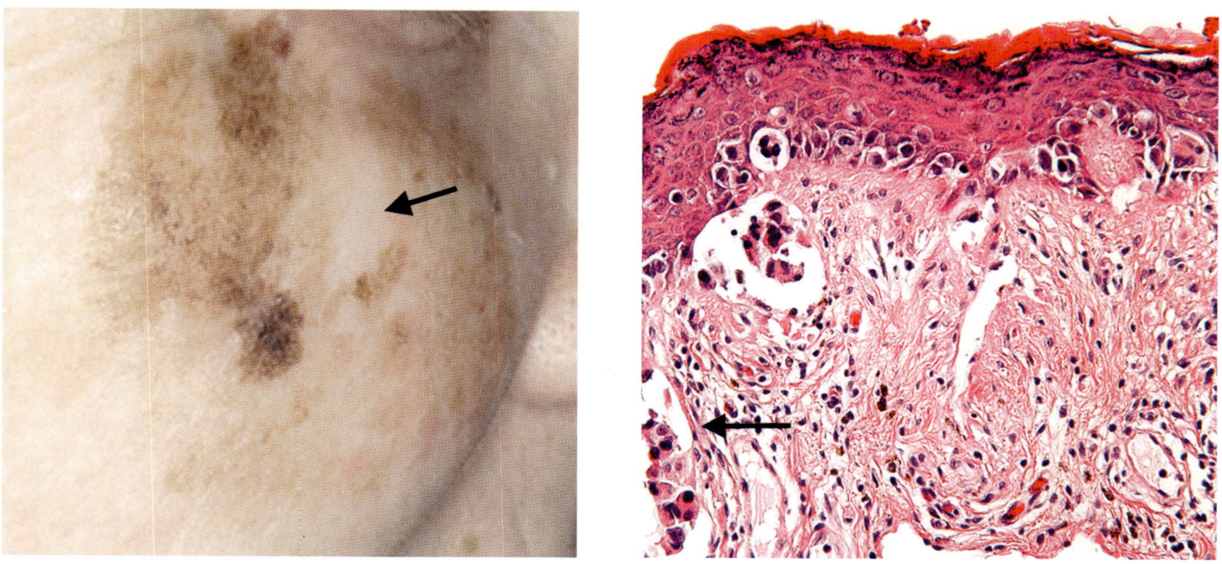

Figure 3-30

LENTIGO MALIGNA

Left: A poorly circumscribed, variably pigmented patch is seen on the upper cheek and lower eyelid of a light-skinned patient. An area of depigmentation (arrow), suggestive of regression, is present.

Right: A contiguous proliferation of atypical melanocytes populates the basal layer of the epidermis, with focal nesting and focal extension into the dermal appendage (arrow) (H&E stain).

8th edition, applies only to primary carcinoma of the eyelid and includes the following: basal cell carcinoma, squamous cell carcinoma, sebaceous carcinoma, and other rare carcinomas, such as all varieties of sweat gland carcinoma. This system classifies carcinomas of the head and neck region from non-eyelid and orbital sites under the rubric of "head and neck cancer." Cutaneous melanoma and Merkel cell carcinoma staging systems apply to lesions arising in the eyelid (11).

The clinical and pathological T classification schemes are the same for primary carcinomas of the eyelid. A precise microscopic description is essential: the surgical procedure performed, specimen size, specific tumor type, and degree of differentiation. Per AJCC manual, greatest tumor dimension and evaluation of the surgical specimen margins are mandatory. Beyond T, N, and M categories, the system does not require the identification of additional prognostic factors for stage groupings (11).

REFERENCES

1. Last RJ. The appendages of the eye. In: Wollf E, Last RJ. Anatomy of the eye and orbit, including the central connections, development, and comparative anatomy of the visual apparatus, 6th ed. London: Lewis; 1968;182-240.
2. Eagle RC Jr. Eyelid. In: Eagle RC Jr. Eye pathology: an atlas and text, 3rd ed. Philadelphia: Wolters Kluwer; 2017;269-96.
3. Font RL, Croxatto JO, Rao NA. Tumors of the eyelids. In: Font RL, Croxatto JO, Rao NA. Tumors of the eye and ocular adnexa. AFIP Atlas of Tumor Pathology, 4th series, Fascicle 5. Washington DC: American Registry of Pathology; 2006; 155-221.
4. Deprez M, Uffer S. Clinicopathological features of eyelid skin tumors. A retrospective study of 5504 cases and review of literature. Am J Dermatopathol 2009;31:256-62.
5. Gundogan FC, Yolcu U, Tas A, et al. Eyelid tumors: clinical data from an eye center in Ankara, Turkey. Asian Pac J Cancer Prev 2015;16:4265-9.
6. Obata H, Aoki Y, Kubota S, Kanai N, Tsuru T. [Incidence of benign and malignant lesions of eyelid and conjunctival tumors.] Nippon Ganka Gakkai Zasshi 2005;109:573-9. [Japanese]
7. Abdi U, Tyagi N, Maheshwari V, Gogi R, Tyagi SP. Tumours of eyelid: a clinicopathologic study. J Indian Med Assoc 1996;94:405-9.
8. Hoguet A, Warrow D, Milite J, et al. Mucin-producing sweat gland carcinoma of the eyelid: diagnostic and prognostic considerations. Am J Ophthalmol 2013;155:585-92.
9. Shields JA, Saktanasate J, Lally SE, Carrasco JR, Shields CL. Sebaceous carcinoma of the ocular region: the 2014 Professor Winifred Mao Lecture. Asia Pac J Ophthalmol (Phila) 2015;4:221-7.
10. Yin VT, Merritt HA, Sniegowski M, Esmaeli B. Eyelid and ocular surface carcinoma: diagnosis and management. Clin Dermatol 2015;33:159-69.
11. Amin MB, Edge SB, Greene FL, et al, eds. AJCC cancer staging manual, 8th ed. Switzerland: Springer; 2017.
12. Milman T, McCormick SA. The molecular genetics of eyelid tumors: recent advances and future directions. Graefes Arch Clin Exp Ophthalmol 2013;251:419-33.
13. Wu A, Sun MT, Huilgol SC, Madge S, Selva D. Histological subtypes of periocular basal cell carcinoma. Clin Exp Ophthalmol 2014;42:603-7.
14. Sun MT, Wu A, Huilgol SC, Selva D. Accuracy of Biopsy in subtyping periocular basal cell carcinoma. Ophthal Plast Reconstr Surg 2015;31:449-51.
15. Pe'er J. Pathology of eyelid tumors. Indian J Ophthalmol 2016;64:177-90.
16. Compton LA, Murphy GF, Lian CG. Diagnostic immunohistochemistry in cutaneous neoplasia: an update. Dermatopathology (Basel) 2015;2:15-42.
17. Yin VT, Esmaeli B. Targeting the hedgehog pathway for locally advanced and metastatic basal cell carcinoma. Curr Pharm Des 2017;23:655-9.
18. Yin VT, Pfeiffer ML, Esmaeli B. Targeted therapy for orbital and periocular basal cell carcinoma and squamous cell carcinoma. Ophthal Plast Reconstr Surg 2013;29:87-92.
19. Sun MT, Andrew NH, O'Donnell B, McNab A, Huilgol SC, Selva D. Periocular squamous cell carcinoma: TNM staging and recurrence. Ophthalmology 2015;122:1512-6.
20. Malhotra R, Huilgol SC, Huynh NT, Selva D. The Australian Mohs database: periocular squamous cell carcinoma. Ophthalmology 2004;111:617-23.
21. Rishi K, Font RL. Sebaceous gland tumors of the eyelids and conjunctiva in the Muir-Torre syndrome: a clinicopathologic study of five cases and literature review. Ophthal Plast Reconstr Surg 2004;20:31-6.
22. Eisen DB, Michael DJ. Sebaceous lesions and their associated syndromes: part I. J Am Acad Dermatol 2009;61:549-60.
23. Singh RS, Grayson W, Redston M, et al. Site and tumor type predicts DNA mismatch repair status in cutaneous sebaceous neoplasia. Am J Surg Pathol 2008;32:936-42.

24. Lee JB, Litzner BR, Vidal CI. Review of the current medical literature and assessment of current utilization patterns regarding mismatch repair protein immunohistochemistry in cutaneous Muir-Torre syndrome-associated neoplasms. J Cutan Pathol. 2017;44:931-7.
25. Kaliki S, Ayyar A, Dave TV, Ali MJ, Mishra DK, Naik MN. Sebaceous gland carcinoma of the eyelid: clinicopathological features and outcome in Asian Indians. Eye (Lond) 2015;29:958-63.
26. Koreen IV, Flint A, Nelson CC, Frueh BR, Elner VM. Nondiagnostic conjunctival map biopsies for sebaceous carcinoma. Arch Ophthalmol 2009;127:961-3.
27. Milman T, Schear MJ, Eagle RC Jr. Diagnostic utility of adipophilin immunostain in periocular carcinomas. Ophthalmology 2014;121:964-71.
28. Rajan Kd A, Burris C, Iliff N, Grant M, Eshleman JR, Eberhart CG. DNA mismatch repair defects and microsatellite instability status in periocular sebaceous carcinoma. Am J Ophthalmol 2014;157:640-7.
29. Tetzlaff MT, Singh RR, Seviour EG, et al. Next-generation sequencing identifies high frequency of mutations in potentially clinically actionable genes in sebaceous carcinoma. J Pathol 2016;240:84-95.
30. Zembowicz A, Garcia CF, Tannous ZS, Mihm MC, Koerner F, Pilch BZ. Endocrine mucin-producing sweat gland carcinoma: twelve new cases suggest that it is a precursor of some invasive mucinous carcinomas. Am J Surg Pathol 2005;29:1330-9.
31. Dhaliwal CA, Torgersen A, Ross JJ, Ironside JW, Biswas A. Endocrine mucin-producing sweat gland carcinoma: report of two cases of an under-recognized malignant neoplasm and review of the literature. Am J Dermatopathol 2013;35:117-24.
32. Abdulkader M, Kuhar M, Hattab E, Linos K. GATA3 positivity in endocrine mucin-producing sweat gland carcinoma and invasive mucinous carcinoma of the eyelid: report of 2 cases. Am J Dermatopathol 2016;38:789-91.
33. Cornejo KM, Hutchinson L, Meng X, O'Donnell P, Deng A. Endocrine mucin-producing sweat gland carcinoma of the eyelid: a report of a case with molecular analysis. Am J Dermatopathol 2016;38:636-8.
34. Romano RC, McDonough P, Salomao DR, Fritchie KJ. Phakomatous choristoma in a 10-week-old boy: a case report and review of the literature. Pediatr Dermatol 2015;32:405-9.
35. Thaung C, Bonshek RE, Leatherbarrow B. Phakomatous choristoma of the eyelid: a case with associated eye abnormalities. Br J Ophthalmol 2006;90:245-6.
36. Dekmezian MS, Cohen PR, Sami M, Tschen JA. Malignancies of the eyelid: a review of primary and metastatic cancers. Int J Dermatol 2013;52:903-26.
37. Kramer TR, Grossniklaus HE, McLean IW, et al. Histiocytoid variant of eccrine sweat gland carcinoma of the eyelid and orbit: report of five cases. Ophthalmology 2002;109:553-9.
37a. Elder DE, Murphy GF. Melanocytic tumors of the skin. AFIP Atlas of Tumor Pathology, 4th Series, Fascicle 12. Washington DC: American Registry of Pathology; 2010.
37b. Patterson JW, Wick MR. Nonmelanocytic tumors of the skin. AFIP Atlas of Tumor Pathology, 4th Series, Fascicle 9. Washington DC: American Registry of Pathology; 2006.
38. Kwiek B, Schwartz RA. Keratoacanthoma (KA): an update and review. J Am Acad Dermatol 2016;74:1220-33.
39. Leibovitch I, Huilgol SC, James CL, Hsuan JD, Davis G, Selva D. Periocular keratoacanthoma: can we always rely on the clinical diagnosis? Br J Ophthalmol 2005;89:1201-4.
40. Gordon S, Fischer C, Martin A, Rosman IS, Council ML. Microcystic Adnexal carcinoma: a review of the literature. Dermatol Surg 2017;43:1012-6.
41. Bostan C, de Souza MB, Zoroquiain P, de Souza LA, Burnier MN Jr. Case series: Merkel cell carcinoma of the eyelid. Can J Ophthalmol 2017;52:e182-5.
42. Raven ML, Selid PD, Lucarelli MJ. Merkel cell carcinoma of the eyelid. Ophthalmology 2016; 123:2126.
43. Kim JW, Lee DK. Unusual eyelid, periocular, and periorbital cutaneous malignancies. Int Ophthalmol Clin 2009;49:77-96.
44. Sanka RK, Eagle RC Jr, Wojno TH, Neufeld KR, Grossniklaus HE. Spectrum of CD30+ lymphoid proliferations in the eyelid lymphomatoid papulosis, cutaneous anaplastic large cell lymphoma, and anaplastic large cell lymphoma. Ophthalmology 2010 ;117:343-51.
45. Franceschini D, Dinulos JG. Dermal melanocytosis and associated disorders. Curr Opin Pediatr 2015;27:480-5.
46. Van Raamsdonk CD, Bezrookove V, Green G, et al. Frequent somatic mutations of GNAQ in uveal melanoma and blue naevi. Nature 2009;457:599-602.

4 TUMORS OF THE ORBIT

ANATOMY AND HISTOLOGY

The orbits are pyramidal structures limited by the bony walls that contain the eyeball and the surrounding supporting soft tissues. The average dimensions of the orbit are 35 mm vertically and 40 mm horizontally at the orbital rim, with a depth of 45 mm and a volume of 30 mL (1).

Seven bones contribute to the four components of the orbital wall: the roof, the medial wall, the inferior wall, and the lateral wall. The frontal and the lesser wing of the sphenoid bone define the orbital roof. The ethmoid and lacrimal bones form the medial wall. The palatine, maxillary, and zygomatic bones contribute to the inferior wall, the orbital floor. The greater wing of the sphenoid and zygomatic bones form the lateral wall.

Several apertures in the orbital bones allow for the passage of blood vessels and nerves. The optic foramen transmits the optic nerve, the ophthalmic artery, and accompanying sympathetic nerve fibers. The superior orbital fissure transmits divisions of the fifth cranial nerve, sympathetic nerve fibers, and the superior ophthalmic vein. The inferior orbital fissure carries the maxillary division of the trigeminal nerve and venous connections between the orbit and the pterygoid fossa (1,2).

The contents of the orbit include the globe, optic nerve with its meningeal sheaths, lacrimal gland, proximal lacrimal drainage apparatus, cartilaginous trochlea, skeletal muscles, and orbital fat, which is subdivided into compartments by the connective tissue fascia. All of the extraocular muscles, except the inferior oblique muscle, originate in the orbital apex and travel anteriorly to insert onto the eye or eyelid.

The four rectus muscles (superior, medial, lateral, and inferior recti) originate at the annulus of Zinn. The levator muscle arises above the annulus on the lesser wing of the sphenoid. The superior oblique muscle originates slightly medial to the levator muscle and travels anteriorly through the trochlea on the superomedial orbital rim, where it turns posterolaterally toward the eye. The inferior oblique muscle originates in the anterior orbital floor lateral to the lacrimal sac and travels posterolaterally within the lower eyelid retractors to insert into the sclera, inferolateral to the macula.

In the anterior portion of the orbit, the rectus muscles are connected by connective tissue, the intermuscular septum, forming a funnel-shaped muscle cone. This subdivides the orbital fat into intraconal fat (central surgical space) and extraconal fat (peripheral surgical space) (1,2). Figure 4-1 illustrates the anatomic structures of the orbit.

CLASSIFICATION

A full spectrum of tumors and tumor-like lesions involve the orbit (Table 4-1). Primary tumors of the orbit arise predominantly from orbital soft tissues and bone. Lacrimal gland lesions are also frequent in the orbit (see chapter 5). Neoplasms from adjacent sites, including the eyeball, optic nerve, and sinonasal sinuses, can secondarily involve the orbit by direct extension. Additionally, various malignancies have the potential to metastasize to the orbit (2,3).

The reported incidence of orbital lesions is significantly variable, influenced by the geographic location, age group, method of examination, source institution, and medical specialty. Despite this variability, the combined evidence from several earlier and recent series indicates that lymphoproliferative tumors, secondary malignancies and metastases, and cystic lesions predominate in the orbital location. When analyzed by age, benign cysts are the most common orbital lesions in children and lymphoproliferative tumors, secondary malignancies, and metastases are the most

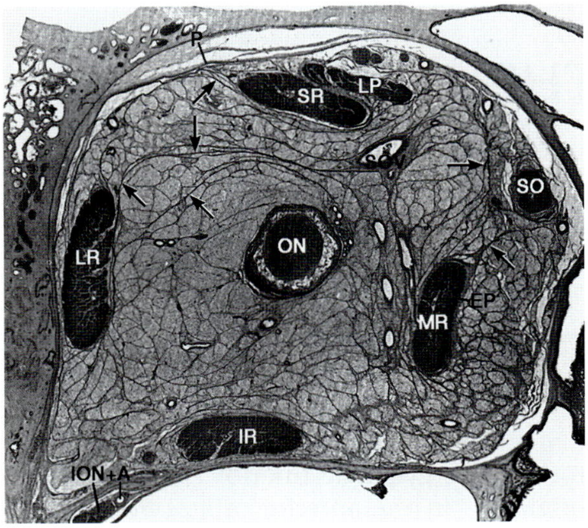

Figure 4-1

CONTENTS OF THE ORBIT

Coronal section of the posterior orbit discloses the orbital bony wall, periosteum (P), adipose tissue and fasciae (arrows), extraocular muscles (SR, superior rectus muscle; LP, levator palpebrae superioris muscle; LR, lateral rectus muscle; MR, medial rectus muscle; SO, superior oblique muscle; IR, inferior rectus muscle), peripheral nerves and blood vessels (ION+A), inferior orbital nerve and artery (SOV, superior orbital vein), and optic nerve (ON).

frequent orbital neoplasms in adults (2,3). The percentage of orbital malignancies steadily rises with age, from around 20 percent in children younger than 19 years, 30 percent in young adults and middle-aged patients, to 60 percent in patients aged 60 and older (3).

HERNIATED ORBITAL FAT

Definition. *Herniated orbital fat* is a benign adipocytic lesion that results from the migration of intraconal orbital fat into subconjunctival space under the influence of gravity and pressure (4).

Clinical Features. Herniated orbital fat is uncommon, occurring mainly in elderly obese men 65 to 72 years of age. Less frequently, orbital fat prolapse is described in younger patients following trauma.

The characteristic clinical presentation is a bilateral, frequently asymmetric, yellow to orange superotemporal subconjunctival mass (fig. 4-2). Prolapsed fat pads can easily be repositioned into the orbit and become more prominent by retropulsion, inward or downward rotation of the globe, aiding in distinction of herniated orbital fat from other similar-appearing lesions,

Table 4-1

TYPES AND FREQUENCY OF ORBITAL TUMORS[a]

Type of Tumor	General	Children	Adults
Cystic lesions	12%	26%	3%
Fibrocytic and adipose lesions	3%	3%	<1%
Rhabdomyosarcoma	1%	9%	NA[b]
Vascular lesions	9%	15%	14%
Peripheral nerve sheath tumors	3%	5%	NA
Optic nerve and meninges	5%	13%	5%
Osseous and cartilaginous lesions	4%	4%	<1%
Primary melanocytic lesions	<1%	<1%	1%
Lacrimal gland lesions	6%	2%	3%
Lymphoproliferative tumors and leukemia	8%	4%	28%
Histiocytic and related lesions	<1%	2%	NA
Secondary and metastatic tumors	14%	5%	26%
Inflammatory lesions	12%	8%	11%
Thyroid-related orbitopathy	17%	NA	NA
Other	8%	5%	NA

[a]Data from reference 2.
[b]NA = not available.

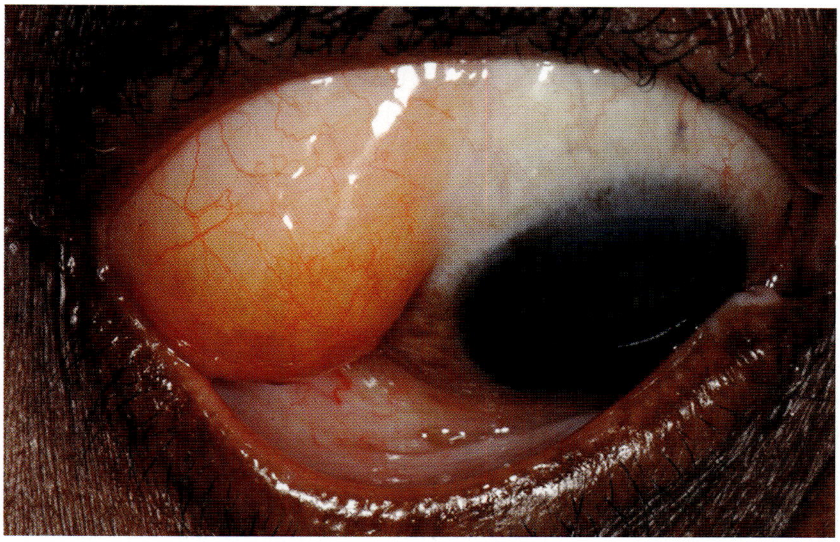

Figure 4-2

HERNIATED ORBITAL FAT

A yellow-orange superotemporal subconjunctival mass is seen.

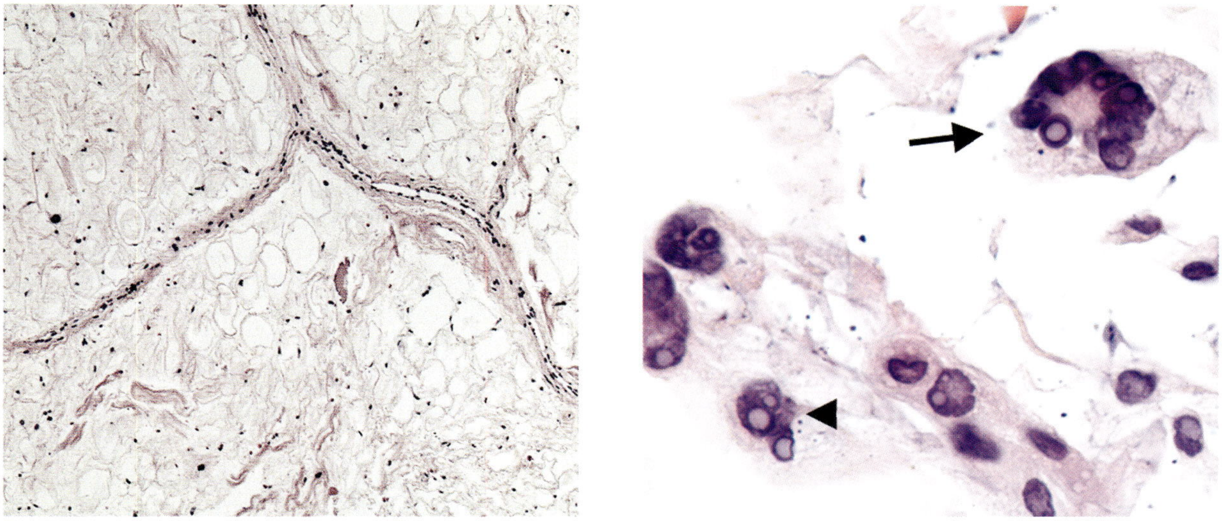

Figure 4-3

HERNIATED ORBITAL FAT

Left: Mature lobulated fibroadipose tissue.

Right: Occasional floret-like giant cells (arrow) and Lochkern cells (arrowhead) (hematoxylin and eosin [H&E] stain). (Courtesy of Dr. R. C. Eagle Jr., Philadelphia, PA.)

such as dermolipoma, lacrimal gland tumors, and lipomatous neoplasms (4).

Microscopic Findings. Microscopically, herniated orbital fat is characterized by the presence of unencapsulated, generally uniformly sized, small adipose tissue lobules separated by thin fibrous septa, the features normally seen in intraconal orbital fat. Additionally, adipocytes with intranuclear pseudovacuoles (Lochkern cells) and multinucleated floret-like giant cells are identified, most frequently within the fibrous septa (fig. 4-3) (4). These multinucleated giant cells lack appreciable nuclear hyperchromasia and atypia, express CD34 and apoptotic markers (5), and lack expression of S-100 protein, CD68, FXIIIa (4), MDM2, and CDK4 (5) compatible with reactive, degenerative changes of fibroblastic cells in the orbital fat (4–6).

The diagnosis of herniated orbital fat is generally made on histopathologic evaluation alone, without the aid of immunohistochemical, or molecular genetic, or cytogenetic techniques.

Recently, array comparative genomic hybridization analysis performed on herniated orbital fat tissue samples demonstrated no copy number alterations in these lesions, further confirming its distinction from lipomatous neoplasms, such as pleomorphic lipoma and atypical lipomatous tumor, and supporting its categorization as a reactive process (6).

Differential Diagnosis. The presence of floret-like giant cells, formerly thought to be pathognomonic of spindle cell/pleomorphic lipoma and frequently present in atypical lipomatous neoplasms, has led in the past to interpretation of herniated orbital fat as a neoplastic process (7). The floret-type giant cells in herniated orbital fat differ from the giant cells in pleomorphic lipoma by their small, uniform, normochromatic nuclei, and frequent (though not invariable) presence within the fibrous septa, rather than within the fat. The wire-like, fibrillar collagen and variably myxoid stroma seen in pleomorphic lipoma is absent in herniated orbital fat. Similarly, differentiation of Lochkern cells from true lipoblasts is essential in avoiding the diagnosis of liposarcoma (4). In challenging cases, immunohistochemical staining for RB1, MDM2, and CDK4, and molecular cytogenetic studies investigating for rearrangements in chromosomes 13q and 16p, and amplification of *MDM2* gene locus (12q15), may help in distinction of herniated orbital fat from pleomorphic lipoma and atypical lipomatous tumor/well-differentiated liposarcoma (6).

Treatment and Prognosis. The treatment of herniated orbital fat is directed toward palliation of discomfort and cosmesis. Herniated orbital fat is a benign lesion, with low recurrence rate following excision (4).

CONGENITAL TUMORS AND TUMOR-LIKE LESIONS

Dermoid Cysts and Variations

Definition. *Dermoid cyst* is a congenital lesion presumed to arise from the sequestration of surface ectoderm at bony suture lines or along lines of embryonic closure (8). Other rare congenital orbital cysts of the surface epithelium include *conjunctival dermoid ("conjunctivoid") cyst, epidermal cyst, conjunctival epithelial cyst, respiratory epithelial cyst*, and *apocrine gland cyst*. Similar to dermoid cyst, these lesions are believed to originate from the sequestration of the surface epithelium in the orbital tissues during fetal development (8).

Clinical Features. Dermoid cyst is a common lesion of the orbital region. In most cases, dermoid cyst is found superotemporally, near the frontozygomatic suture. Patients typically present in childhood with a slowly progressive, painless subcutaneous mass that is firm to palpation and may be attached to the underlying bone. Pain and inflammatory presentation may occur if the lesion spontaneously ruptures. Conjunctival dermoid has a predilection for the superonasal aspect of the orbit and frequently presents in this location in older patients as a subcutaneous mass (8).

Microscopic Findings. A typical dermoid cyst is lined by stratified squamous keratinized epithelium with dermal appendages (hair follicles, sebaceous glands, and sweat glands), luminal keratin, and hair shafts (fig. 4-4). The conjunctival dermoid cyst is lined by conjunctival epithelium (stratified squamous nonkeratinized epithelium with goblet cells) with adnexal structures in its wall, and is filled with clear fluid. As their names imply, epidermal cyst, conjunctival epithelial cyst, respiratory epithelial cyst, and apocrine gland cyst are lined by their respective epithelia (8).

Differential Diagnosis. Secondary cysts, arising from traumatic implantation of epithelium into the stroma or from extension from adjacent sites (such as mucocele) are the primary clinical and histopathologic diagnostic considerations (8). The distinction between primary and secondary etiologies rests on correlation of microscopic findings with the clinical history and imaging studies (7), and on detection of tissue changes compatible with trauma.

Treatment and Prognosis. Ophthalmologists manage symptomatic orbital cysts by complete excision. There are no reports in the literature of progression to malignancy (8).

Orbital Teratoma and Other Germline Tumors

Definition. *Teratoma* is a congenital, multicystic tumor that may originate from ectopic germ cells in the orbital area. Orbital teratomas are primary (localized to the orbit) or combined (involving periorbital and intracranial space). Although infrequent, primary orbital teratoma is

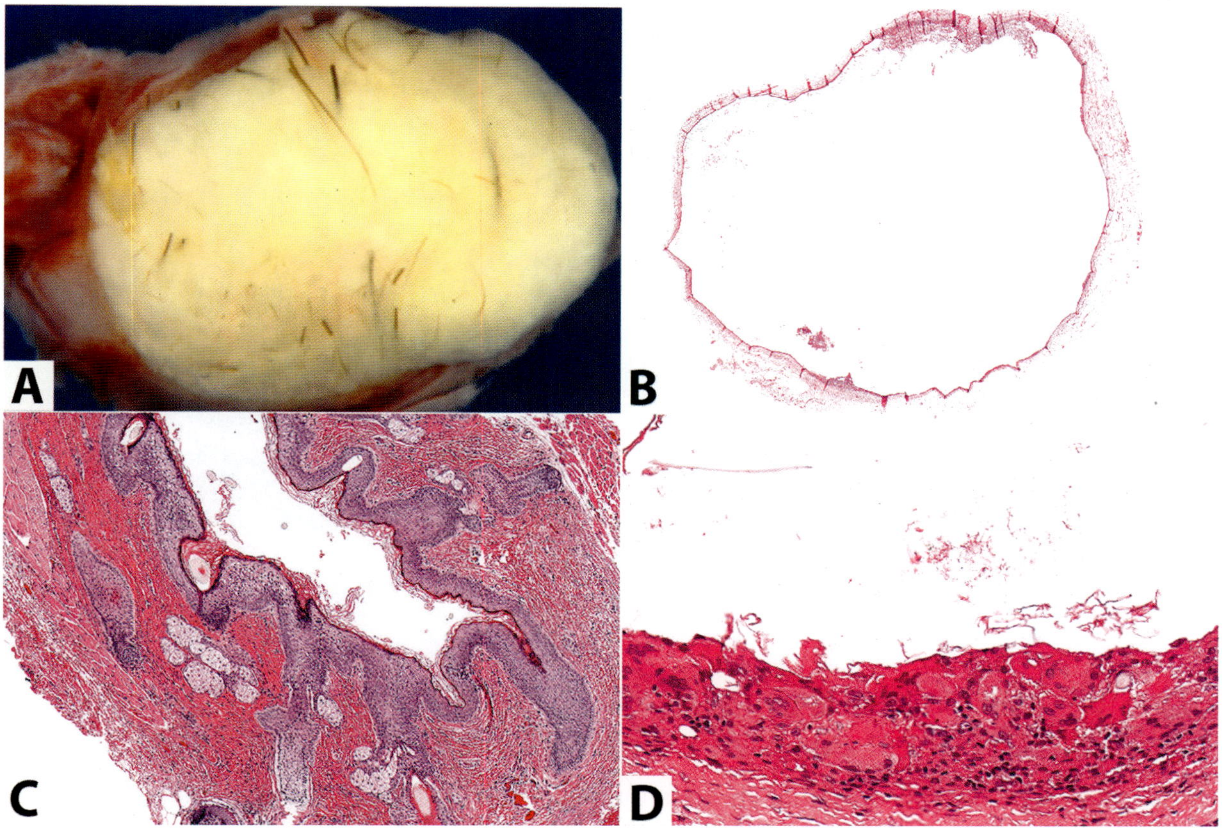

Figure 4-4

DERMOID CYST

A: Cystic mass contains yellow cheesy keratin material and hair shafts.
B: Scanning magnification demonstrates a cystic mass (B–D: H&E stain).
C: The cyst is lined by the keratinizing epithelium (epidermis), with dermal appendages, luminal keratin, and hair.
D: Foreign body giant cell reaction to keratin and hair in ruptured dermoid cyst.

the most common germ cell tumor in the orbit. Other primary orbital germ cell tumors, such as *yolk sac (endodermal sinus) tumor,* are rare (9,10).

Clinical Features. Orbital teratoma is a sporadic congenital neoplasm with a female predilection. This tumor typically presents at birth with severe unilateral proptosis, which may be progressive in the first few weeks of life (fig. 4-5) (8). Large lesions are associated with facial dysmorphism and nasopharyngeal (oropharyngeal) and intracranial involvement, which can cause significant morbidity (11).

Microscopic Findings. In its strict definition, an orbital teratoma is characterized by the presence of tissue derived from all three embryonic layers (ectoderm, mesoderm, and endoderm), including cysts lined by surface, gastrointestinal, and respiratory epithelia (fig. 4-5). Cartilage and central nervous tissue may be observed. Some orbital teratomas even contain formed parts of a fetus. Occasionally, the eye is disorganized and is incorporated in the teratomatous mass (8,10).

Differential Diagnosis. The differential diagnosis includes other congenital cystic orbital lesions, such as a dermoid cyst. The presence of three embryonic germ cell layers is diagnostic of orbital teratoma (8). Although primary malignant germ cell tumors in the orbit are rare, careful evaluation for signs of malignancy is prudent. Germ cell tumors can involve the orbit secondarily by direct extension and as metastasis (9). Correlation with clinical presentation, radiographic features, and oncologic findings is essential.

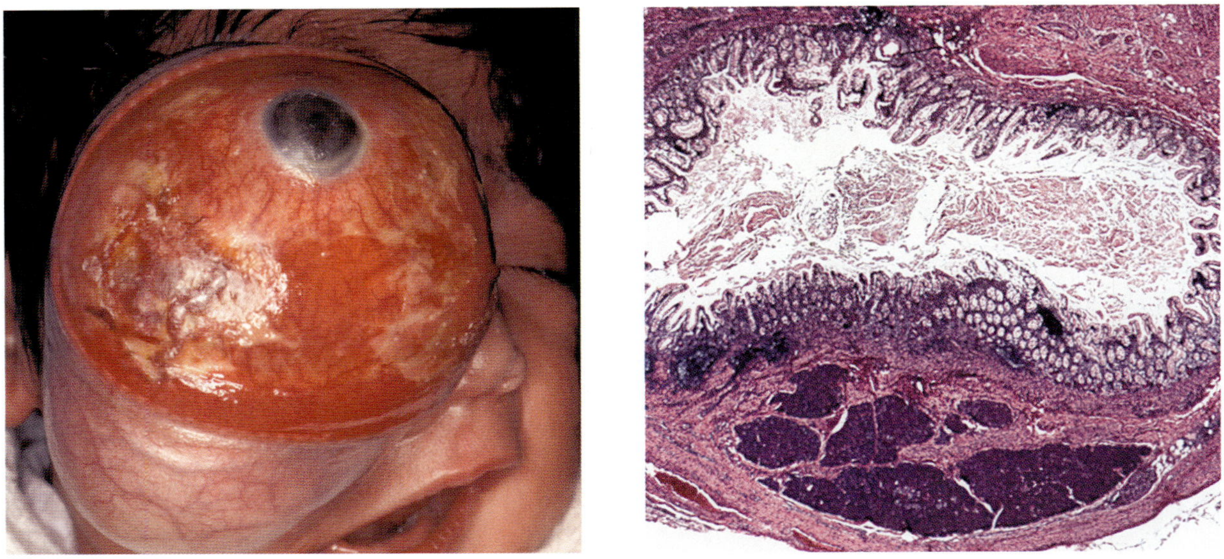

Figure 4-5

ORBITAL TERATOMA

Left: A large orbital teratoma in a child exposes the globe.
Right: The cavity is lined by multilayered epithelium of the intestinal tract and lobules of pancreatic tissue (bottom).

Treatment and Prognosis. Although orbital teratomas carry low malignant potential, several reports of malignant transformation in incompletely excised tumors exist in the literature (10). Treatment consists of complete resection with a goal of preservation of visual function. Combined orbital teratomas and secondary teratomas require multidisciplinary surgical approaches (11).

PRIMARY ORBITAL MELANOMA

Definition. *Primary orbital melanoma* is a malignant neoplasm arising in the orbital soft tissues.

Clinical Features. Primary orbital melanoma is exceptionally uncommon, with approximately 50 cases documented in the literature and, as a consequence, not well-defined epidemiologically. Based on the available evidence, this neoplasm manifests at an average age of 50 years (12) and may develop in the context of the nevus of Ota (oculodermal melanocytosis) and its variants (fig. 4-6) (12–14) (see chapter 3).

Microscopic Findings. Primary orbital melanomas feature a proliferation of atypical, mitotically active melanocytes with a variable degree of pigmentation and variable morphology, ranging from spindle to epithelioid melanoma cells. In tumors originating from nevus of Ota and cellular blue nevus, a precursor lesion can be identified (fig. 4-6) (12–14).

Molecular Genetic Findings. The molecular genetic data on primary orbital melanoma is sparse. Recent studies have demonstrated *GNAQ* and *GNA11* mutations in nevi of Ota, blue nevi, and uveal melanoma, suggesting that these presumably initiating mutational events explain the link between these neoplasms (15,16). One molecular genetic and cytogenetic analysis of primary orbital melanoma developing in the nevus of Ota demonstrated chromosomal copy number changes typical of melanoma in the morphologically malignant areas (14). This finding illustrated the relationship between tumor progression and the progressive acquisition of genetic aberrations but did not identify mutations in *GNAQ*, *BRAF*, *NRAS*, and *KIT* genes.

Differential Diagnosis. Orbital involvement by melanoma occurs much more frequently as secondary disease (local invasion from a uveal, conjunctival, or eyelid primary tumor), or as metastasis from other sites, such as skin or uvea. It is important to exclude these etiologies by a thorough clinical and imaging evaluation, and (when appropriate) by histologic examination of both uveal tracts (12).

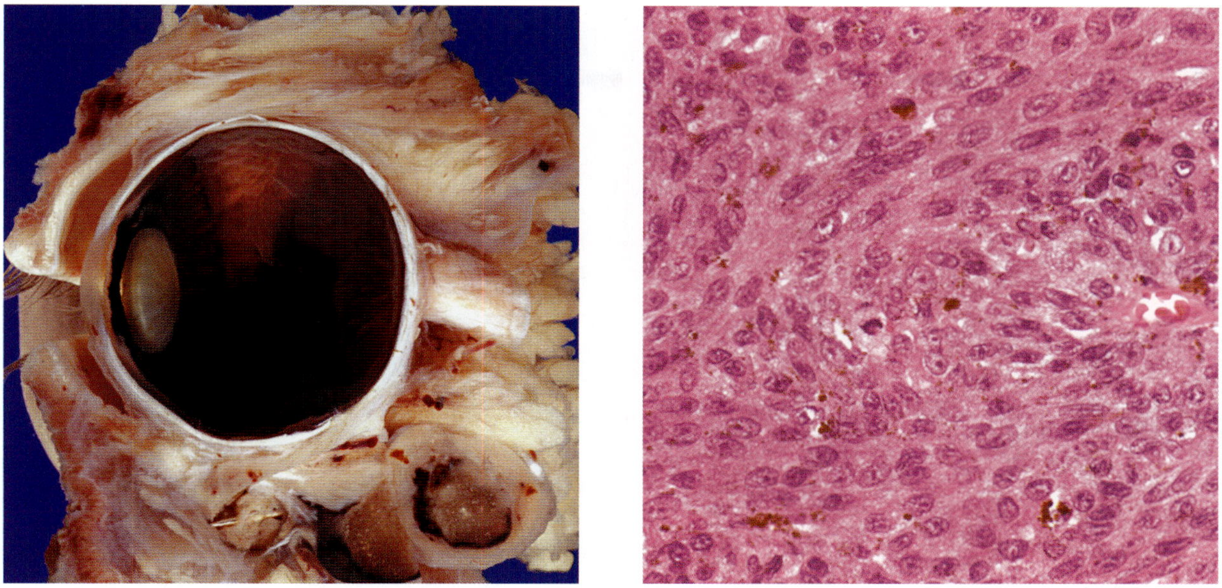

Figure 4-6

PRIMARY ORBITAL MELANOMA

Left: A multinodular pigmented tumor is in the posterior orbit inferiorly.

Right: The tumor depicted on the left was composed of variable pigmented, plump, spindle-shaped cells with ovoid nuclei and prominent nucleoli.

Treatment and Prognosis. Because of the limited number of patients with primary orbital melanoma described in the literature, and the continually evolving therapeutic modalities, the prognosis and management of this rare orbital neoplasm are not well defined. Based on the published evidence, primary orbital melanoma has a propensity for regional lymph node and distant (hepatic) metastases and significant mortality. Therapeutic options have traditionally included orbital exenteration, which has been more recently replaced by radiotherapy to the orbit and targeted therapies (12).

METASTASES TO ORBIT

Definition. *Metastatic disease* of the orbit refers to malignant orbital tumors involving the orbit from distant organs or contralateral eye or orbit (2).

Clinical Features. Metastases to the orbit are among the most common orbital malignancies in adults (2,3,17). Approximately half of these metastases are from a breast primary and frequently occur in a setting of widespread metastatic disease (fig. 4-7). Other common primary tumors that metastasize to the orbit are prostate and lung (adeno)carcinomas, uveal melanoma, renal cell carcinoma, and gastrointestinal tract carcinoid tumors (fig. 4-8). In 5 to 15 percent of the patients, the primary site cannot be determined (3,17).

Metastases to the orbit typically present as a rapidly progressive orbital mass, associated with displacement of the eyeball, restriction of ocular motility, and double vision. A rare presentation with enophthalmos (posterior displacement of the eyeball) occurred with scirrhous carcinoma of the breast and prostatic adenocarcinoma. Breast carcinoma and melanoma have a predilection for extraocular muscles and orbital fat, while prostate gland and thyroid gland carcinomas have tendency to metastasize to bone (2).

Microscopic Findings. Orbital metastases may adopt an infiltrative, nodular, or mixed growth pattern. Metastatic tumors tend to retain the morphology and immunophenotype of the tumor of origin (figs. 4-7, 4-8). Characteristic involvement of the orbital stroma in a single file arrangement has been observed in metastatic carcinomas (fig. 4-7) (2).

Differential Diagnosis. Metastatic tumors to the orbit must be distinguished from primary

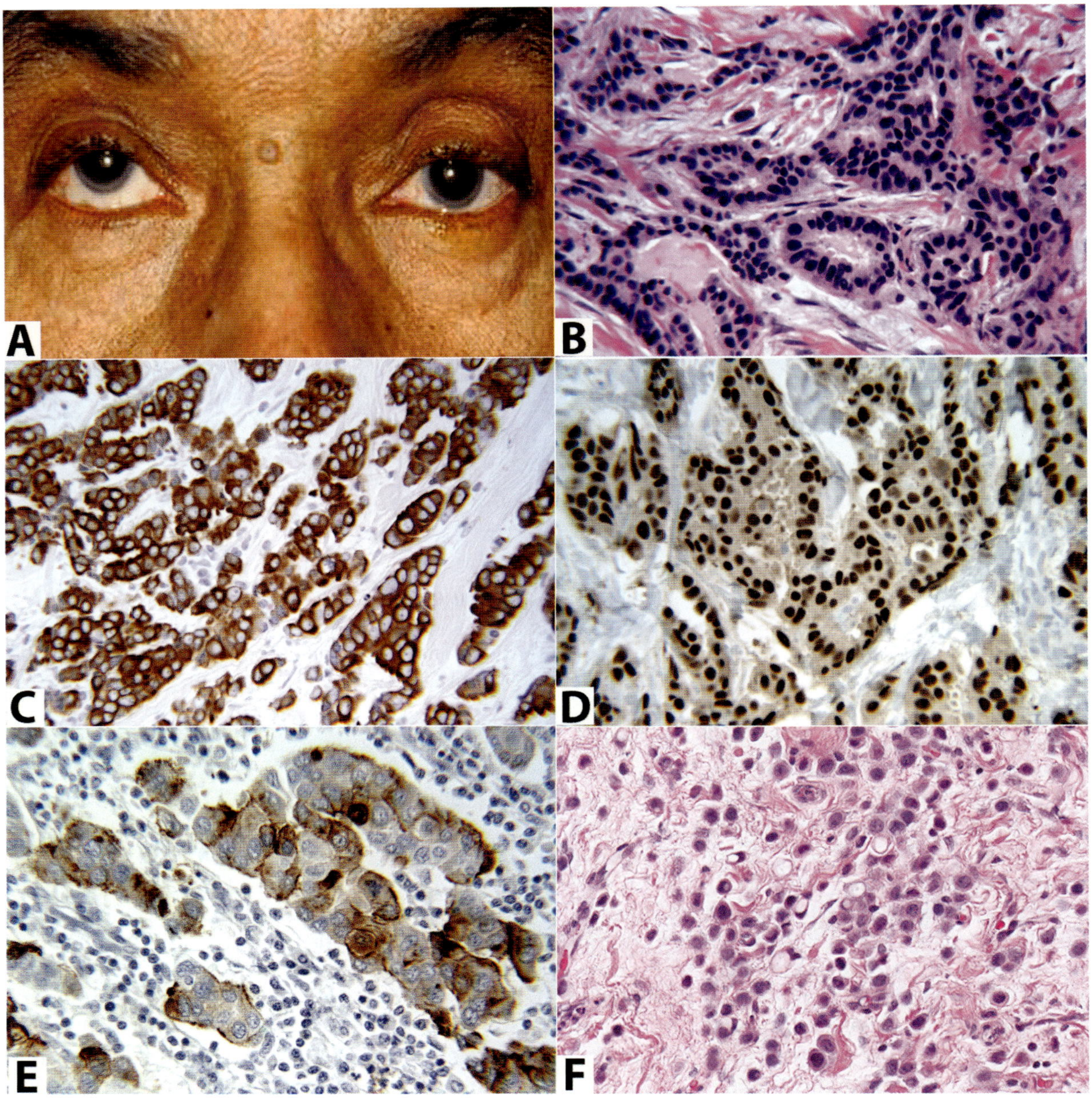

Figure 4-7

BREAST CARCINOMA METASTATIC TO ORBIT

A: Orbital metastasis presents with superior displacement of the right eye (hyperglobus).

B–E: Metastatic ductal carcinoma (B), composed of neoplastic cells infiltrating fibrotic orbital tissue as strands and ductules expresses CK7 (C), estrogen receptors (ER) (D), and BRST-2 (E).

F: Lobular carcinoma metastatic to the orbit presents as infiltrate of histiocytoid cells with occasional intracytoplasmic vacuoles (signet ring morphology).

orbital neoplasms. Observation of an in situ component and correlation with clinical/oncologic history help establish the diagnosis. Immunohistochemical panels to elucidate or confirm primary site and correlation with prior surgical material are helpful in identifying the tumor of origin (2).

Treatment and Prognosis. Survival and treatment of patients with orbital metastases depend on the type of primary tumor and

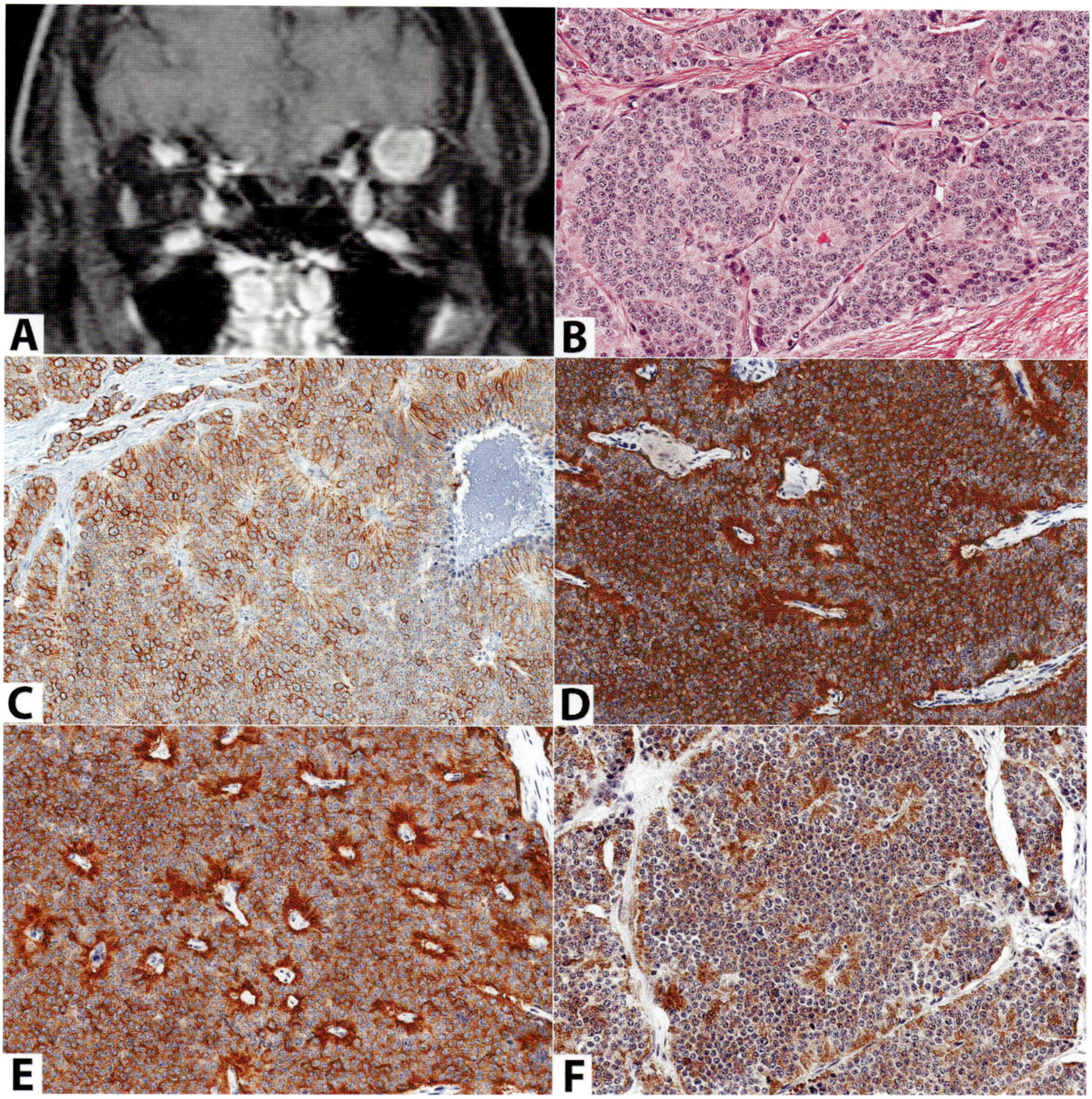

Figure 4-8

ILEAL CARCINOID TUMOR METASTATIC TO ORBIT

A: Well-circumscribed extraconal orbital enhancing mass on coronal magnetic resonance imaging (MRI).
B–F: Minimally pleomorphic neoplastic cells with finely granular (salt and pepper) nuclear chromatin in a nested (organoid) arrangement (B) express cytokeratins AE1/AE3 (C), chromogranin (D), synaptophysin (E), and serotonin (F).

oncologic staging. In disseminated malignancies, management of metastatic disease is generally directed toward the systemic oncologic control of the tumor. Surgery and orbital radiotherapy are used as palliative modalities and, in select cases, therapeutically (2).

SOFT TISSUE TUMORS

The discussion of lesions mentioned in this section is limited to the site-specific information. Most of the orbital soft tissue tumors are clinically, microscopically, and genetically

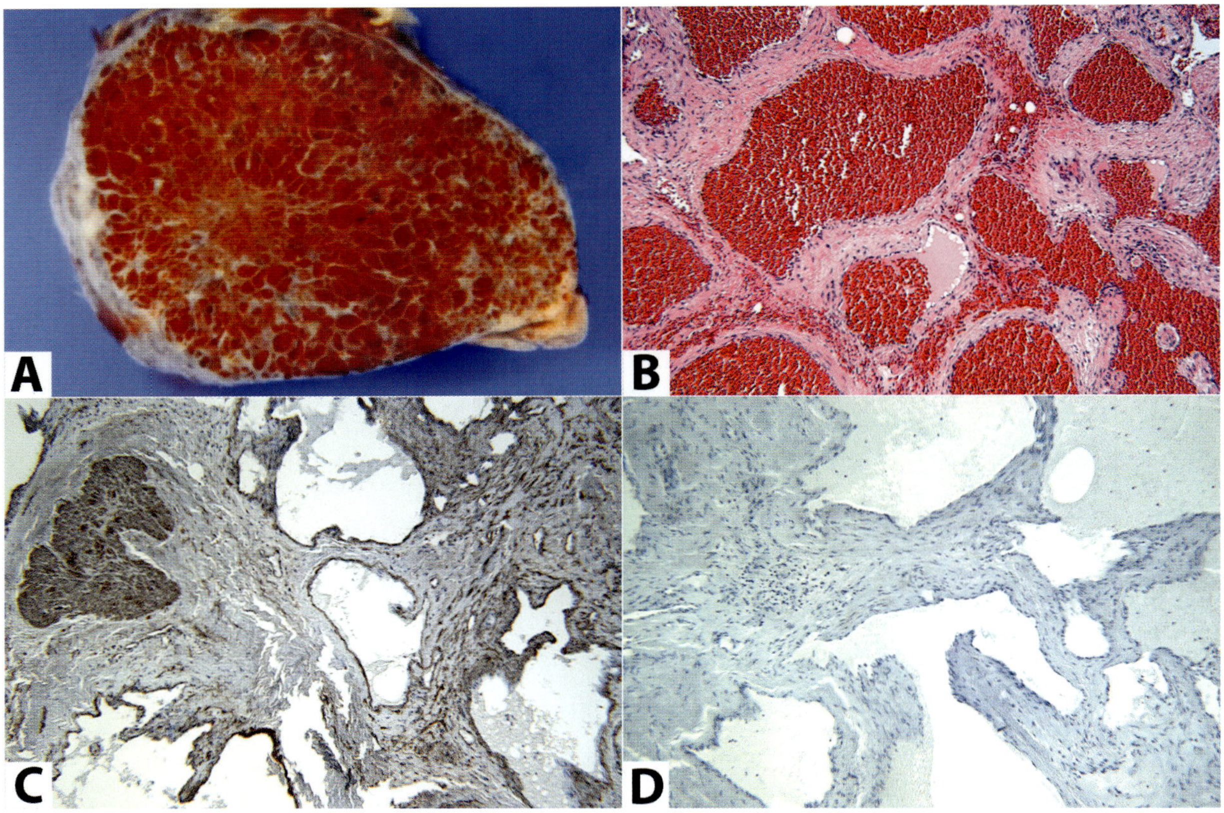

Figure 4-9

ORBITAL CAVERNOUS HEMANGIOMA (CAVERNOUS VENOUS MALFORMATION)

A: Ovoid encapsulated mass, composed of large, uniform, blood-filled vascular channels separated by thick fibrous septa.

B: The vascular septa contain smooth muscle, but lack well-defined elastic lamina, compatible with venous derivation. The focal serum-red blood cell separation reflects a low-flow lesion.

C: The smooth muscle in the walls of the lesion is highlighted with smooth muscle actin (SMA).

D: There is no expression of lymphatic endothelial marker D2-40, supporting presence of vascular endothelium.

similar to the same lesions in extraocular locations. For additional information on these lesions, please see corresponding Fascicles in this series.

The location of the lesion within the orbit influences the management and prognosis of locally aggressive and malignant primary orbital soft tissue tumors. Despite the surgical challenges of complete resection of infiltrative orbital lesions imposed by the proximity of the orbital walls, cranium, paranasal sinuses, and eyeball, some primary orbital sarcomas are associated with a favorable prognosis when compared with the extraorbital sites. Recognition of these site-specific differences in management and prognosis is reflected in the ophthalmic site-specific TNM staging system for most primary orbital sarcomas in the current edition of American Joint Committee on Cancer (AJCC) manual (17).

VASCULAR TUMORS AND TUMOR-LIKE LESIONS

A variety of vascular lesions involve the orbital tissues. Of these, *cavernous hemangioma (orbital cavernous venous malformation)* is the most frequent, emerging as one of the most common primary orbital tumors in the adult population (3,18). This vascular lesion characteristically presents in adult females with slowly progressive, painless proptosis (anterior displacement of the eyeball) and visual loss. Resection is performed for symptomatic tumors and is generally easily accomplished due to circumscription or encapsulation of the tumor (fig. 4-9).

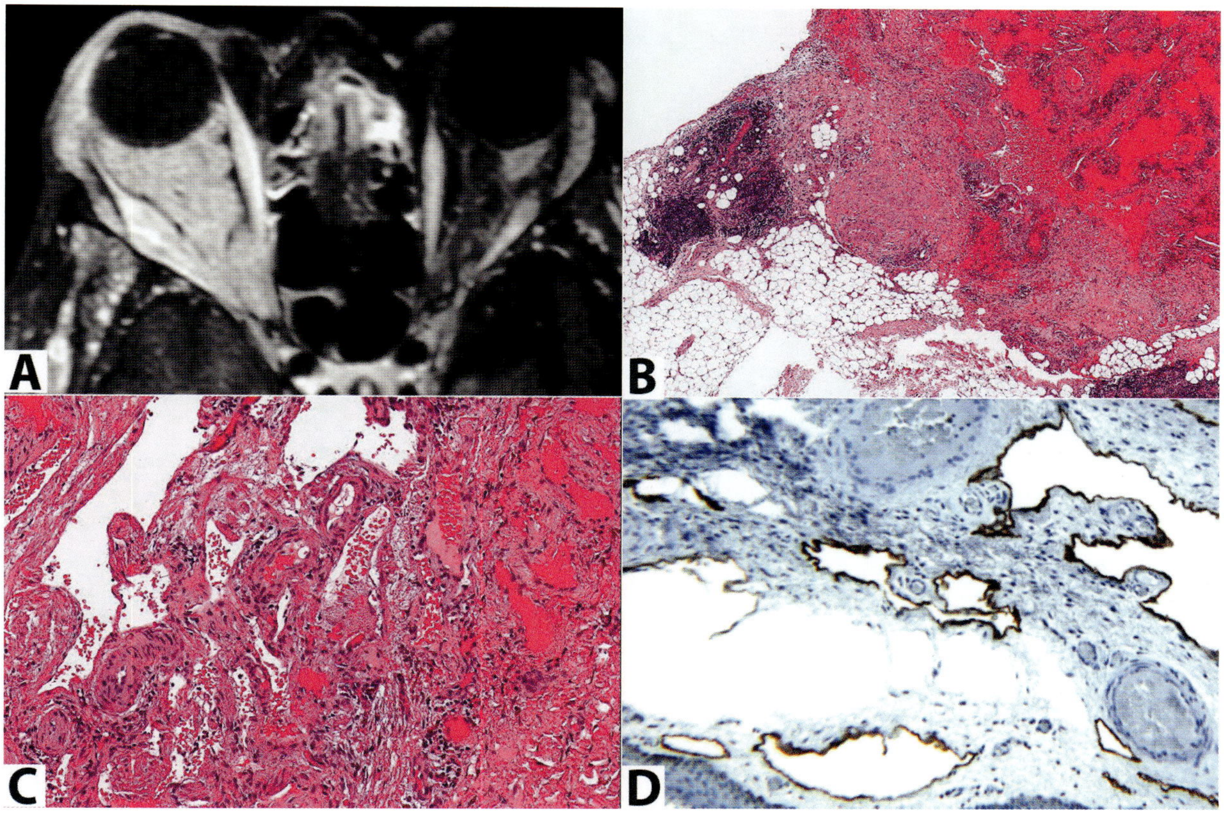

Figure 4-10

ORBITAL LYMPHANGIOMA (LYMPHATIC-VENOUS MALFORMATION)

A: Poorly circumscribed, infiltrative right orbital mass causes anterior displacement of the eyeball (proptosis).
B: The lesion infiltrates orbital fat and is associated with a lymphoid aggregate.
C: Higher magnification identifies irregular, focally collapsed, blood-filled lymphatic channels and few thick-walled venous channels, with hemosiderin deposition in the walls.
D: The lymphatic endothelium expresses D2-40.

Lymphangioma (lymphatic-venous malformation) presents in young patients with fluctuating proptosis related to intralesional hemorrhage or in a context of upper respiratory illness. The infiltrative nature of this lesion, with frequent intraoperative hemorrhage, makes surgical resection challenging (fig. 4-10). As a result, surgery is reserved for lesions associated with severe visual compromise.

Other vascular lesions described in the orbital region include *arteriovenous malformation, epithelioid hemangioma, angiosarcoma*, and *Kaposi sarcoma*. These orbital vascular lesions have a histopathologic appearance similar to the tumors in extraocular sites (2). Primary epithelioid hemangioendotheliomas and angiosarcomas of the orbit are staged following ophthalmic site-specific protocols in the current AJCC staging manual (17).

FIBROCYTIC AND MYOFIBROBLASTIC TUMOR-LIKE LESIONS AND NEOPLASMS

The literature cites examples of various fibrocytic and myofibroblastic tumors in the orbit, including *nodular fasciitis, solitary fibrous tumor, deep fibrous histiocytoma*, and *fibrosarcoma*. Similar to extraorbital sites, the classification of orbital fibrocytic lesions has undergone evolution with the advent of refined modern immunohistochemical methods and molecular genetic techniques. The past decades have witnessed a decrease in frequency of orbital fibrous histiocytoma and the disappearance of hemangiopericytoma and malignant fibrous histiocytoma, with the corresponding

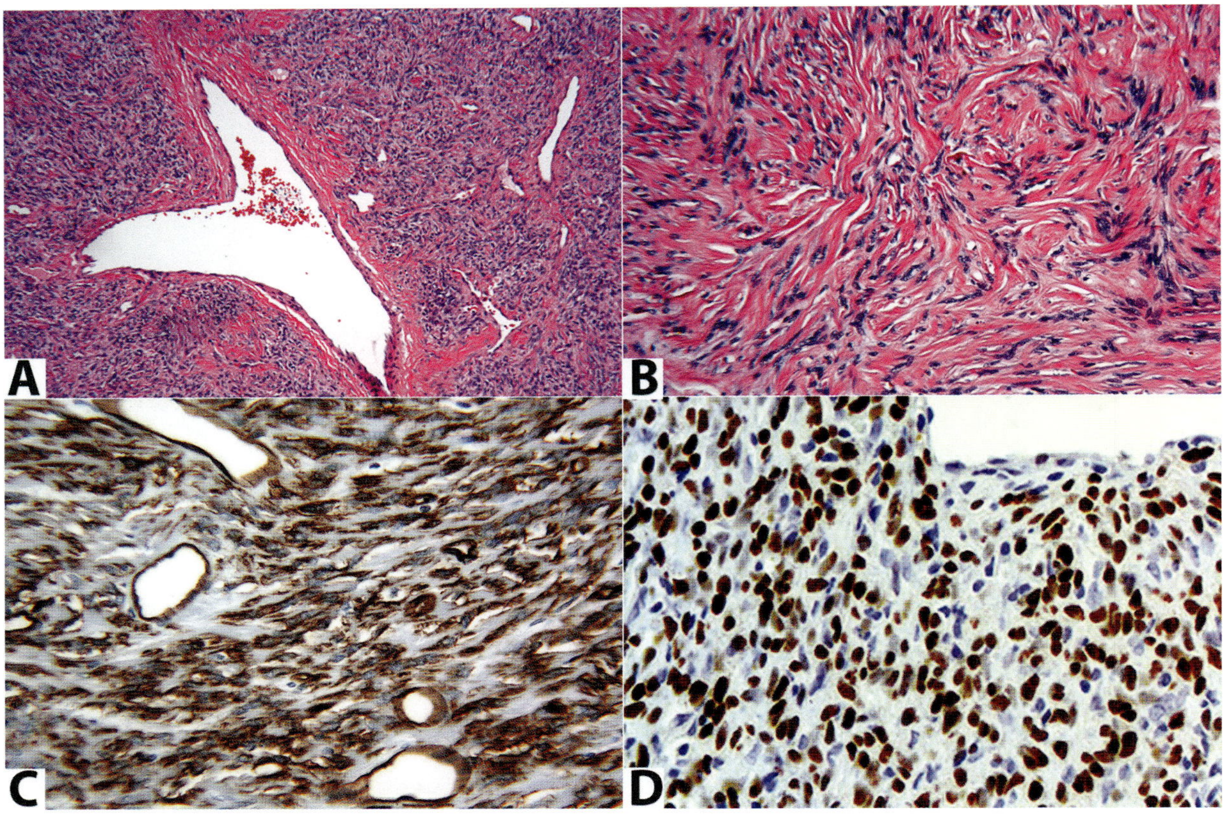

Figure 4-11

SOLITARY FIBROUS TUMOR OF ORBIT

A: Cellular proliferation of spindle to ovoid cells are separated by large angulated thick-walled vascular channels in a solitary fibrous tumor that was previously designated as hemangiopericytoma.

B: Less cellular areas in the tumor are characterized by a haphazard, "patternless" arrangement of ovoid and spindle cells separated by coarse collagen bundles (A,B: H&E stain).

C: The neoplastic cells diffusely express CD34.

D: Strong nuclear staining for STAT-6.

increase in the frequency of solitary fibrous tumors (fig. 4-11) and specific sarcoma subtypes.

Orbital fibrocytic and myofibroblastic tumors have morphologic and molecular genetic features (when elucidated) similar to those of extraorbital counterparts. The location of locally aggressive lesions in the orbit may influence the clinical behavior, management, and prognosis because of limited options to achieve a complete excision. Primary sarcomas of the orbit with fibrocytic and myofibroblastic differentiation are staged following ophthalmic site-specific protocols in the current AJCC staging manual (17).

TUMORS WITH MUSCLE DIFFERENTIATION

Benign and malignant neoplasms with muscle differentiation, including *leiomyoma, leiomyosarcoma*, and *rhabdomyosarcoma* can involve the orbital soft tissues. Of these, orbital rhabdomyosarcoma commands particular attention, because of its status as the most frequent primary malignant orbital neoplasm in the pediatric population and because of its management and prognostic implications.

Orbital rhabdomyosarcoma comprises approximately 10 percent of all rhabdomyosarcoma cases. It typically presents at the average age of 6 to 8 years, with a male predilection, as rapidly progressive proptosis and lid swelling without inflammatory signs (fig. 4-12) (2,19). Similar to other rhabdomyosarcomas of the head and neck, the embryonal subtype is the most frequent in the orbit, followed by alveolar rhabdomyosarcoma (19–21).

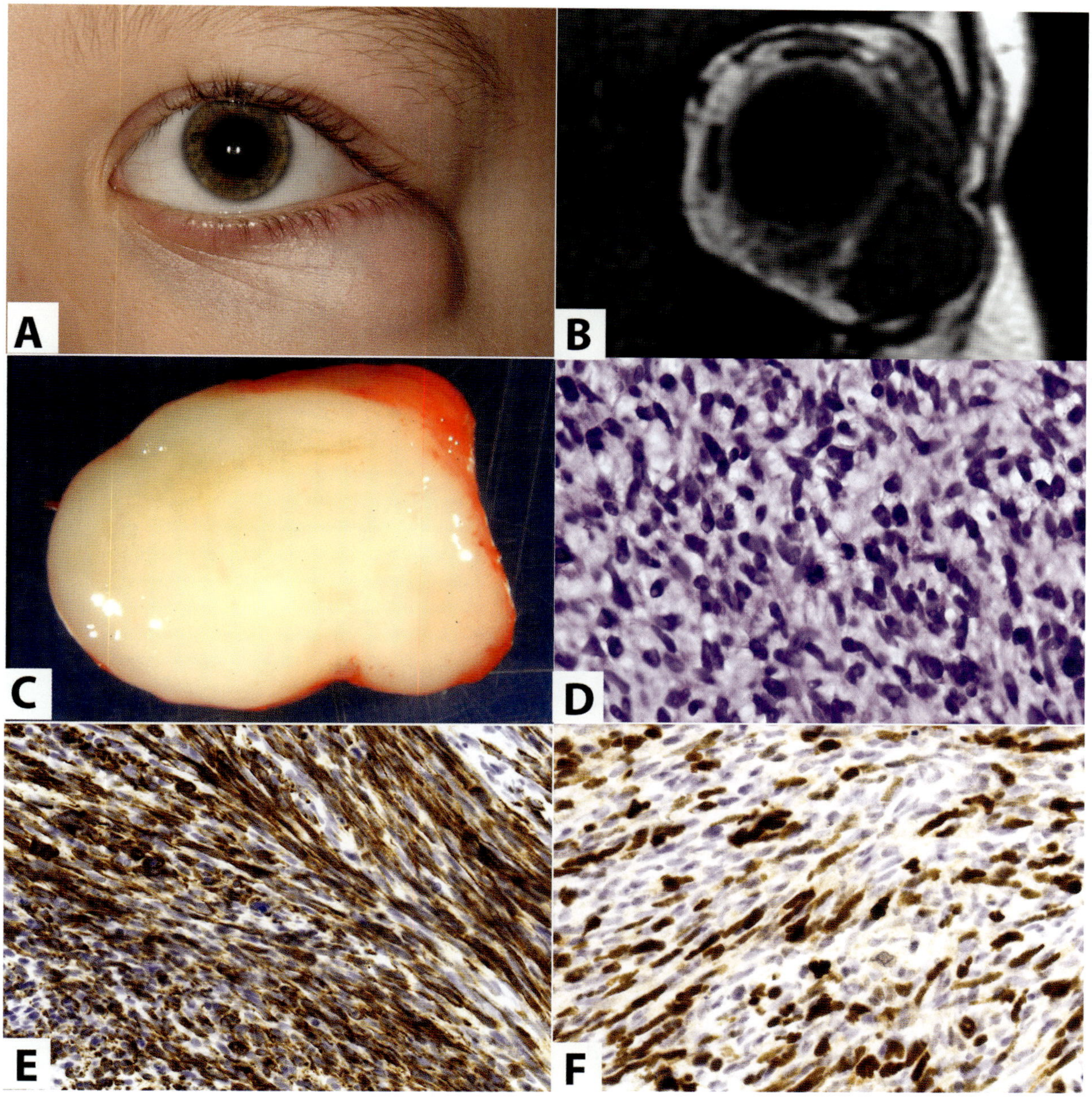

Figure 4-12

EMBRYONAL RHABDOMYOSARCOMA OF ORBIT

A: A mass in the inferior orbit of a young child, without inflammatory signs.
B: Coronal MRI delineates the well-circumscribed extraconal mass in the inferior orbit.
C: Gelatinous cut surface in the tumor reflects the presence of myxoid stroma.
D: The neoplasm is composed of primitive mesenchymal cells in an abundant myxoid matrix.
E: Neoplastic cells diffusely express desmin.
F: There is focal nuclear expression of myogenin.

The diagnosis of orbital rhabdomyosarcoma is established following the current morphologic, immunohistochemical, and molecular genetic guidelines established for all rhabdomyosarcomas (fig. 4-12) (20,21). The orbit is a favorable site. As a result, incompletely resected tumors belong in the prognostically favorable category. Pathologic staging of orbital

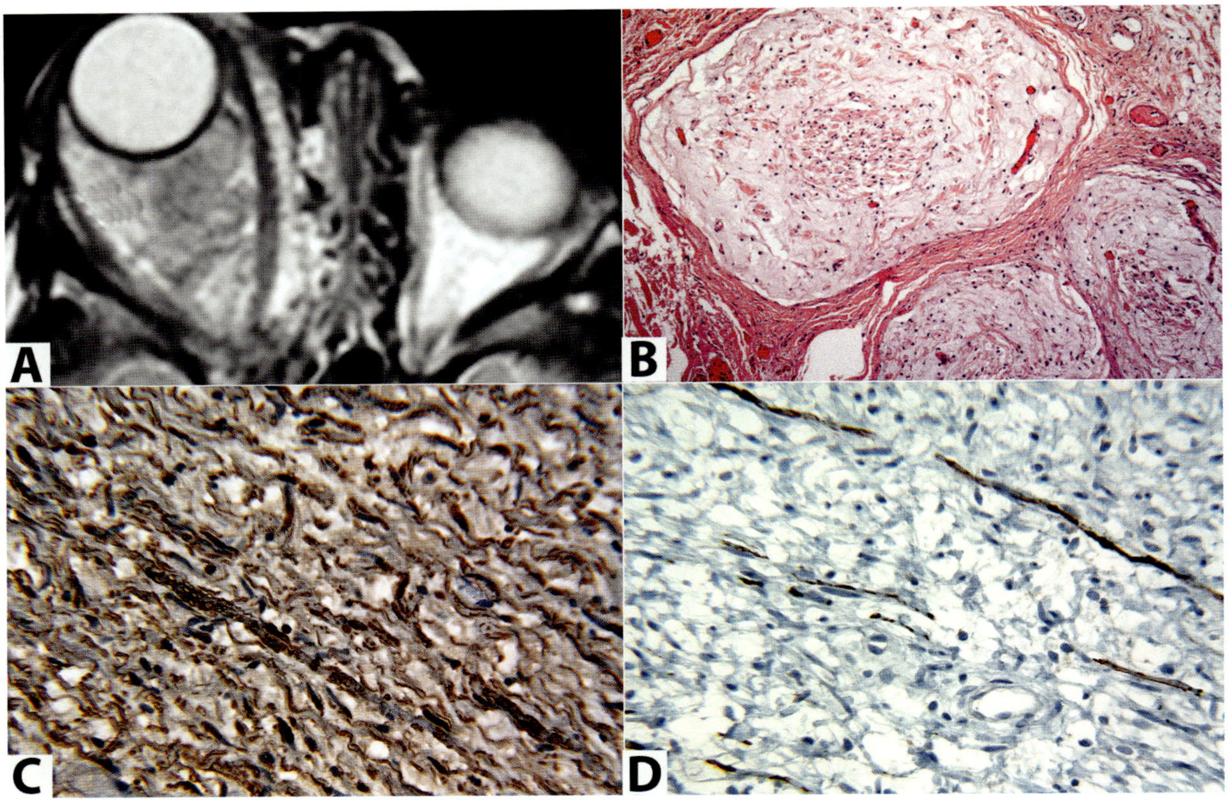

Figure 4-13

PLEXIFORM NEUROFIBROMA OF ORBIT

A: Poorly circumscribed, unencapsulated undulating mass in the intraconal orbital space seen with MRI.

B: Characteristic expanded neural bundles caused by proliferation of fibroblasts and Schwann cells in the background of ropy collagen, myxoid stroma, and few entrapped axons.

C: There is patchy expression of S-100 protein.

D: The axons are highlighted by neurofilament (NF) immunostain.

rhabdomyosarcoma traditionally uses the pretreatment TNM staging system established by the Intergroup Rhabdomyosarcoma Study Group (20,21). Additionally, primary orbital rhabdomyosarcomas can be staged following the ophthalmic site-specific AJCC classification guidelines for orbital sarcoma (17).

LIPOMATOUS TUMORS AND TUMOR-LIKE LESIONS

Various lipomatous neoplasms, ranging from benign *lipoma* and its variants to *liposarcoma* have been described in an orbital location, where they are vanishingly rare. Care should be taken not to misinterpret the reactive changes in orbital fat as a neoplastic process (see Herniated Orbital Fat section above) (4). Primary liposarcomas of the orbit are staged following ophthalmic site-specific protocols for orbital sarcoma in the current AJCC staging manual (17).

PERIPHERAL NERVE SHEATH TUMORS

The orbital region, rich in peripheral nerves, can spawn a variety of peripheral nerve sheath tumors, including *neurofibroma, schwannoma, malignant peripheral nerve sheath tumor*, and *granular cell tumor*. These have clinical features, systemic associations, histopathologic appearances, and biologic behavior similar to extraorbital counterparts (figs. 4-13, 4-14) (2). Primary malignant peripheral nerve sheath tumors of the orbit are staged by ophthalmic site-specific protocols for orbital sarcoma in the current AJCC staging manual (18).

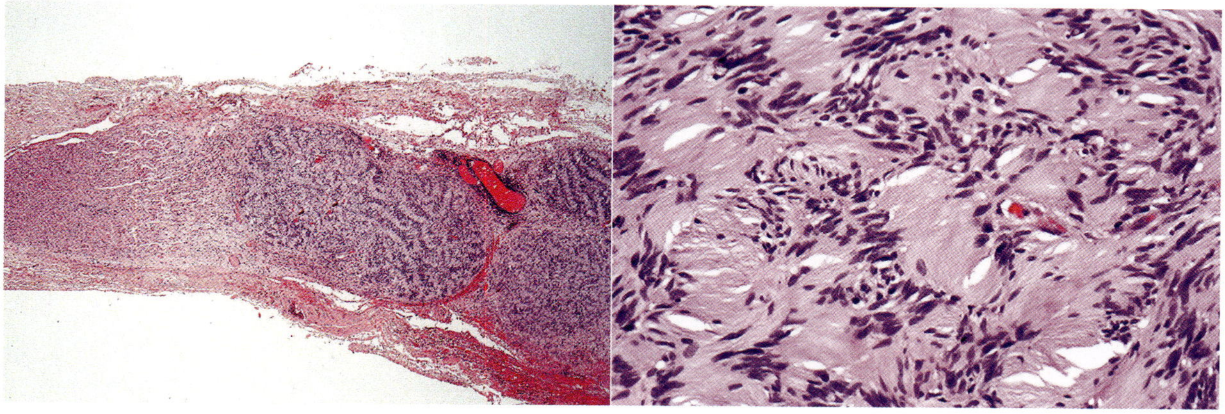

Figure 4-14

SCHWANNOMA OF ORBIT

Left: Cellular fascicles of schwannoma arise from the peripheral nerve.

Right: At higher magnification, the characteristic palisaded arrangement of Schwann cells (Antoni A pattern) with Verocay body formation is visualized (H&E stain).

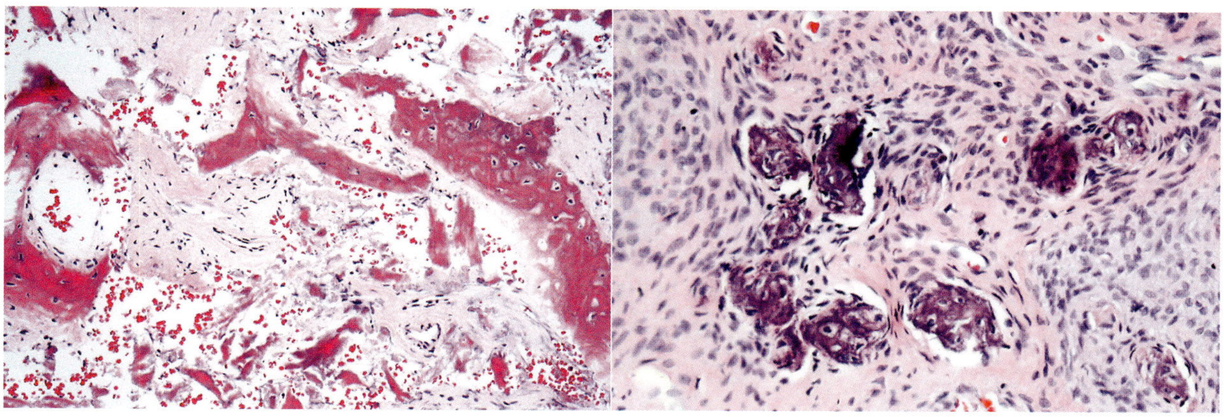

Figure 4-15

FIBRO-OSSEOUS LESIONS OF ORBIT

Left: Fibrous dysplasia is characterized by irregular, geographic, "Chinese character-like" arrangement of woven bone trabeculae, without appreciable osteoblastic rimming in a background of loose fibrovascular stroma.

Right: Ossifying fibroma is characterized by densely mineralized ossicles with resident osteocytes and osteoblastic rimming, in a background of dense fibrous stroma (H&E stain).

TUMORS AND TUMOR-LIKE LESIONS OF ORBITAL BONES

Primary lesions of the bony orbit are rare, comprising 0.6 to 2.0 percent of all orbital tumors. While uncommon, the lesions arising within the walls of the bony orbit are diverse and include a variety of reactive processes, benign neoplasms, and malignant tumors.

Fibro-osseous, osseous, and cartilaginous tumors (*fibrous dysplasia, ossifying fibroma, osteoblastoma, osteoid osteoma, aneurysmal bone cyst, osteosarcoma*, and *chondrosarcoma*) account for approximately half of all primary orbital bone lesions (figs. 4-15, 4-16) (3,22). The higher frequency of fibrous dysplasia, ossifying fibroma, and osteoma in the orbital and craniofacial location, as compared to the extracranial skeleton, is believed to reflect the underlying developmental differences between these body sites. Historically, orbital osteosarcoma was more common in retinoblastoma survivors exposed

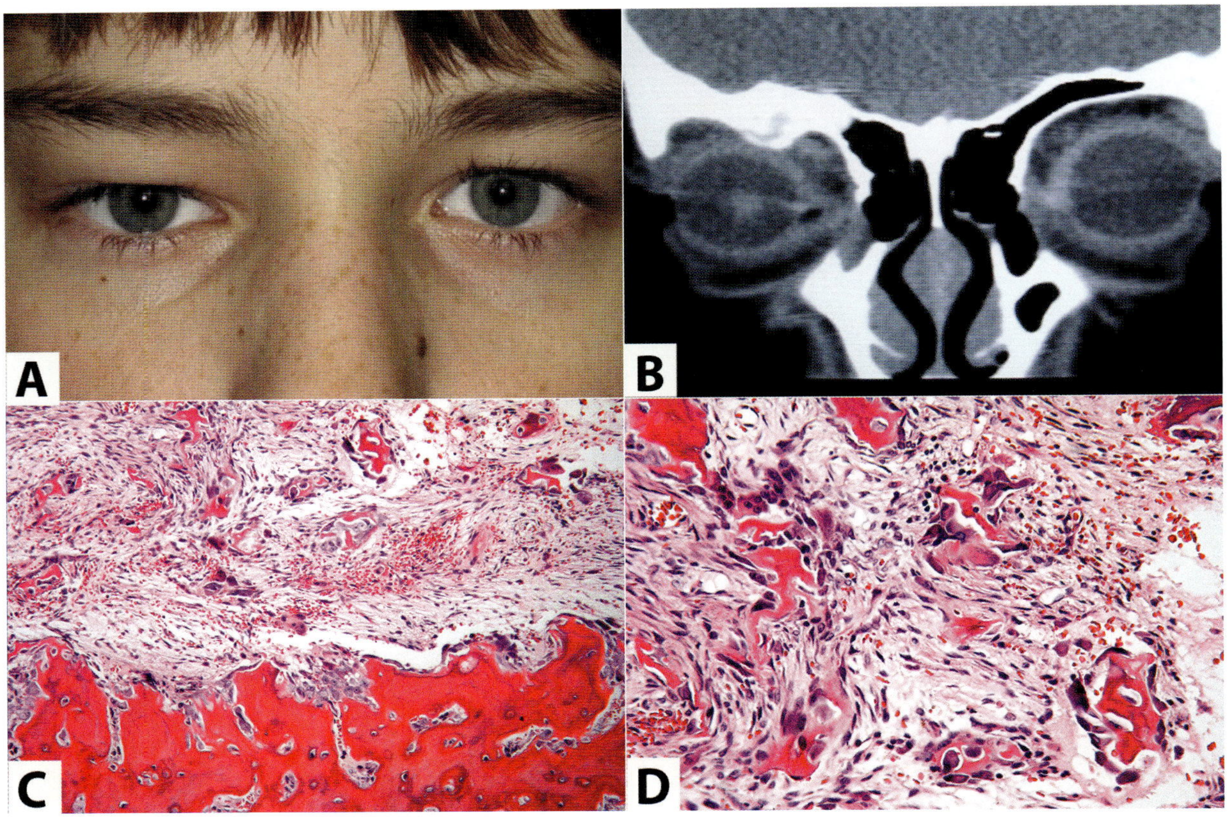

Figure 4-16

OSTEOBLASTOMA OF ORBIT

A: There is right superior orbital fullness with mild downward displacement of the eyeball.

B: Axial computed tomography scan shows a well-circumscribed mass with sclerotic border and central lucent "nidus" in the superior nasal orbital rim.

C: The center of the nidus of osteoblastoma is composed of immature trabeculae of woven bone, lined by plump osteoblasts in a background of richly vascularized stroma. The periphery of the tumor is lined by mature sclerotic lamellar bone.

D: Higher magnification of the nidus highlights the plump osteoblasts and osteoclasts rimming the immature bone trabeculae and the osteoid.

to external beam radiotherapy (2). This association has been decreasing with the advent of other treatment modalities for retinoblastoma.

The close association of the orbital bones with cranial nerves and the eyeball significantly affects the surgical approach, which, in turn, affects the nature of specimens submitted for pathologic evaluation. In contrast with similar lesions of long bones, orbital lesions are frequently curetted, limiting the contribution of gross evaluation to pathologic assessment and contributing to the microscopic evaluation challenges (23,24). An orbital location influences the clinical behavior, management, and prognosis of locally aggressive lesions because the location may limit the ability to resect the lesion completely (23). Osseous and cartilaginous tumors arising in bone are staged according to the general classification for tumors of the bone (18).

PATHOLOGIST REPORT FOR ORBITAL TUMORS

The system presented in Appendix 4 from the Taskforce for Staging of Cancer of the Eye of the American Joint Committee on Cancer (AJCC), 8th edition, applies only to primary orbital sarcoma, except for lacrimal gland sarcomas and osseous and cartilaginous tumors arising in the bone. This system does not apply to secondary tumors that arise in adjacent periorbital sites, with orbital invasion and metastases (18).

REFERENCES

1. Wolff E, Last RJ. The appendages of the eye. In: Wolff E, Last RJ. Anatomy of the eye and orbit, 6th ed. London: Lewis; 1968:182-240.
2. Font RL, Croxatto JO, Rao NA. Tumors of the orbit. AFIP Atlas of Tumor Pathology, 4th series, Fascicle 5. Washington DC: American Registry of Pathology; 2006: 265-319.
3. Shields JA, Shields CL, Scartozzi R. Survey of 1264 patients with orbital tumors and simulating lesions: the 2002 Montgomery lecture, part 1. Ophthalmology 2004;111:997-1008.
4. Schmack I, Patel RM, Folpe AL, et al. Subconjunctival herniated orbital fat: a benign adipocytic lesion that may mimic pleomorphic lipoma and atypical lipomatous tumor. Am J Surg Pathol 2007;31:193-8.
5. Ebrahimi KB, Ren S, Green WR. Floretlike cells in in situ and prolapsed orbital fat. Ophthalmology 2007;114:2345-9.
6. Stacy RC, Bernardo LA, Nielsen GP. Absence of chromosomal abnormalities in herniated orbital fat. Histopathology 2014;65:273-7.
7. Daniel CS, Beaconsfield M, Rose GE, Luthert PJ, Heathcote JG, Clark BJ. Pleomorphic lipoma of the orbit: a case series and review of literature. Ophthalmology 2003;110:101-5.
8. Shields JA, Shields CL. Orbital cysts of childhood—classification, clinical features, and management. Surv Ophthalmol 2004;49:281-99.
9. Kamal S, Kaliki S, Sreedhar A, Mishra DK. Primary orbital yolk sac tumor: report of a case and review of literature. Int Ophthalmol 2016;36:435-44.
10. Kivelä T, Tarkkanen A. Orbital germ cell tumors revisited: a clinicopathological approach to classification. Surv Ophthalmol 1994;38:541-54.
11. Moreddu E, Pereira J, Vaz R, Lena G, Triglia JM, Nicollas R. Combined endonasal and neurosurgical resection of a congenital teratoma with pharyngeal, intracranial and orbital extension: case report, surgical technique and review of the literature. Int J Pediatr Otorhinolaryngol 2015; 79:1991-4.
12. Rose AM, Luthert PJ, Jayasena CN, Verity DH, Rose GE. Primary orbital melanoma: presentation, treatment, and long-term outcomes for 13 patients. Front Oncol 2017;7:316.
13. Buntinx-Krieg T, Ouyang J, Cartwright M. An orbital malignant melanoma arising in cellular blue nevus in a patient with nevus of Ota. Cureus 2016;8:e698.
14. Gerami P, Pouryazdanparast P, Vemula S, Bastian BC. Molecular analysis of a case of nevus of Ota showing progressive evolution to melanoma with intermediate stages resembling cellular blue nevus. Am J Dermatopathol 2010;32:301-5.
15. Franceschini D, Dinulos JG. Dermal melanocytosis and associated disorders. Curr Opin Pediatr 2015;27:480-5.
16. Van Raamsdonk CD, Bezrookove V, Green G, et al. Frequent somatic mutations of GNAQ in uveal melanoma and blue naevi. Nature 2009;457:599-602.
17. Amin MB, Edge SB, Greene FL, et al., eds. AJCC cancer staging manual, 8th ed. Switzerland: Springer; 2017.
18. Bonavolontà G, Strianese D, Grassi P, Comune C, Tranfa F, Uccello G, Iuliano A. An analysis of 2,480 space-occupying lesions of the orbit from 1976 to 2011. Ophthal Plast Reconstr Surg 2013; 29:79-86.
19. Shields JA, Shields CL. Rhabdomyosarcoma: review for the ophthalmologist. Surv Ophthalmol 2003;48:39-57.
20. Jurdy L, Merks JH, Pieters BR, et al. Orbital rhabdomyosarcomas: a review. Saudi J Ophthalmol 2013;27:167-75.
21. Radzikowska J, Kukwa W, Kukwa A, Czarnecka A, Krzeski A. Rhabdomyosarcoma of the head and neck in children. Contemp Oncol (Pozn) 2015; 19:98-107.
22. Selva D, White VA, O'Connell JX, Rootman J. Primary bone tumors of the orbit. Surv Ophthalmol 2004;49:328-42.
23. Wenig BM, Mafee MF, Ghosh L. Fibro-osseous, osseous, and cartilaginous lesions of the orbit and paraorbital region. Correlative clinicopathologic and radiographic features, including the diagnostic role of CT and MR imaging. Radiol Clin North Am 1998;36:1241-59.
24. El-Mofty SK. Fibro-osseous lesions of the craniofacial skeleton: an update. Head Neck Pathol 2014;8:432-44.

5 TUMORS OF THE LACRIMAL GLAND

ANATOMY AND HISTOLOGY

The lacrimal gland, a serous gland that contributes to the aqueous component of the tear film, is situated in the superolateral aspect of the orbit. It is separated anatomically into two lobes by the aponeurosis of the levator palpebrae superioris. The larger orbital lobe lies in the lacrimal fossa, behind the orbital septum and above the levator aponeurosis. The smaller palpebral lobe lies more inferiorly, below the levator aponeurosis, adjacent to the superior and lateral conjunctival fornix (1). It may be visible beneath the palpebral conjunctiva superolaterally when the eyelid is everted. The orbital and palpebral lobes are continuous where the gland bends around the edge of the levator aponeurosis (1,2). The lacrimal gland measures approximately 20 x 12 x 5 mm (2).

The lacrimal gland is not encapsulated. It is composed of multiple lobules separated by loose connective tissue septa (fig. 5-1). Each lobule is composed of acini and intralobular ducts. Connective tissue septa contain interlobular ducts, blood vessels, nerves, fibroblasts, and some plasma cells and lymphocytes. Interlobular ducts join to form approximately 12 excretory ducts, which open into the conjunctival fornix (1). Acini are composed of pyramid-shaped cells with numerous periodic acid–Schiff (PAS)-positive secretory granules and a basally located nucleus. Mucus-secreting cells are not present.

The acinus is surrounded by myoepithelial cells, which are often inconspicuous in hematoxylin and eosin (H&E)-stained sections (fig. 5-1, left), but may be highlighted with alpha-smooth muscle actin (α-SMA) and cytokeratin (CK)5/6.

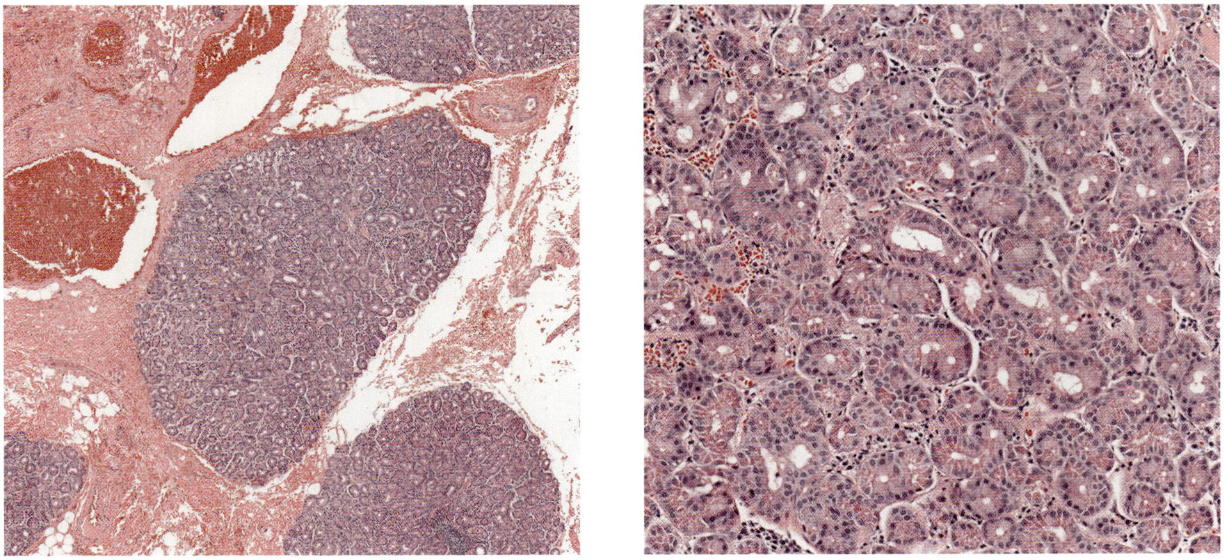

Figure 5-1

NORMAL LACRIMAL GLAND

Left: The parenchymal tissue has a lobular arrangement, supported by fibrovascular connective tissue (hematoxylin and eosin [H&E] stain).

Right: At higher magnification, the histology is identical to that of a serous accessory salivary gland. A duct is in the center of the figure. It is considered normal to identify scattered lymphocytes in the parenchymal stroma.

Table 5-1

FREQUENCY OF LACRIMAL GLAND TUMORS

Tumors	Shields et al. (4) (2004) No. (%)	Lacrimal Gland Tumor Study Group (2a) (2005) No. (%)	Bona-volonta et al. (3) (2013) No. (%)	Teo et al. (3a) (2013) No. (%)	Von Holstein et al. (5) (2013) No. (%)	Andrew et al. (6) (2015) No. (%)
Inflammatory	37 (32)	50 (16)	54 (22)	32 (46)	62 (27)	134 (50)
Lymphoid	26 (23)	87 (28)	29 (12)	26 (38)	35 (15)	86 (32)
Epithelial (includes dacryops)	49 (43)	146 (47)	158 (66)	9 (13)	87 (38)	40 (15)
Benign	11 (10)	87 (28)	85 (35)	7 (10)	31 (13)	21 (8)
Pleomorphic adenoma	11 (10)	86 (28)	78 (32)	7 (10)	31 (13)	21 (8)
Other	0	1 (<1)	7 (3)	0	0	0
Malignant	19 (17)	37 (12)	69 (29)	1 (1)	32 (14)	11 (4)
Adenoid cystic carcinoma	14 (12)	19 (6)	44 (18)	1 (1)	20 (9)	6 (2)
Adenocarcinoma	0	11 (4)	8 (3)	0	2 (1)	3 (1)
Carcinoma ex pleomorphic adenoma	4 (4)	7 (2)	7 (3)	0	5 (2)	1 (<1)
Other	1 (<1)	0	10 (4)	0	5 (2)	1 (<1)
Dacryops	19 (17)	22 (7)	4 (2)	1 (1)	24 (10)	8 (3)
Miscellaneous	2 (2)	28 (9)	0	2 (3)	48 (21)	8 (3)
Total	**114**	**311**	**241**	**69**	**232**	**268**

Ducts have 2 to 3 layers of epithelial cells and lack myoepithelial cells (1). Plasma cells present within the lacrimal gland produce immunoglobulin (Ig) A, which is secreted in the tear film and provides mucosal immunity.

CLASSIFICATION

Lacrimal gland lesions comprise approximately 10 percent of orbital tumors studied (3,4). These lesions are usually biopsied for suspicion of tumor. Proper diagnosis is essential for appropriate management and assessment of prognosis. Unnecessary removal of the lacrimal gland causes dry eye, which can lead to scarring and vision loss. On the other hand, improper treatment of a malignant neoplasm can result in vision loss and death (5).

Lacrimal gland tumors are commonly grouped into epithelial and nonepithelial categories. Between 55 and 80 percent of biopsied lacrimal gland lesions are nonepithelial (Table 5-1). These nonepithelial lesions include inflammatory entities, such as chronic dacryoadenitis and IgG4-related inflammation, benign and atypical lymphoid hyperplasias, and lymphomas.

The reported incidence of epithelial tumors ranges from 13 to 66 percent, of which 11 to 44 percent are malignant epithelial tumors. The most commonly encountered epithelial tumors are pleomorphic adenoma and adenoid cystic carcinoma. Classification of lacrimal gland tumors is based on the World Health Organization (WHO) histologic classification of salivary gland tumors, 2017 (Table 5-2).

This chapter highlights lacrimal gland lesions of special ophthalmic interest, including chronic dacryoadenitis, IgG4-related inflammation, lacrimal gland ductal cyst, pleomorphic adenoma, the most common benign epithelial neoplasm of the lacrimal gland, and adenoid cystic carcinoma, the most common malignant epithelial neoplasm of the lacrimal gland. The most common lymphoma of the lacrimal gland, extranodal marginal zone lymphoma, is discussed in Tumors of the Lymph Nodes and Spleen (5a).

CHRONIC DACRYOADENITIS

Inflammatory lesions comprise a significant proportion of lacrimal gland biopsies, up to 50 percent in some series (5,6). The main causes of lacrimal gland inflammation are idiopathic, autoimmune (e.g., immunoglobulin G4 (IgG4)-related, sarcoidosis, Sjogren syndrome, granulomatosis with polyangiitis), and infectious (e.g., Epstein-Barr virus, *Staphylococcus aureus* infection).

Table 5-2

CLASSIFICATION OF LACRIMAL GLAND TUMORS[a]

Benign Epithelial Tumors
Pleomorphic adenoma
Myoepithelioma
Oncocytoma
Malignant Epithelial Tumors
Adenoid cystic carcinoma
Carcinoma ex pleomorphic adenoma
Adenocarcinoma, NOS
Mucoepidermoid carcinoma
Polymorphous carcinoma
Epithelial-myoepithelial carcinoma
Acinic cell carcinoma
Basal cell adenocarcinoma
Mucinous adenocarcinoma
Ductal adenocarcinoma
Squamous cell carcinoma
Sebaceous adenocarcinoma
Myoepithelial carcinoma
Carcinosarcoma
Soft Tissue Tumors
Hematolymphoid Tumors
Extranodal Marginal Zone B-Cell Lymphoma

[a]Data from reference 33.

Idiopathic Chronic Dacryoadenitis

Nonspecific chronic inflammation of the lacrimal gland represents the most common histologic diagnosis in lacrimal gland lesions (7). Clinically, the onset is acute to subacute. Patients typically present with pain and eyelid erythema, and 50 percent present with dry eye. In 20 percent of patients, the condition is bilateral. Imaging studies show diffuse enlargement of the lacrimal gland with contrast enhancement and possible extension into the adjacent extraocular muscles or orbital fat (7).

Microscopically, there is a variable amount of periductal and interlobular fibrosis, and a lymphoplasmacytic infiltrate with or without lymphoid follicles with germinal centers admixed with neutrophils, eosinophils, and occasional histiocytes. Acini are atrophic (fig. 5-2).

Most patients are treated with corticosteroids and orbital radiotherapy may be reserved for relapsing or refractory cases. Surgical excision or debulking is also effective (8). Other treatments include methotrexate and rituximab.

IgG4-Related Inflammation

Definition. *IgG4-related disease* is a recently described systemic condition underlying a significant proportion of cases of idiopathic dacryoadenitis. It is characterized by abundant tissue infiltration by IgG4-positive plasma cells and elevated serum levels of IgG4 (9,10).

Clinical Features. The orbit, particularly the lacrimal gland, is a frequently involved site. A significant proportion of patients with orbital involvement have disease elsewhere, either at presentation or developing over time (11). Involvement of the lacrimal gland may be unilateral or bilateral, and may be associated with salivary gland enlargement (previously designated as Mikulicz disease) (12).

The mean age at diagnosis is 50 to 60 years, and there is no sex predominance. Patients typically present with eyelid swelling and other ophthalmic complaints, such as proptosis, double vision, ptosis, eye pain, decreased visual acuity, dry eye, and epiphora. Bilateral lacrimal gland swelling is seen in 50 to 70 percent of patients. On imaging studies, lesions are well defined and have homogeneous attenuation/signal intensities on computerized tomography (CT)/magnetic resonance imaging (MRI), with homogeneous enhancement after injection of contrast (13). Some patients have involvement of other orbital tissues, such as extraocular muscles, orbital soft tissue, or optic nerve sheath (12).

Microscopic Findings. Histologic diagnosis of IgG4-related dacryoadenitis is based on identifying the characteristic histopathologic features and increased numbers of IgG4-positive plasma cells within the tissue (fig. 5-3) (10). Characteristic histologic features are: 1) dense lymphoplasmacytic infiltrate; 2) fibrosis, usually storiform; and 3) obliterative phlebitis. In the lacrimal gland, obliterative phlebitis is typically absent, therefore, only features 1 and 2 are required for diagnosis (11). The lymphocytic infiltrate is usually composed of reactive lymphoid follicles with germinal centers. Scattered eosinophils may be seen. In the lacrimal gland, more than 100 IgG4-positive plasma cells per high-power field is required for the diagnosis, and the IgG4- to IgG-positive plasma cell ratio must be greater than 40 percent. The constellation of these features is considered "histologically highly suggestive of IgG4-related disease" (10).

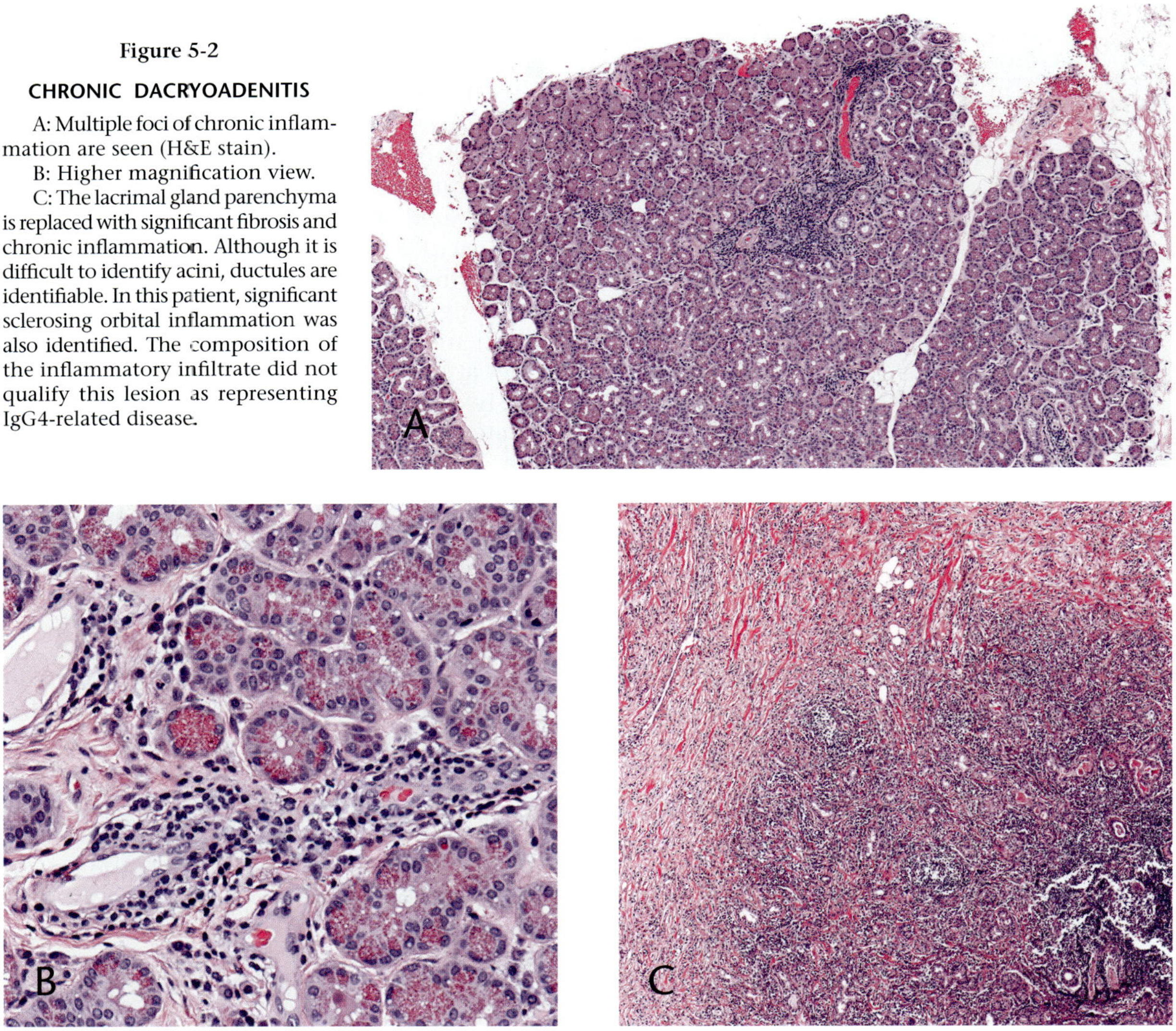

Figure 5-2

CHRONIC DACRYOADENITIS

A: Multiple foci of chronic inflammation are seen (H&E stain).

B: Higher magnification view.

C: The lacrimal gland parenchyma is replaced with significant fibrosis and chronic inflammation. Although it is difficult to identify acini, ductules are identifiable. In this patient, significant sclerosing orbital inflammation was also identified. The composition of the inflammatory infiltrate did not qualify this lesion as representing IgG4-related disease.

The authors of the consensus statement (10) suggest three categories of diagnostic terminology: 1) histologically highly suggestive of IgG4-related disease; 2) probable histologic features of IgG4-related disease; and 3) insufficient histologic evidence of IgG4-related disease. Probable histologic features of IgG4-related disease usually involves cases in which a single histopathologic feature and the required number of IgG4-positive plasma cells are present. Additional evidence to confirm the diagnosis of IgG4-related disease includes serum IgG4 greater than 135 mg/dL and other organ involvement, as demonstrated by radiologic or pathologic examination.

Treatment and Prognosis. Corticosteroids are the mainstay of treatment, usually with excellent response. Relapse is common. Serum IgG4 levels may decrease with treatment, but may not return to normal. Patients with higher serum IgG4 levels after treatment are at increased risk of relapse (14). Other immunosuppressants, such as azathioprine, methotrexate, and mycophenolate, have been used, typically when tapering or stopping corticosteroids. Good clinical responses have been documented with the anti-CD20 monoclonal antibody Rituximab (15). Low-dose radiotherapy can be used, and the response is variable.

Sarcoidosis Dacryoadenitis

Sarcoidosis may affect the palpebral and orbital lobes of the lacrimal gland and presents as a painless, palpable mass or upper eyelid swelling, sometimes associated with dry eye. Imaging

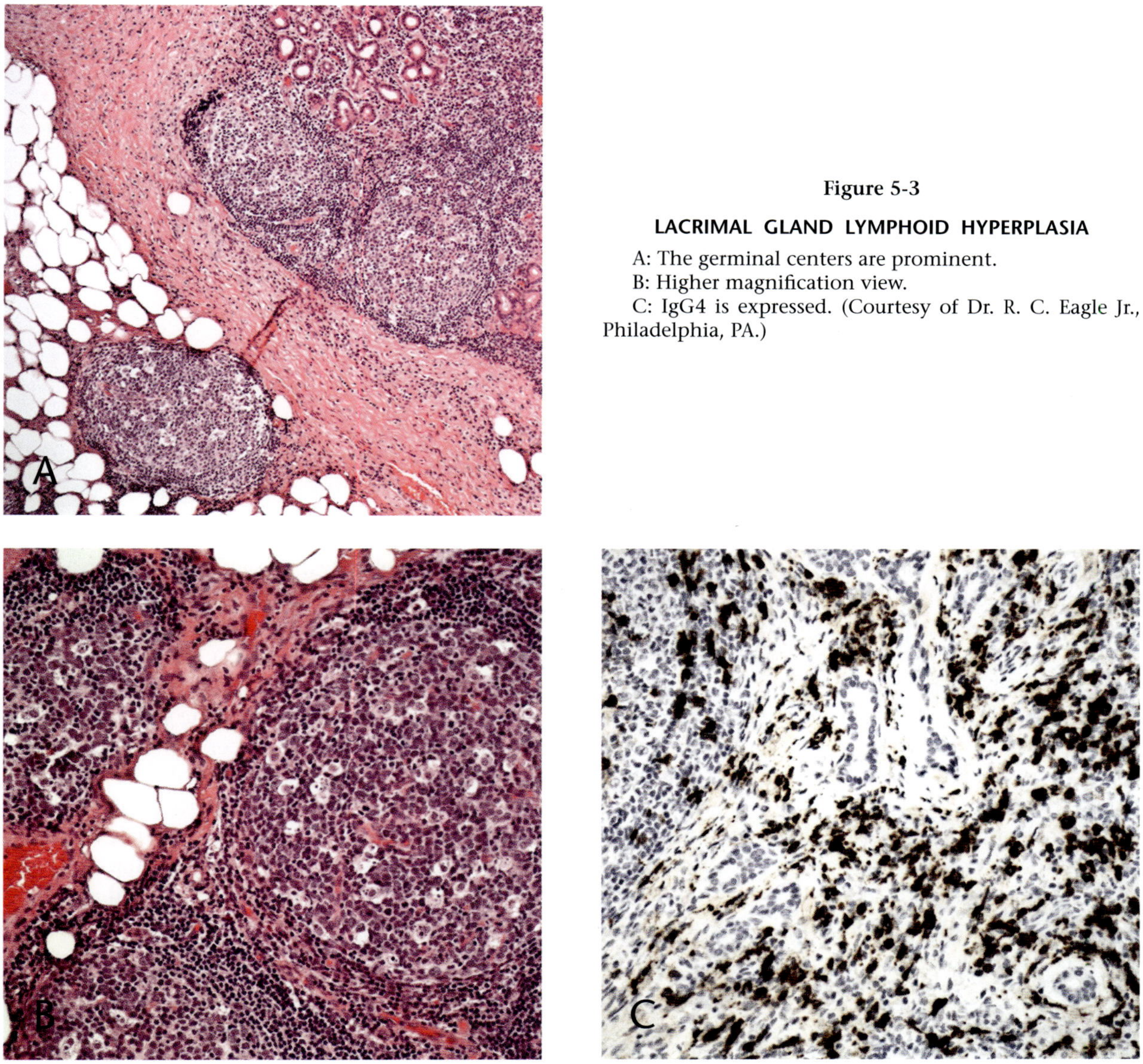

Figure 5-3

LACRIMAL GLAND LYMPHOID HYPERPLASIA

A: The germinal centers are prominent.

B: Higher magnification view.

C: IgG4 is expressed. (Courtesy of Dr. R. C. Eagle Jr., Philadelphia, PA.)

studies show diffuse enlargement of the lacrimal gland with homogeneous enhancement, occasionally involving adjacent orbital fat or lateral rectus muscle.

Histologically, discrete, well-formed, non-necrotizing granulomas, composed of epithelioid and multinucleated histiocytes and surrounded by little to no inflammation, are typical. There is little fibrosis (fig. 5-4). Sarcoid-like granulomas can also be seen in the lacrimal gland of patients with regional enteritis (Crohn disease) (16), or those treated with interferon-α and ribavirin for hepatitis C infection (17,18).

For disease localized to the lacrimal gland, observation, intralesional corticosteroid injection, surgical excision, or debulking are used. For systemic disease, treatment consists of systemic corticosteroid therapy with methotrexate as a steroid-sparing agent (7).

LACRIMAL GLAND DUCTAL CYST (DACRYOPS)

Lacrimal gland ductal cyst (*dacryops*) is an uncommon cystic lesion that can occur anywhere lacrimal gland tissue is present (19). Bullock et al. (20) classified lacrimal duct cysts based on the location of the lacrimal gland tissue: 1) *palpebral lobe cyst (simple dacryops)*, 2) *orbital lobe cyst*, 3) *cyst of accessory lacrimal glands of Krause and Wolfring*, and 4) *cysts of ectopic lacrimal glands*.

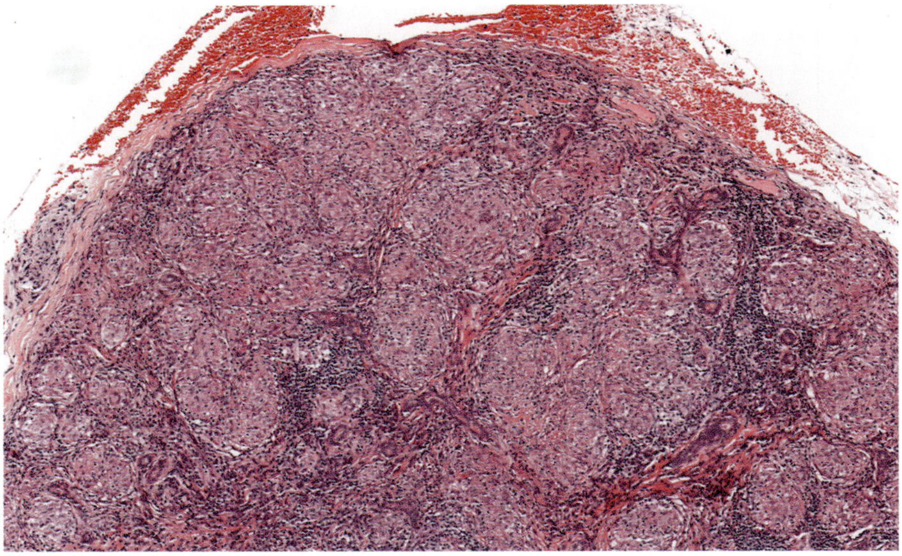

Figure 5-4

SARCOID AFFECTING THE LACRIMAL GLAND

Although it is difficult to identify acini, it is still possible to identify ductules (H&E stain).

Most instances of dacryops arise in the palpebral lobe of the lacrimal gland.

Symptoms include eyelid fullness, mild eye irritation (21), or mechanical ptosis (22). The lesion may be seen on eversion of the upper eyelid as a transparent or translucent, vascularized cyst that may appear multiloculated and have a bluish tint. These cysts are fluctuant and freely movable.

Microscopically, the cyst is lined by a bilayer of cuboidal to columnar cells with goblet cells (fig. 5-5). Focally, adlumenal cells have apical snouts (21). Treatment is complete removal.

BENIGN EPITHELIAL TUMORS

Benign epithelial tumors that have been described in the lacrimal gland include pleomorphic adenoma, myoepithelioma, and oncocytoma. Of these, pleomorphic adenoma is the most common, with only a handful of case reports of myoepithelioma (23–28) and oncocytoma (29–31) in the literature (32). Pleomorphic adenoma of the lacrimal gland is described briefly here; additional information, as well as myoepithelioma and oncocytoma, is found in the Fourth Series Fascicle, *Tumors of the Salivary Glands* (32a).

Pleomorphic Adenoma

Definition. *Pleomorphic adenoma* is a well-circumscribed, benign tumor with a variable cytomorphologic and architectural appearance. Epithelial and myoepithelial elements are essential for the diagnosis (33).

Clinical Features. Pleomorphic adenoma is the most common epithelial tumor of the lacrimal gland, accounting for more than 50 percent of primary epithelial tumors (34), but is uncommon overall, accounting for 0.6 percent of orbital tumors. The orbital lobe of the lacrimal gland is involved in 80 to 90 percent of cases and the remainder involve the palpebral lobe (34,35).

Well-circumscribed enlargement of the lacrimal gland is a characteristic feature. These lesions may indent the globe or remodel adjacent bone (34,35). In one study (36), the internal architecture in over 80 percent of cases was homogeneous on CT and heterogeneous on T2-weighted MRI. Pleomorphic adenoma is isointense on T1-weighted images and hyperintense and heterogeneous on T2-weighted images, with moderate contrast enhancement. Some tumors may contain cysts. Focal calcifications, when present, are more common in tumors with cysts (35).

Gross Findings. Pleomorphic adenoma is pseudoencapsulated, and has a smooth surface and bosselated appearance. The cut surface is light tan to gray, lobulated, and variegated, with mucinous and fibrous areas, degenerative and cystic changes, and focal hemorrhage.

Microscopic Findings. Pleomorphic adenoma is composed of epithelium-lined duct-like structures surrounded by myoepithelial cells that stream or feather away into the chondromyxoid stroma (37). Myoepithelial cells have fusiform, round, stellate, plasmacytoid, oncocytoid, and polygonal shapes. Occasionally, clear cells and

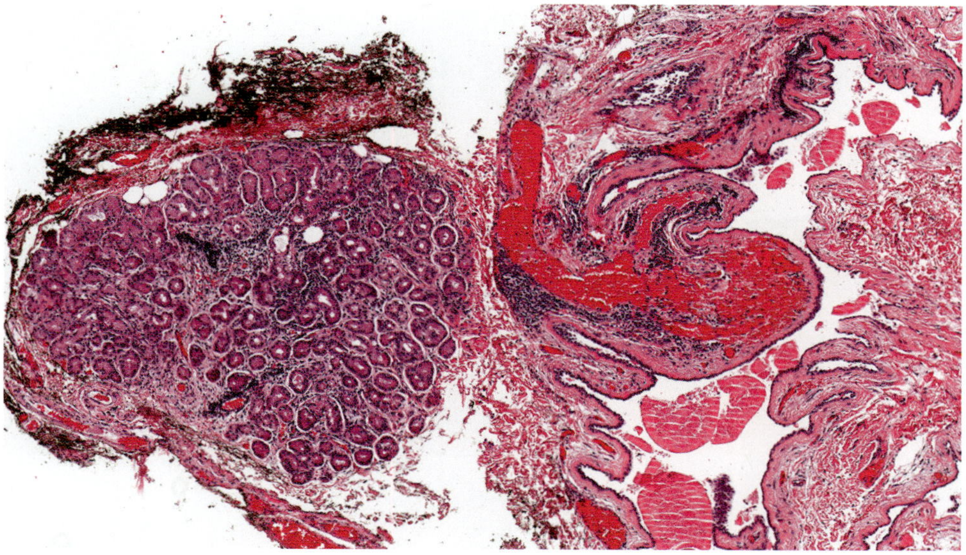

Figure 5-5

DACRYOPS

A large dilated duct is present to the right of the lacrimal gland tissue (H&E stain).

squamous elements are present. Myxoid, hyaline, cartilaginous, or osseous stromal elements may be identified. Mitotic figures are rarely detected.

Tumor pseudocapsules vary in thickness, up to a few hundred micrometers (32). The pseudocapsule should be carefully examined for tumor invasion and the status of the capsular integrity (fig. 5-6).

By immunohistochemistry, epithelial and myoepithelial cells express cytokeratin. Myoepithelial cells may also express muscle-specific actin and glial fibrillary acid protein (GFAP) (38).

Cytologic Findings. Variable amounts of epithelial, myoepithelial, and stromal cells, associated with extracellular, chondromyxoid stroma may be detected in fine-needle biopsies (39). The typically fibrillary stroma characteristically appears magenta on air-dried Romanowsky preparations (33).

Molecular Genetic Findings. Lacrimal gland pleomorphic adenomas are genetically stable tumors with no or comparatively few genomic imbalances. In one study, 10 of 17 primary lacrimal gland pleomorphic adenomas had normal genomic profiles by array comparative genomic hybridization (array CGH) (40). Losses involving 1p, 6q, 8q, and 13q and gains of 9p are recurrent alterations (40). Salivary gland pleomorphic adenomas have a highly specific pattern of chromosomal rearrangements with translocations involving breakpoints in 8q12 and 12q14-15: target genes of the 8q12 and 12q14-15 rearrangements are the transcription factor genes *PLAG1* and *HMGA2*. However, Von Holstein et al. (40) did not find any genomic imbalances with breakpoints in *PLAG1* or *HMGA2* in their series of 31 lacrimal gland pleomorphic adenomas. Nevertheless, a significant percentage of pleomorphic adenomas show nuclear staining for PLAG1, indicating overexpression (37,40). Overexpression of HMGA2 was found in 2 pleomorphic adenomas, consistent with prior studies demonstrating that fewer than 10 percent of pleomorphic adenomas have activation of *HMGA2* (40).

Differential Diagnosis. A variety of lesions may mimic the clinical-radiologic features of lacrimal gland pleomorphic adenoma, including lymphoma, chronic dacryoadenitis, adenoid cystic carcinoma, Sjogren syndrome, cavernous hemangioma, benign lymphoid hyperplasia, and granulomatous dacryoadenitis (34). These lesions are easily distinguished histopathologically.

Treatment and Prognosis. The recommended treatment is complete excision with a margin of normal tissue. Tumors of the palpebral lobe are excised using a transseptal approach, whereas tumors of the orbital lobe are removed via lateral orbitotomy (41).

Incomplete excision leads to recurrence in a significant number of cases. Longstanding, recurrent, and incompletely excised tumors may undergo malignant transformation. Incisional biopsy prior to excision is typically not recommended because of the belief that there is an increased risk of recurrence from disruption of the pseudocapsule and tumor spillage (42).

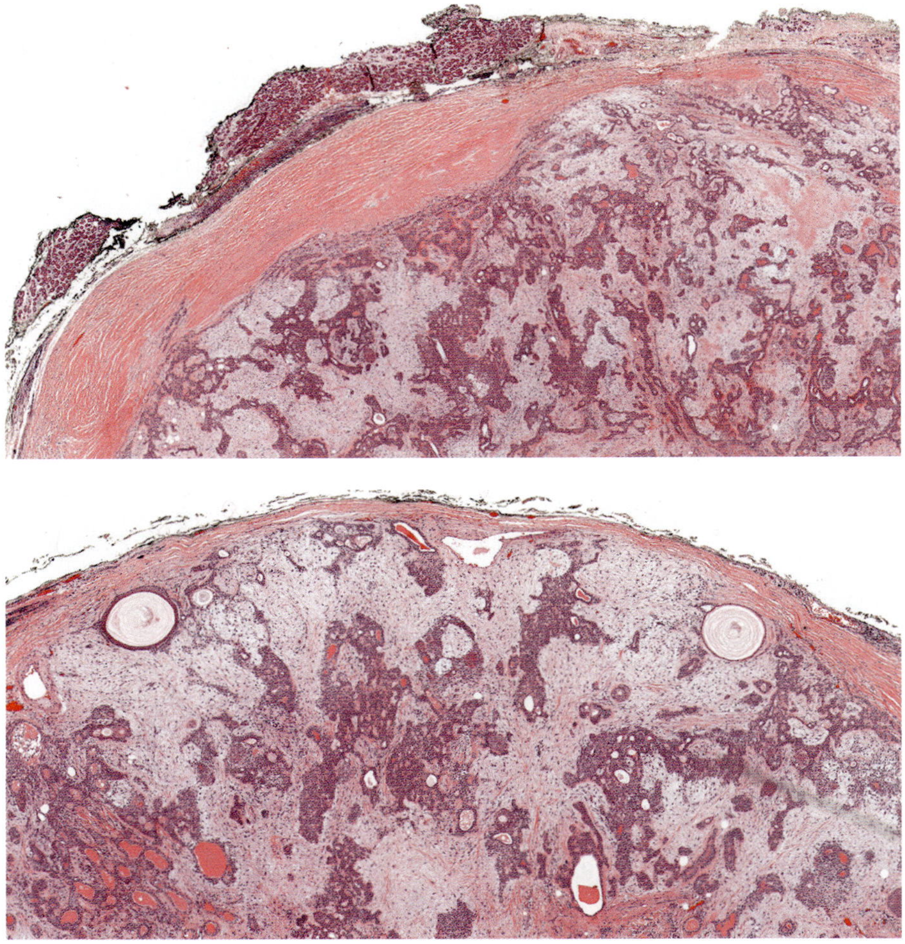

Figure 5-6

PLEOMORPHIC ADENOMA

Top: Compressed and atrophic lacrimal gland tissue is present outside of the capsule (upper left). The histology of pleomorphic adenoma in the lacrimal gland is identical to that of its namesake in salivary gland tissue (H&E stain).

Bottom: The capsule in this case is exquisitely thin.

However, there is no evidence in the literature to show that incisional biopsy or fine-needle aspiration biopsy, in themselves, lead to a greater risk of recurrence or malignant transformation of lacrimal gland pleomorphic adenoma. If a preoperative or intraoperative biopsy is performed, complete excision of the tumor and biopsy tract is recommended (42).

MALIGNANT EPITHELIAL TUMORS

Table 5-3 lists the *malignant epithelial tumors* that have been described in the lacrimal gland. Of these, *adenoid cystic carcinoma* is the most common, followed by *carcinoma ex pleomorphic adenoma, adenocarcinoma*, and *mucoepidermoid carcinoma*, with scattered case reports of the others listed. Adenoid cystic carcinoma of the lacrimal gland will be described briefly here; additional information about this and other malignant epithelial tumors listed in Table 5-3 can be found in *Tumors of the Salivary Glands* (32a).

Adenoid Cystic Carcinoma

Definition. *Adenoid cystic carcinoma* (ACC) is a malignant epithelial tumor composed of epithelial and myoepithelial cells that form various patterns, including tubular, cribriform, and solid forms (33)

Clinical Features. ACC is the most common epithelial malignancy of the lacrimal gland, but overall is rare. There is a 15-year survival rate of 20 percent. Although the mean age at presentation is 40 years, ACC may develop in the lacrimal gland at any age (43). Patients typically present with inferior and nasal displacement of the globe and proptosis (43). The lesion may be painful because of nerve invasion, and firm to hard by palpation (44). Patients may have hyperesthesia of the skin in the area of the first and second divisions of the trigeminal nerve.

CT and MRI are important diagnostic tools. Contrast-enhanced MRI is the preferred imaging modality to assess for perineural spread. CT is used to assess bone destruction and calcification (44). On CT images, ACC is solid, usually with irregular margins and infiltration into adjacent tissues. On MRI, the tumor is isointense to muscle on T1- and T2-weighted images, with moderate enhancement and perineural spread (44).

Gross Findings. ACC is a firm, gray-white tumor that may appear circumscribed.

Microscopic Findings. ACC is seen in tubular, cribriform, and solid patterns (fig. 5-7). The cribriform pattern is the most common: it consists of sharply punched-out spaces filled with basophilic matrix. With the tubular and solid patterns, basaloid myoepithelial cells are dominant, and there is scarcity or absence of matrix-filled spaces (38). The tumor is nonencapsulated and can infiltrate surrounding fat, periorbita, and muscle (32). Perineural invasion may be identified. Tumors with a solid component have higher mitotic activity than those with tubular and cribriform patterns (32). Immunohistochemical staining for CD117 is usually positive in inner epithelial cells. P63 and smooth muscle actin are positive in peripheral myoepithelial cells (33).

Cytologic Findings. Numerous fragments of epithelial cells and individual cells may be identified in samples obtained by fine-needle biopsies. The cells are compact and basaloid, with round to oval nuclei, high nuclear to cytoplasmic ratio, and minimal, easily disrupted cytoplasm. Cells may surround balls or globs of basement membrane-like material (39).

Table 5-3

CELLULAR CLASSIFICATION OF MALIGNANT EPITHELIAL LACRIMAL GLAND TUMORS[a]

Low-Grade Tumors
Carcinoma ex pleomorphic adenoma
Polymorphous low-grade adenocarcinoma
Mucoepidermoid carcinoma, grades 1 and 2
Epithelial-myoepithelial carcinoma
Acinic cell carcinoma
Basal cell adenocarcinoma
Mucinous adenocarcinoma
High-Grade Tumors
Carcinoma ex pleomorphic adenoma (malignant mixed tumor) that includes adenocarcinoma and adenoid cystic carcinoma arising in a pleomorphic adenoma
Adenoid cystic carcinoma
Adenocarcinoma, not otherwise specified
Mucoepidermoid carcinoma, grade 3
Ductal adenocarcinoma
Squamous cell carcinoma
Sebaceous adenocarcinoma
Myoepithelial carcinoma
Lymphoepithelial carcinoma
Carcinosarcoma
Other Rare and Unclassifiable Carcinomas

[a]Data from reference 46.

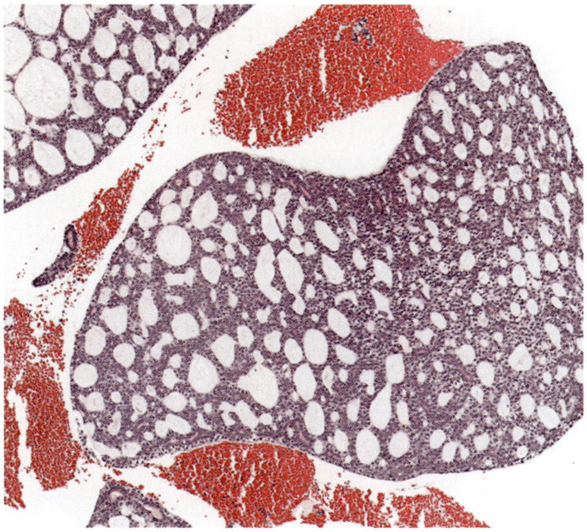

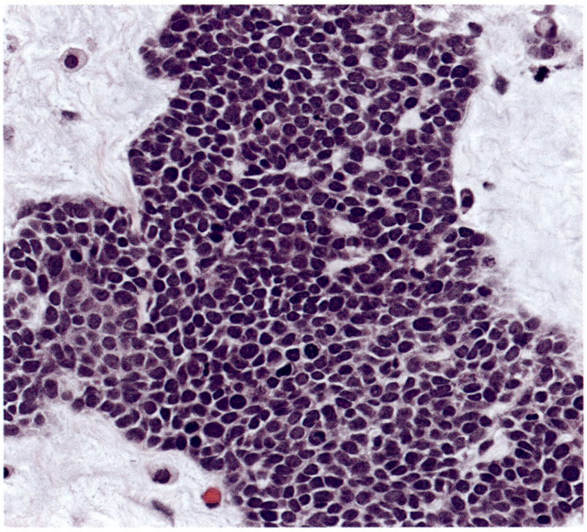

Figure 5-7

ADENOID CYSTIC CARCINOMA OF THE LACRIMAL GLAND

Left: Characteristic Swiss-cheese pattern.
Right: Solid pattern.

Molecular Genetic Findings. ACC is characterized by overexpression of the oncogene *MYB*, which encodes a transcription factor that regulates other target genes. In the majority of cases, *MYB* activation is caused by t(6;9)(q22-23;p23-24) chromosomal translocation, which results in *MYB-NFIB* fusion gene. Immunohistochemistry for MYB may help in diagnosis. Mendoza et al. (37) found that six of seven cases of ACC showed strong, positive staining for MYB. ACC is also characterized by recurrent copy number alterations involving segmental losses of 1p, 5q, 6q, 11q, 12q, 14q, and 17q, and segmental gains of 1q, 8q, 11q, 19q, and 22q (38).

Differential Diagnosis. ACC may mimic pleomorphic adenoma. The basaloid variant of ACC has lobules as seen in basal cell carcinoma of the skin, but lacks peripheral palisading (43). The basaloid variant should also be differentiated from basal cell adenocarcinoma of the lacrimal gland because it has a worse prognosis. Alcian blue and smooth muscle actin are positive in ACC in contrast to basal cell adenocarcinoma.

Treatment and Prognosis. Optimal treatment of ACC is controversial. The type of surgery, whether globe-sparing or exenteration, has not been demonstrated to correlate with prognosis. Use of adjuvant treatments is also debated (38). Patients whose tumors have a basaloid component have a poorer prognosis. In a study of 54 cases of ACC, Gamel and Font (45) showed that patients whose tumors showed a basaloid pattern had a 21 percent 5-year survival rate and 3-year median survival period, compared with 71 percent 5-year survival rate and 8-year median survival period for patients whose tumors did not have a basaloid pattern.

PATHOLOGIST REPORT FOR LACRIMAL GLAND TUMORS

The system presented in Appendix 5, from the Task force for Staging of Cancer of the Eye of the American Joint Committee on Cancer (AJCC), 8th edition, is used to stage carcinomas of the lacrimal gland (46). The 8th edition staging system is different from prior editions because it is designed to determine which factor (tumor size, periosteal invasion, or bone invasion) is most important in determining outcome. The staging system applies only to primary carcinomas of the lacrimal gland, and does not include mesenchymal, hematologic, or melanocytic tumors of the lacrimal gland, or carcinomas of the lacrimal sac or nasolacrimal duct system.

Clinical staging should include a complete history (with emphasis on duration of symptoms, pain, or dysesthesia) and physical examination (including globe displacement or distortion, palpation, and sensory and motor examination) (46,47). Orbital imaging, including CT and MRI, should be performed to evaluate the size, shape, extent of the lesion, and invasion of adjacent structures, including periosteum, bone, skull base, and periorbital areas. Evaluation of cervical lymph nodes, lungs, and bone should be performed to stage the extent of disease.

The tumor should be sampled thoroughly to evaluate histologic type and grade, size, presence of preexisting pleomorphic adenoma, presence of perineural invasion, presence of high-grade transformation, and surgical margins. For carcinomas arising in pleomorphic adenoma, extent of invasion beyond the capsule should be described. For ACCs, the approximate percentage of the basaloid pattern present on histopathologic examination should be reported. Any and all bone removed should be fully examined pathologically for evidence of periosteum only or periosteum and bone involvement. The pathologist should perform Ki-67 immunohistochemistry on all lacrimal gland carcinomas and report the percentage of positive tumor cells (Ki-67 growth fraction) (46).

REFERENCES

1. Obata H. Anatomy and histopathology of the human lacrimal gland. Cornea 2006;25(Suppl 1):S82-9.
2. Dutton JJ. The lacrimal system. In: Dutton JJ. Atlas of clinical and surgical orbital anatomy, 2nd ed. Chapel Hill: Elsevier Saunders; 2011:164-74.

2a. Lacrimal Gland Tumor Study Group. An epidemiological survey of lacrimal fossa lesions in Japan: number of patients and their sex ratio by pathological diagnosis. Jpn J Ophthalmol 2005;49:343-8.

3. Bonavolontà G, Strianese D, Grassi P, et al. An analysis of 2,480 space-occupying lesions of the orbit from 1976 to 2011. Ophthal Plast Reconstr Surg 2013;29:79-86.

3a. Teo L, Seah LL, Choo CT, Chee SP, Chee E, Looi A. A survey of the histopathology of lacrimal gland lesions in a tertiary referral centre. Orbit 2013;32:1-7.

4. Shields JA, Shields CL, Scartozzi R. Survey of 1264 patients with orbital tumors and simulating lesions: The 2002 Montgomery Lecture, part 1. Ophthalmology 2004;111:997-1008.
5. von Holstein SL, Therkildsen MH, Prause JU, Stenman G, Siersma VD, Heegaard S. Lacrimal gland lesions in Denmark between 1974 and 2007. Acta Ophthalmol 2013;91:349-54.

5a. Medeiros L J, O'Malley DP, Caraway PN, Vega F, Elenitoba-Johnson KS, Lim MS. Tumors of the lymph nodes and spleen. AFIP Atlas of Tumor Pathology, 4th Series, Fascicle 25. Washington DC: American Registry of Pathology; 2017.

6. Andrew NH, McNab AA, Selva D. Review of 268 lacrimal gland biopsies in an Australian cohort. Clin Exp Ophthalmol 2015;43:5-11.
7. Mombaerts I. The many facets of dacryoadenitis. Curr Opin Ophthalmol 2015;26:399-407.
8. Mombaerts I, Cameron JD, Chanlalit W, Garrity JA. Surgical debulking for idiopathic dacryoadenitis: a diagnosis and a cure. Ophthalmology 2014;121:603-9.
9. Kubota T, Moritani S. Orbital IgG4-related disease: clinical features and diagnosis. ISRN Rheumatol 2012;2012:412896.
10. Deshpande V, Zen Y, Chan JK, et al. Consensus statement on the pathology of IgG4-related disease. Mod Pathol 2012;25:1181-92.
11. McNab AA, McKelvie P. IgG4-related ophthalmic disease. Part I: background and pathology. Ophthalmic Plast Reconstr Surg 2015;31:83.
12. Plaza JA, Garrity JA, Dogan A, Ananthamurthy A, Witzig TE, Salomão DR. Orbital inflammation with IgG4-positive plasma cells: manifestation of IgG4 systemic disease. Arch Ophthalmol 2011;129:421-8.
13. Song YS, Choung H, Park S, Kim J, Khwarg SI, Jeon YK. Ocular adnexal IgG4-related disease: CT and MRI findings. Br J Ophthalmol 2013;97:412-8.
14. McNab AA, McKelvie P. IgG4-related ophthalmic disease. Part II: clinical aspects. Ophthalmic Plast Reconstr Surg 2015;31:167.
15. Wallace ZS, Khosroshahi A, Jakobiec FA, et al. IgG4-related systemic disease as a cause of "idiopathic" orbital inflammation, including orbital myositis, and trigeminal nerve involvement. Surv Ophthalmol 2012;57:26-33.
16. Jakobiec FA, Rashid A, Lane KA, Kazim M. Granulomatous dacryoadenitis in regional enteritis (Crohn disease). Am J Ophthalmol 2014;158:838-44.
17. Hwang CJ, Gausas RE. Sarcoid-like granulomatous orbital inflammation induced by interferon-alpha treatment. Ophthal Plast Reconstr Surg 2008;24:311-3.
18. Morley AM, O'Sullivan E, Thaung C, Malhotra R. Sarcoid-related dacryoadenitis following treatment with interferon alpha and ribavarin for hepatitis C. Orbit 2011;30:27-9.
19. Lam K, Brownstein S, Jordan DR, Jastrzebski A. Dacryops: a series of 5 cases and a proposed pathogenesis. JAMA Ophthalmol 2013;131:929-32.
20. Bullock JD, Fleishman JA, Rosset JS. Lacrimal ductal cysts. Ophthalmology 1986;93:1355-60.
21. Jakobiec FA, Zakka FR, Perry LP. The cytologic composition of dacryops: an immunohistochemical investigation of 15 lesions compared to the normal lacrimal gland. Am J Ophthalmol 2013;155:380-96.
22. Galindo-Ferreiro A, Alkatan HM, Muinos-Diaz Y, Akaishi PM, Galvez-Ruiz A, Cruz AV. Accessory lacrimal gland duct cyst: 23 years of experience in the Saudi population. Ann Saudi Med 2015;35:394-9.
23. Bolzoni A, Pianta L, Farina D, Nicolai P. Benign myoepithelioma of the lacrimal gland: report of a case. Eur Arch Otorhinolaryngol 2005;262:186-8.
24. Pasquale S, Strianese D, Mansueto G, Tranfa F. Epithelioid myoepithelioma of lacrimal gland. Virchows Arch 2005;446:97.
25. Font RL, Garner A. Myoepithelioma of the lacrimal gland: report of a case with spindle cell morphology. Br J Ophthalmol 1992;76:634-6.
26. Grossniklaus HE, Wojno TH, Wilson MW, Someren AO. Myoepithelioma of the lacrimal gland. Arch Ophthalmol 1997;115:1588-90.

27. Heathcote JG, Hurwitz JJ, Dardick I. A spindle-cell myoepithelioma of the lacrimal gland. Arch Ophthalmol 1990;108:1135-9.
28. Okudela K, Ito T, Iida MI, Kameda Y, Furuno K, Kitamura H. Myoepithelioma of the lacrimal gland: report of a case with potentially malignant transformation. Pathol Int 2000;50:238-43.
29. Calle CA, Castillo IG, Eagle RC, Daza MT. Oncocytoma of the lacrimal gland: case report and review of the literature. Orbit 2006;25:243-7.
30. Economou MA, Seregard S, Sahlin S. Oncocytoma of the lacrimal gland. Acta Ophthalmol Scand 2007;85:576-7.
31. Kim J, Park H, Paik J, Kim D, Yang S. Oncocytoma of the lacrimal gland: an Asian case. Jpn J Ophthalmol 2010;54:239-41.
32. Weis E, Rootman J, Joly TJ, et al. Epithelial lacrimal gland tumors: pathologic classification and current understanding. Arch Ophthalmol 2009;127:1016-28.
32a. Ellis GL, Auclair PL. Tumors of the salivary glands. AFIP Atlas of Tumor Pathology, 4th Series, Fascicle 9. Washington DC: American Registry of Pathology; 2008.
33. El-Naggar AK, Chank JK, Grandis JR, Takata T, Slootweg PJ. WHO classification of head and neck tumours, 4th ed. Lyon: IARC Press; 2017.
34. Prabhakaran VC, Cannon PS, McNab A, et al. Lesions mimicking lacrimal gland pleomorphic adenoma. Br J Ophthalmol 2010;94:1509-12.
35. Watanabe A, Andrew N, Ueda K, et al. Clinico-radiological features of primary lacrimal gland pleomorphic adenoma: an analysis of 37 cases. Jpn J Ophthalmol 2016;60:286-93.
36. Gündüz K, Shields C, Günalp I, Shields J. Magnetic resonance imaging of unilateral lacrimal gland lesions. Graefes Arch Clin Exp Ophthalmol 2003;241:907-13.
37. Mendoza PR, Jakobiec FA, Krane JF. Immunohistochemical features of lacrimal gland epithelial tumors. Am J Ophthalmol 2013;156:1147.
38. von Holstein SL, Rasmussen PK, Heegaard S. Tumors of the lacrimal gland. Semin Diagn Pathol 2016;33:156-63.
39. Sturgis CD, Silverman JF, Kennerdell JS, Raab SS. Fine-needle aspiration for the diagnosis of primary epithelial tumors of the lacrimal gland and ocular adnexa. Diagn Cytopathol 2001;24:86-9.
40. von Holstein SL, Fehr A, Persson M, et al. Lacrimal gland pleomorphic adenoma and carcinoma ex pleomorphic adenoma: genomic profiles, gene fusions, and clinical characteristics. Ophthalmology 2014;121:1125-33.
41. Rose GE, Wright JE. Pleomorphic adenoma of the lacrimal gland. Br J Ophthalmol 1992;76:395-400.
42. Lai T, Prabhakaran VC, Malhotra R, Selva D. Pleomorphic adenoma of the lacrimal gland: is there a role for biopsy? Eye 2008;23:2-6.
43. Shields JA, Shields CL, Epstein JA, Scartozzi R, Eagle RC. Review: primary epithelial malignancies of the lacrimal gland: the 2003 Ramon L. Font lecture. Ophthal Plast Reconstr Surg 2004;20:10-21.
44. Vaidhyanath R, Kirke R, Brown L, Sampath R. Lacrimal fossa lesions: pictorial review of CT and MRI features. Orbit 2008;27:410-8.
45. Gamel JW, Font RL. Adenoid cystic carcinoma of the lacrimal gland: the clinical significance of a basaloid histologic pattern. Histopathology 1982;13:219-25.
46. Amin MB, Edge SB, Greene FL, et al., eds. AJCC cancer staging manual, 8th ed. Switzerland: Springer; 2017.
47. Rootman J, White VA. Changes in the 7th edition of the AJCC TNM classification and recommendations for pathologic analysis of lacrimal gland tumors. Arch Pathol Lab Med 2009;133:1268-71.

6 TUMORS OF THE LACRIMAL DRAINAGE SYSTEM

ANATOMY AND HISTOLOGY

The upper and lower puncta, canaliculi, common canaliculus, lacrimal sac, and nasolacrimal duct are components of the lacrimal drainage system (fig. 6-1). Tears move toward the medial canthus and enter the puncta. The puncta are 0.5 to 1.5 mm in diameter. From the puncta, tears flow into the canaliculi, which have an initial vertically oriented segment for 2 mm, then bend at a right angle medially as an 8-mm horizontal segment (1).

The upper and lower canaliculi merge into the common canaliculus, which enters the lacrimal sac through the valve of Rosenmuller. In some individuals, the upper and lower canaliculi may not join, and enter the lacrimal sac separately. The superiorly located lacrimal sac fundus, which extends 3 to 5 mm superior to the medial canthal tendon, and the body inferiorly, which extends approximately 10 mm to the nasolacrimal duct, are the two components of the sac. The nasolacrimal duct is approximately 17 mm in length, with a 12-mm intraosseous and 5-mm membranous component. The duct opens onto the lateral wall of the inferior nasal meatus (1).

The canaliculi are lined by nonkeratinizing stratified squamous epithelium with collagenous and elastic tissue within the wall (fig. 6-2). The lacrimal sac is lined by pseudostratified columnar epithelium with scattered goblet cells (figs. 6-3). The epithelium rests on a basement membrane. The underlying connective tissue stroma contains scattered lymphocytes. Similar to the lacrimal sac, the lacrimal duct is lined by pseudostratified columnar epithelium with cilia. An associated venous plexus connects to the vascular tissue of the inferior turbinate (1).

CLASSIFICATION

Lacrimal sac tumors are uncommon, with approximately 775 cases reported in the literature (2). Primary tumors are broadly classified into epithelial and nonepithelial types. Approximately 70 percent of lacrimal sac tumors are epithelial (Table 6-1). Benign epithelial tumors include papillomas, oncocytomas, and benign mixed tumors (Table 6-2). Malignant epithelial tumors can arise within papillomas or de novo, and most commonly are squamous cell carcinoma or transitional cell carcinoma. Other malignant epithelial tumors are mucoepidermoid carcinoma, adenoid cystic carcinoma, adenocarcinoma, oncocytic adenocarcinoma, and poorly differentiated carcinoma (Table 6-2).

Nonepithelial tumors are further subclassified into mesenchymal, hematolymphoid, melanocytic, and other. Mesenchymal tumors comprise

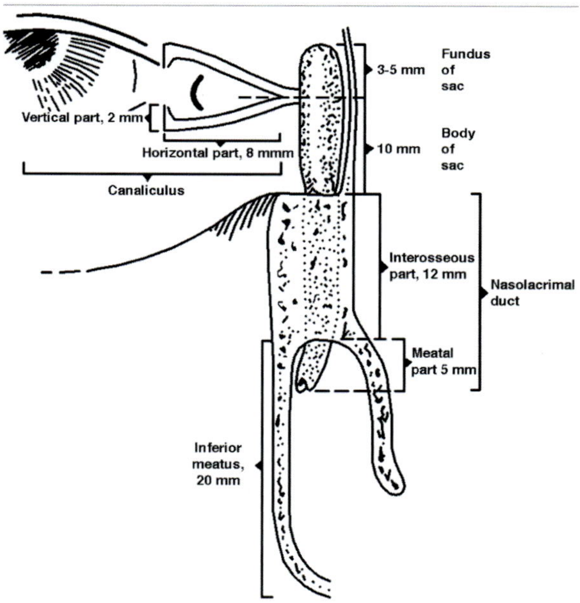

Figure 6-1

LACRIMAL DRAINAGE SYSTEM

The measurements of the adult lacrimal drainage system. (Fig. 60 from Jones LT. The lacrimal apparatus. In: Jones LT, Reeh MJ, Wirtschapter JD, eds. Ophthalmic anatomy: a manual with some clinical application. Rochester: American Academy of Ophthalmology and Otolaryngology; 1970:225.)

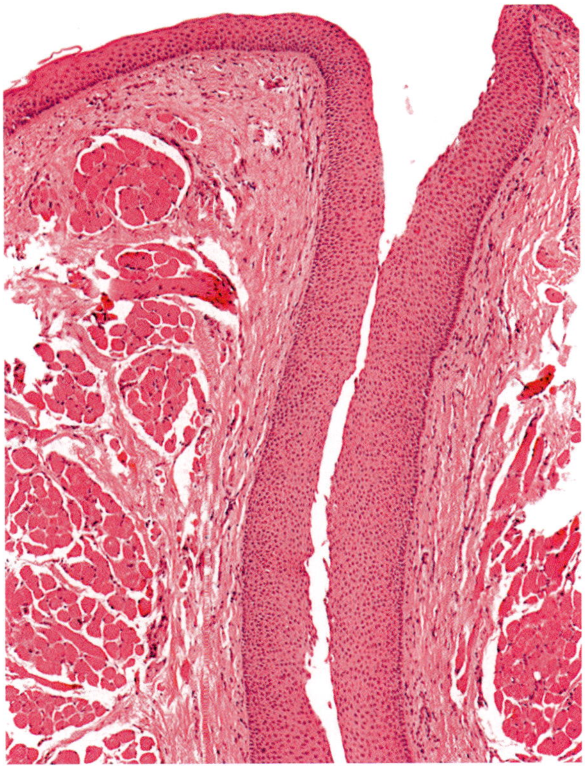

Figure 6-2

NORMAL CANALICULUS

The canaliculus is lined by stratified nonkeratinized epithelium.

12 to 14 percent of lacrimal sac tumors. Fibrous histiocytoma is the most common. The literature also cites reports of other tumors, such as hemangiopericytoma, hemangioma, lipoma, osteoma (3), rhabdomyosarcoma (4,5), and Kaposi sarcoma (6). Hematolymphoid lesions account for approximately 11 percent of lacrimal drainage system tumors, and include diffuse large B-cell lymphoma (DLBCL) (7,8), extranodal marginal zone lymphoma of mucosa-associated lymphoid tissue (MALT lymphoma) (7–9), transitional MALT lymphoma with features between MALT lymphoma and DLBCL (8), unclassified B-cell lymphoma (8), plasmacytoma (9), granulocytic sarcoma (10), and chronic lymphocytic leukemia (9,11,12). Melanoma is rare, comprising approximately 4 percent of lacrimal sac tumors (2,13,14). Secondary tumors of the lacrimal drainage system arise from invasion from adjacent structures, such as the nose, paranasal sinuses, orbit, conjunctiva, and skin (2,13).

Most patients with lacrimal sac neoplasms present with signs and symptoms of acquired nasolacrimal duct obstruction, such as epiphora and palpable medial canthal mass, possibly leading to a misdiagnosis of chronic dacryocystitis. Features that should raise suspicion for

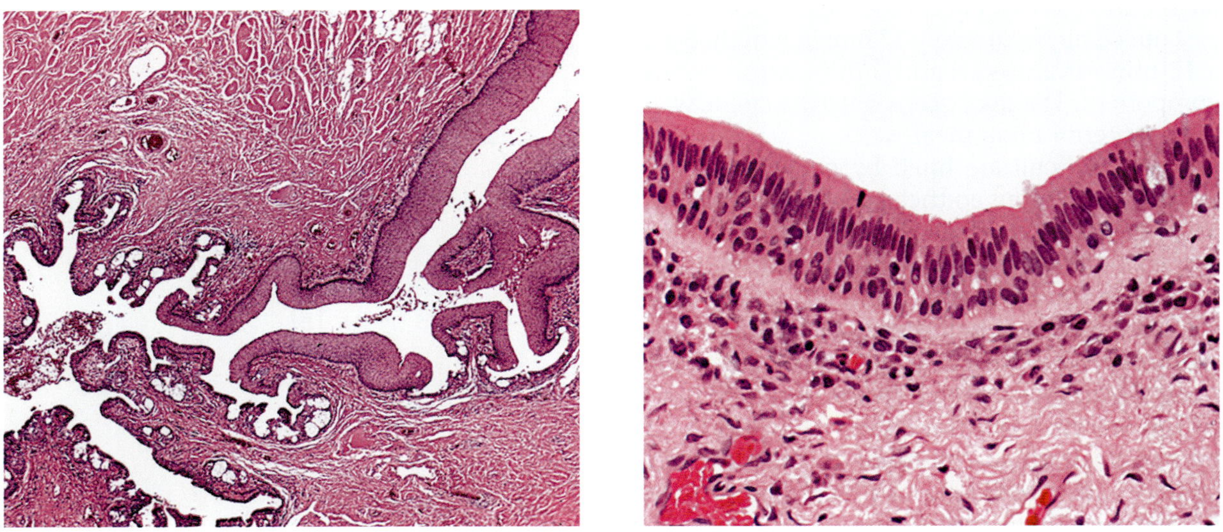

Figure 6-3

LACRIMAL SAC AND CANALICULUS

Left: The canaliculus on the right is lined by nonkeratinized squamous epithelium and the sac on the left, stratified columnar epithelium containing goblet cells.

Right: Higher magnification of the sac. Stratified columnar epithelium contains goblet cells.

Table 6-1

FREQUENCY OF PRIMARY LACRIMAL SAC TUMORS (REVIEW OF MAJOR CASE SERIES)

	Epithelial		Mesenchymal		Hemato-	Melano-	
	Benign	Malignant	Benign	Malignant	lymphoid	cytic	Other
Ashton (68)	11	27	30	0	5	0	0
Radnot (69)	22	64	5	26	15	6	0
Ryan (39)	11	16	0	0	0	0	0
Schenck (70)	4	14	1	2	0	3	1
Ni (41)	6	71	1	2	1	0	1
Stefanyszyn (15)	38	44	15	0	11	8	1
Anderson (9)	6	4	0	0	13	0	0
Parmar (7)	1	8	1	0	5	0	0
Sjo (8,71)	5	6	0	0	15	0	0
Bi (59)	0	83	3	0	5	3	2
TOTAL (%)	104 (17)	337 (54)	56 (9)	30 (5)	70 (11)	20 (3)	5 (1)

malignancy include a palpable, firm, incompressible, immobile mass above the medial canthal tendon; chronic dacryocystitis which irrigates or partially irrigates; telangiectasias or ulceration of the skin overlying the mass; serosanguineous discharge or bloody reflux on irrigation; and proptosis or displacement of the eye (2,13).

Table 6-2

CLASSIFICATION OF LACRIMAL SAC TUMORS

Epithelial Tumors
- Benign epithelial tumors
  - Papilloma (exophytic, inverted)
  - Oncocytoma
  - Benign mixed tumor
- Malignant epithelial tumors
  - Squamous cell carcinoma
  - Transitional cell carcinoma
  - Mucoepidermoid carcinoma
  - Adenoid cystic carcinoma
  - Adenocarcinoma
  - Oncocytic adenocarcinoma
  - Poorly differentiated carcinoma

Nonepithelial Tumors
- Mesenchymal
- Hematolymphoid
- Melanoma
- Other

BENIGN EPITHELIAL TUMORS

Exophytic Papilloma

Definition. *Exophytic papilloma* is a benign epithelial neoplasm with finger-like fronds of proliferating epithelium growing in an outward direction toward the lumen (2,13). The epithelium may be squamous, transitional, or mixed cell. *Squamous papillomas* are composed of acanthotic stratified squamous epithelium with occasional foci of dyskeratosis (13). *Transitional cell papillomas* are composed of stratified columnar epithelium with scattered goblet cells, and sometimes cilia. *Mixed cell papillomas* are composed of a mixture of both squamous and transitional epithelium.

Clinical Features. In an extensive series of lacrimal sac tumors reported by Stefanyszyn et al. (15), the mean age of patients with benign papillomas was 44 years, 47 years for patients with squamous papillomas, and 40 years for patients with transitional cell papillomas. Patients with exophytic papillomas typically present with epiphora, recurrent dacryocystitis, and a palpable mass in the medial canthal region. Some patients have bloody discharge spontaneously or after irrigation.

Lesions can recur (16) or undergo malignant transformation, although malignant transformation occurs less frequently in exophytic papillomas than inverted papillomas (15). Recurrent tumors may cause ulceration of the overlying skin (fig. 6-4). Imaging studies, such as dacryocystography and computerized tomography (CT) or magnetic resonance imaging (MRI) of

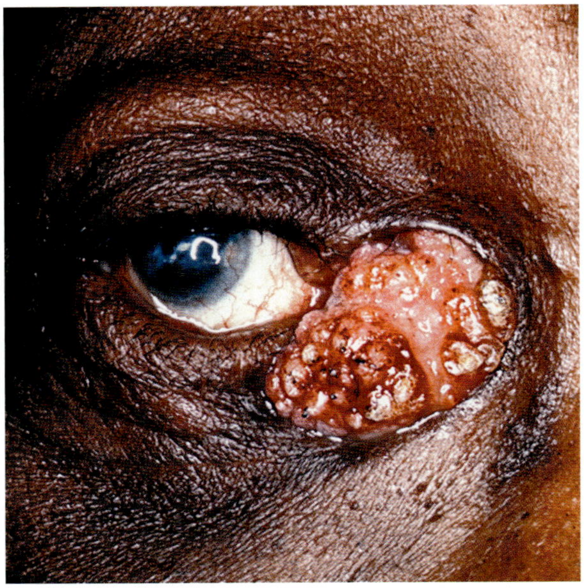

Figure 6-4

RECURRENT PAPILLOMA

Recurrent papilloma of the lacrimal sac ulcerates the overlying skin.

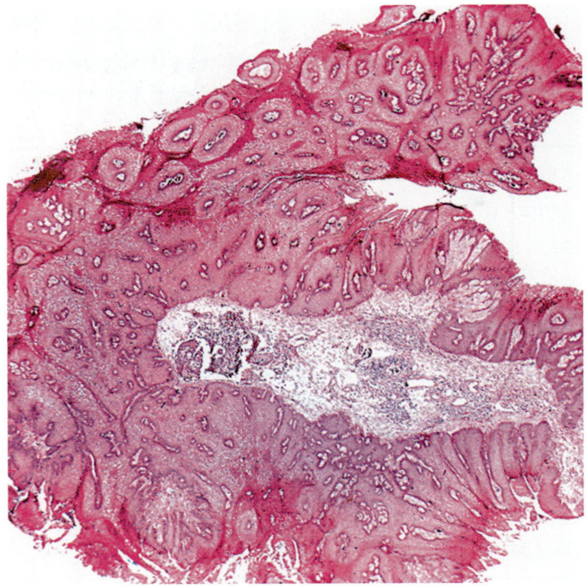

Figure 6-5

SQUAMOUS CELL PAPILLOMA

Squamous cell papilloma with an exophytic pattern.

the orbit and sinuses may be helpful in the workup of these patients. The dacryocystogram characteristically shows a filling defect with an uneven or mottled density of contrast and delayed drainage of contrast media (7,17). CT or MRI are used to characterize the extent of the tumor and relationship with surrounding bony and soft tissue structures.

Microscopic Findings. Exophytic papillomas are fungiform masses, with finger-like projections of acanthotic epithelium covering fibrovascular stalks. The epithelium may be squamous (fig. 6-5), transitional (fig. 6-6), or a mixture of both. The epithelium contains goblet cells or columnar ciliated respiratory epithelial cells. Cytologic atypia/dysplasia or foci of carcinoma may be present (fig. 6-7).

Immunohistochemical, in situ hybridization, and polymerase chain reaction studies show evidence of human papillomavirus (HPV) types 6 and 11. This suggests that HPV may be an etiologic factor (18,19).

Differential Diagnosis. The principal entities in the differential diagnosis are primary lacrimal gland carcinomas, most commonly squamous cell carcinoma and transitional cell carcinoma. It is essential to perform a thorough microscopic examination of the lesion to look for areas of dysplasia and foci of carcinoma.

Treatment and Prognosis. Surgeons usually manage exophytic papillomas by dacryocystorhinostomy and complete excision of the mass (13,15). The recurrence rate after surgery is 10 to 40 percent and recurrence is more frequent with inverted papillomas. Malignant transformation is seen with tumor recurrence, and may be more likely in cases demonstrating dysplasia (17).

Inverted Papilloma

Definition. *Inverted papilloma* is a benign polypoid neoplasm arising from sinonasal respiratory epithelium or Schneiderian epithelium; it has an endophytic growth pattern (20). Similar to exophytic papillomas, the epithelium may be stratified squamous, transitional, or a mixture of both.

Clinical Features. As with most patients with lacrimal sac tumors, patients with inverted papilloma typically present with epiphora and a medial canthal mass. Inverted papilloma is locally aggressive, with a tendency to recur, and a capacity for bony destruction and intraorbital and intracranial extension (21–23).

CT is the preferred imaging modality for inverted papillomas to look for bony erosions and

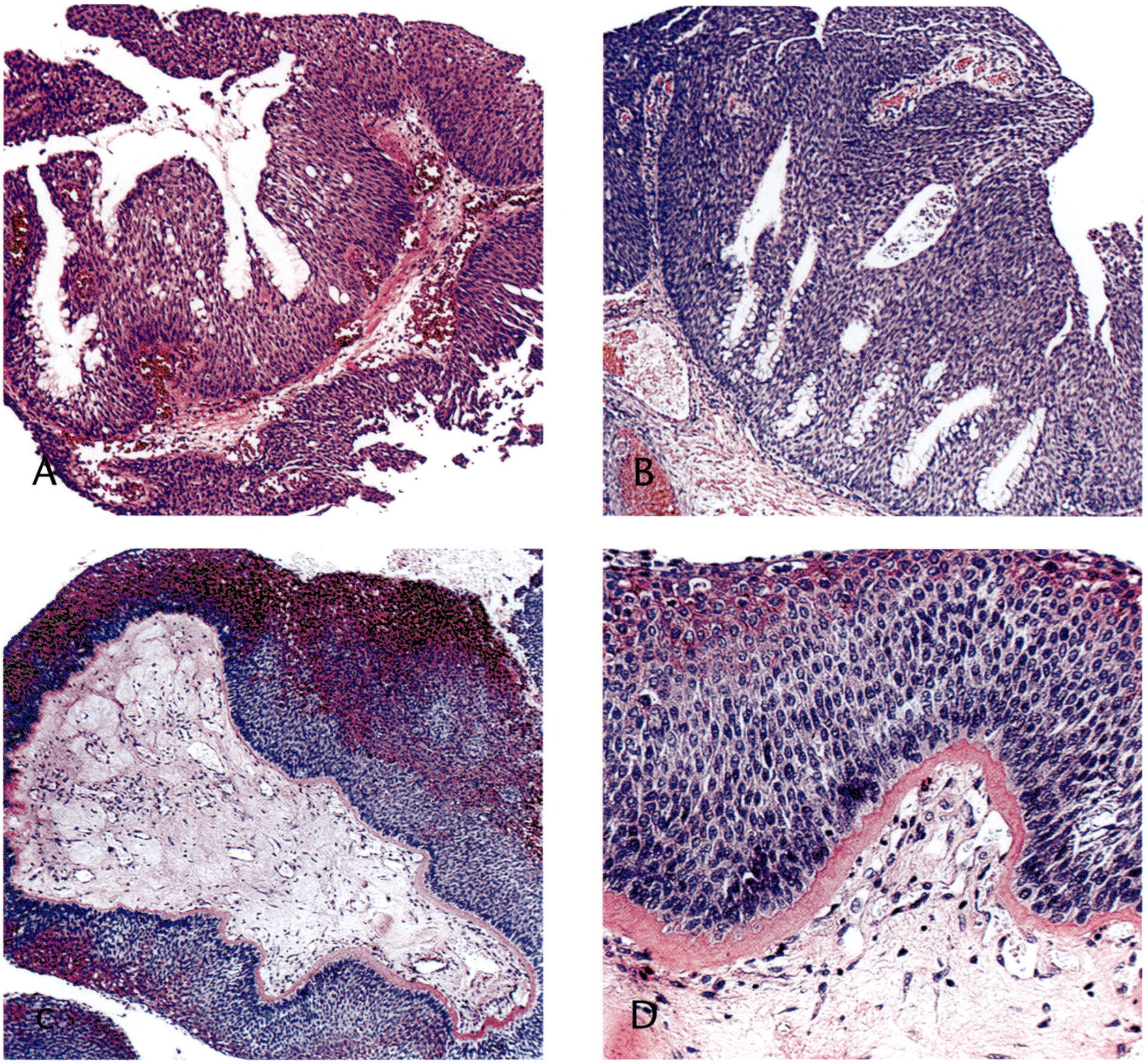

Figure 6-6

TRANSITIONAL CELL PAPILLOMA

A: Transitional cell papilloma of the lacrimal sac with focal goblet cell hyperplasia.

B: In another example, the cystic spaces accompany goblet cell hyperplasia.

C: Transitional cell papilloma with an exophytic pattern. The periodic acid–Schiff (PAS)-positive material in the epithelium is glycogen.

D: The basement membrane is thickened (PAS positive).

confirm the extent of the tumor. The tumor is usually well-defined and homogeneous without contrast enhancement (24).

Microscopic Findings. Hyperplastic, stratified epithelium is surrounded by basement membrane with an endophytic growth pattern and no evidence of invasion. The epithelium may be composed of squamous epithelial cells (fig. 6-8), transitional epithelium (fig. 6-9), or a mixture. An inflammatory cell infiltrate may be present within the epithelium (24). There may be areas of dysplasia or foci of carcinoma (25).

Immunohistochemistry for p16 may be positive in cases caused by HPV infection. HPV, identified by immunohistochemistry or in situ hybridization, is found in some cases (18,24).

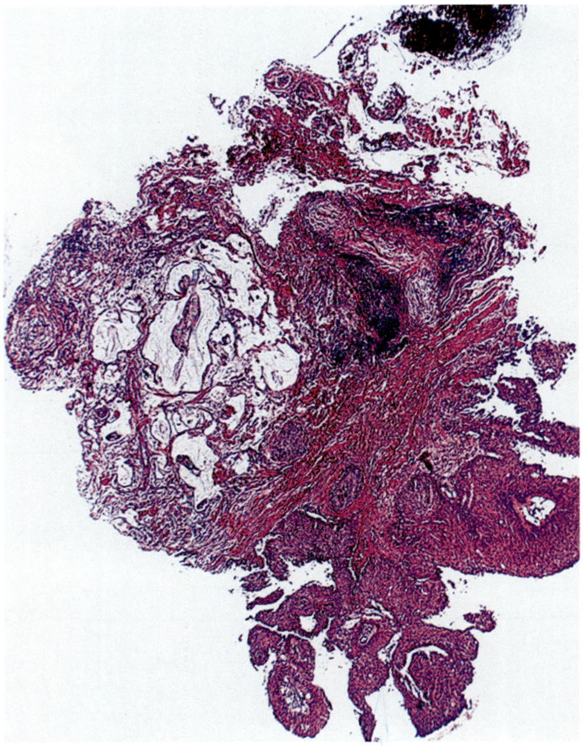

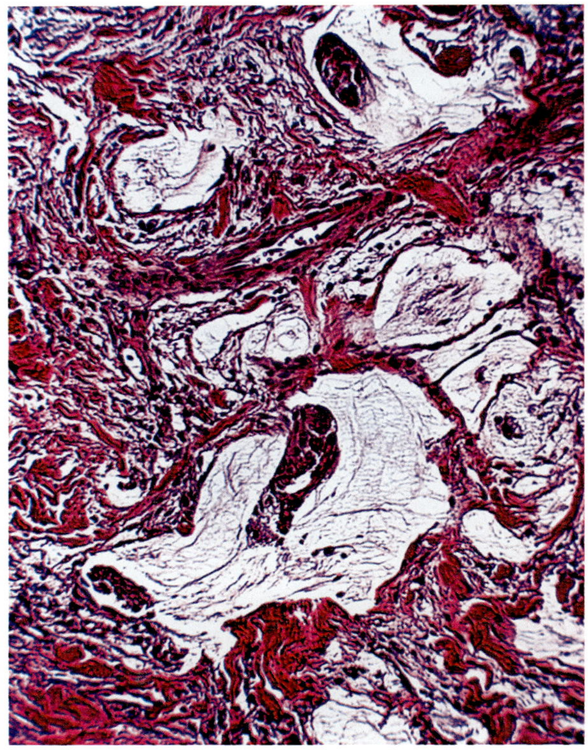

Figure 6-7

EXOPHYTIC PAPILLOMA WITH ADENOCARCINOMA

Left: An infiltrating mucus-secreting adenocarcinoma is at the base of an exophytic papilloma. Right: High-power view shows mucin-filled pools containing epithelial islands.

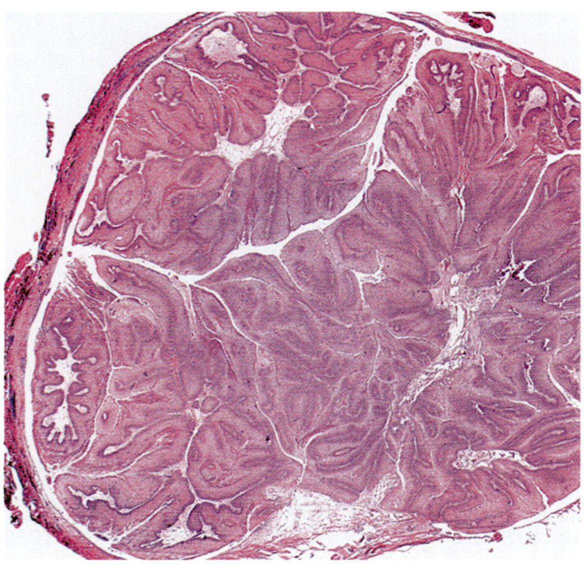

Figure 6-8

INVERTED SQUAMOUS CELL PAPILLOMA

Squamous cell papilloma with an inverted pattern fills the lacrimal sac.

Differential Diagnosis. The principal differential diagnoses include squamous cell carcinoma and transitional cell carcinoma of the lacrimal drainage system and secondary involvement by inverted papilloma or carcinoma from the nasal cavity (24).

Treatment and Prognosis. Inverted papillomas are locally aggressive and have a propensity to recur (reported recurrence rates, 27 to 71 percent) (23,26). There may be bony destruction and intraorbital and intracranial extension (22,23). Foci of carcinoma may be present or develop, especially in recurrences or cases with orbital invasion, and are usually squamous cell carcinoma or transitional cell carcinoma (21,22,27).

Treatment usually consists of dacryocystectomy with complete excision of the tumor with clear margins (20,28). Radiotherapy may be used to treat patients with positive surgical margins (29). Mitomycin C (MMC) has potential as an adjunct therapy due to its effectiveness in the treatment of superficial bladder cancer and

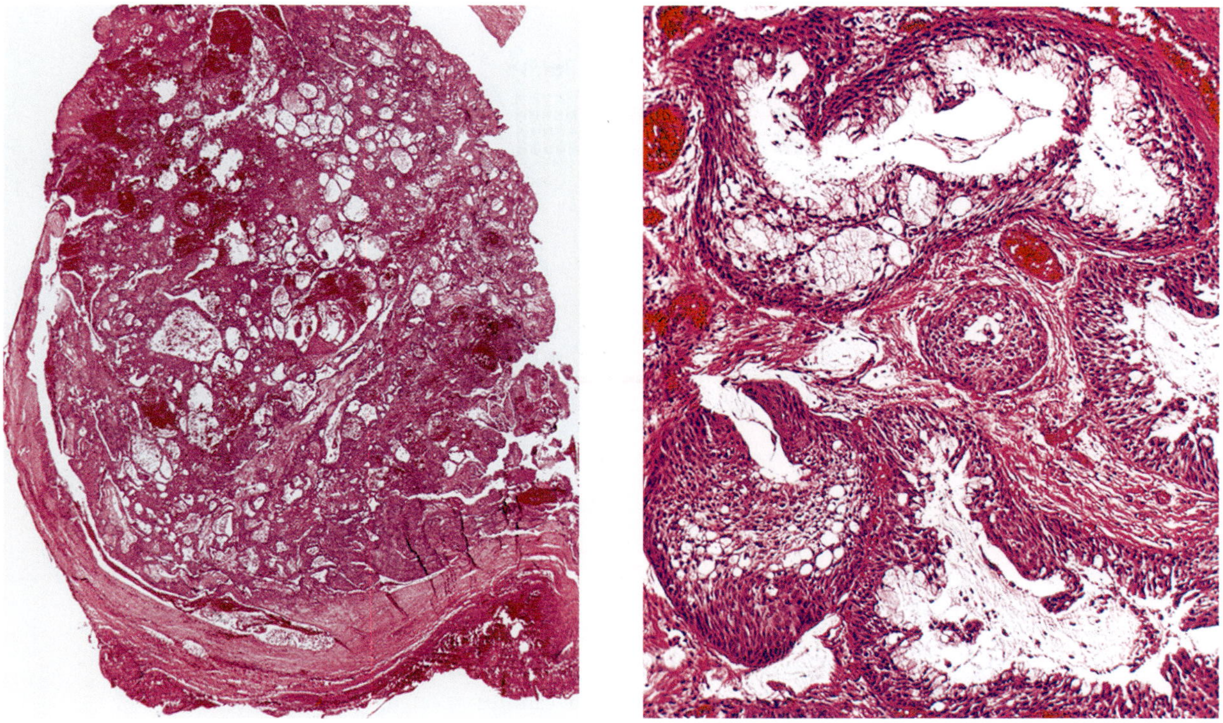

Figure 6-9

INVERTED TRANSITIONAL CELL PAPILLOMA

Left: The papilloma has an inverted pattern and goblet cell hyperplasia with mucus-filled cystic spaces. Right: Inverted pattern and areas of goblet cell hyperplasia.

inverted papillomas of the bladder. Woodcock et al. (28) used MMC after a recurrence of inverted papilloma, and the patient had no recurrence after 18 months of follow-up.

MALIGNANT EPITHELIAL TUMORS

Squamous Cell Carcinoma

Definition. *Squamous cell carcinoma* is a malignant epithelial neoplasm that shows features of squamous differentiation, such as the presence of keratinization and intercellular bridges.

Clinical Features. Squamous cell carcinoma is the most common malignant epithelial tumor of the lacrimal sac. Patients are generally older than those with benign tumors, and typically present with tearing and a palpable mass in the medial canthal area (fig. 6-10). In late-stage tumors, patients may have proptosis or nonaxial globe displacement, skin ulceration over the tumor, invasion of adjacent structures, and involvement of regional lymph nodes (7). Tumors may develop from squamous papilloma or de novo.

Thin-section (1.25 mm) CT with contrast is recommended as the first-line imaging study to evaluate the extent of the tumor, especially for expansion or erosion of bony lacrimal canal. MRI with contrast may be a helpful follow-up study if postobstructive sinus secretions are present because it is better than CT for distinguishing tumor from secretions in the sinonasal cavity. Most tumors demonstrate moderate contrast enhancement on CT and MRI and are isointense on T1-weighted and T2-weighted imaging (30).

Gross and Microscopic Findings. Grossly, squamous cell carcinomas are firm, white masses (fig. 6-11) that may have a papillary architecture. Microscopically, they range from well-differentiated with keratin pearls to poorly differentiated. The histologic features include atypical squamous cells with abundant eosinophilic cytoplasm, nuclear pleomorphism, and prominent nucleoli (fig. 6-12) (2,13). Tumors may have a papillary growth pattern (fig. 6-13) or a nonpapillary, infiltrative pattern.

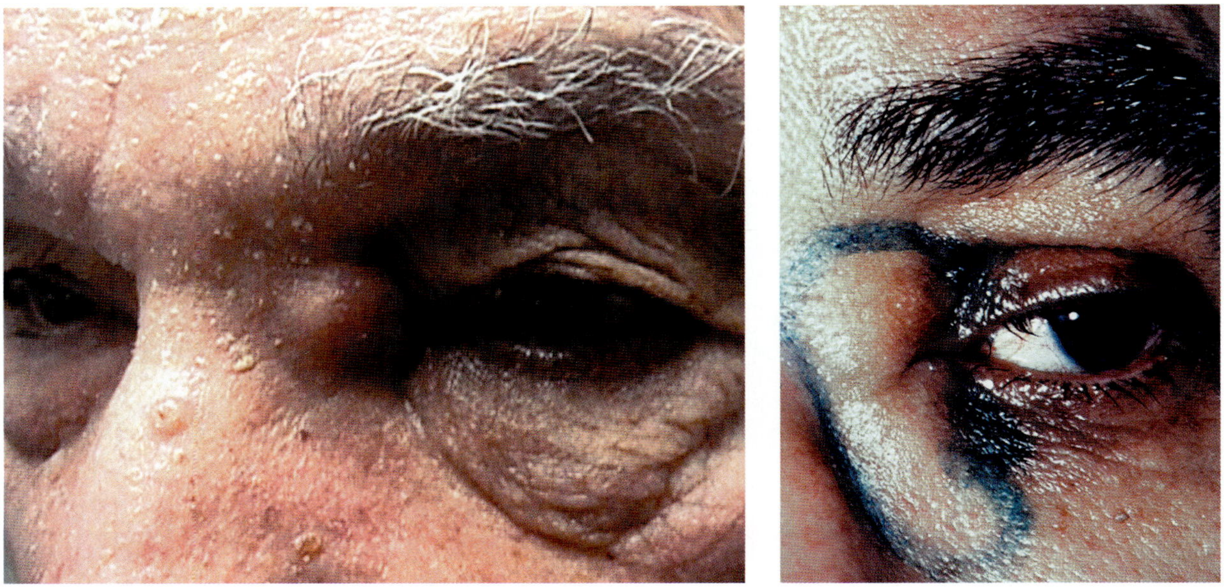

Figure 6-10

SQUAMOUS CELL CARCINOMA

Left: The tumor is located in the left medial canthus.
Right: The tumor is localized mainly above the left medial canthal ligament.

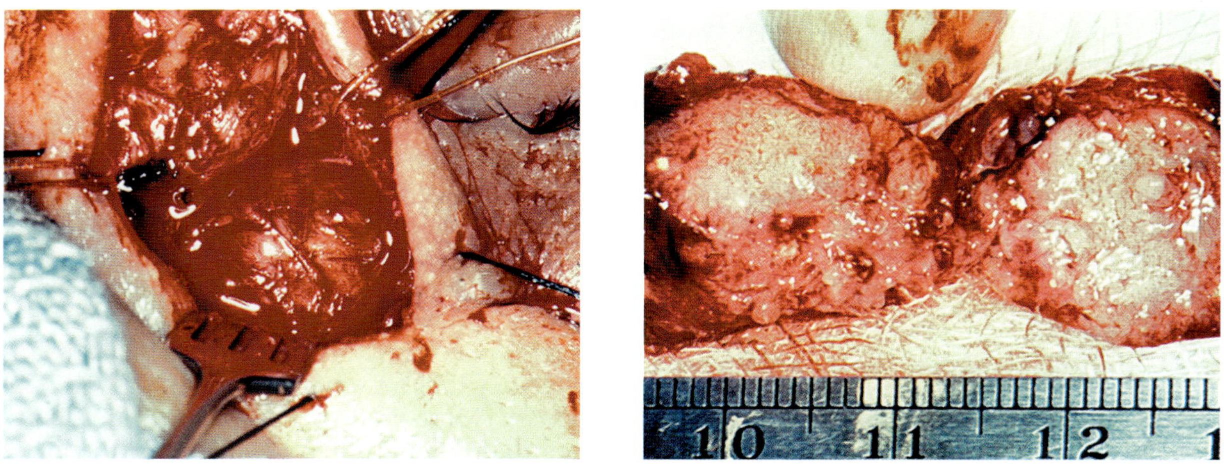

Figure 6-11

SQUAMOUS CELL CARCINOMA

Left: Surgically exposed mass at the medial canthus.
Right: Gross appearance of the sectioned mass.

High-risk human papillomavirus (HR-HPV) has been detected in a majority of squamous cell carcinomas of the lacrimal sac, suggesting the possibility of an etiologic role for HR-HPV in the development of these tumors. In one study, 8 of 9 (89 percent) lacrimal sac squamous cell carcinomas were positive for HR-HPV by p16 immunohistochemistry and HPV DNA in situ hybridization or p16 immunohistochemistry and polymerase chain reaction (PCR) (31).

Differential Diagnosis. The principal differential diagnoses include benign epithelial tumors, such as exophytic squamous papilloma and inverted papilloma, and other malignant

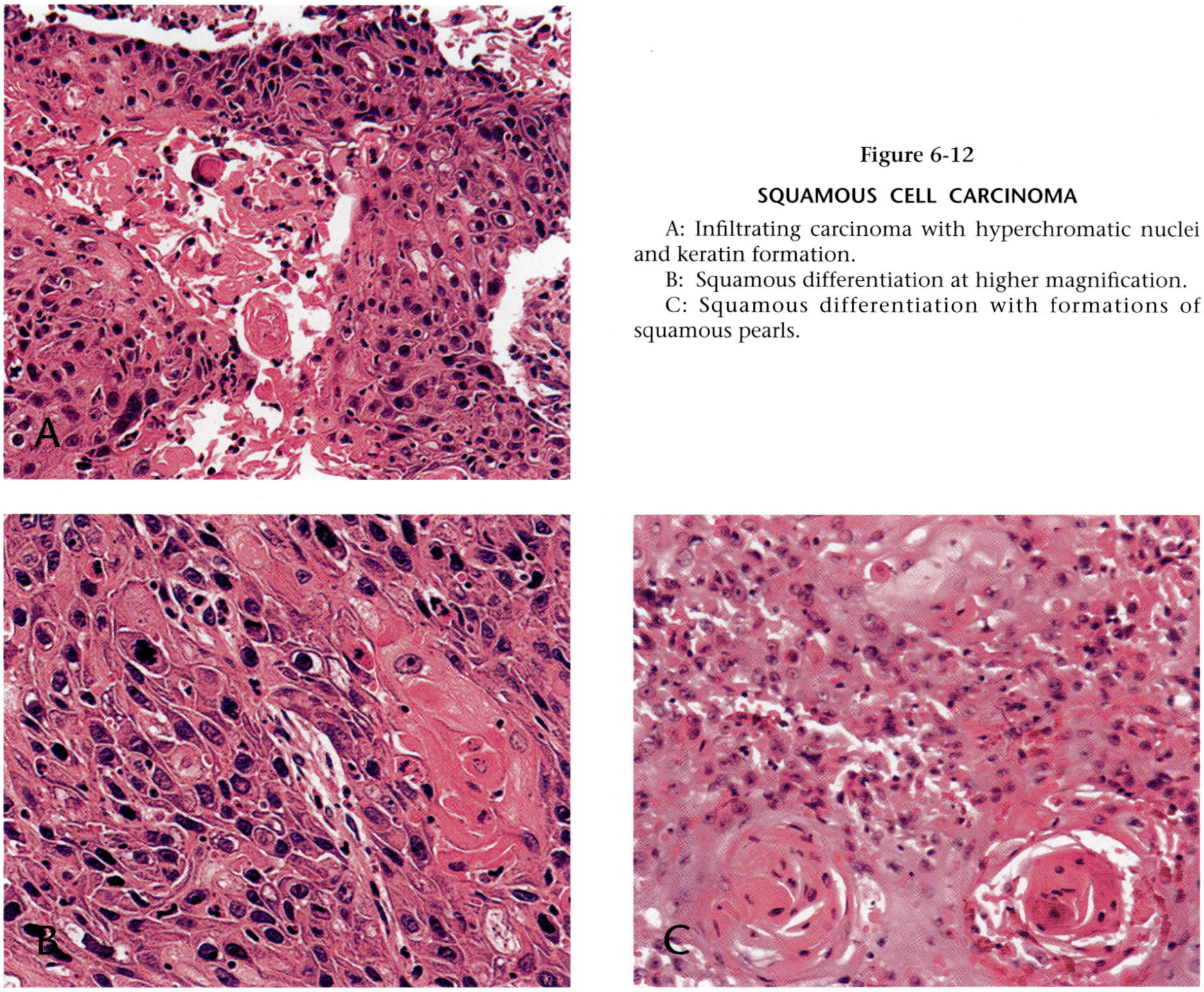

Figure 6-12

SQUAMOUS CELL CARCINOMA

A: Infiltrating carcinoma with hyperchromatic nuclei and keratin formation.

B: Squamous differentiation at higher magnification.

C: Squamous differentiation with formations of squamous pearls.

epithelial tumors, particularly mucoepidermoid carcinoma. Invasive squamous cell carcinoma may be challenging to differentiate from the endophytic nests of inverted papilloma. Invasive nests of squamous cell carcinoma are more irregular than inverted papilloma, basement membrane does not surround the invasive component, and a desmoplastic stromal reaction is identified. Squamous cell carcinoma is distinguished from mucoepidermoid carcinoma by the absence of intermediate and mucous cells.

Treatment and Prognosis. Treatment is the complete surgical removal of the tumor. Frozen sections help confirm clear resection margins (34,35). Cases with positive surgical margins or that involve adjacent structures may benefit from adjuvant radiotherapy (33). An en bloc excision technique includes extirpation of the canaliculi, lacrimal sac, nasolacrimal duct, and adjacent bone (32,35). For more advanced cases, orbital exenteration, medial maxillectomy, paranasal sinus resection, and lymph node dissection may be needed (7,15,36).

A significant concern for squamous cell carcinomas is local recurrence (32,33). Lymph node metastasis or distant metastasis to the lungs can occur but is rare (34).

Transitional Cell Carcinoma

Definition. *Transitional cell carcinoma* is a malignant epithelial neoplasm that arises from transitional epithelium in the lacrimal sac.

Clinical Features. Transitional cell carcinoma is the second most common malignant epithelial neoplasm of the lacrimal sac. The mean age at diagnosis is younger than that for squamous cell carcinoma (47 years versus 63 years) (26,37). There is no sex predilection (38).

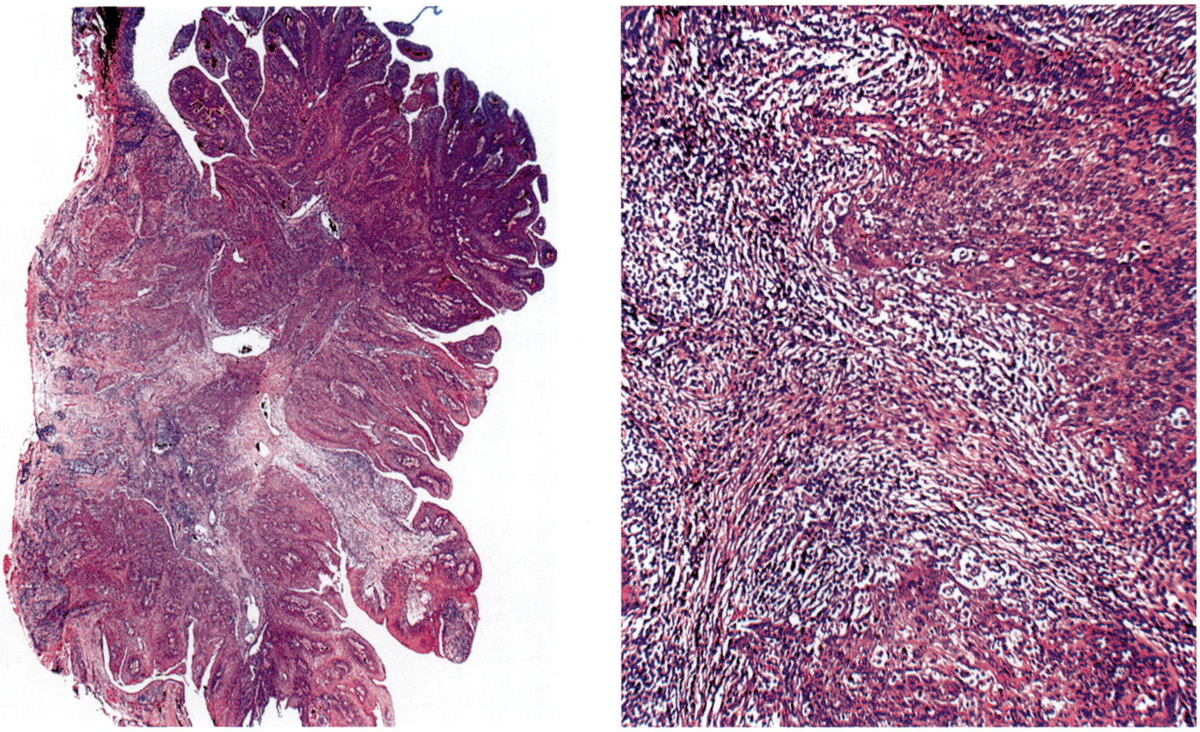

Figure 6-13

PAPILLARY SQUAMOUS CELL CARCINOMA

Left: Invasion at the base of a papillary carcinoma.

Right: Higher magnification of left figure reveals glands of infiltrating squamous cell carcinoma. The stroma shows scarring and chronic inflammatory cell infiltration.

Like other lacrimal sac neoplasms, patients with transitional cell carcinoma typically present with signs of chronic dacryocystitis, epiphora, and a medial canthal mass, which may extend superiorly to the medial canthal tendon (fig. 6-14).

Thin-section (1.25 mm) CT with contrast is the first-line imaging study to evaluate the extent of the tumor, especially for expansion or erosion of bony lacrimal canal (6-15). The tumor may appear heterogeneous on CT (37).

Gross and Microscopic Findings. Transitional cell carcinomas resemble those of the urinary bladder and may be friable and discohesive. Microscopically, the tumor is composed of irregular, stratified epithelium with pleomorphic cells that have prominent nucleoli, conspicuous mitotic activity, and invasive growth. Foci of squamous metaplasia may be present (13,26).

Low- and high-risk HPV types have been demonstrated in transitional cell carcinomas, suggesting a role for HPV in the development of this tumor. In a series of lacrimal sac papillomas and carcinomas, Sjo et al. (19) had two cases of transitional cell carcinoma with concomitant infection by HPV 6 or 11 and 16, one case positive for HPV 11 and one case positive for HPV16.

Differential Diagnosis. The main differential diagnoses are inverted transitional cell papilloma, transitional cell carcinoma in situ, invasive squamous cell carcinoma, and secondary involvement of transitional cell carcinoma from the sinonasal tract.

Treatment and Prognosis. Primary treatment is wide local excision with clear margins for disease control, with or without adjuvant radiotherapy. An en bloc excision technique includes extirpation of the canaliculi, lacrimal sac, nasolacrimal duct, and adjacent bone (32,35). With this approach, Alabiad et al. (32) found that all four of their patients with transitional cell carcinoma had no recurrence at 16 to 62 months follow-up. For more advanced cases, orbital exenteration, medial maxillectomy, paranasal sinus resection, and lymph node

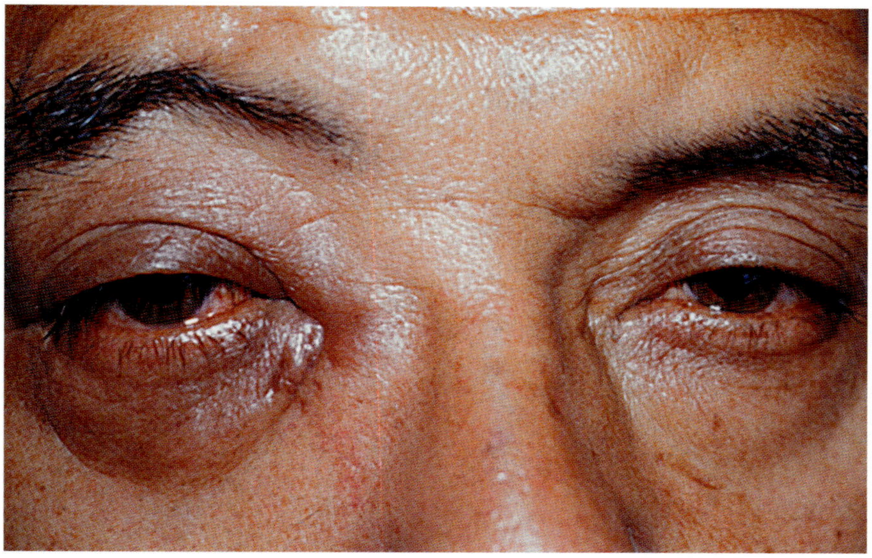

Figure 6-14

TRANSITIONAL CELL CARCINOMA

The mass is located at the right medial canthus.

dissection are treatment options (7,15,36). Postoperative radiotherapy is for patients with clear but narrow margins or positive surgical margins (36).

Transitional cell carcinoma is reported to have a high recurrence rate of 50 percent (39) and average mortality of 44 percent, ranging from 0 to 100 percent (40,41). Lymph node metastasis and hematogenous metastases (fig. 6-16) to lungs or esophagus occur in advanced cases (2).

Mucoepidermoid Carcinoma

Definition. *Mucoepidermoid carcinoma* is a malignant tumor that contains both neoplastic mucin-producing and epidermoid cells (42,43).

Clinical Features. Mucoepidermoid carcinoma is a rare malignancy of the lacrimal sac, with 15 reported cases in the literature (42–53). The mean age at diagnosis is 46 years (range, 19 to 78), and there is no sex predilection. Patients usually present with epiphora and a palpable, firm, solid medial canthal mass that is painless (42,43,48–51,53). Occasionally, skin telangiectasias or ulceration may appear over the tumor site. Two cases arose in the same area as a previously resected squamous (50) or transitional squamous (49) papilloma of the lacrimal sac. CT and MRI with contrast usually show a well-enhanced soft tissue mass in the lacrimal fossa (52,54).

Gross, Microscopic, and Cytologic Findings. Grossly, mucoepidermoid carcinomas are firm white masses that may be adherent to adjacent tissues, such as bone, orbital soft tissue, or overlying skin. Microscopically, the tumor is made up of islands and nests of intermediate, epidermoid, and mucous cells within a fibrosclerotic stroma (43). Intermediate cells are oval with scant cytoplasm; epidermoid cells are polygonal with abundant eosinophilic cytoplasm and may form keratin pearls; and mucous cells have small eccentric nuclei and intracytoplasmic mucin. In some cases, cystic spaces are filled with mucinous material or pools of mucin dissecting into the stroma (55).

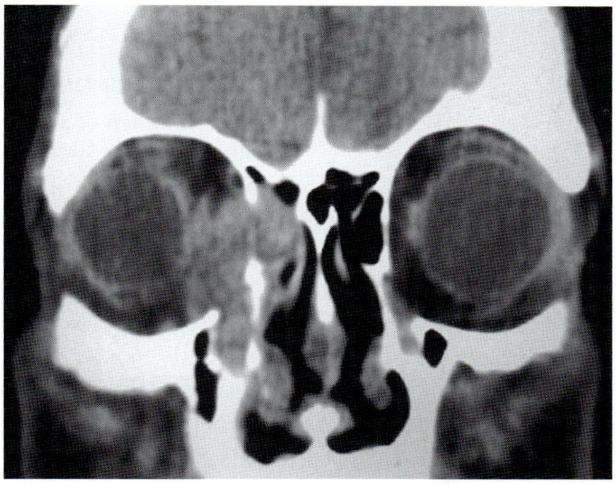

Figure 6-15

TRANSITIONAL CELL CARCINOMA

Computerized tomography (CT) shows a right medial canthal mass with destruction of the medial orbital bony wall.

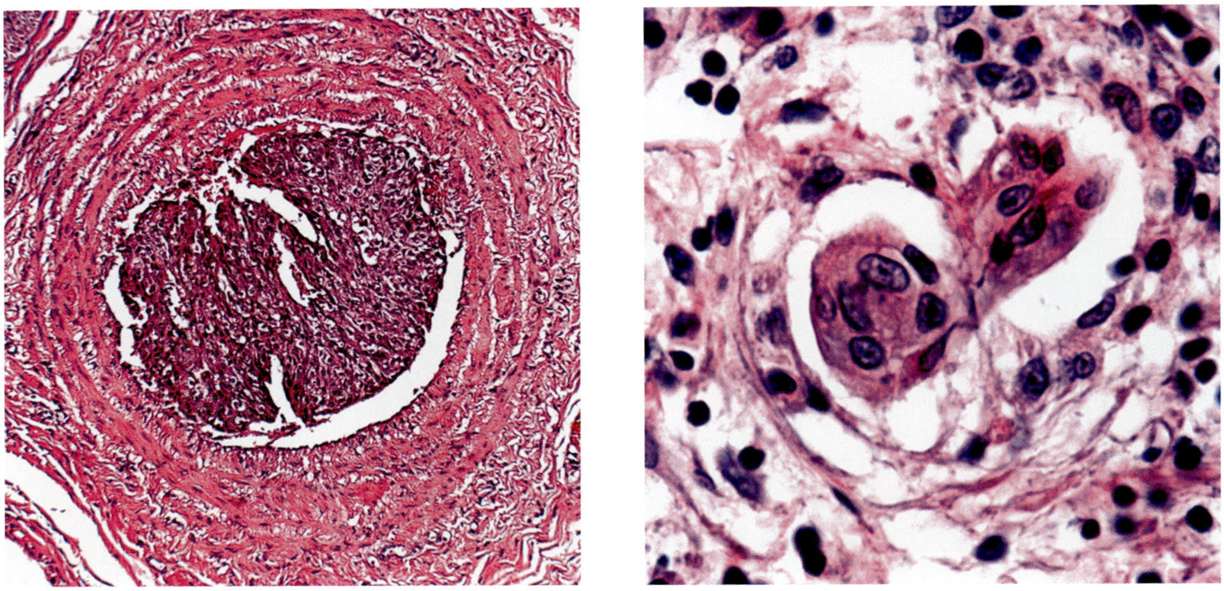

Figure 6-16

CARCINOMA OF THE LACRIMAL SAC

Left: The lumen of a vein is invaded by carcinoma.
Right: Emboli of tumor cells are seen in lymphatic vessels.

Mucinous material stains for Alcian blue, mucicarmine, periodic acid–Schiff (PAS), and PAS with diastase. Tumor cells are positive for cytokeratin AE1/3, CK 5/6, epithelial membrane antigen, CK7, and p63 (51); focally positive for vimentin and carcinoembryonic antigen (CEA); and negative for CK20, smooth muscle actin (SMA), and S-100 protein (43).

The diagnosis of mucoepidermoid carcinoma by cytology is challenging because it requires adequate cellularity, identifiable cytologic atypia, and the presence of all three cell types: epidermoid cells, intermediate cells, and mucin-secreting cells in a fine-needle aspiration specimen (56). Williams et al. (48) suggest that the presence of abundant mucin and epithelial cells in a fine-needle aspiration biopsy of a mass lesion from the lacrimal sac should raise suspicion for mucoepidermoid carcinoma.

Differential Diagnosis. The principal differential diagnoses include benign epithelial tumors, such as transitional/squamous papilloma and inverted papilloma, and other malignant epithelial tumors, particularly squamous cell carcinoma. There was one reported case of mucoepidermoid carcinoma with an overlying papilloma (51). Adequate sampling is required to avoid missing areas with intermediate and mucous cells that differentiate this tumor from squamous cell carcinoma.

Treatment and Prognosis. In the initial case reports, surgeons treated patients by orbital exenteration (44,57). Now, more conservative, globe-sparing approaches, typically surgical resection in combination with preoperative or postoperative radiotherapy, are used. Mucoepidermoid carcinoma is locally aggressive, but the rate of recurrence is low. Regional lymph node involvement occurs late in the course, and distant metastases are rare (54), with one reported case of metastatic mucoepidermoid carcinoma to the cavernous sinus (53).

MELANOMA

Definition. Primary *malignant melanoma* is a form of mucosal melanoma of the head and neck that arises in the lacrimal drainage system. Melanocytes do not appear in normal lacrimal sac epithelium. This tumor is thought to arise from either ectopic melanocytic cells deposited during embryonic migration of neural crest cells or migration from the embryologic conjunctival sac (58–60).

Clinical Features. Primary melanoma of the lacrimal sac is rare, with 31 cases reported world-

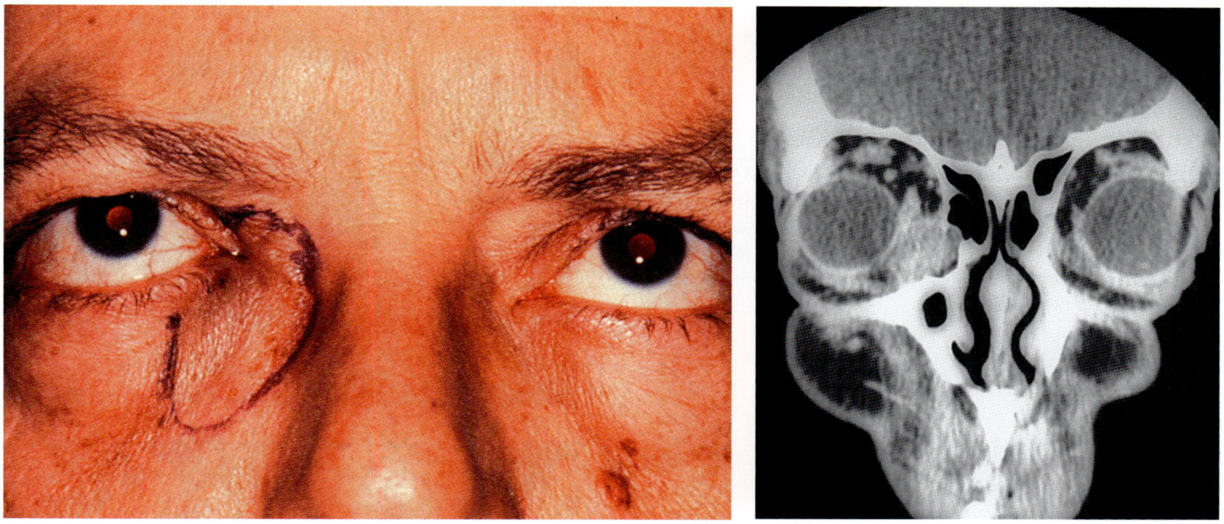

Figure 6-17

PRIMARY MELANOMA

Left: The edges of the tumor are outlined.
Right: CT shows the mass involving the right lacrimal sac.

wide (61). Melanomas comprise 3 to 13 percent of primary lacrimal sac tumors in various series (13,59). The mean age at diagnosis is 60 years (range, 38 to 81 years). There is a slight female predilection of 1.7 to 1.0; however, the mean age at diagnosis is older in men than women (65 and 58 years, respectively) (61).

Similar to other lacrimal sac neoplasms, patients with melanoma present with signs of obstruction, such as epiphora or dacryocystitis. Patients may experience bloody discharge, either from the punctum or paranasal sinuses (61–63). As the tumor grows, patients may develop a mass in the medial canthal region that extends above the medial canthal tendon (fig. 6-17, left) (15,62). Orbital imaging studies are crucial for patients with the above symptoms to rule out a tumor. CT or MRI help identify the extent of the lesion (fig. 6-17, right), in particular, to determine whether bony destruction, infiltration of surrounding soft tissue, or extension into the orbit is present. CT of the chest, abdomen, and pelvis, as well as positron emission tomography (PET)-CT are useful studies for assessing metastatic disease (61,62).

Microscopic Findings. Like melanomas elsewhere, sheets of epithelioid and spindle cells with a variable amount of melanin pigmentation are identified (fig. 6-18). Some cells have hyperchromatic nuclei with prominent nucleoli. A variable number of mitotic figures is seen.

Differential Diagnosis. The primary histologic differential diagnosis is a poorly differentiated carcinoma, which can be distinguished from melanoma by immunohistochemical studies. Melanoma cells are immunoreactive to S-100 protein, Melan A, and HMB-45, and negative for cytokeratin, in contrast to carcinoma, which is immunoreactive for cytokeratin.

Treatment and Prognosis. Due to the rarity of this tumor, there is no consensus regarding the treatment of lacrimal sac melanoma, and surgeons may use therapies designed to treat head and neck mucosal malignancies (60,62). Treatment strategies consist of varying combinations of surgery, radiotherapy, chemotherapy, and immunotherapy (60,62,65). The goal of surgery is wide excision with clear margins (at least 5 mm clearance) (61). Depending on the extent of the tumor, this can range from dacryocystectomy to orbit-naso-maxillectomy and orbital exenteration (61). Postoperative radiotherapy may be given to reduce the risk of local recurrence (35,61,62). Currently, chemotherapy and immunotherapy are used as palliative treatments for metastatic disease or unresectable tumors (62,66). Experts do not recommend neck dissection (66,67).

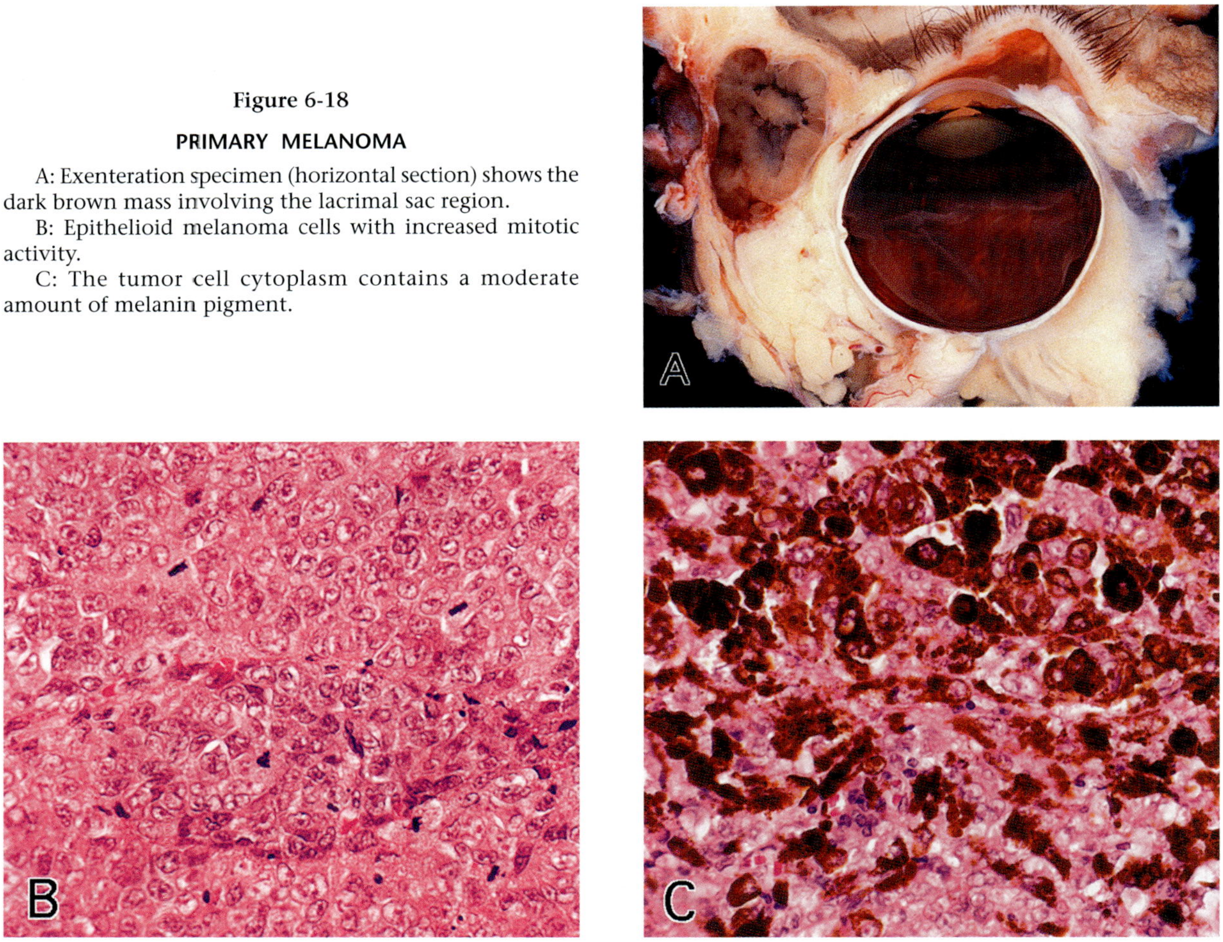

Figure 6-18

PRIMARY MELANOMA

A: Exenteration specimen (horizontal section) shows the dark brown mass involving the lacrimal sac region.

B: Epithelioid melanoma cells with increased mitotic activity.

C: The tumor cell cytoplasm contains a moderate amount of melanin pigment.

A poor prognosis reflects a typically insidious onset and proximity to an intricate vascular and lymphatic network (62). The recurrence rate is high at 69 percent, and median time to recurrence is eight months (64).

PATHOLOGIST REPORT FOR LACRIMAL SAC TUMORS

Due to the rarity of lacrimal sac tumors and lack of reports with long-term follow-up, there are no evidence-based guidelines for therapy or staging of these lesions (13,32). Nevertheless, the histopathologic diagnosis and anatomic extent of the lesion should be reported, including involvement of adjacent structures, resection margin status, and presence of metastases (2,13).

The goal of treatment is complete removal of the tumor. Surgical excision may range from dacryocystectomy to en bloc resection with lateral rhinotomy/paranasal sinus resection to orbital exenteration, without or with cervical lymph node resection (2,13,32,35). Treatments are often multidisciplinary and may include varying combinations of surgery, radiotherapy, chemotherapy, and immunotherapy. It is important to individualize treatment strategies to each patient based on tumor type and extent, margin status, and patient health status and preferences.

REFERENCES

1. Marchioni D, Bettini M, Soloperto D. Anatomy of the lacrimal drainage 1 system. In: Presutti L, Mattioli F, eds. Endoscopic surgery of the lacrimal drainage system. Switzerland: Springer; 2016:1-12.
2. Krishna Y, Coupland S. Lacrimal sac tumors—a review. Asia Pac J Ophthalmol 2017;6:173-8.
3. Kim JY, Kwon JH. An isolated nasolacrimal duct osteoma. J Craniofac Surg 2013;24:319-20.
4. Burkat CN, Lucarelli MJ. Rhabdomyosarcoma masquerading as acute dacryocystitis. Ophthalmic Plast Reconstr Surg 2005;21:456-8.
5. Neffendorf JE, Bagdonaite L, Mudhar HS, Pearson AR. Adult alveolar rhabdomyosarcoma of the lacrimal sac. Orbit 2014;33:468-70.
6. Khan MA, Dhillon B. Epiphora due to Kaposi's sarcoma of the nasolacrimal duct. Br J Ophthalmol 1999;83:501-2.
7. Parmar DN, Rose GE. Management of lacrimal sac tumours. Eye (Lond) 2003;17:599-606.
8. Sjö LD, Ralfkiaer E, Juhl BR, et al. Primary lymphoma of the lacrimal sac: an EORTC ophthalmic oncology task force study. Br J Ophthalmol 2006;90:1004-9.
9. Anderson NG, Wojno TH, Grossniklaus HE. Clinicopathologic findings from lacrimal sac biopsy specimens obtained during dacryocystorhinostomy. Ophthalmic Plast Reconstr Surg 2003;19:173-6.
10. Pe'er JJ, Stefanyszyn M, Hidayat AA. Nonepithelial tumors of the lacrimal sac. Am J Ophthalmol 1994;118:650-8.
11. Litschel R, Siano M, Tasman A, Cogliatti S. Nasolacrimal duct obstruction caused by lymphoproliferative infiltration in the course of chronic lymphocytic leukemia. Allergy Rhinol (Providence) 2015;6:191-4.
12. Krishna Y, Irion LD, Karim S, Dharmsena A, McCormick A, Coupland SE. Chronic lymphocytic leukaemia/small-cell lymphocytic lymphoma of the lacrimal sac: a case series. Ocul Oncol Pathol 2017;3:224-8.
13. Heindl LM, Jünemann AGM, Kruse FE, Holbach LM. Tumors of the lacrimal drainage system. Orbit 2010;29:298-306.
14. Heindl LM, Schick B, Kampgen E, Kruse FE, Holbach LM. [Malignant melanoma of the lacrimal sac.] Ophthalmologe 2008;105:1146-9. [German]
15. Stefanyszyn MA, Hidayat AA, Pe'er JJ, Flanagan JC. Lacrimal sac tumors. Ophthal Plast Reconstr Surg 1994;10:169-84.
16. Hung SL, Ma L. Recurrent lacrimal sac papilloma: case report. Chang Gung Med J 2000;23:113-7.
17. Jang JH, Chang SD, Choe MS. A case of recurrent schneiderian papilloma of the lacrimal sac invading the nasal cavity. Korean J Ophthalmol 2009;23:100-3.
18. Nakamura Y, Mashima Y, Kameyama K. Human papilloma virus DNA detected in case of inverted squamous papilloma of the lacrimal sac. Br J Ophthalmol 1995;79:392-3.
19. Sjo NC, von Buchwald C, Cassonnet P, et al. Human papillomavirus: cause of epithelial lacrimal sac neoplasia? Acta Ophthalmol Scand 2007;85:551-6.
20. Hardy AW, Dwivedi RC, Masterson L, et al. Inverted papilloma of lacrimal sac invading into the orbit: case report and review of literature. J Cancer Res Ther 2015;11:238-40.
21. Gustafson C, Einhorn E, Scanlon MH, Morgenstern KE, Howlett PJ, Cohen NA. Synchronous verrucous carcinoma and inverted papilloma of the lacrimal sac: case report and clinical update. Ear Nose Throat J 2013;92:E1.
22. Elner VM, Burnstine MA, Goodman ML, Dortzbach RK. Inverted papillomas that invade the orbit. Arch Ophthalmol 1995;113:1178-83.
23. Golub JS, Parikh SL, Budnick SD, Bernardino CR, DelGaudio JM. Inverted papilloma of the nasolacrimal system invading the orbit. Ophthalmic Plast Reconstr Surg 2007;23:151-3.
24. Raemdonck TY, Van den Broecke, Claerhout I, Decock CE. Inverted papilloma arising primarily from the lacrimal sac. Orbit 2009;28:181-4.
25. Walijee HZ, Berry S, Quine S, Lane C, Morris DS, Bowman B. Inverted papilloma originating primarily from the nasolacrimal duct: a case report and review of the pertinent literature. Case Rep Otolaryngol 2015;2015:123694.
26. Heathcote JG. Transitional neoplasms of the naso-lacrimal system: a review of the histopathology and histogenesis. Saudi J Ophthalmol 2012;26:125-31.
27. Islam S, Eisenberg RL, Hoffman GR. Malignant transformation of metachronous bilateral schneiderian inverted papilloma of the lacrimal sac: management considerations and the contentious issue of orbital exenteration. Eur Arch Otorhinolaryngol 2014;271:1857-60.
28. Woodcock M, Mollan SP, Harrison D, Taylor D, Lecuona K. Mitomycin C in the treatment of a Schneiderian (inverted) papilloma of the lacrimal sac. Int Ophthalmol 2010;30:303-5.

29. Anderson KK, Lessner AM, Hood I, Mendenhall W, Stringer S, Warren R. Invasive transitional cell carcinoma of the lacrimal sac arising in an inverted papilloma. Arch Ophthalmol 1994;112:306-7.
30. Kumar VA, Esmaeli B, Ahmed S, Gogia B, Debnam JM, Ginsberg LE. Imaging features of malignant lacrimal sac and nasolacrimal duct tumors. AJNR Am J Neuroradiol 2016;37:2134-7.
31. Afrogheh AH, Jakobiec FA, Hammon R, et al. Evaluation for high-risk HPV in squamous cell carcinomas and precursor lesions arising in the conjunctiva and lacrimal sac. Am J Surg Pathol 2016;40:519-28.
32. Alabiad CR, Weed DT, Walker TJ, et al. En bloc resection of lacrimal sac tumors and simultaneous orbital reconstruction: surgical and functional outcomes. Ophthal Plast Reconstr Surg 2014;30:459-67.
33. Montalban A, Liétin B, Louvrier C, et al. Malignant lacrimal sac tumors. Eur Ann Otorhinolaryngol Head Neck Dis 2010;127:165-72.
34. Salib RJ, Afoakwah EO. De novo papillary squamous carcinoma of the lacrimal sac. Auris Nasus Larynx 2003;30:205-8.
35. Valenzuela AA, Selva D, McNab AA, Simon GB, Sullivan TJ. En bloc excision in malignant tumors of the lacrimal drainage apparatus. Ophthal Plast Reconstr Surg 2006;22:356-60.
36. Islam S, Thomas A, Eisenberg RL, Hoffman GR. Surgical management of transitional cell carcinoma of the lacrimal sac: is it time for a new treatment algorithm? J Plast Reconstr Aesthet Surg 2012;65:33-6.
37. Azari AA, Kanavi MR, Saipe N, et al. Transitional cell carcinoma of the lacrimal sac presenting with bloody tears. JAMA Ophthalmology 2013;131:689-90.
38. Karim R, Ghabrial R, Lin B. Transitional cell carcinoma of the nasolacrimal sac. Clin Ophthalmol 2009;3:587-91.
39. Ryan SJ, Font RL. Primary epithelial neoplasms of the lacrimal sac. Am J Ophthalmol 1973;76:73-88.
40. Krishna Y, Coupland SE. Lacrimal sac tumors—a review. Asia Pac J Ophthalmol (Phila) 2017;6:173-8.
41. Ni C, D'Amico DJ, Fan CQ, Kuo PK. Tumors of the lacrimal sac: a clinicopathological analysis of 82 cases. Int Ophthalmol Clin 1982;22:121-40.
42. Blake J, Mullaney J, Gillan J. Lacrimal sac mucoepidermoid carcinoma. Br J Ophthalmol 1986;70:681-5.
43. Bianchi FA, Tosco P, Campisi P, Namsyl-Kaletka A, Munoz F, Ramieri G. Mucoepidermoid carcinoma of the lacrimal sac masquerading as dacryocystitis. J Craniofac Surg 2010;21:797-800.
44. Bambirra EA, Miranda D, Rayes A. Mucoepidermoid tumor of the lacrimal sac. Arch Ophthalmol 1981;99:2149-50.
45. Ni C, Wagoner MD, Wang WJ, Albert DM, Fan CO, Robinson N. Mucoepidermoid carcinomas of the lacrimal sac. Arch Ophthalmol 1983;101:1572-4.
46. Khan JA, Sutula FC, Pilch BZ, Joseph MP. Mucoepidermoid carcinoma involving the lacrimal sac. Ophthal Plast Reconstr Surg 1988;4:153-7.
47. Fliss DM, Freeman JL, Hurwitz JJ, Heathcote JG. Mucoepidermoid carcinoma of the lacrimal sac: a report of two cases, with observations on the histogenesis. Can J Ophthalmol 1993;28:228-35.
48. Williams JD, Agrawal A, Wakely PE Jr. Mucoepidermoid carcinoma of the lacrimal sac. Ann Diagn Pathol 2003;7:31-4.
49. Brar ST, Meyer D. Diagnosis and management of mucoepidermoid carcinoma of the lacrimal duct. Orbit 2011;30:34-6.
50. Lee SB, Kim KN, Lee SR, Bernardino CR. Mucoepidermoid carcinoma of the lacrimal sac after dacryocystectomy for squamous papilloma. Ophthalmic Plast Reconstr Surg 2011;27:e44-6.
51. Iordanous Y, Belrose JC, Cadieux DC, Chakrabarti S, Farmer JP, Allen LH. P63 positive mucoepidermoid tumor of the lacrimal sac with associated papilloma. Orbit 2015;34:220-2.
52. Yuksel D, Kosker M, Saribas F, Simsek S. Surgical treatment of mucoepidermoid carcinoma of the lacrimal sac. Semin Ophthalmol 2014;29:70-2.
53. Roos JC, Beigi B. Lacrimal sac mucoepidermoid carcinoma with metastases to the cavernous sinus following dacryocystorhinostomy treated with stereotactic radiotherapy. Case Rep Ophthalmol 2016;7:274-8.
54. Bianchi FA, Tosco P, Campisi P, Namsyl-Kaletka A, Munoz F, Ramieri G. Mucoepidermoid carcinoma of the lacrimal sac masquerading as dacryocystitis. J Craniofac Surg 2010;21:797-800.
55. Williams JD, Agrawal A, Wakely PE Jr. Mucoepidermoid carcinoma of the lacrimal sac. Ann Diagn Pathol 2003;7:31-4.
56. Kumar N, Kapila K, Verma K. Fine needle aspiration cytology of mucoepidermoid carcinoma. A diagnostic problem. Acta Cytol 1991;35:357-9.
57. Ni C, Wagoner MD, Wang WJ, Albert DM, Fan CO, Robinson N. Mucoepidermoid carcinomas of the lacrimal sac. Arch Ophthalmol 1983;101:1572-4.
58. Fishman G, Ophir D. Malignant melanoma of the lacrimal sac: a case study. Am J Otolaryngol 1999;20:336-9.
59. Bi YW, Chen RJ, Li XP. [Clinical and pathological analysis of primary lacrimal sac tumors.] Zhonghua Yan Ke Za Zhi 2007;43:499-504. [Chinese]

60. Li Y, Zhu S, Yan H, Han J, Wang D, Xu S. Primary malignant melanoma of the lacrimal sac. BMJ Case Rep 2012;2012.
61. Subramaniam SS, Anand R, Mellor TK, Brennan PA. Primary lacrimal sac melanoma with metastatic cervical disease: a review of the literature and case report. J Oral Maxillofac Surg 2017;75:1438-41.
62. Sitole S, Zender CA, Ahmad AZ, Hammadeh R, Petruzzelli GJ. Lacrimal sac melanoma. Ophthal Plast Reconstr Surg 2007;23:417-9.
63. Dugguid IM. Malignant melanoma of the lacrimal sac. Br J Ophthalmol 1964;48:394-8.
64. Gleizal A, Kodjikian L, Lebreton F, Beziat JL. Early CT-scan for chronic lacrimal duct symptoms - case report of a malignant melanoma of the lacrimal sac and review of the literature. J Craniomaxillofac Surg 2005;33:201-4.
65. Lewis AM, Clarke PM, Olver JM. Primary nasolacrimal duct malignant mucosal melanoma. Br J Ophthalmol 2006;90:114-5.
66. Breik O, Sim F, Wong T, Nastri A, Iseli TA, Wiesenfeld D. Survival outcomes of mucosal melanoma in the head and neck: case series and review of current treatment guidelines. J Oral Maxillofac Surg 2016;74:1859-71.
67. Lloyd WC, Leone CR. Malignant melanoma of the lacrimal sac. Arch Ophthalmol 1984;102:104-7.
68. Ashton N, Choyce DP, Fison LG. Carcinoma of the lacrimal sac. Br J Ophthalmol 1951;35:366-76.
69. Radnót M, Gáll J. [Tumors of the lacrimal sac]. Ophthalmologica 1966;151:2-22. [German]
70. Schenck NL, Ogura JH, Pratt LL. Cancer of the lacrimal sac. Presentation of five cases and review of the literature. Ann Otol Rhinol Laryngol 1973;82:153-61.
71. Sjö NC, von Buchwald C, Cassonnet P, et al. Human papillomavirus: cause of epithelial lacrimal sac neoplasia? Acta Ophthalmol Scand 2007;85:551-6.

7 OCULAR ADNEXAL HEMATOLYMPHOID NEOPLASMS AND TUMOR-LIKE LESIONS

The ocular adnexa, consisting the eyelids, conjunctiva, lacrimal apparatus, and orbital soft tissue, is frequently involved by hematolymphoid neoplasms and tumor-like lesions (1–3) (Tables 2-1, 3-1, and 4-1). The contribution of neural crest to the microenvironment of ocular adnexa and the function of lacrimal gland and conjunctiva as mucosa-associated lymphoid tissue (MALT) are believed to play a role in the site-specific histologic, cytogenetic, molecular genetic, and clinical features of ocular adnexal hematolymphoid lesions (3).

Similar to what is encountered in extraocular locations, the spectrum of lymphoproliferative orbital and ocular adnexal lesions ranges from reactive lymphoid hyperplasia to lymphoma. Reactive lymphoid hyperplasia accounts for 8 to 27 percent of all lymphoproliferative tumors of the ocular adnexa (4). Primary ocular adnexal lymphoma (POAL) is lymphoma of the ocular adnexa without evidence of concurrent systemic lymphoma and no prior history of lymphoma. One percent of all lymphomas and approximately 8 percent of all extranodal lymphomas arise in the ocular adnexa, and the incidence is increasing. Of the ocular adnexal sites, the orbit is most frequently affected by POAL (37 percent), where lymphoma is the most common orbital malignancy, followed by the conjunctiva (29 percent) and the lacrimal apparatus (20 percent). The eyelids are least frequently involved in primary lymphoma (14 percent) and are, conversely, the most common site for secondary lymphoma (3).

POALs are classified following the World Health Organization Classification (WHO) of the Tumors of Hematopoietic and Lymphoid Tissues (5). They are staged following the TNM classification devised for POAL (Appendix 6) (6), which has been shown to be a more accurate predictor of patient survival than the Ann Arbor system (7).

REACTIVE AND ATYPICAL LYMPHOID HYPERPLASIA

Reactive lymphoid hyperplasia is common in the lacrimal gland and conjunctival regions, as expected from their roles as MALT (3). Despite an extensive search for self-peptides and foreign microbes that may play a role in the development of ocular adnexal lymphoid hyperplasia, no causative agent has been convincingly identified (4). Similar to nonocular adnexal sites, the diagnosis of reactive lymphoid hyperplasia is made by exclusion of a clonal lymphoid cell population, in accordance with the established morphologic, immunohistochemical, flow cytometric, and molecular genetic guidelines (fig. 7-1) and in conjunction with systemic clinical evaluation (3–5).

Lesions that morphologically and immunophenotypically appear polyclonal and demonstrate clonality on molecular genetic testing, and lesions with morphologic atypia suspicious for lymphoma, with no evidence of clonality, are assigned the diagnosis of *atypical lymphoid hyperplasia*. With the advent of refined ancillary diagnostic studies, there has been a shift toward the diagnosis of lymphoma for lesions that formerly would have been classified as hyperplasia (3,4).

IgG4-related disease (see chapter 5) may mimic lymphoid hyperplasia both clinically and histologically and IgG4 and IgG immunostaining is recommended to rule out this diagnosis (4,8). Idiopathic orbital inflammatory disease, a diagnosis of exclusion, can feature reactive follicles analogous to reactive lymphoid hyperplasia. This condition is distinguished from lymphoid hyperplasia by a mixed cellular infiltrate consisting of small lymphocytes, plasma cells, neutrophils, eosinophils, and, occasionally, macrophages in a background of edematous, variably fibrotic stroma. Other

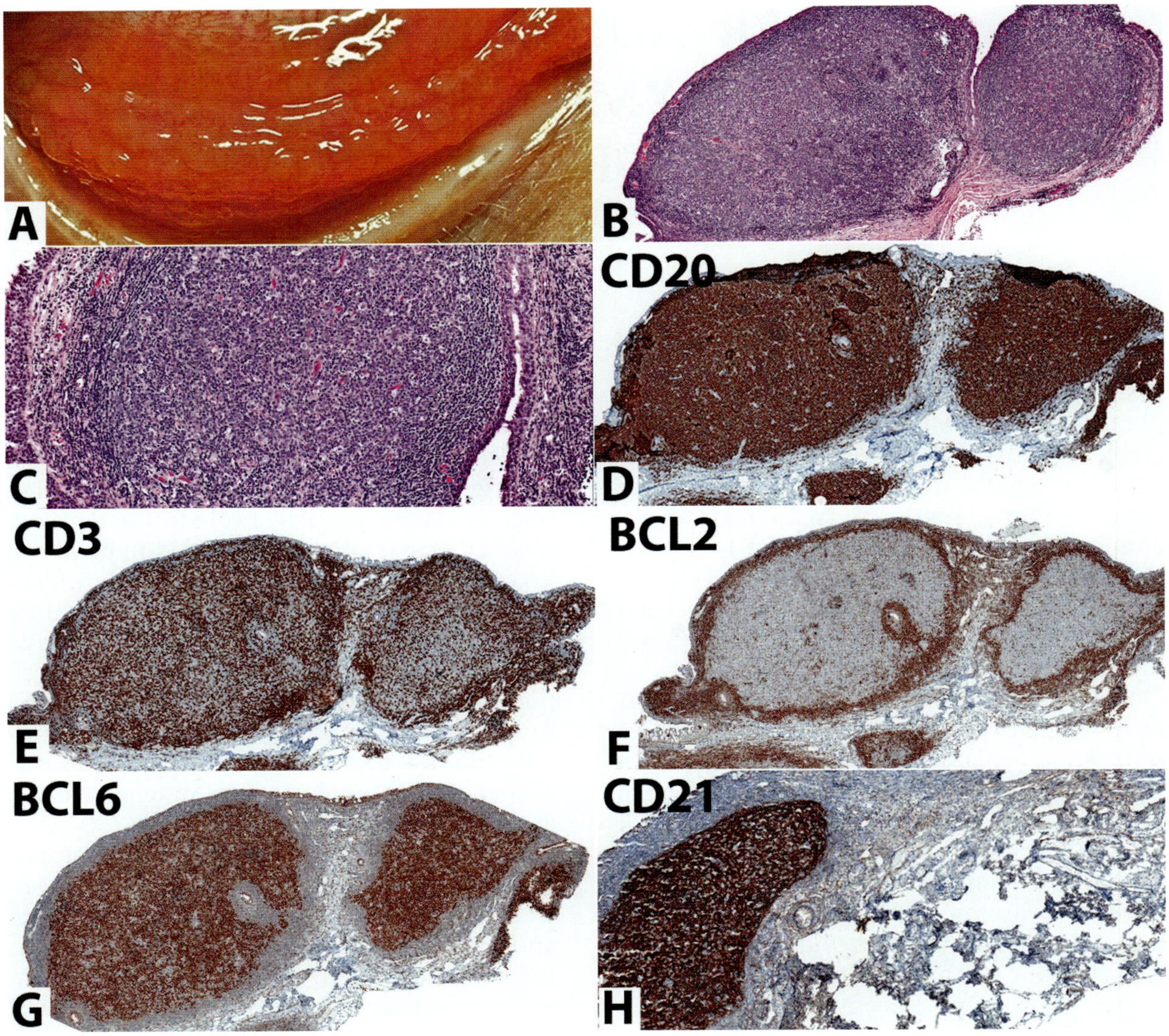

Figure 7-1

REACTIVE LYMPHOID HYPERPLASIA: CONJUNCTIVA

A: A salmon-colored forniceal conjunctival mass has numerous dome-shaped lymphoid follicles.

B: Scanning magnification shows the well-delineated nodular architecture of individual reactive lymphoid follicles.

C: At higher magnification, the typical reactive follicle morphology with a central germinal center and peripheral mantle zone is highlighted.

D–H: The reactive follicles appropriately express CD20 (D). CD3 (E) is predominantly expressed in the interfollicular T cells. BCL6 expression (G) is confined to the germinal center cells and BCL2 expression (F) in the mantle zones. CD21 (H) highlights the intact follicular dendritic meshworks.

conditions that feature reactive follicles, such as epithelioid hemangioma, xanthogranuloma, and Castleman disease, must also be considered in the histologic differential diagnosis of reactive lymphoid hyperplasia (4).

Ocular adnexal lymphoid hyperplasia is managed successfully by a variety of treatment modalities including observation, corticosteroids, rituximab, and low-dose radiotherapy (4). Reactive lymphoid hyperplasia follows an indolent course with potential for recurrence (20 to 30 percent) and low risk of progression to lymphoma (15 percent). The risk of progression to lymphoma is higher in lesions classified as atypical lymphoid hyperplasia. The most common lymphoma arising in ocular adnexal lymphoid hyperplasia is extranodal marginal zone B-cell lymphoma (EMZL) of MALT type (MALT lymphoma) which carries minimal risk of dissemination and disease-related mortality of 1 to 2 percent (4).

PRIMARY OCULAR ADNEXAL LYMPHOMA

In the ocular adnexal region, indolent mature small *B-cell lymphomas* predominate, characteristically presenting in older patients as painless gradual displacement of the eyeball, orbital and eyelid fullness, and a fleshy "salmon-patch"-like conjunctival mass (fig. 7-2). Lymphoid orbital tumors frequently "mold" to the orbital tissues and the globe without indenting these structures and with no associated bone destruction, radiographic features helpful in distinguishing lymphoproliferative processes from other space-occupying lesions.

EMZL/MALT lymphoma is by far the most frequent lymphoma type, accounting for more than half of all POALs (fig. 7-2). Other frequent POAL types are *follicular lymphoma* (fig. 7-3) and *diffuse large B-cell lymphoma* (DLBCL) (fig. 7-4). *Primary mantle cell lymphoma* (MCL), *small lymphocytic lymphoma*, and *T-cell lymphomas* are encountered infrequently in the ocular adnexa (3).

EMZL/MALT lymphoma deserves special consideration because of its site-specific clinical, histologic, and cytogenetic/molecular genetic features. EMZL may arise in ocular adnexal MALT following chronic antigenic stimulation from persistent inflammation, as exemplified by the higher incidence of ocular adnexal MALT lymphomas in patients with autoimmune disorders (3,9). Similarly, infectious agents, most notably *Chlamydia psittaci*, are associated with ocular adnexal MALT lymphomas (3,9,10) but their role in the pathogenesis of this neoplasm has not been established firmly (3).

Ocular adnexal MALT lymphoma is associated with prominent plasmacytic differentiation when compared with nonocular adnexal EMZL. Cytogenetically, ocular adnexal MALT lymphomas more frequently harbor t(14;18)(q32;q21), trisomy 3, and trisomy 18 than gastric MALT lymphomas, while t(11;18)(q21;q21), common in gastric and lung MALT lymphomas, is seldom encountered in ocular adnexal lymphoid lesions (9,11).

The management of ocular adnexal lymphomas is heterogeneous and tailored to the type of lymphoma, degree of dissemination, and patient-specific factors. Management options include regional radiotherapy (an especially attractive option for localized MALT lymphomas), chemotherapy, and targeted therapies. When confined to the orbit, patients with POAL have a favorable prognosis when compared with respective non-ocular adnexal lymphoma types: 5-year overall survival rates over 90 percent for MALT lymphoma (9) and follicular lymphoma (12). DLBCL and MCL are associated with much higher rates of systemic dissemination and patients have been documented to have 5-year survival rates of 20 to 60 percent (3,13) and 40 percent (14), respectively.

PRIMARY OCULAR ADNEXAL PLASMACYTOMA

Primary extramedullary orbital plasmacytoma is extremely rare, with fewer than 30 cases reported in the literature. This neoplasm tends to present in older patients with a male predilection, as painless globe displacement and double vision (15–18). Other less common presentations include a cellulitis-type feature (fig. 7-5) (16) and a fungating mass (18).

The diagnosis of primary ocular adnexal plasmacytoma is made following the current WHO classification guidelines for tumors of hematopoietic and lymphoid tissues and rests on the exclusion of systemic plasma cell neoplasia and B-cell lymphomas with plasmacytic differentiation (5,15–18).

XANTHOGRANULOMATOUS DISORDERS

Definition. *Ocular adnexal xanthogranulomatous disorders* are a heterogeneous group of diseases that span all ages, presenting in childhood as *juvenile xanthogranuloma* (JXG) or in adulthood as an *adult orbital xanthogranulomatous disease* (AOXGD).

Pathogenesis. The pathogenesis of xanthogranulomatous disease is not fully understood. Various mechanisms have been proposed, including infection-driven proliferation of histiocytic and dendritic cells and the underlying genetic defects (19). Recent studies have highlighted the emerging importance of the RAF/MEK/ERK (MAPK) pathway, particularly *BRAF* V600E activating mutations in the pathogenesis of these diseases, supporting the concept of inflammatory myeloid neoplasia (5,20,21).

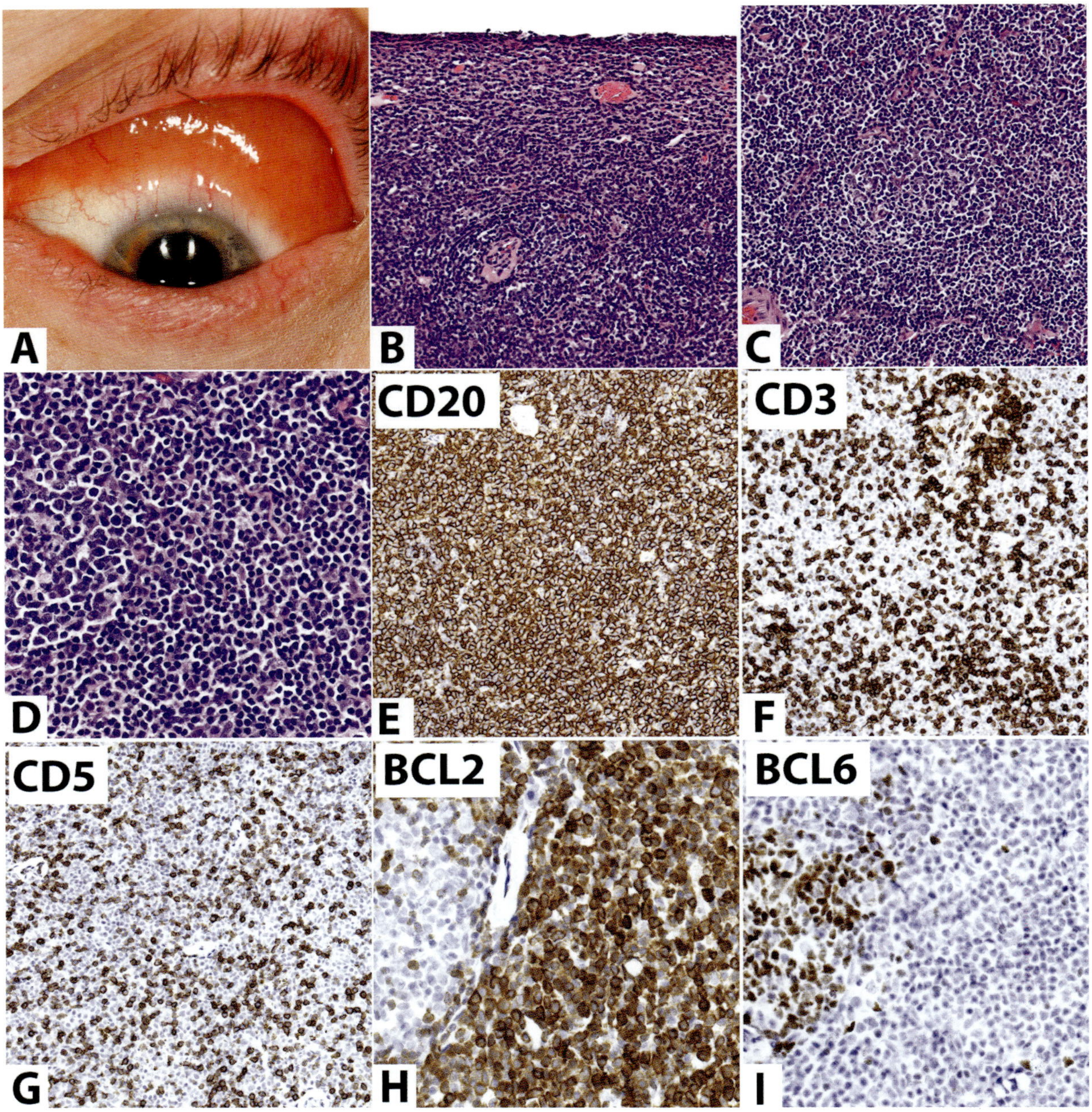

Figure 7-2

OCULAR ADNEXAL MUCOSA-ASSOCIATED LYMPHOID TISSUE (MALT) LYMPHOMA

A: A fleshy orange-salmon conjunctival forniceal mass is seen.

B: There is a diffuse sheet-like infiltrate of small lymphocytes in the conjunctival substantia propria with intraepithelial migration (B–D: H&E stain).

C: Poorly defined germinal centers in the background of lymphoid proliferation lend a vaguely nodular appearance to the lesion.

D: Higher magnification highlights the monomorphic small monocytoid cells with scant cytoplasm, inconspicuous nucleoli, and inconspicuous mitotic figures.

E–L: The neoplastic cells diffusely express CD20 and BCL2 and lack expression of CD5, CD10, and CD23, with a Ki-67 proliferative index of less than 5 percent outside of germinal centers. The germinal center B cells in the background of neoplastic proliferation appropriately express CD10, BCL6, and Ki-67. The follicular dendritic cell meshworks are highlighted with CD23. The few scattered small T cells express CD3 and CD5.

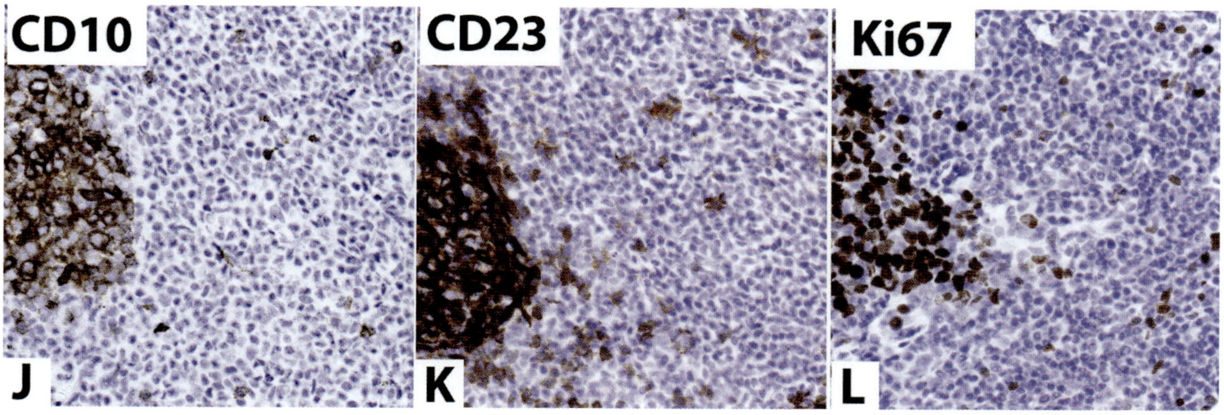

Figure 7-2, continued

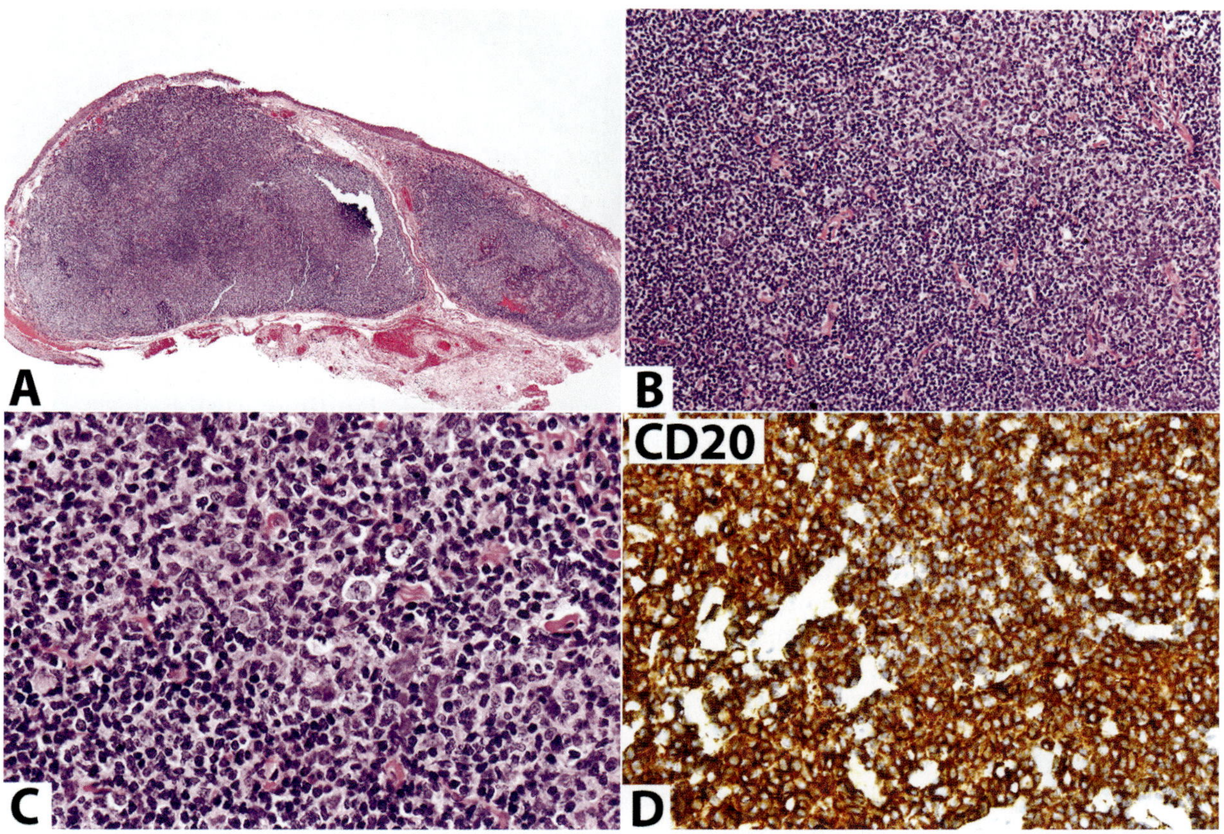

Figure 7-3

LOW-GRADE FOLLICULAR LYMPHOMA: CONJUNCTIVA

A: A nodular lymphoid infiltrate is seen in the conjunctival substantia propria (A–C: H&E stain).

B,C: The neoplastic cells are predominantly small to medium-sized, with irregular nuclear contours and inconspicuous nucleoli (centrocytes). Rare centroblasts are identified.

D–H: The neoplastic cells diffusely express CD20 (D), CD10 (G), BCL2, and BCL6 (F). CD5 (E) is confined to the few scattered small T cells.

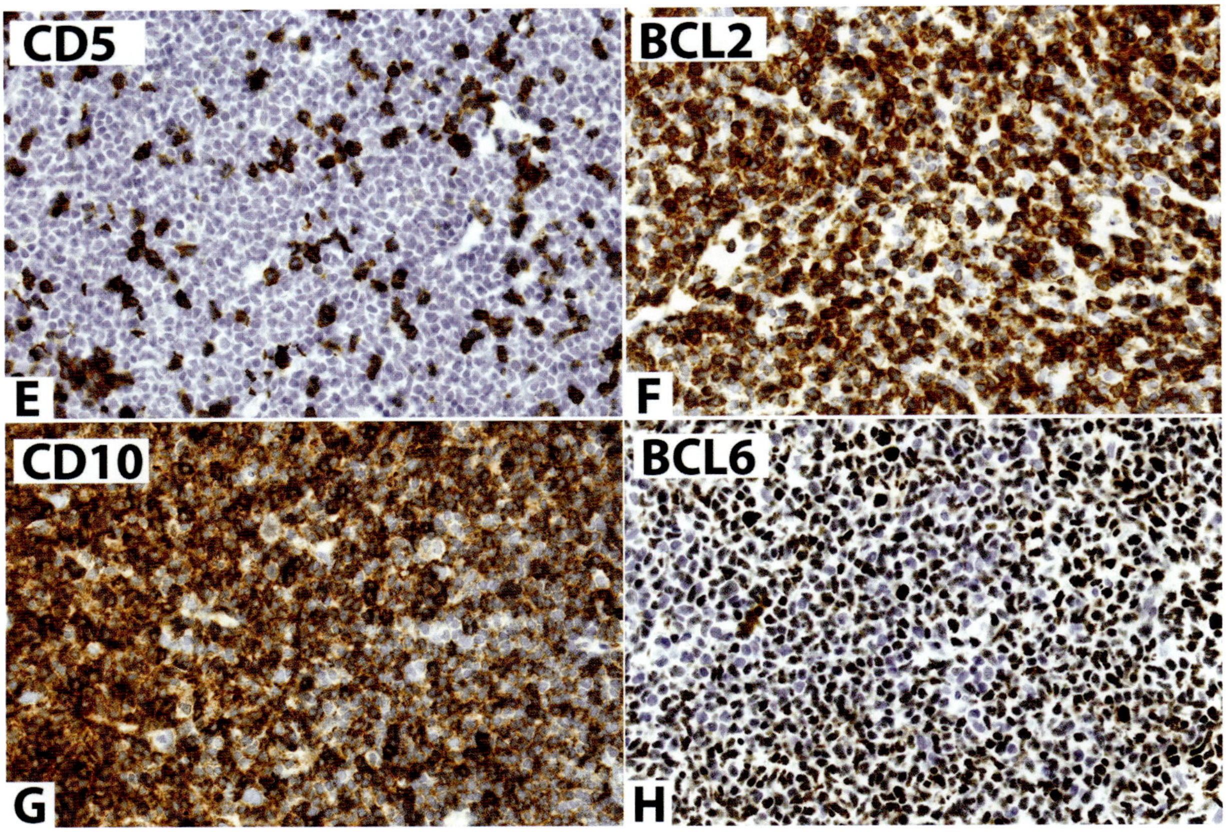

Figure 7-3, continued

Clinical Features. JXG is a self-limited disorder of infants and children and, less frequently, older patients. Skin, particularly the head and neck region, is primarily affected by the reddish to yellowish papules and nodules (fig. 7-6). The eye, especially the uveal tract, is the most frequent site of extracutaneous disease (22).

AOXGD consists of four entities, classified by degree of systemic involvement: *adult-onset xanthogranuloma* (AOX), *necrobiotic xanthogranuloma* (NXG), *Erdheim-Chester disease* (ECD), and *adult-onset asthma and periocular xanthogranuloma* (AAPOX). AOXGD is associated with a wide range of manifestations in the orbital and ocular adnexal region, presenting as yellow-orange to red-violaceous, elevated, indurated xanthomatous eyelid plaques, nodules, and papules or as orbital mass lesions with variable involvement of the extraocular muscles, orbital fat, and lacrimal gland (fig. 7-7). AOX is an isolated and frequently self-limited condition. In contrast, ECD frequently affects bone and viscera and NXG and AAPOX are both associated with paraproteinemias and lymphoproliferative/plasma cell proliferative disorders (19).

Microscopic Findings. Histologically, xanthogranulomatous diseases are characterized by infiltration of foamy histiocytes (xanthoma cells) and variable numbers of Touton-type giant cells, in a background of variable fibrosis and lymphoplasmacytic infiltrate. Foamy histiocytes and Touton-type giant cells may be inconspicuous in early JXG lesions in which fibroblast-like cells predominate (fig. 7-6).

Immunohistochemically, the foamy histiocytes are strongly positive for CD68 and CD163, frequently express factor XIIIa, and are usually negative for CD21, CD35, S-100 protein, CD1a, and langerin. Although particular histopathologic features are associated with individual adult xanthogranulomatous disorders, such as necrobiosis in NXG, reactive lymphoid follicles in AAPOX, and fibrosis in ECD (19), there is a significant histologic overlap between these conditions. It is, therefore, recommended that a systemic workup is undertaken for all of AOXGD entities (23).

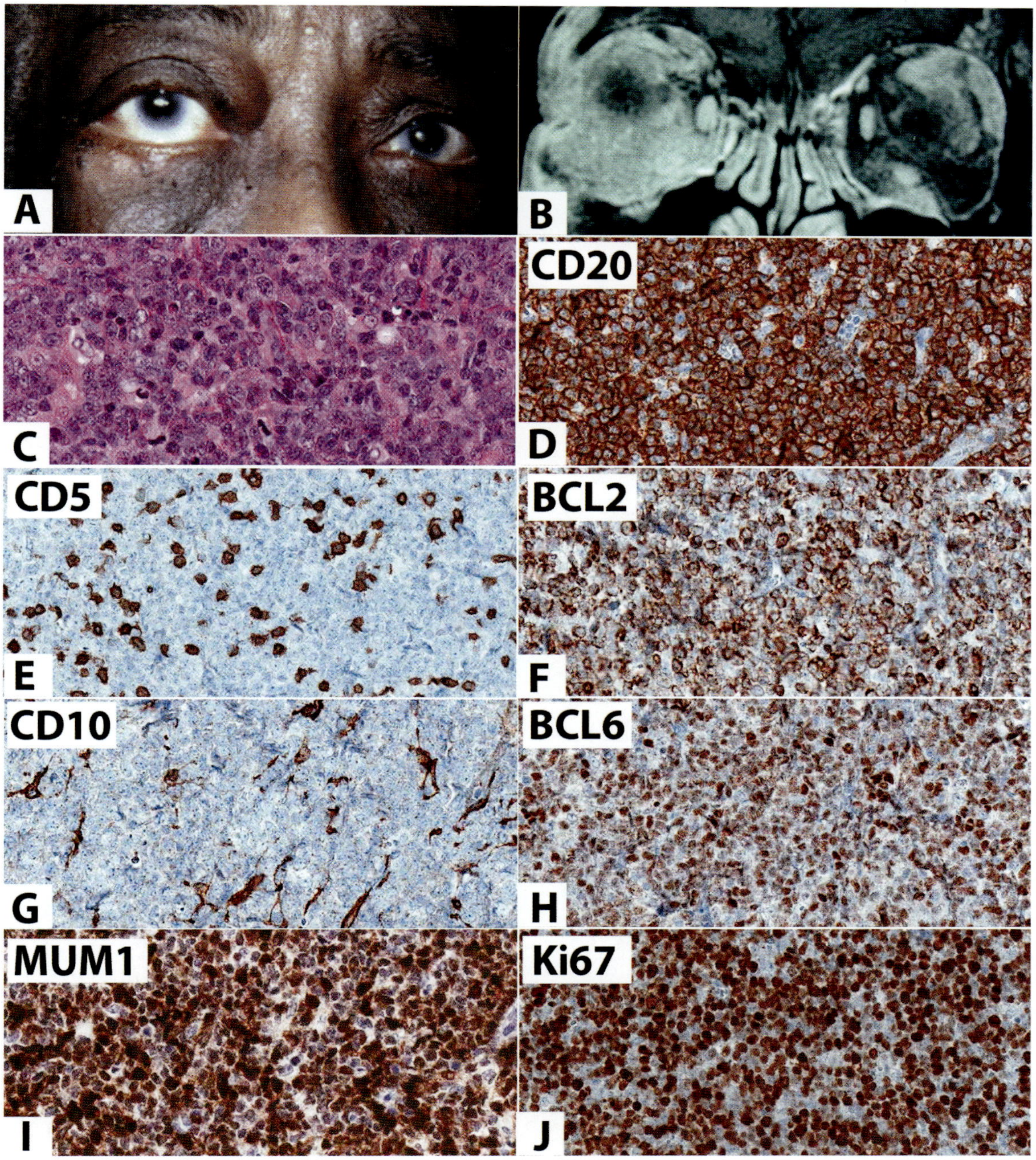

Figure 7-4

HIGH-GRADE B-CELL LYMPHOMA: ORBIT

A: Eyelid and orbital fullness with displacement of the eyeball.

B: Bilateral asymmetric infiltration of the orbital tissues by enhancing mass lesions (diffusely on the right and predominantly in the lacrimal gland region on the left).

C: Diffuse proliferation of neoplastic cells with large nuclei, irregular nuclear contours, prominent nucleoli, scant to moderate amount of cytoplasm, and brisk mitotic activity.

D–I: The malignant cells express diffusely CD20 (D), BCL2 (F), BCL6 (H), and MUM1 (I) and are negative for CD10 (G), compatible with a nongerminal center cell phenotype (Hans algorithm).

J: The Ki-67 proliferative index is high, at greater than 90 percent. CD5 is expressed in few scattered small B cells. FISH cytogenetic studies (not shown here) demonstrated rearrangements in *MYC* and *BCL6* genes, compatible with a high-grade B-cell lymphoma with rearrangement of *MYC* and *BCL6* genes (double-hit lymphoma).

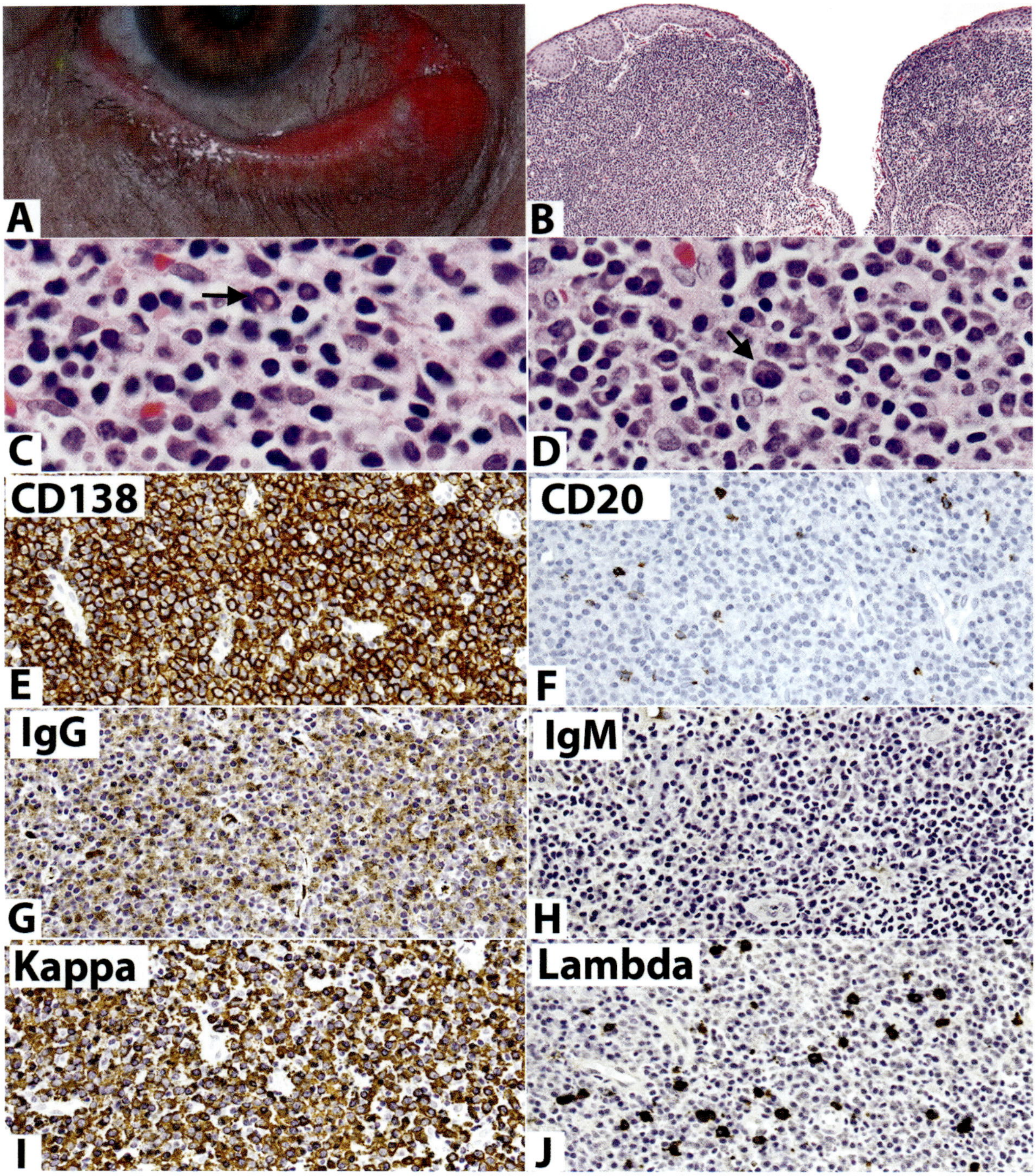

Figure 7-5

PRIMARY OCULAR ADNEXAL PLASMACYTOMA

A: Pseudoinflammatory presentation with canaliculitis-type features.

B–D: Diffuse infiltrate of plasma cells in the stroma with occasional Dutcher body formation (arrow, C) and focally prominent central nucleoli (arrow, D) (B–D: H&E stain).

E–J: The neoplastic cells diffusely express CD138 (E) and demonstrate IgG kappa restriction. Rare scattered small CD20-positive B cells (F) are present in the background.

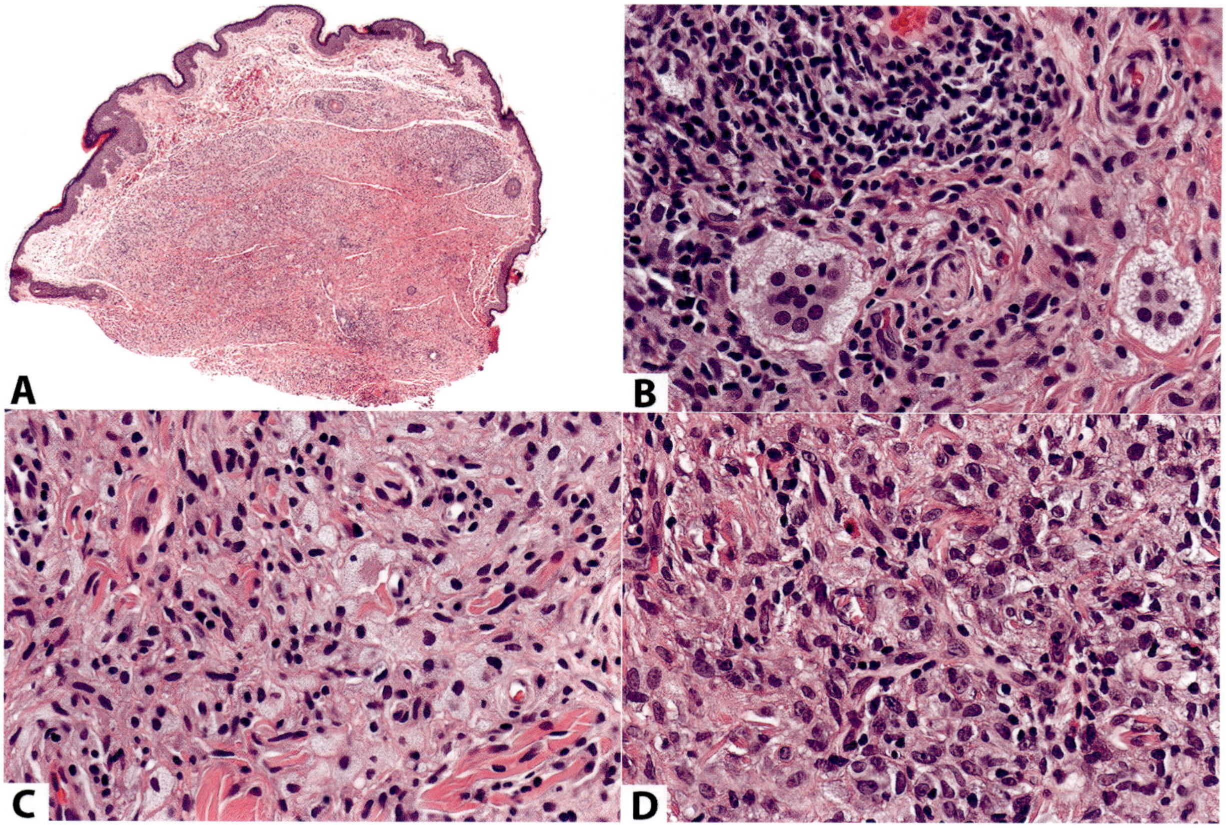

Figure 7-6

JUVENILE XANTHOGRANULOMA: EYELID SKIN

A: A circumscribed nodular mixed cellular proliferation is in the dermis.

B,C: an infiltrate of foamy histiocytes (xanthoma cells) and few multinucleated Touton-type giant cells are in a background of lymphocytes and rare eosinophils.

D: Focally, the lesional cells have a spindle, fibroblast-like morphology without appreciable cytoplasmic lipidization.

Several studies have described elevated numbers of IgG4-positive plasma cells in AOXGD (fig. 7-7), which in itself is not diagnostic of IgG4-related disease. For an independent diagnosis of IgG4-related disease, documentation of characteristic histopathologic findings of IgG4-related disease in a site other than the orbital tissue involved by xanthogranuloma is recommended (24). Marked elevation of serum IgG4 also supports the diagnosis, but this study is limited by population-dependent variation and suboptimal sensitivity and specificity (24,25).

Molecular Genetic Findings. Mutation analysis for key genes involved in the MAPK pathway, particularly *BRAF* and *NRAS*, can be considered for patients with systemic disease that are candidates for targeted therapy (see below).

Differential Diagnosis. Xanthelasma palpebrarum, a common xanthomatous lesion localized to the eyelids, must be distinguished clinically and histopathologically from xanthogranuloma. Unlike xanthogranuloma, xanthelasma typically lacks induration, orange or violaceous hue, and is not associated with eyelid fullness or globe displacement indicative of the involvement of deep ocular adnexal and orbital tissues (fig. 7-8). Xanthelasma histologically appears as a circumscribed lesion without infiltration of deep dermis or orbit, composed mostly of xanthoma cells surrounding dermal capillaries with a sparse lymphocytic background and absent fibrosis. Touton-type giant cells are infrequent (fig. 7-8) (26).

Treatment and Prognosis. There is no consensus on the optimal management of orbital

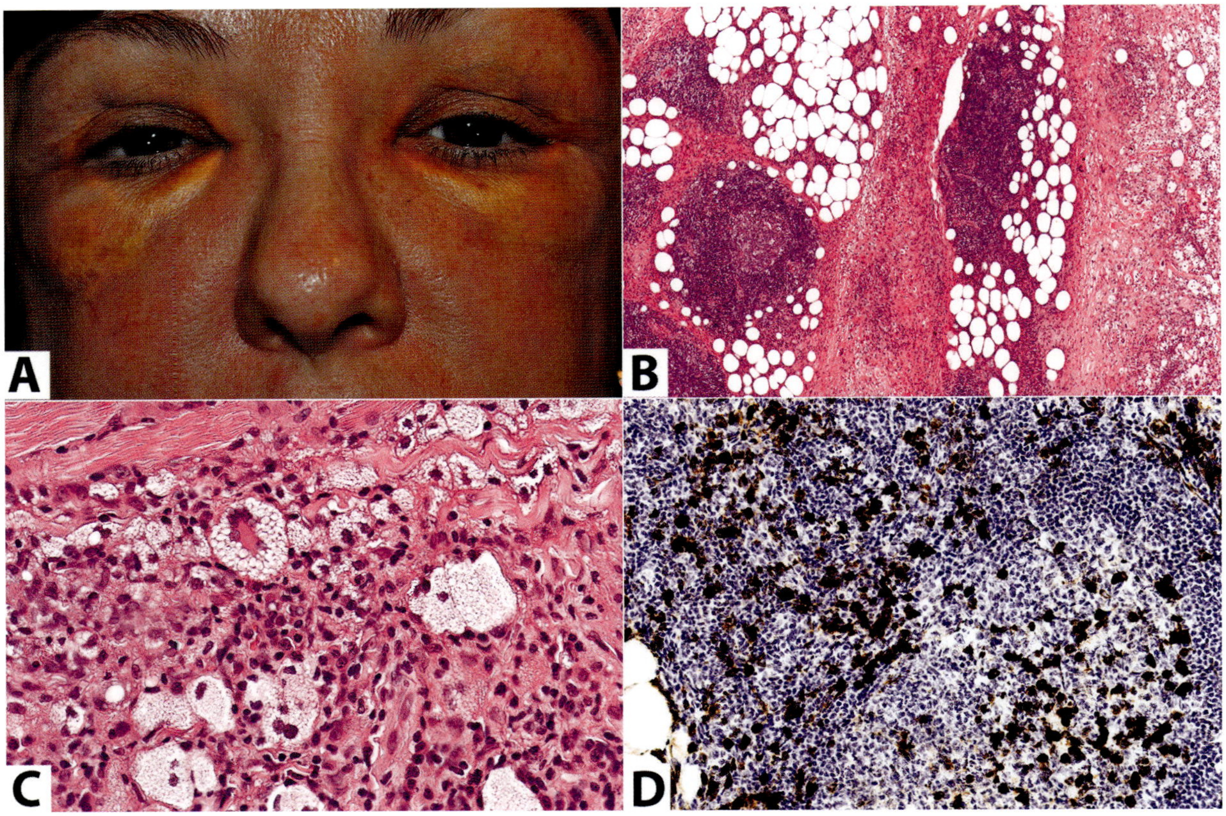

Figure 7-7

ADULT ORBITAL XANTHOGRANULOMATOUS DISORDER IN PATIENT WITH LONGSTANDING AUTOIMMUNE THYROIDITIS

A: Bilateral yellow-orange, indurated, subcutaneous and anterior orbital mass lesions.

B: Infiltration of the deep orbital tissues by lipidized histiocytes (right) in a background of variable fibrosis and reactive follicles (left).

C. Touton-type multinucleated giant cells and foamy histiocytes in a background of lymphoplasmacytic infiltrate.

D: Numerous IgG4-expressing plasma cells, most prominent in the lymphoid follicles, are associated with the lesion.

xanthogranulomatous disorders. Management options depend on the disease entity and include observation, excision, intralesional or systemic corticosteroids, radiotherapy (19), rituximab and, more recently, chemotherapy and targeted therapy (BRAF inhibitors) (27). The systemic associations of ocular adnexal-orbital xanthogranulomatous disorders should guide the workup and systemic therapy (27). The prognosis ranges from excellent for self-limited conditions (JXG and AOX) to guarded with a high rate of morbidity and mortality for diseases with the sizeable systemic burden (ECD and NXG) (19).

ROSAI-DORFMAN DISEASE

Rosai-Dorfman disease (RDD), also known as *sinus histiocytosis with massive lymphadenopathy* (SHML), is a rare histiocytic proliferative disorder. The average age of onset is 20 years, but the age distribution varies greatly. Eye and ocular adnexal involvement occurs in approximately 10 percent of patients with RDD, most commonly manifesting as an orbital or epibulbar mass, uveitis, scleritis, and choroidal mass. Patients with ocular involvement have more severe systemic manifestations than patients with orbital disease (28).

Histopathologically, ophthalmic RDD is similar to extranodal RDD in nonocular sites, characterized by heterogeneous infiltrate of lymphocytes, plasma cells, and large histiocytes with abundant, eosinophilic cytoplasm and intracytoplasmic vacuoles that contain phagocytized cells (emperipolesis). The histiocytes in

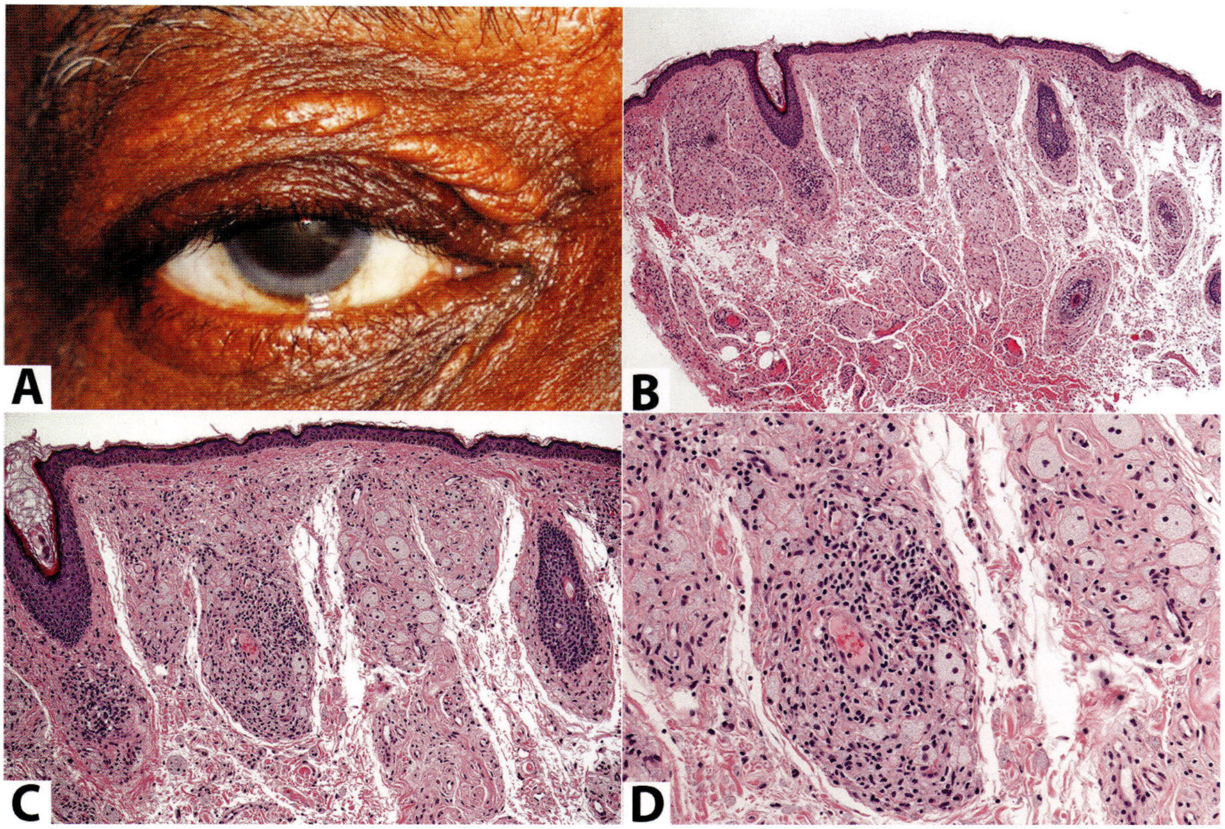

Figure 7-8

XANTHELASMA PALPEBRARUM

A: Yellowish, circumscribed, subcutaneous plaques are most prominent in the superonasal ocular adnexal region. Note also the brown conjunctival pigment in this patient. This is *conjunctival skin-tone associated melanosis* (*C-STAM*) (see chapter 2).

B,C: A dominant infiltrate of xanthoma cells is localized to the superficial and mid-dermis, with minimal lymphocytic component and with no appreciable fibrosis.

D: The xanthoma cells and lymphocytes tend to cluster around dermal vascular channels.

RDD express S-100 protein in addition to the histiocytic markers CD68 and CD163, and are negative for CD1a and langerin (fig. 7-9) (28).

The etiology of RDD is currently poorly understood and debates on its pathogenesis center around a reactive immunologic process versus a neoplastic process. Although *BRAF* mutations are not to date identified in RDD (28,29), mutations in other components of the MAPK pathway have been documented in several studies (29). Management strategies include chemotherapy, radiation, rituximab, and possible surgical excision with variable response (28).

LANGERHANS CELL HISTIOCYTOSIS

Langerhans cell histiocytosis (LCH) is currently recognized as a neoplastic clonal proliferation of myeloid-dendritic lineage–derived Langerhans cells, frequently driven by MAPK pathway mutations (5,20). The most common ophthalmic manifestation is a solitary lytic lesion of orbital bone (*eosinophilic granuloma*), typically in superior or superotemporal location (30).

Histologically, orbital eosinophilic granuloma is characterized by a bone-destructive infiltrate of neoplastic Langerhans cells with elongated, folded (coffee bean-like) nuclei in a background of osteoclast-like giant cells, eosinophils, neutrophils, plasma cells, and lymphocytes. Langerhans cells express S-100 protein, CD1a, and langerin (fig. 7-10) (30). This expression distinguishes cutaneous LCH from early nonlipidized JXG and lipidized LCH from xanthogranulomatous disorders. Recent

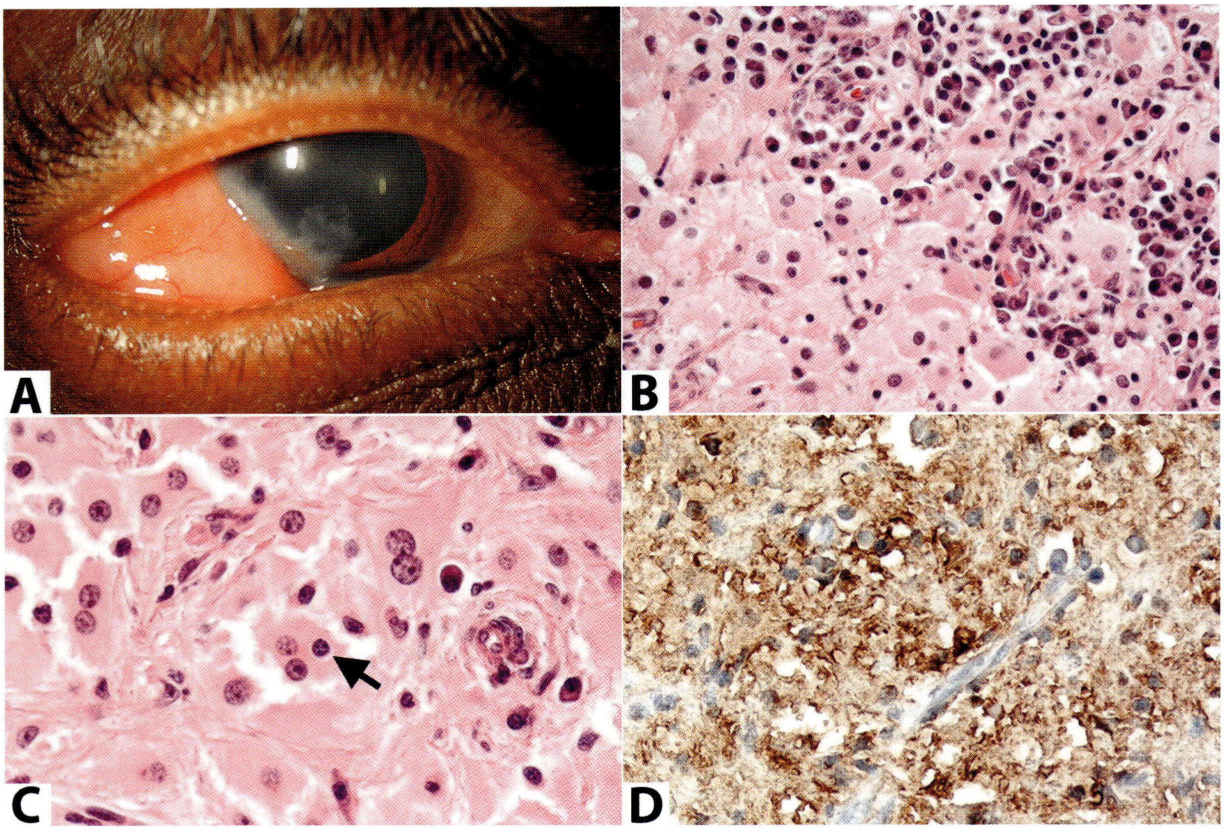

Figure 7-9

OCULAR ROSAI-DORFMAN DISEASE

A: Epibulbar fleshy, salmon-colored, nodular mass.

B:. An infiltrate of large histiocytes with round nuclei, central nucleoli, and abundant eosinophilic cytoplasm, is interspersed with aggregates of plasma cells.

C: Higher magnification highlights emperipolesis.

D: The histiocytes express diffusely S-100 protein.

studies have also documented BRAF expression by some LCH lesions, suggesting that BRAF immunohistochemistry may be a useful screening test for underlying potentially actionable MAPK pathway alterations (31).

Localized orbital and ocular adnexal LCH typically responds to minimally invasive therapy, such as curettage or excision. In more widespread cases, chemotherapy and radiotherapy are traditionally employed (30). Recent recognition of high frequency of *BRAF* mutations in LCH has raised an interest in targeted therapies for this disorder (20).

SECONDARY HEMATOLYMPHOID NEOPLASMS

Hematolymphoid neoplasms from nonocular sites may occasionally disseminate to involve the orbit and ocular adnexa. The frequency of secondary hematolymphoid neoplasms ranges from approximately 50 percent of all lymphoproliferative lesions in the eyelid region to around 25 percent in the remaining ocular adnexa and the orbit (3). Higher-grade B-cell lymphomas frequently cause orbital and ocular adnexal involvement in systemic disease. Therapy focuses on the control of the systemic disease (31).

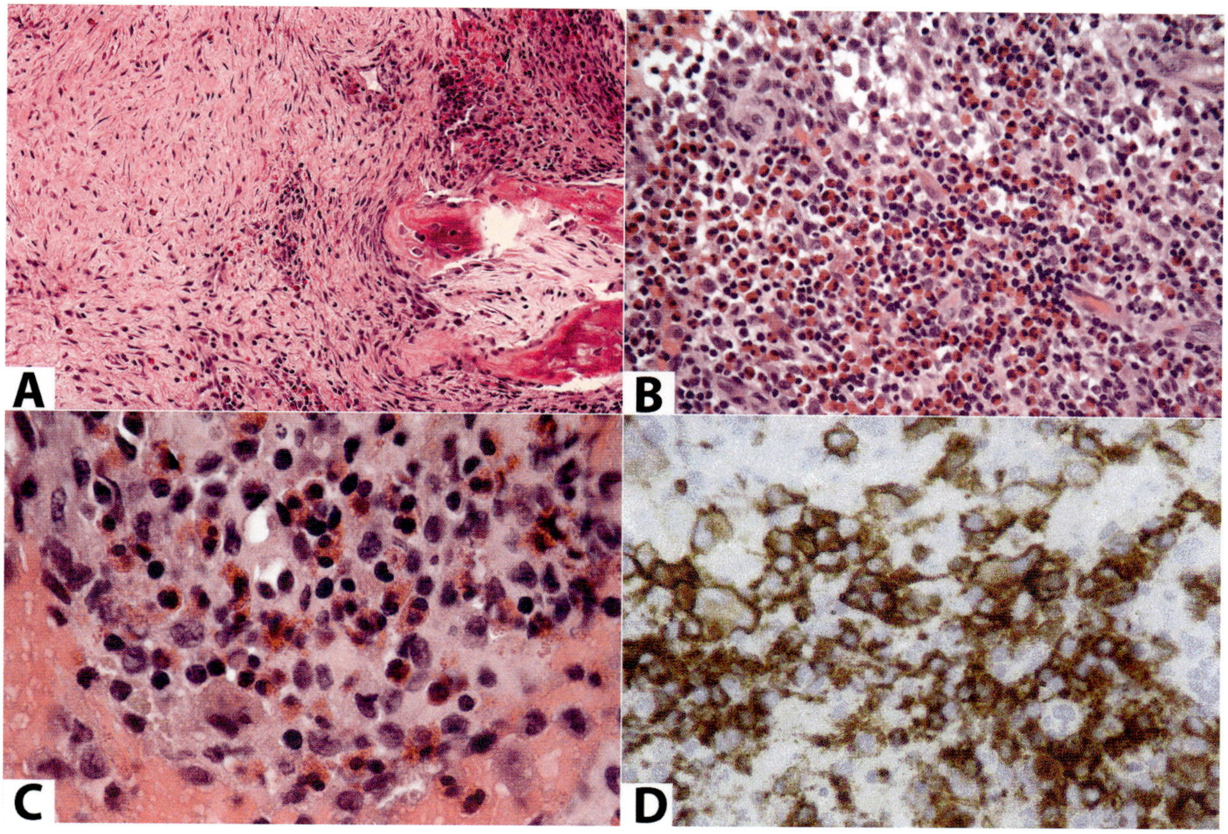

Figure 7-10

EOSINOPHILIC GRANULOMA: ORBIT

A: Reactive bone trabeculae (right) are associated with a cellular infiltrate and fibrosis.

B,C: The infiltrate consists of histiocyte-like cells with elongated nuclei and occasional nuclear grooves in a background of numerous eosinophils.

D: The neoplastic cells express CD1a.

REFERENCES

1. Shields JA, Shields CL, Scartozzi R. Survey of 1264 patients with orbital tumors and simulating lesions: the 2002 Montgomery Lecture, part 1. Ophthalmology 2004;111:997-1008.
2. Shinder R, Al-Zubidi N, Esmaeli B. Survey of orbital tumors at a comprehensive cancer center in the United States. Head Neck 2011;33:610-4.
3. Verdijk RM. Lymphoproliferative tumors of the ocular adnexa. Asia Pac J Ophthalmol (Phila) 2017;6:132-42.
4. Andrew NH, Coupland SE, Pirbhai A, Selva D. Lymphoid hyperplasia of the orbit and ocular adnexa: a clinical pathologic review. Surv Ophthalmol 2016;61:778-90.
5. Swerdlow SH, Campo E, Harris NL, et al. WHO classification of tumours of haematopoietic and lymphoid tissues, 4th ed. Lyon: IARC Press; 2017.
6. Amin MB, Edge SB, Greene FL, et al., eds. AJCC cancer staging manual, 8th ed. Switzerland: Springer; 2017.
7. Rath S, Connors JM, Dolman PJ, Rootman J, Rootman DB, White VA. Comparison of American Joint Committee on Cancer TNM-based staging system (7th edition) and Ann Arbor classification for predicting outcome in ocular adnexal lymphoma. Orbit 2014;33:23-8.

8. Heidari P, Verdijk RM, van den Bosch WA, Paridaens D. Biopsy-proven recurrence of unilateral IgG4-related orbital inflammation after 20 years. Orbit 2014;33:388-91.
9. Sassone M. Ponzoni M, Ferreri AJ. Ocular adnexal marginal zone lymphoma: clinical presentation, pathogenesis, diagnosis, prognosis, and treatment. Best Pract Res Clin Haematol 2017;30:118-30.
10. Usui Y, Rao NA, Takase H, et al. Sci Rep. Comprehensive polymerase chain reaction assay for detection of pathogenic DNA in lymphoproliferative disorders of the ocular adnexa. Sci Rep 2016;6:36621.
11. Takahashi H, Usui Y, Ueda S, et al. Genome-wide analysis of ocular adnexal lymphoproliferative disorders using high-resolution single nucleotide polymorphism array. Invest Ophthalmol Vis Sci 2015;56:4156-65.
12. Rasmussen PK, Coupland SE, Finger PT, et al. Ocular adnexal follicular lymphoma: a multicenter international study. JAMA Ophthalmol 2014;132:851-8.
13. Rasmussen PK, Ralfkiaer E, Prause JU, et al. Diffuse large B-cell lymphoma of the ocular adnexal region: a nation-based study. Acta Ophthalmol 2013;91:163-9.
14. Knudsen MK, Rasmussen PK, Coupland SE, et al. Clinicopathological features of ocular adnexal mantle-cell lymphoma in an international multicenter cohort. JAMA Ophthalmol 2017;135:1367-74.
15. Romero IL, Campos FA, Damasceno RW, Vital Filho J, Bison SH. Plasmacytoma of the orbit involving lacrimal gland with secondary transformation into multiple myeloma: case report. Arq Bras Oftalmol 2009;72:236-8.
16. You JY, Glassman ML, McCormick SA, Milman T. Plasmacytoma associated with canaliculitis. Ophthal Plast Reconstr Surg 2012;28:e60-2.
17. Uceda-Montañés A, Blanco G, Saornil MA, Gonzalez C, Sarasa JL, Cuevas J. Extramedullary plasmacytoma of the orbit. Acta Ophthalmol Scand 2000;78:601-3.
18. Sen S, Kashyap S, Betharia S. Primary orbital plasmacytoma: a case report. Orbit 2003;22:317-9.
19. Kerstetter J, Wang J. Adult orbital xanthogranulomatous disease: a review with emphasis on etiology, systemic associations, diagnostic tools, and treatment. Dermatol Clin 2015;33:457-63.
20. Allen CE, Parsons DW. Biological and clinical significance of somatic mutations in Langerhans cell histiocytosis and related histiocytic neoplastic disorders. Hematology Am Soc Hematol Educ Program 2015;2015:559-64.
21. Haroche J, Abla O. Uncommon histiocytic disorders: Rosai-Dorfman, juvenile xanthogranuloma, and Erdheim-Chester disease. Hematology Am Soc Hematol Educ Program 2015;2015:571-8.
22. Pajaziti L, Hapçiu SR, Pajaziti A. Juvenile xanthogranuloma: a case report and review of the literature. BMC Res Notes 2014;7:174-7.
23. Davies MJ, Whitehead K, Quagliotto G, Wood D, Patheja RS, Sullivan TJ. Adult orbital and adnexal xanthogranulomatous disease. Asia Pac J Ophthalmol (Phila) 2017;6:435-43.
24. McKelvie P, McNab AA, Hardy T, Rathi V. Comparative study of clinical, pathological, radiological, and genetic features of patients with adult ocular adnexal xanthogranulomatous disease, erdheim-chester disease, and IgG4-related disease of the orbit/ocular adnexa. Ophthal Plast Reconstr Surg 2017;33:112-9.
25. Qi R, Chen LYC, Park S, et al. Utility of serum IgG4 levels in a multiethnic population. Am J Med Sci 2018;355:61-6.
26. Baykal C, Polat Ekinci A, Yazganoglu KD, Buyukbabani N. The clinical spectrum of xanthomatous lesions of the eyelids. Int J Dermatol 2017;56:981-92.
27. Goyal G, Shah MV, Call TG, Litzow MR, Hogan WJ, Go RS. Clinical and radiologic responses to cladribine for the treatment of Erdheim-Chester disease. JAMA Oncol 20171;3:1253-6.
28. Choi MB, Salomão DR, Smith WM, Pulido JS, Garrity JA. Ophthalmic findings of Rosai-Dorfman disease. Am J Ophthalmol 2018;188:164-72.
29. Shanmugam V, Margolskee E, Kluk M, Giorgadze T, Orazi A. Rosai-Dorfman disease harboring an activating KRAS K117N missense mutation. Head Neck Pathol 2016:10:394-9.
30. Herwig MC, Wojno T, Zhang Q, Grossniklaus HE. Langerhans cell histiocytosis of the orbit: five clinicopathologic cases and review of the literature. Surv Ophthalmol 2013;58:330-40.
31. Loo E, Khalili P, Beuhler K, Siddiqi I, Vasef MA. BRAF V600E Mutation across multiple tumor types: correlation between DNA-based sequencing and mutation-specific immunohistochemistry. Appl Immunohistochem Mol Morphol 2018;26:709-13.
32. Esmaeli B, Ahmadi MA, Manning J, McLaughlin PW, Ginsberg L. Clinical presentation and treatment of secondary orbital lymphoma. Ophthal Plast Reconstr Surg 2002;18:247-53.

8 TUMORS OF THE UVEA

ANATOMY AND HISTOLOGY

The iris (the diaphragm that encircles the pupil) connects to the ciliary body, and the ciliary body blends into choroid posteriorly. Together, these anatomic zones define the uvea (fig. 8-1).

Melanocytes derived from the neural crest populate the mesenchymal tissue of the iris, ciliary body, and choroid. The neural crest also gives rise to smooth muscle in the iris and ciliary body, a feature that is critical to understanding the histogenesis of rare tumors of smooth muscle origin in these locations.

The ciliary body is described morphologically as two zones. A flat zone, the pars plana, is contiguous with the choroid posteriorly and a corrugated zone, the pars plicata, bridges the pars plana to the iris. The outermost layers of the ciliary body and choroid are adjacent to the sclera, and the ciliary muscle inserts into a collagenous protuberance of the sclera known as the scleral spur (fig. 8-2).

The choroid is highly vascular, especially posteriorly, explaining the tendency for metastases to the eye to be identified at this site. The choroid abuts the sclera. The choroid is separated from the overlying retinal pigment epithelium by Bruch membrane (fig. 8-3). The neurosensory retina is situated internal to the retinal pigment epithelium. Both the neurosensory retina and the retinal pigment epithelium derive embryologically from neuroectoderm.

There are two intraocular pigmented cells: the pigment epithelia of the retina, the ciliary body, and the iris and neural crest-derived melanocytes that populate the choroid, ciliary body, and iris stroma. The pigmented epithelia are capable of an array of histologic responses to injury including hyperplasia and metaplasia, and these tissues are assumed to be capable of generating intraocular fibrous membranes and even bone. Reports of benign pigment epithelial neoplasms are rare. There are reports

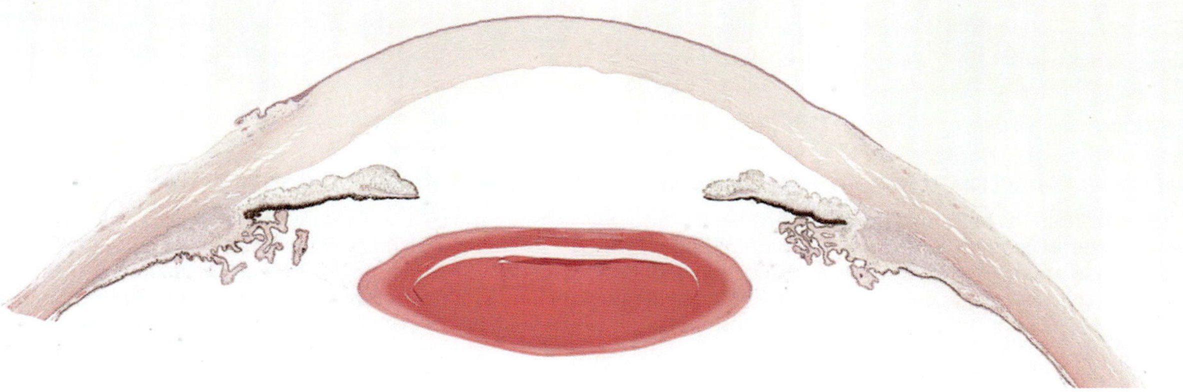

Figure 8-1

ANTERIOR SEGMENT OF NORMAL ADULT EYE

Beginning from the left lower corner and working clockwise, there is the flat portion of the ciliary body, the pars plana, which merges imperceptibly into the pars plicata, just behind the iris. The surface of the iris undulates. Flattening of the surface of the iris at this magnification suggests that a membrane has contracted.

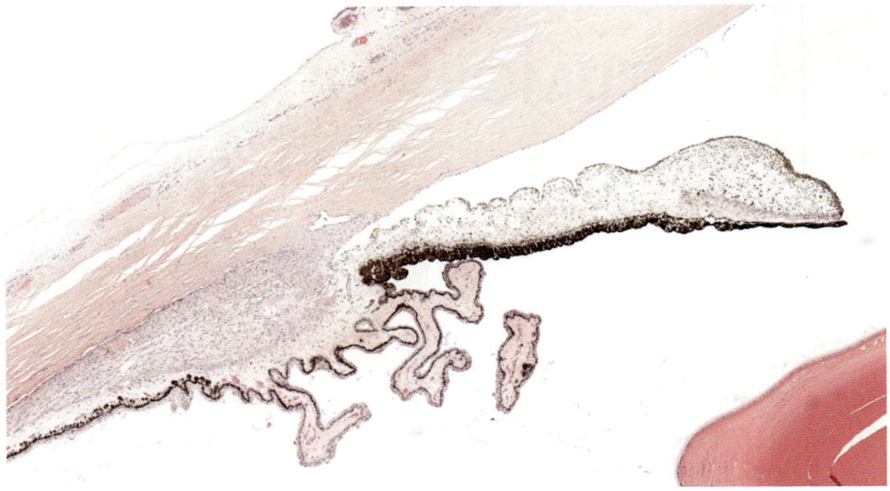

Figure 8-2

CILIARY BODY AND IRIS OF NORMAL ADULT EYE

The same tissue as illustrated in figure 8-1, at 4X magnification shows that the iris, ciliary body (attached to the sclera), and cornea outline the filtration angle. The irregularly shaped triangular space in the sclera just above the anteriormost extent of the ciliary body smooth muscle is the canal of Schlemm.

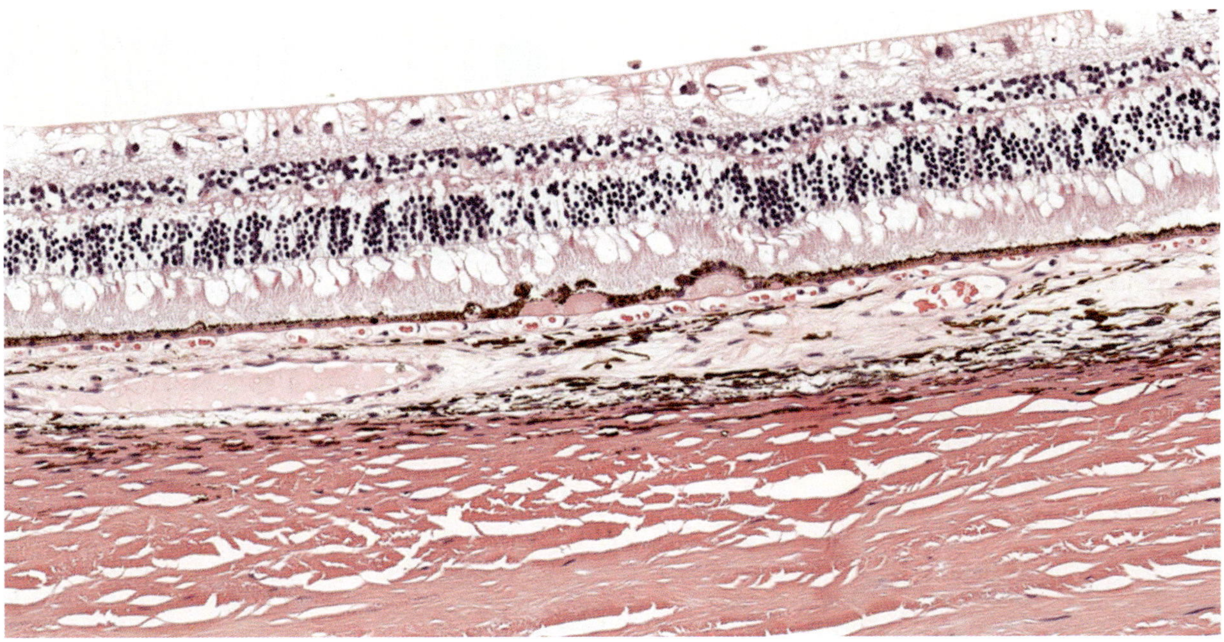

Figure 8-3

SCLERA, CHOROID, RETINAL PIGMENT EPITHELIUM, AND NEUROSENSORY RETINA

Starting from the bottom of the figure, first seen is the densely eosinophilic sclera. The loose fibrovascular connective tissue between the sclera and the unilayered ribbon of pigmented cells (the retinal pigment epithelium) is the choroid. Spindled melanocytes are present in the choroid. These pigmented cells are derived embryologically from the neural crest and give rise to choroidal nevi and melanomas. The retinal pigment epithelium is derived embryologically from neuroectoderm and is a neuroepithelium. Toward the center two eosinophilic deposits are seen between the retinal pigment epithelium and the inner choroid. These deposits are known as drusen of the pigment epithelium. The neurosensory retina is positioned above the retinal pigment epithelium. The space at the top is the vitreous. By convention, tissues and pathologic changes located closer to the sclera are referred to as "outer" while tissues and pathologic conditions closer to the vitreous are described as "inner." Starting from outer to inner, there are three layers of nuclei: the outer nuclear layer, the inner nuclear layer, and only a few large ganglion cells. The eosinophilic line separating the neurosensory retina from the vitreous is the internal limiting membrane (ILM) of the retina. Note the rarefaction of the retina just beneath the ILM, a zone that is occupied normally by the retinal nerve fiber layer, which is atrophic in this case. The dropout of retinal ganglion cells and atrophy of the nerve fiber layer most likely resulted from the effects of longstanding glaucoma. A tumor affecting the filtration angle in this eye, not illustrated here, contributed to the pathogenesis of glaucoma.

of malignant pigment epithelial tumors, but malignancy is manifested by local invasion and not the tendency to generate metastases and contribute to mortality.

Uveal melanocytes have a much more limited biologic repertoire. There is no mention in the literature of "reactive" uveal melanocytic lesions. The literature is silent about "uveal melanocytic hyperplasia." Dermatopathologists may select from a library of terms to describe nevi (common acquired nevus, congenital nevus, Spitz nevus, blue nevus and more) while the ophthalmic pathologist is limited to the terms nevus and melanocytoma (a magnocellular nevus) and, of course, melanoma.

Except for a lymphatic-like circulation in the ciliary body, there are no lymphatics within the uvea. Thus, uveal melanomas do not commonly spread to regional lymph nodes (with the rare exception of some ciliary body melanomas that erode through the sclera to gain access to lymphatic channels in the overlying conjunctival substantia propria). Uveal melanoma tends to spread first and almost exclusively to the liver, and is, therefore, a human model of pure hematogenous dissemination of cancer and organ-targeted metastasis. Also, the uvea, and the interior of the eye, is an immune-privileged site. Therefore, the appearance of tumor-infiltrating lymphocytes within a uveal melanoma may indicate systemic immune contact with tumor cells and may be an indicator of subclinical metastasis.

CLASSIFICATION

Tumors of the uvea are classified into the following broad categories: lesions arising from neural crest-derived melanocytes (nevi, melanomas, and the rare paraneoplastic bilateral diffuse uveal melanocytic proliferation syndrome), mesenchymal neoplasms (hemangioma and leiomyoma), neural and peripheral nerve sheath tumors (neurofibroma, ganglioneuroma, and neurilemmoma), metastases to the uvea, tumor-like lesions (choroidal osteoma, xanthogranuloma), and secondary and metastatic lesions.

It is unlikely that pathologists will encounter any of the entities described in this chapter. Few iris nevi are excised unless there is clinical concern about the rare progression of these lesions to melanoma. It is not possible to excise a "suspicious" choroidal nevus without disrupting vision. Needle biopsies may be performed on suspicious choroidal nevi and the material obtained is typically sent for gene expression profiling or other cytogenetic and molecular testing, with or without cytopathologic review. Pathologists usually see a uveal nevus in the context of the removal of an eye for conditions unrelated to melanocytic neoplasia, the blind and painful eye, for example.

It used to be the case that the preferred primary treatment for choroidal and ciliary body melanomas was enucleation, removal of the eye. With the discovery that vision-sparing treatments such as brachytherapy and proton beam therapy yield the same survival outcome as enucleation (1), fewer enucleations are performed now. Enucleations are often reserved for patients whose tumors cannot be treated successfully by radiation without compromising vision or for complications of radiation therapy. Some ciliary body melanomas are amenable to excisional biopsy through a procedure designated as an "eyewall resection," but such procedures are performed at few institutions, and the indications for such procedures are restrictive (2).

Cancers that metastasize to the eye are infrequently subjected to a fine-needle aspiration biopsy if the primary source of the malignancy is unknown. Typically, metastasis to the eye occurs late in the course of disseminated cancer and life expectancy after this ocular complication is short. Treatment of metastatic cancer to the eye is typically palliative, addressing the desire to restore and preserve vision through a phase of terminal care (3). Eyes harboring metastatic cancers are not removed unless the eye is blind or painful, or the source of the primary cancer cannot be identified by any means other than tissue examination.

The remainder of the lesions described in this section are rare and seen as postmortem specimens or as incidental findings in eyes removed for non-neoplastic conditions. This chapter is divided into sections based on histogenesis instead of anatomy. Therefore, nevi of the iris, ciliary body, and choroid are presented as a unit rather than as separate entities. Hopefully, this presents an opportunity for fresh insights into these lesions.

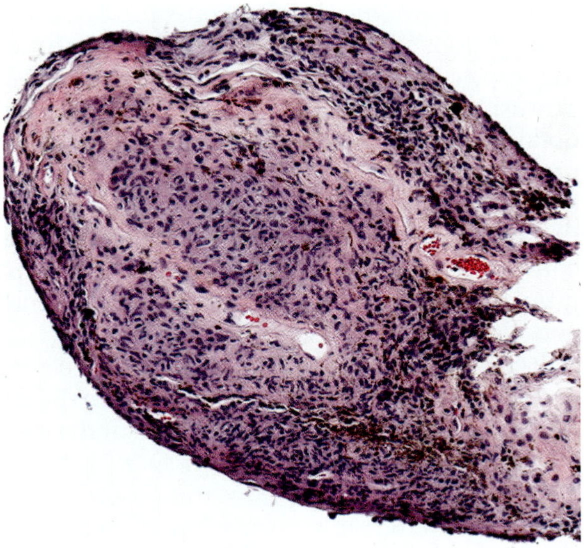

Figure 8-4

IRIS NEVUS

The iris is flimsy and the tissue often folds back upon itself. A small zone of pigment epithelium (which normally lines the posterior surface of the iris, see figs. 8-1, 8-2), is visible to the left. The stroma is occupied focally by a cohesive collection of cells that appears to surround blood vessels that are widely patent; the vessels are not compressed or distorted by the lesion, characteristic of benign lesions in general at scanning magnification. The surface of the iris is smooth, not undulating, suggesting a membrane of nevus cells on the surface.

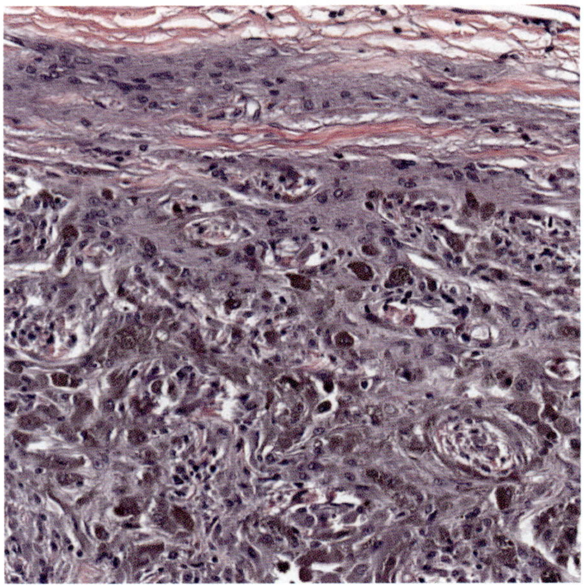

Figure 8-5

MELANOCYTOMA OF THE OPTIC NERVE

At the top are the characteristic large cells with cytologically bland nuclei that feature small or barely perceptible and uniformly sized and shaped nucleoli. Some spindled cells are also identifiable and the nuclear features associated with malignancy are not seen.

BENIGN MELANOCYTIC LESIONS

Uveal Nevi

Definition. The common *uveal nevus* is characterized histologically as a spindled melanocyte that lacks evidence of a nucleolus or in which nucleoli are inconspicuous and without nucleolar pleomorphism (fig. 8-4). A nevus of the iris or choroid is a benign, generally circumscribed lesion composed of histologically of nevus cells. Nevi of the ciliary body are rare, except for the nevus-variant of melanocytoma (see below). A *uveal melanocytoma*, also known as a *magnocellular nevus*, is populated by polygonal cells or spindled cells with centrally placed nuclei and inconspicuous nucleoli (fig. 8-5).

Nevi are acquired or congenital. Most uveal nevi are circumscribed and are assumed to be acquired. Within the realm of congenital nevi, there are small circumscribed nevi with limited potential for progression to melanoma and large congenital nevi with some risk of progression to malignancy. One variant of a congenital nevus is *congenital melanosis oculi* or *congenital ocular melanocytosis* (fig. 8-6). There may be a risk of progression to melanoma in eyes with congenital melanosis oculi in the white population (4).

Clinical Features. Nevi of the iris tend to develop in the inferior half of lightly pigmented irides. Clinically, a pigmented lesion is termed an iris freckle (perhaps a marker of chronic sun exposure) (5) and the distinction between nevus and freckle is a distortion of the iris stroma in nevi. This distortion is manifest by effacement of the highly textured iris surface that is so distinctive for its crypts that iris scans may form the basis for one type of biometric security. Iris nevi may distort the pupil (corectopia). Progression to melanoma is uncommon (6).

Most iris nevi are not resected unless there is clinical evidence of growth documented through serial photography over time. The development of feeder vessels or encroachment of the lesion into the filtration angle may prompt therapy

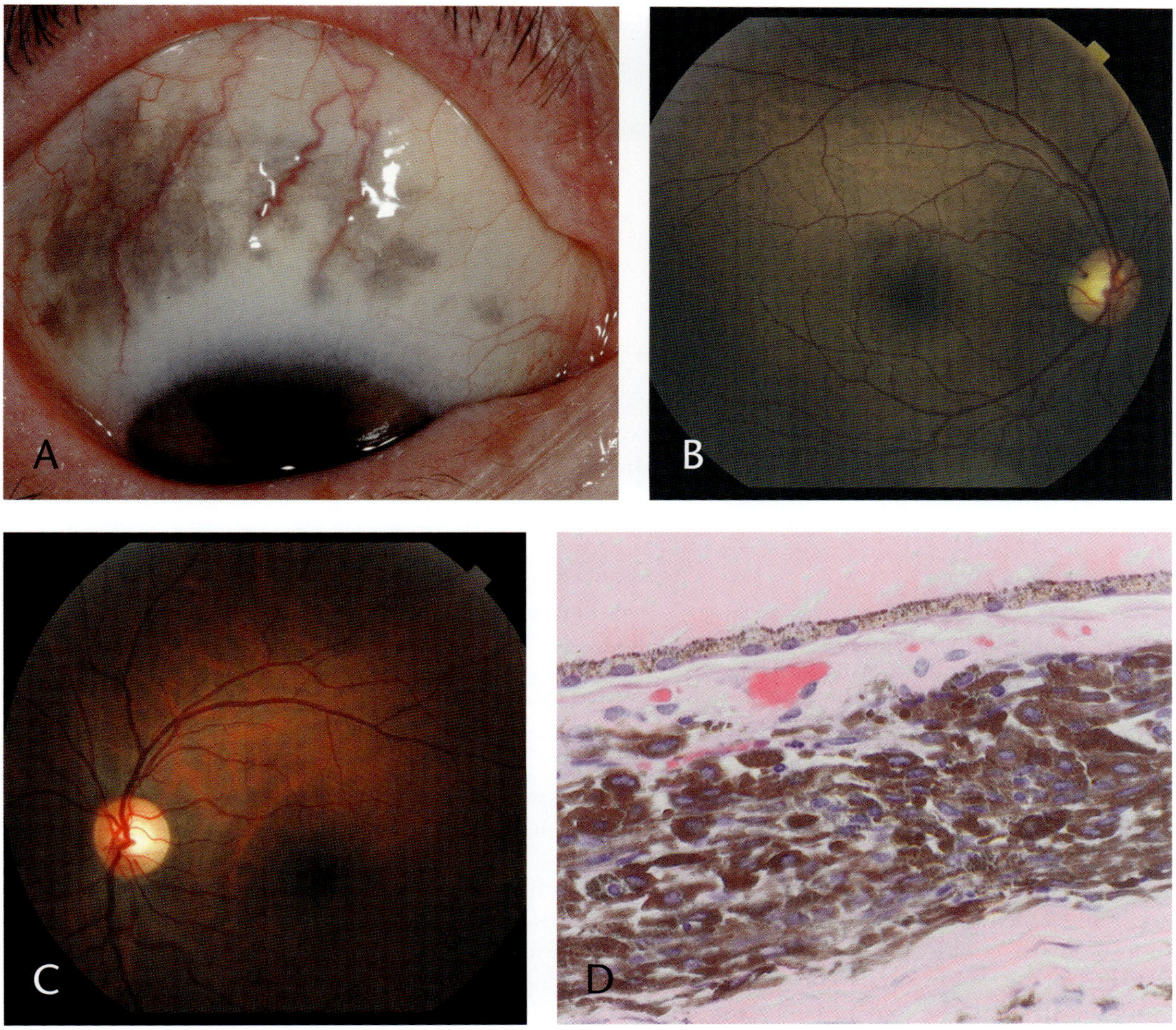

Figure 8-6

CONGENITAL MELANOSIS OCULI

A: The sclera is gray to blue. The conjunctival vessels are superficial to the pigmented zones. The scleral discoloration results from the presence of pigmented melanocytes in the sclera beneath the surface of the eye, associated with hyperpigmentation of the underlying uvea.

B,C: The dark appearance of the posterior pole (B) is compared with the unaffected normal eye shown in C. (A–C: courtesy of Drs. J. H. Francis and D. H. Abramson, New York, NY.)

D: The choroid is packed with cytologically benign melanocytes oriented along the same plane. Compare the appearance of the choroid in this eye with congenital melanosis oculi with the density of choroidal melanocytes in an eye not affected by this condition in figure 8-3.

Iris nevi may be discrete (sometimes referred to as solitary [fig. 8-7]) or diffuse (sometimes referred to as "sectoral" [7,8]). Iris nevi tend not to extend to the filtration angle, and if they do, then there is concern that the lesion may be a melanoma. Some iris nevi efface of the surface texture beyond the borders of the visible lesion. This effacement, in this clinical context, may be the result of a clinically transparent membrane of amelanotic or lightly pigmented iris nevus cells that can extend across the surface of the iris and generate adhesions to the cornea or trabecular meshwork (peripheral anterior synechias), thereby contributing to an elevation in intraocular pressure (9).

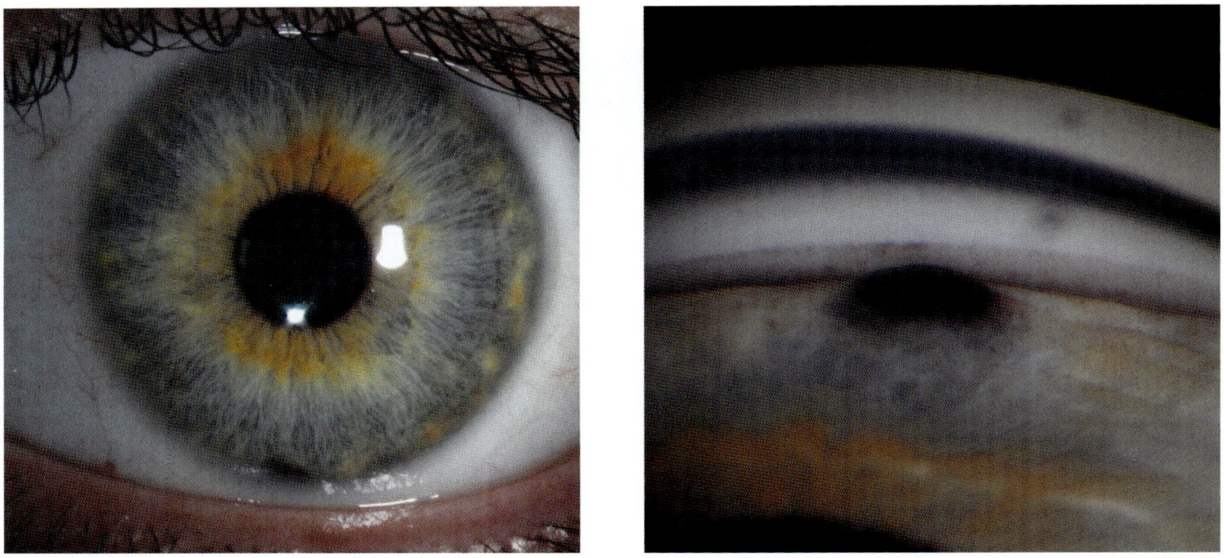

Figure 8-7

IRIS NEVUS

Left: A pigmented lesion is present inferiorly in this blue iris. The distinctive crypts and folds of the iris surface are preserved, indicating an absence of a membrane lining the iris.

Right: Gonioscopy shows that this discrete lesion is elevated. (Left, right: courtesy of Drs. J. H. Francis and D. H. Abramson, New York, NY.)

Most nevi of the ciliary body are subclinical with one exception. Melanocytomas (magnocellular nevus) may affect the iris and ciliary body and erode through the sclera even though they are benign. A lesion in this area that has been biopsied with a clinical diagnosis of extraocular extension by melanoma may represent rare benign extraocular extension of a melanocytoma (10).

The choroid is the most common location for uveal nevi, which are present in nearly 5 percent of the population. Choroidal nevi are small and circumscribed. The presence of drusen (histologically, excrescences over Bruch membrane deposited by the retinal pigment epithelium) indicates chronicity (fig. 8-8) but is not definite evidence of benignity: overlying drusen may accompany melanomas. Choroidal nevi are typically devoid of orange pigment overlying the lesion (lipofuscin in the retinal pigment epithelium). Typically, there is no fluid between the nevus and the overlying neurosensory retina. Nevi that are thick with overlying orange pigment and subretinal fluid, especially in the absence of pigment epithelial drusen, may be classified clinically as atypical nevi or perhaps small melanomas. The overall risk of progression to melanoma is about 1 percent (11).

Melanocytomas appear anywhere in the choroid, but are commonly located adjacent to the optic nerve head. These lesions are jet black. Progression to melanoma is rare (figs. 8-9, 8-10).

Microscopic Findings. Iris nevi are composed of spindled cells that typically lack nucleoli. If nucleoli are present, they are small and indistinct, without evidence of nucleolar pleomorphism. The nucleus may be hollowed-out, and some cells feature nuclear folds along the long nuclear axis, typical of spindle A melanoma cells. Uveal melanocytic lesions composed exclusively of spindle A cells are usually classified as nevi. Mitotic figures are not seen in iris nevi, and the presence of mitoses should suggest iris melanoma.

Membranes do not line the surface of the healthy iris. The undulating histologic surface corresponds to the iris crypts and ridges seen clinically (figs. 8-1, 8-2). When a flat anterior surface is identified at low magnification, indicating effacement of these crypts, then the presence of a membrane on the iris surface may be inferred. An uncommon cause of such a membrane is the migration of benign nevus

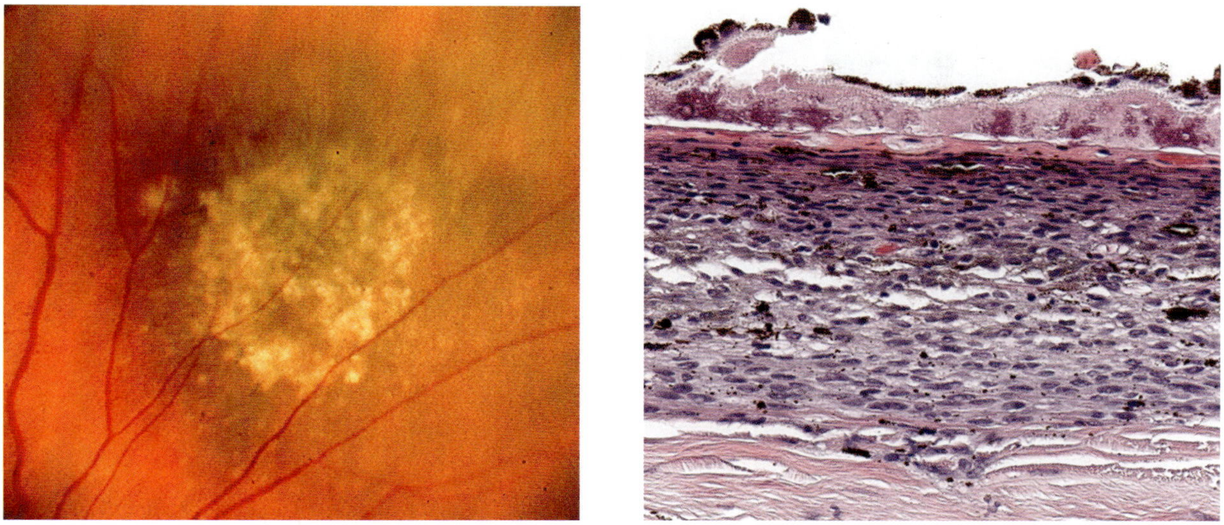

Figure 8-8

CHOROIDAL NEVUS WITH DRUSEN

Left: The lesion is small and circumscribed. The presence of drusen overlying the lesion suggests chronicity, and orange pigment (lipofuscin) is absent. Orange pigment overlying a choroidal pigmented lesion is suggestive clinically of malignancy. (Courtesy of Drs. J. H. Francis and D. H. Abramson, New York, NY.)

Right: Compared with the normal choroid accompanied by two isolated drusen that was illustrated in figure 8-3, the choroid here is thickened by cytologically benign spindled cells that are oriented in one plane parallel to the sclera. Nucleoli are either absent or inconspicuous.

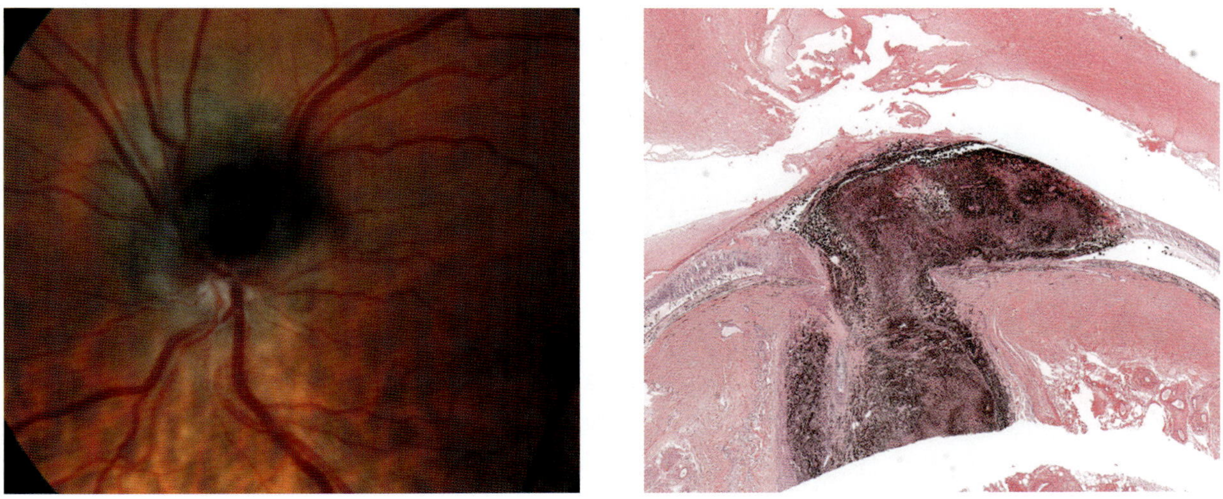

Figure 8-9

MELANOCYTOMA OF THE OPTIC NERVE HEAD

Left: The jet-black lesion is characteristic of melanocytoma. (Courtesy of Drs. J. H. Francis and D. H. Abramson, New York, NY.)

Right: The presence of intense melanin pigmentation in these lesions typically necessitates bleaching.

cells across the iris surface. The presence of a nevus with a surface membrane (plaque) should be noted in the pathology report because these clinically transparent membranes may progress across the surface of the iris and contribute to the formation of adhesions between the iris and cornea or the iris and the trabecular meshwork (peripheral anterior synechias (fig. 8-11).

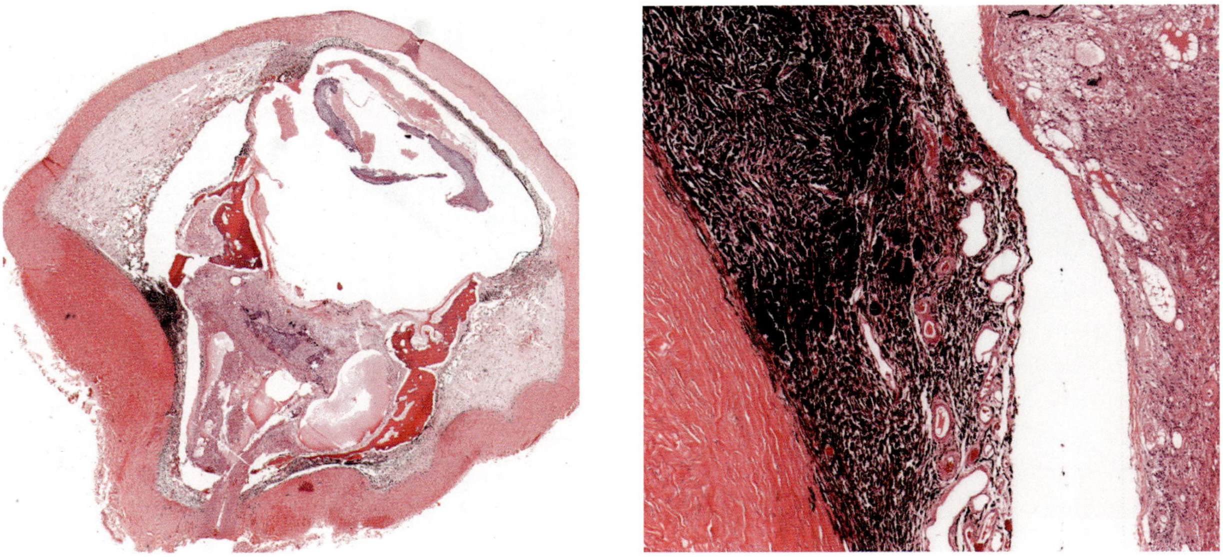

Figure 8-10

MELANOCYTOMA OF THE CHOROID

Left: Melanocytomas, although characteristically associated with a location at or near the optic nerve, may develop anywhere in the uvea. This choroidal lesion was identified as an incidental lesion in a blind and painful eye.

Right: Higher magnification of the lesion.

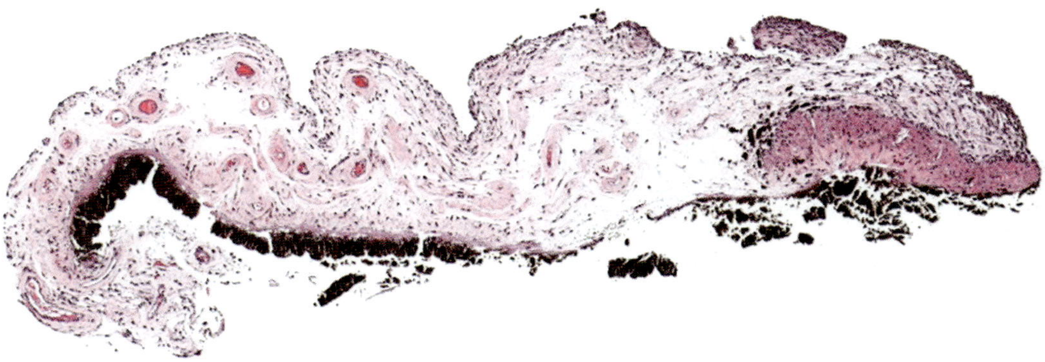

Figure 8-11

IRIS NEVUS WITH SURFACE PLAQUE

The thick iris sphincter muscle is at the right margin of the tissue sample. A small theque of nevus cells is on the surface of the iris just above the left-most extent of the sphincter muscle. At low magnification, a thin layer of pigmented cells on the surface of the iris is seen. It effaces the iris crypts for about 50 percent of the surface of this biopsy sample, from right to left. This thin, darkly pigmented layer is absent over the portion of the iris from the middle of this sample to the left-most margin and here, the undulating surface of the normal iris architecture is appreciated.

Choroidal nevi are typically incidental findings in an eye removed for other reasons. Spindled melanocytes, again without nucleoli or with small nucleoli without nucleolar pleomorphism, populate choroidal nevi. The architecture of a choroidal nevus is such that cells align typically in a single plane (fig. 8-12, top). Melanomas, by contrast, are populated by cells with prominent nucleoli and considerable variation in nucleolar size and shape. The orientation of melanoma cells to each other is chaotic and not regular (fig. 8-12, bottom).

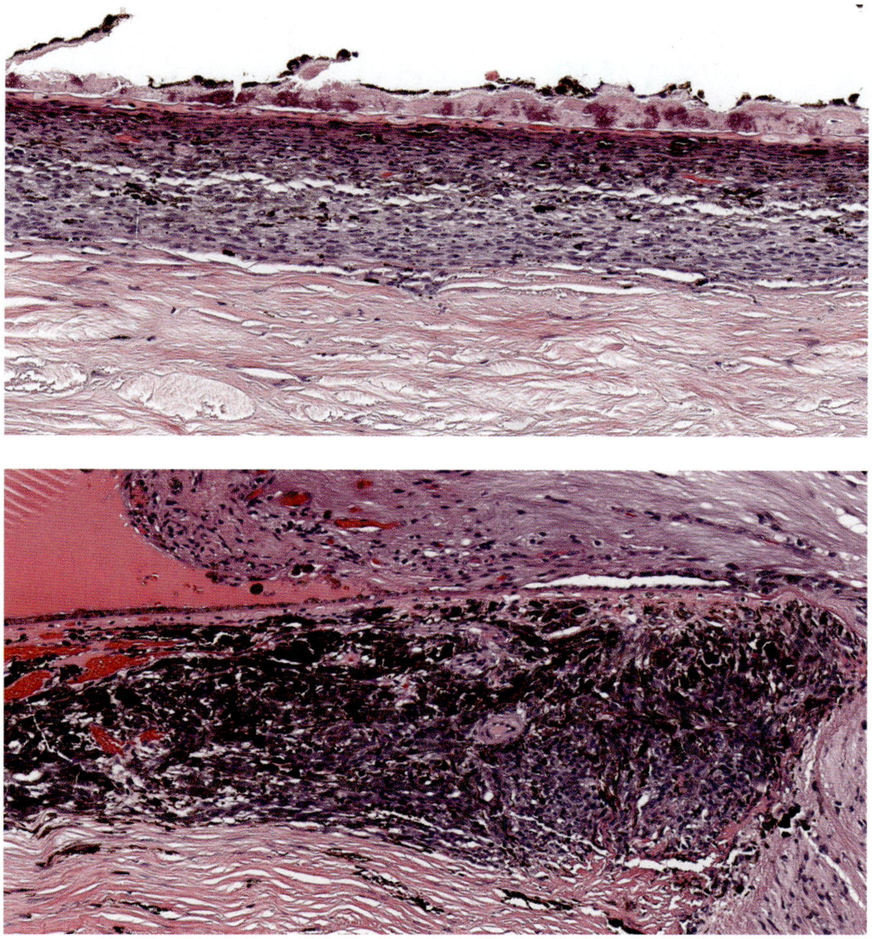

Figure 8-12

CHOROIDAL NEVUS COMPARED TO CHOROIDAL MELANOMA

Top: A choroidal nevus with overlying drusen is seen.

Bottom: A choroidal spindle cell melanoma shows the chaotic arrangement of the spindled cells compared with the nevus.

Drusen of the retinal pigment epithelium may overlie choroidal nevi (see fig. 8-8). These are periodic acid–Schiff (PAS)-positive excrescences between the surface of Bruch membrane and the overlying retinal pigment epithelium.

Melanocytoma, the magnocellular nevus, is heavily pigmented (fig. 8-13). Melanin bleaching is required to visualize polygonal or spindled cells (12) with uniformly sized nuclei and small, uniformly sized and shaped nucleoli (see fig. 8-5).

Immunohistochemical Findings. Although it is true that "nevi grow," the presence of cells expressing proliferation markers such as Ki-67 should shift the diagnostic focus toward melanoma. Ki-67 stains do not need to be routine. The use of proliferation markers may be relevant when there is doubt about benignity.

Special stains are not required to establish the diagnosis. If there is uncertainty about the histogenesis of the lesion, then markers for melanocytes such as HMB-45, S-100 protein, SOX10, and MITF may be helpful. Some melanocytic lesions will not stain for one marker or another and if immunohistochemistry is required, and this is not the case with most such lesions, more than one marker should be used.

Cytologic Findings. A suspicious iris lesion may be sampled by fine-needle aspiration biopsy before attempting resection. The cytologic differentiation between nevus and melanoma (or atypical nevus) again focuses upon the general absence of nucleoli in nevi and the presence of prominent, pleomorphic nucleoli in more aggressive lesions.

Melanocytomas and even necrotic melanomas may generate melanophages in the anterior chamber that are detectable upon aspiration of aqueous humor. The accumulation of melanophages in the anterior chamber may contribute to the development of glaucoma (melanomalytic or melanocytomalytic glaucoma). The volume of aqueous in the anterior chamber is very small

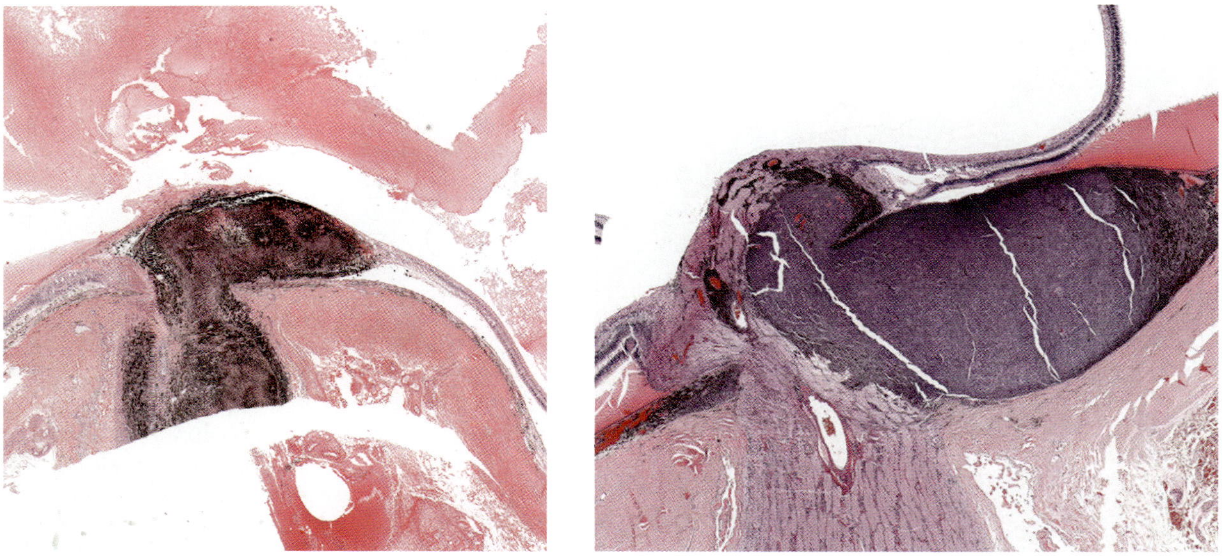

Figure 8-13

MELANOCYTOMA COMPARED WITH MELANOMA OF THE OPTIC NERVE

Left: This is a heavily pigmented melanocytoma.

Right: The malignant melanoma is adjacent to the optic nerve head, a juxtapapillary melanoma. The component of the lesion to the right of the insertion of the optic nerve has ruptured Bruch membrane, imparting a mushroom configuration to the lesion. The component of this tumor to the left of the insertion of the optic nerve is the area from which figure 8-12, bottom was taken.

and cytologic aspirations of fluid here must be handled in the laboratory as one would process cytology samples of cerebrospinal fluid.

Nevi of the choroid are not amenable to incisional or excisional biopsy, and unless these lesions are clinically "atypical" (large size, overlying orange pigment, or subretinal fluid) they are likely to be followed by serial photography for evidence of change. A histologic encounter with a choroidal nevus may occur during the examination of an eye removed for other reasons.

If a choroidal nevus is clinically atypical or in the clinical gray zone between nevus and melanoma, a fine-needle aspiration biopsy of the lesion may be helpful. Historically, these lesions were cytologically classified according to the modified Callender classification (see below) (13) but more recently, cells extracted by fine-needle aspiration biopsies are sent directly for cytogenetic analysis or gene expression profiling. Unless specifically ordered, the cytopathologist may not examine cells microscopically.

Molecular Genetic Findings. *GNAQ* mutations have been detected in a minority of ordinary choroidal nevi (13 percent) (14) and in some melanocytomas (15). Screening for this mutation is at this time purely investigational.

Differential Diagnosis. Most iris nevi are solitary lesions. If iris nevi are multiple, a different diagnosis should be considered. Lisch nodules, lesions that form in neurofibromatosis (NF)1, are seldom biopsied. In one study, the histopathology of a Lisch nodule was likened to a neurofibroma (16).

The *Cogan-Reese syndrome* (also known as the *iris-nevus syndrome*) is a component of a spectrum of disorders characterized by the extension of corneal endothelial cells over the surface of the iris, with the deposition of a neo-Descemet membrane over the iris. The spectrum of disorders is known as the *iridocorneal-endothelial* (ICE) *syndrome*. The iris stroma prolapses through defects in the neo-Descemet membrane and the concentration of stromal elements in the prolapsed stroma may give the clinical appearance of a nevus (17). Usually, there are multiple prolapses, and the clinical distinction between these lesions and a nevus of melanocytic origin is clear. Lesions that are part of the ICE syndrome are seldom biopsied.

The histologic differential diagnosis of melanocytoma includes an aggregate of melanophages and melanoma. As in any other location, the aggregation of melanin into intracytoplasmic packets of varying shapes and sizes in melanophages distinguishes these cells from those of melanocytic origin, where the dispersion of melanin is uniform. The primary diagnostic concern in the histologic examination of a choroidal nevus is the possibility of melanoma.

Treatment and Prognosis. Most uveal nevi are observed and not treated. Serial imaging technologies allow documentation of any growth in iris nevi. Fine-needle aspiration biopsy is rarely performed for iris melanocytic lesions but can identify melanoma in a clinically suspicious lesion (18). Suspicious iris nevi may be resected.

In small choroidal lesions where the clinical diagnosis is uncertain (suspicious nevus versus small melanoma), fine-needle aspiration biopsy of the lesion may provide a diagnosis. Gene expression profiling (GEP) (19) may separate lesions with an indolent course from those with an anticipated aggressive course. Nonmelanocytic lesions of the choroid may also be segregated into GEP prognostic classes (20) and confirmation of melanocytic histogenesis by cytopathology (and immunohistochemistry if necessary) should precede or accompany GEP testing. The application of molecular markers to prognosis is discussed in the section related to uveal melanoma.

The prevalence of iris nevi in the white population is about 4 percent (8). The risk of progression to melanoma is low. The prevalence of choroidal nevi in the white population in the United States ranges from 4.6 to 7.9 percent and the risk of progression to melanoma is extremely low (21). In one recent study, the overall risk of progression to melanoma was calculated at about 1 percent (11).

Congenital Melanosis Oculi (Congenital Ocular Melanocytosis)

There are patients for whom the white of the eye, the sclera, appears clinically gray or blue and for whom the condition is present at birth or becomes evident shortly after that (see fig. 8-6A). The clinical differential of "blue sclera" is broad. In *congenital melanosis oculi* or *congenital ocular melanocytosis*, the sclera appears to be irregularly rather than uniformly blue. The ocular lesion may be associated with a blue lesion of the periocular skin (nevus of Ota or oculodermal melanocytosis) or more uncommonly with the nevus of Ito. Usually, only one eye is affected and when the ophthalmologist looks into an affected eye, the reflexive color of the fundus appears dark compared with the orange "red reflex" noted in the unaffected eye (see fig. 8-6B,C).

Some assert that among patients with this condition who are not of Asian or African descent, there is a risk of progression to melanoma. It may be that congenital ocular melanosis is the uveal counterpart of congenital nevus of the skin where such lesions, when large, may predispose a patient to cutaneous melanoma.

The risk of a uveal melanoma developing in an eye with congenital melanosis oculi is 1 in 400 patients. The risk of metastasis from a melanoma developing in the context of congenital melanosis oculi is estimated to be 1.6 times greater than when congenital melanosis oculi is absent (4).

In an eye removed for uveal melanoma, diffuse choroidal pigmentation may be seen. After melanin is bleached out with permanganate, there may be bland spindled melanocytes, regularly oriented, generally along one plane, with nuclei that lack nucleoli or that feature small nucleoli. Typically, in such eyes, heavily pigmented cytologically unremarkable melanocytes are detected in the sclera. The presence of such heavily pigmented melanocytes is responsible for the clinical appearance of "blue sclera," with the Tyndall effect accounting for the clinical change in color from the brown of melanin to the clinically evident blue (fig. 8-6D). Suspected congenital melanosis oculi, if present, in eyes removed for uveal melanoma, should be mentioned.

Bilateral Diffuse Uveal Melanocytic Proliferation

Bilateral diffuse uveal melanocytic proliferation (BDUMP) is a paraneoplastic syndrome that is associated with ovarian cancer and other abdominal malignancies, especially of endodermal origin (22). In contrast to the lesions of congenital melanosis oculi, BDUMP is acquired and bilateral. The pathology is that of diffuse pigmentation of the choroid, sometimes with focal necrosis.

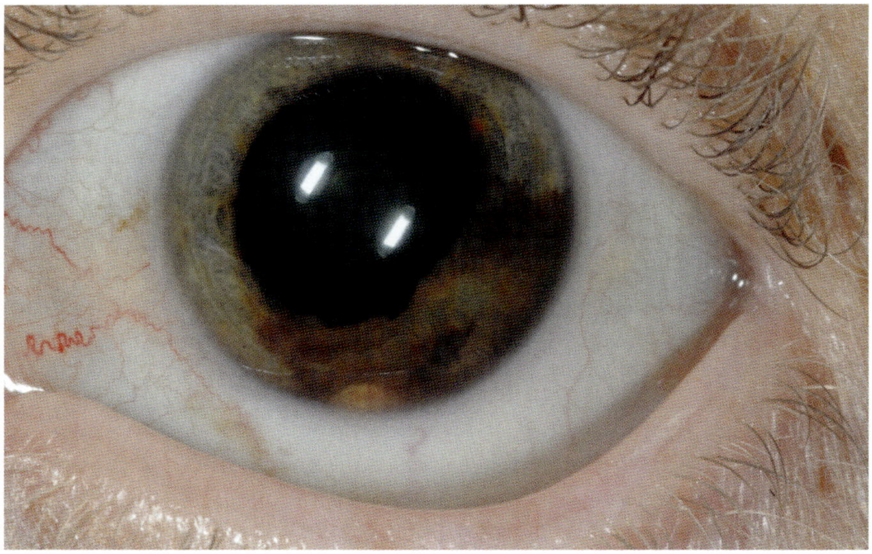

Figure 8-14

IRIS MELANOMA

The elevated, sectoral lesion has distorted the pupillary margin. The highly textured surface of the iris is completely effaced in the region of the melanoma. (Courtesy of Drs. J. H. Francis and D. H. Abramson, New York, NY.)

UVEAL MELANOMA

Definition. *Uveal melanoma* is a melanocytic malignancy arising in the iris, ciliary body, choroid, or a combination of these components of the eye.

General Features. The most common primary intraocular malignancy in adults, the reported of incidence of uveal melanoma is 1/100,000 in white Western populations (23). Unlike cutaneous or even conjunctival melanomas, melanomas of the uvea do not express *BRAF* mutations (24). They tend to develop in patients with a blue iris, and the tumors are located inferiorly, leading some to speculate that sun exposure is related to the pathogenesis of iris melanoma. The association between sun exposure and the pathogenesis of uveal melanoma is tenuous (25), although recent studies suggest that the genetic progression pathway points to the possibility of light in the pathogenesis of some uveal melanomas (26).

Patients at some risk of developing uveal melanoma include those with congenital melanosis oculi (27) (also known as congenital ocular melanocytosis, see below) and the BAP1 cancer syndrome (28). There appears to be no gender predilection (29). Nearly 50 percent of patients develop metastatic disease and currently there is no effective treatment for metastatic uveal melanoma (30).

There are four critical biologic characteristics associated with uveal melanoma. Uveal melanomas develop in an immune-privileged compartment, the interior of the eye. They disseminate almost exclusively by a hematogenous route in the absence of an intraocular lymphatic circulation. Ciliary body melanomas with extraocular extension into the overlying conjunctival substantia propria may provide access to lymphatics. Uveal melanomas metastasize almost exclusively to the liver first, a human model of organ-targeted metastasis (31). Metastases may appear years, and even decades, after the initial diagnosis and treatment (32) and thus become a human model of tumor dormancy.

Clinical Features. Melanomas confined to the iris may be asymptomatic and come to attention primarily because of a change in the appearance of the eye. Involvement of the trabecular meshwork may lead to an elevation in intraocular pressure (fig. 8-14). Melanomas of the ciliary body that grow circumferentially around the major arterial circle of the iris are also asymptomatic unless the patient develops glaucoma (33). Small choroidal melanomas may be asymptomatic and are detected on examination of the fundus. Clinical features such as documented growth by serial photography, the presence of orange pigment over the surface of the lesion, an absence or paucity of overlying drusen of the retinal pigment epithelium, and the presence of subretinal fluid as detected by imaging studies such as optic coherence tomography (OCT) point to a lesion that may represent a small choroidal melanoma. With time, some choroidal melanomas rupture through Bruch membrane to assume a mushroom shape.

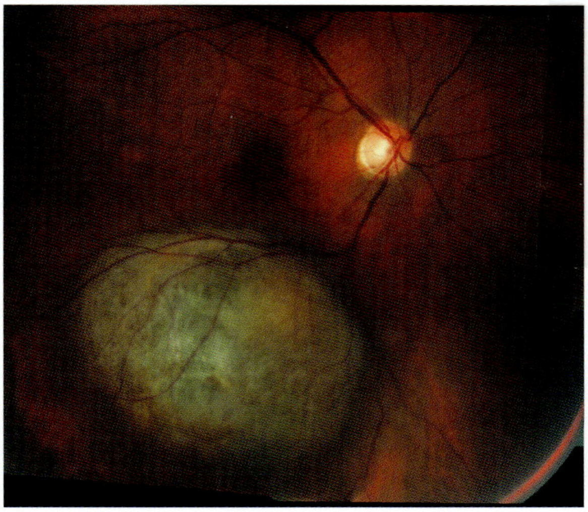

Figure 8-15

CHOROIDAL MELANOMA

This pigmented mass arises from the choroid. The dome-shaped lesion is confined to the choroid and has not ruptured through Bruch membrane. (Courtesy of Drs. J. H. Francis and D. H. Abramson, New York, NY.)

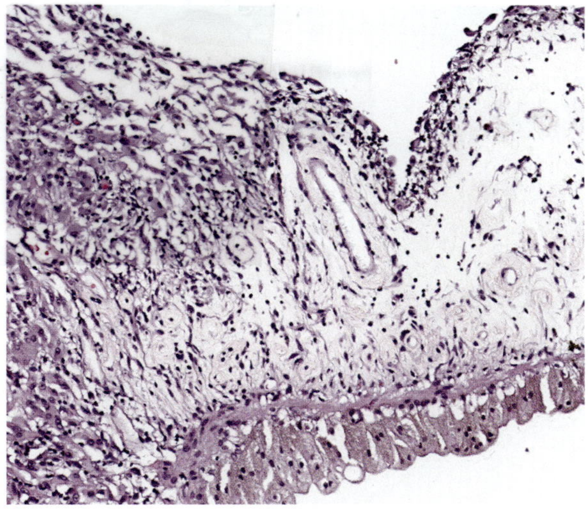

Figure 8-16

IRIS AND CILIARY BODY MELANOMA

The tumor involves the ciliary body (not shown) and the root of the iris. Tumor cells migrate over the surface of the iris. This surface membrane of tumor cells has not yet contracted.

Ciliary body melanomas may become symptomatic by a variety of mechanisms including encroachment upon the lens and the formation of a sectoral cataract. Ciliary body melanomas may lead to occlusion of the filtration angle by mechanical compression or by infiltration, leading to elevated intraocular pressure. Some ciliary body and iris melanomas become necrotic, and melanophages from the necrotic tumor accumulate in the filtration angle and trabecular meshwork and lead to elevated intraocular pressure (melanomalytic glaucoma [34]).

Some ciliary body melanomas have a component in the iris, some extend posteriorly into the choroid, and some involve all three uveal compartments: iris, ciliary body, and choroid. Visual impairment usually accompanies choroidal melanoma because of retinal detachment (fig. 8-15).

Gross and Microscopic Findings. The microscopic findings of iris, ciliary body, and choroidal melanomas are linked tightly to the identification of histologic characteristics associated with outcome. These findings are dependent upon how the specimen was obtained and grossed, and although chapter 1 in this volume is dedicated to the gross examination of ocular tissues, we mention here gross features that are relevant to the histologic examination of uveal melanomas.

Iris melanoma is often seen after local resection. It is essential to assess the lateral margins of the resection and to comment upon the presence or absence of a surface membrane of neoplastic melanocytes that may extend to the resection margins. Melanomas confined to the iris tend to follow an indolent clinical course, with few lesions resulting in metastasis. Melanomas that involve the iris and ciliary body tend to follow a more aggressive clinical course (fig. 8-16).

A ciliary body melanoma may be seen in a local resection of the iris and ciliary body, following a procedure known as an iridocyclectomy. Uveal melanomas are encountered in the context of enucleation, the removal of an eye. Although it is common practice to open such eyes in a plane that captures the pupil and the optic nerve, there are times when an alternative approach to the gross examination is helpful. If a ring melanoma, a melanoma that grows circumferentially around the major arterial circle of the iris, is suspected, then it may be prudent to detach the anterior segment from the posterior segment and to obtain radial sections through the ciliary body, much in the same manner as a cervical cone biopsy.

Most commonly, ciliary body and choroidal melanomas become the focus of histologic

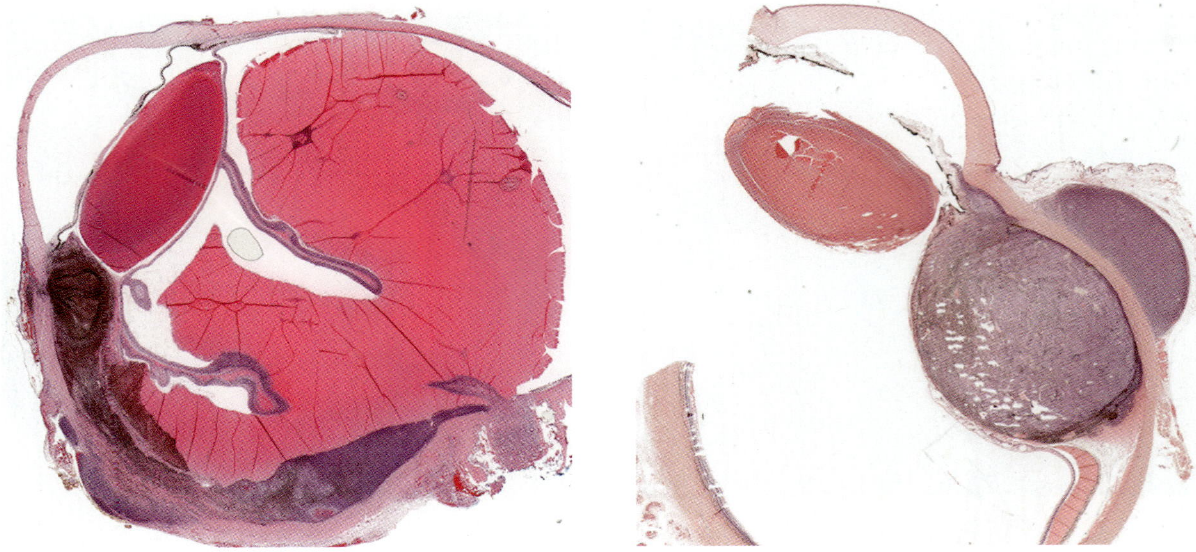

Figure 8-17

UVEAL MELANOMA WITH EXTRAOCULAR EXTENSION

Left: This uveal melanoma involves the choroid and ciliary body, with multiple areas of extraocular extension through the sclera. The retina is detached, with a protein-rich eosinophilic exudate in the subretinal space. There is a space between the optic nerve and the dura posteriorly. The optic nerve is atrophic.

Right: This tumor involves the choroid, ciliary body, and iris root. Tumor is detected beneath a flap of conjunctival tissue. It is rare for uveal melanomas to spread to lymph nodes because of the absence of major lymphatics within the eye. However, in a case like this, melanoma may gain access to a rich plexus of lymphatic vessels in the conjunctival substantia propria.

examination following enucleation. It is important to realize that there is no difference in survival in patients treated for choroidal melanoma with enucleation and those treated by radiation therapy in an attempt to treat the tumor while sparing vision. Therefore, at this time, enucleation specimens are only obtained from patients who express a clear preference for removal of the eye and those whose tumors cannot be treated by radiation without compromising vision. If it is likely that treatment of a choroidal melanoma will almost certainly result in vision compromise, then there may be no advantage to radiation in such cases over the removal of the eye.

Pathologists may encounter eyes that are removed after treatment by radiation, either brachytherapy or proton beam radiation, if the patient experiences vision loss and the eye becomes painful. One complication of radiation treatment is the development of glaucoma secondary to vascularization of the iris and adhesions that form between the iris and the trabecular meshwork and cornea, a condition known as neovascular glaucoma. Eyes with melanomas treated by radiation may also show evidence of tumor recurrence, perhaps with spread outside of the eye (extraocular extension).

During the gross examination, evidence of extraocular extension should be sought, and if this is suspected, then sections should include this focus for microscopic confirmation. Extraocular extension that exceeds the size of a microscopic focus is a histologic feature associated with adverse outcome (fig. 8-17).

Pathology Report. The histologic examination of choroidal and ciliary body melanomas should include a careful examination of tissue for the following characteristics. This description follows the outline of the 8th edition of the American Joint Committee on Cancer (AJCC) guidelines for reporting these tumors, with recommendations for modifying and simplifying these guidelines for clinical relevance. Some histologic features that are omitted by the AJCC recommendations may become relevant as new therapeutic strategies emerge from the laboratory into clinical practice.

Location. The precise location of the tumor is noted, whether involving the ciliary body only, choroid only, or both. Location is related to prognosis: melanomas confined to the iris have the most favorable prognosis, and melanomas involving any component of the ciliary body have the least favorable prognosis. Melanomas confined to the choroid have a prognosis positioned between tumors confined to iris and ciliary body.

Growth Pattern. Mention should be made if a ciliary body melanoma is a ring melanoma or if a melanoma of the choroid is diffuse, affecting multiple quadrants of the eye. An adverse outcome accompanies the identification of these patterns.

Size. Whereas the relationship between size and outcome is the depth of invasion, the dimension most critical to the pathologist's report is the largest diameter of the tumor in contact with the sclera, known as the largest basal diameter. Clinically, tumors are classified by the largest basal diameter and the elevation. Some choroidal melanomas rupture through Bruch membrane in a mushroom configuration, but rupture or the invasion of the overlying neurosensory retina does not point to an adverse outcome.

The AJCC classification and of choroidal melanomas appears in Appendix 7.

Intrascleral Invasion and Extraocular Extension. Tumor may infiltrate into the sclera along the pathway of ciliary arteries or within vortex veins. It may also infiltrate along the pathway of intrascleral ciliary nerves. Sclera may be invaded without following the course of any scleral emissary nerve or vessel. Extension outside the eye, extraocular extension, is associated with adverse outcome, especially if the extension is anything larger than a microscopic focus (fig. 8-17).

Cell Type: the Modified Callender Classification. The cytologic composition of uveal melanomas by cell type has been associated with outcome since the first classification described by Callender in the 1930s. A modification of the Callender classification (fig. 8-18) is presented next, followed by a short discussion of the limitations of applying this in practice.

Spindled melanoma cells lacking nucleoli may feature a nucleus with a central fold along the long axis, and such cells are classified as spindle A cells. Spindled cells with prominent nucleoli that are pleomorphic are designated as spindle B cells. Round cells with round nuclei and large, irregular nucleoli, are designated as epithelioid cells.

Lesions composed exclusively of spindle A cells are classified by many as nevi. Tumors composed of spindle B cells, with or without spindle A cells, but lacking epithelioid cells, are classified as *spindle cell melanomas*. Tumors composed exclusively (or nearly exclusively) of epithelioid cells are classified as *epithelioid melanomas*. Tumors composed of a mixture of spindled and epithelioid cells are classified as *melanomas of mixed cell type*.

The more epithelioid cells, the more adverse the clinical course. Therefore, spindle cell melanomas are associated with a relatively indolent course compared with melanomas of the epithelioid cell type and melanomas of the mixed cell type are associated with an intermediate clinical course.

Some melanomas are entirely necrotic, and under the modified Callender classification, the term *necrotic melanomas* describes these lesions. The prognosis of necrotic melanomas by cell type lies between spindle cell and epithelioid melanomas.

The Callender classification and its modification, although useful, is subject to limitations. Experts debate the number or percentage of epithelioid cells present that would shift the diagnosis from spindle cell melanoma to melanoma of the mixed cell type. Cross sections of spindle B cells may be mistaken for epithelioid melanoma cells. Intraobserver and interobserver reliability may not be high. Nevertheless, it is customary to include mention of the cell type in a pathology report.

Mitotic Figures and Proliferation Index. It is customary to count the number of mitotic figures per 40 high-power fields and to record this number in pathology reports (fig. 8-19, left). In larger tumors, it is worthwhile searching multiple fields until one mitotic figure is seen and to begin counting figures from this location. The identification of mitoses in uveal melanomas is associated with an adverse outcome.

The application of a formal proliferation index as measured by the percentage of cells expressing Ki-67 is not employed routinely. In highly pigmented tumors, it is necessary to use a chromogen that is not brown and to bleach sections gently to allow for detection of signal expression, and this is not practical in many cases since tissue begins to fall off slides after even mild

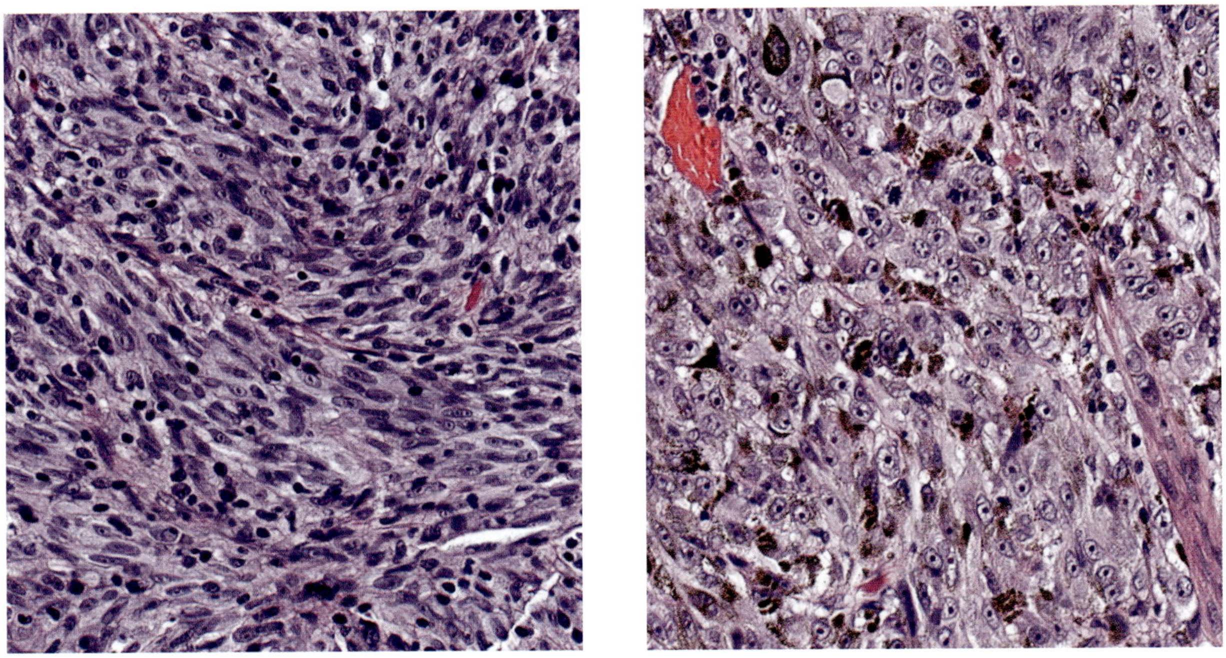

Figure 8-18

THE CALLENDER CLASSIFICATION

Left: The tumor cells in this section are clearly spindled. Cells lacking nucleoli and featuring folds along the longitudinal axis of the cell are designated as spindle A melanoma cells. These are found throughout this field and are especially prominent toward the left. Spindled cells with prominent nucleoli are designated as spindle B melanoma cells.

Right: These are epithelioid melanoma cells. The nuclei are round, the nucleoli are pleomorphic and large.

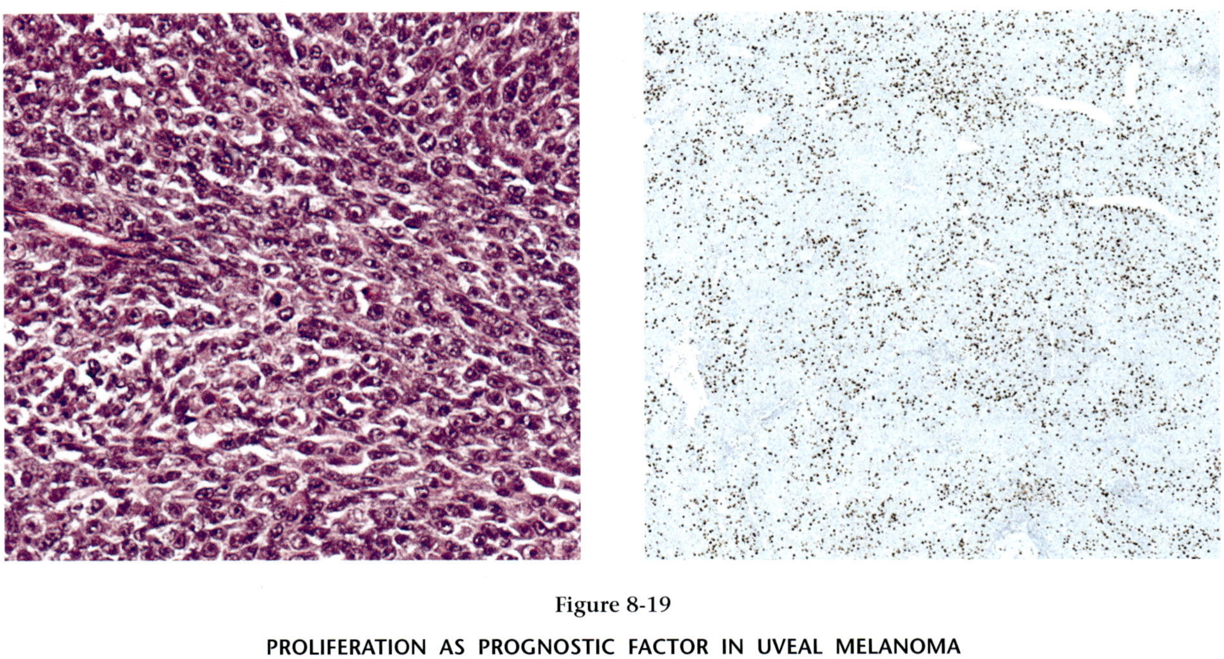

Figure 8-19

PROLIFERATION AS PROGNOSTIC FACTOR IN UVEAL MELANOMA

Left: It is customary to report the number of mitotic figures identified in 40 high-power fields at X40 magnification. A mitotic figure is seen in the center.

Right: This tumor section is stained for the expression of Ki-67. Calculation of proliferation indices based on Ki-67 expression is not performed routinely for cases of uveal melanoma.

bleaching. Nevertheless, comparison of mitotic figures and Ki-67 staining in an amelanotic melanoma illustrates the reality that the detection of mitotic figures may underestimate the population of proliferating cells (fig. 8-19, right).

Vasculogenic Mimicry. Vasculogenic mimicry (VM), the formation of a nonvascular microcirculation by aggressive tumor cells, was described first in uveal melanomas and has now been identified in many other cancers. There are several reasons for noting the absence or presence of VM in uveal melanoma tissues: the presence of VM in uveal melanomas is strongly associated with an adverse outcome, the histologic detection of VM is associated with an aggressive gene expression profile, and therapeutic strategies targeting VM are under investigation.

VM is a microcirculation generated by aggressive tumor cells consisting either of tubes that are not lined by endothelial cells, or highly patterned extracellular matrix material that may contain red blood cells. Loops and networks of back-to-back loops are common VM patterns; endothelial cell-lined vessels do not form closed loops in two-dimensional histologic sections. The extracellular matrix material is composed of an amalgam of proteins, including laminin and fibronectin, and VM is therefore much different from fibrovascular septa which have no prognostic significance. Because most laboratories do not use antibodies to fibronectin and laminin, PAS stain is used to highlight these patterns and the omission of hematoxylin in the counterstain aids in pattern recognition, even after bleaching of tissue to remove melanin that may obscure the identification of VM (fig. 8-20).

Tumor-Infiltrating Lymphocytes. The detection of at least 100 tumor-infiltrating lymphocytes per 20 high-power fields is associated with adverse outcome. Evidence of a lymphocytic infiltrate within the body of the tumor should be sought (fig. 8-21), avoiding the recording of this feature if it appears only at the interface between the tumor and normal adjacent choroid. In some cases, the inflammatory infiltrate is enriched by many plasma cells. This histologic feature was listed among prognostically significant histologic parameters, but recently, this tumor characteristic has fallen out of prognostic favor. Uveal melanoma arises in the context of an immune-privileged site, and the presence of tumor-infiltrating lymphocytes may reflect the presence of subclinical metastasis. It is important to record this histologic parameter, especially if promising immunotherapeutic approaches to the treatment of metastatic uveal melanoma emerge (fig. 8-21).

Immunohistochemical Findings. It is seldom necessary to confirm the melanocytic origin of uveal melanoma with immunohistochemistry. The expression of S-100 protein, MART1/Melan-A, HMB-45, MITF, and SOX10 have all been used to validate the histogenesis of a lesion that appears to be a uveal melanoma from tissue sections stained with hematoxylin and eosin (H&E). Immunohistochemistry to confirm histogenesis is neither expected nor required. The counting of mitotic figures, although less sensitive than the calculation of a proliferation index using the expression of Ki-67, is the method used most commonly to reflect proliferation for reasons articulated above (fig. 8-19, right).

There is some emerging interest in staining uveal melanomas for the expression of BAP1. The loss of expression of BAP1 may emerge eventually as a surrogate marker for the gene expression profile associated with an adverse clinical course.

Molecular Genetic Findings. There is no effective treatment for metastatic uveal melanoma at this time. Nevertheless, it is a high priority to classify patients into risk categories for metastasis with the hope that these stratifications can be used in clinical trials to develop effective treatment. These stratifications may also be used to generate sensitive means for the early detection of subclinical metastases.

There is no difference in survival between patients whose choroidal melanomas are treated with radiation therapy when compared with enucleation (35). The diagnosis of choroidal melanoma from clinical features is high, and patients may be treated with radiation therapy as a primary modality without tissue confirmation of the diagnosis.

Cytogenetic (36,37) and gene expression profiling studies (19,38) may stratify patients reliably into risk categories for metastasis. After a decision has been made to remove an eye, it is possible, even likely, that tissue may be harvested at the time of enucleation for these studies by the surgeon by fashioning a window

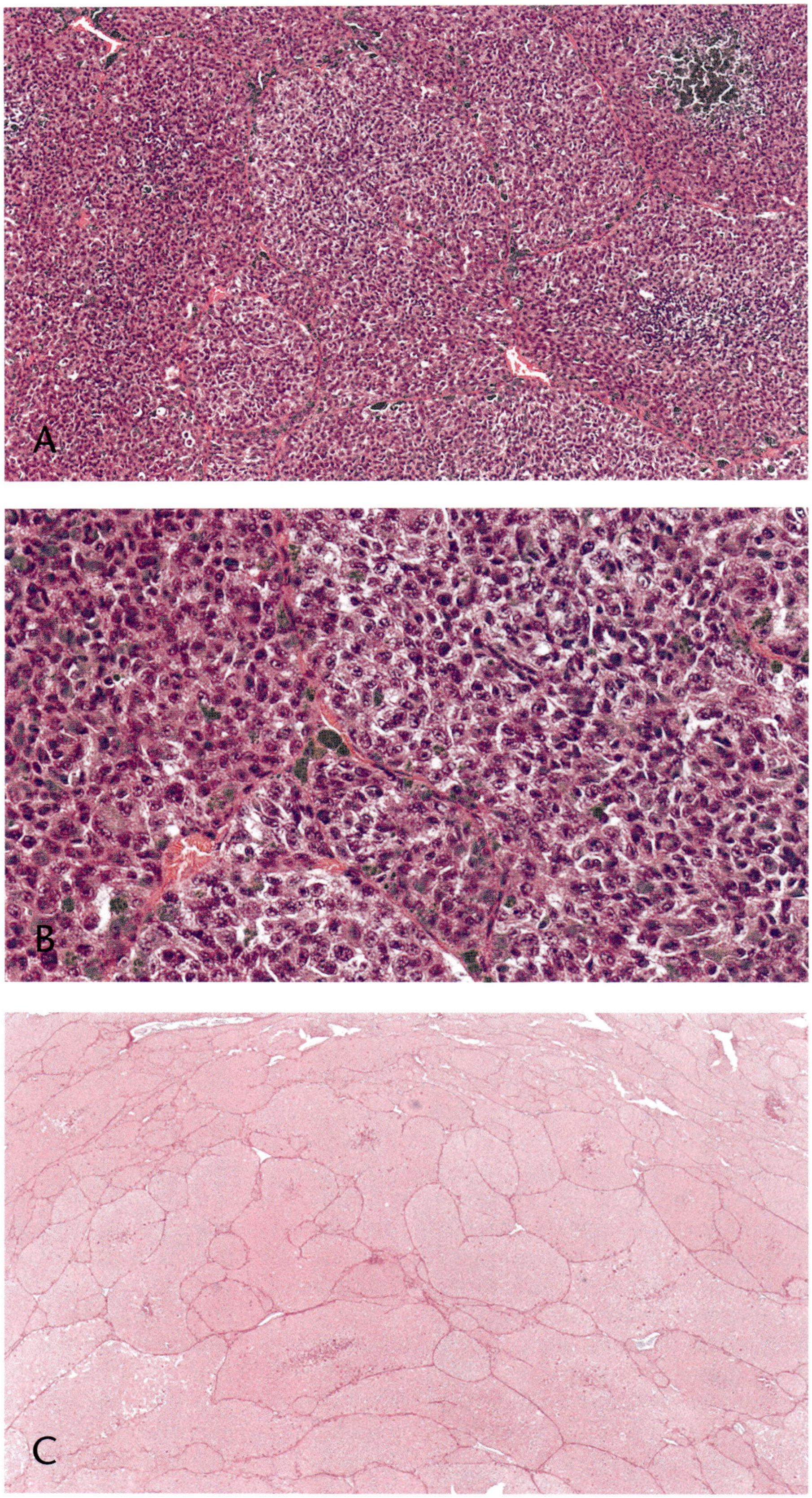

Figure 8-20

VASCULOGENIC MIMICRY

A: Vasculogenic mimicry (VM) at scanning magnification. Packets of tumor cells appear to be encircled and within the delimiting "loop," and a patchy distribution of red blood cells is detected around the circumference of the circle. VM has been shown to be an independent risk factor for metastasis in multiple studies.

B: At higher magnification, the rouleaux formation along the circumference of the delimiting "loops" is seen. Endothelial cells are not identified. The "circle" is composed of an amalgam of extracellular matrix proteins that are generated by genetically dysregulated melanoma cells, and these matrix patterns connect with the vasculature, permitting the matrix to be perfused. These are not blood vessels or components of fibrovascular septa.

C: Most laboratories do not stock antibodies to laminin subtypes or fibronectin. Staining tissue with the periodic acid-Schiff (PAS) stain and omitting counterstaining with hematoxylin generated the tissue section in this photomicrograph. In heavily pigmented tumors, this stain can be performed with permanganate bleaching to remove melanin. The looping patterns connect to blood vessels throughout the section.

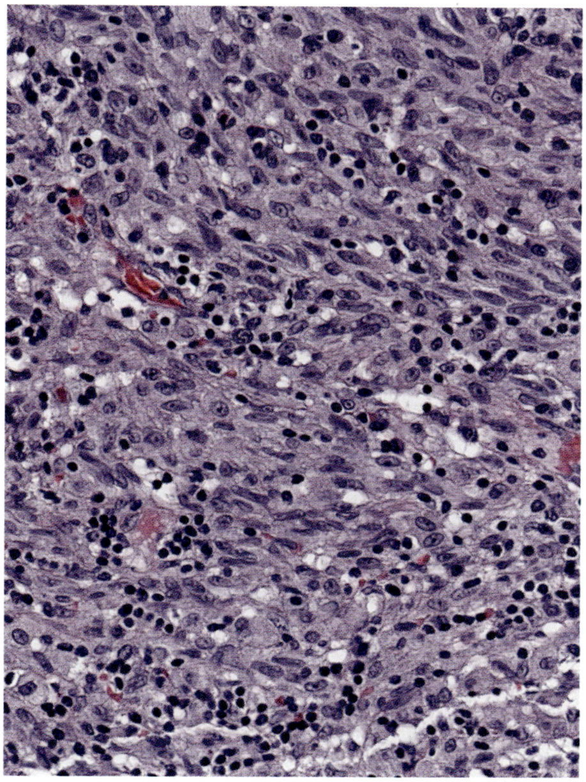

Figure 8-21

TUMOR-INFILTRATING LYMPHOCYTES

These are seen in primary uveal melanoma.

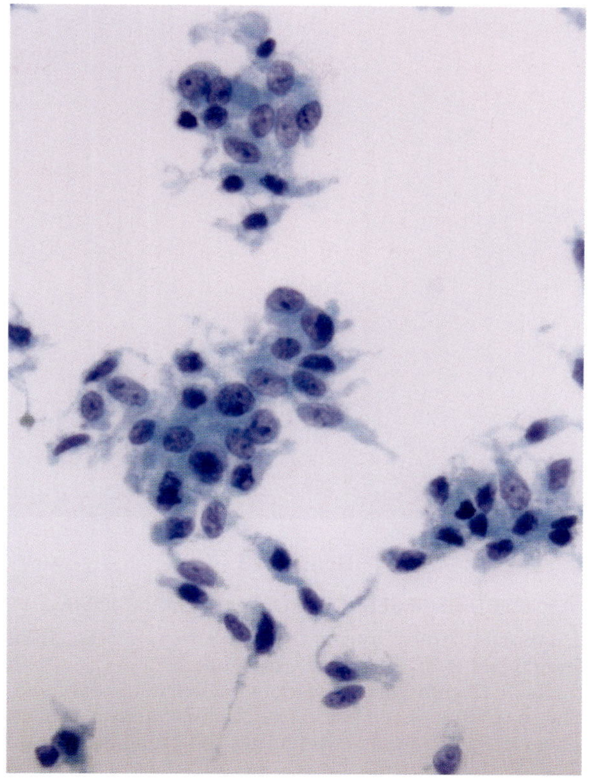

Figure 8-22

CHOROIDAL MASS

Cells obtained from a fine-needle aspiration biopsy of a choroidal mass. Pleomorphic nucleoli are prominent.

in the sclera to extract fresh tissue for analysis. Fine-needle aspiration biopsy is used to sample tumors in the eyes of patients before treatment with radiation therapy (brachytherapy or proton beam radiation). Cells removed may be sent for gene expression profiling or cytogenetic analysis before commencing treatment (19). As mentioned above, in exceptional cases, it is possible to stratify patients into risk categories by gene expression profiling even if the tumor is something other than a uveal melanoma (20). Therefore, practice patterns may evolve in which a fine-needle aspiration biopsy (fig. 8-22) is segmented into samples for cytology (with possible immunohistochemistry) to confirm the diagnosis of uveal melanoma and material for gene expression, cytogenetic studies, or both.

There are four major approaches to incorporating cytogenetic and molecular data into pathology reporting if these data are available. Often, these data become available after the pathologist issues a report. The pathologist should be aware of these tests and request the test results for updating the pathology report. Other classification systems may emerge over time.

Although several chromosomal abnormalities are prognostically significant in uveal melanomas, the dominant prognostic marker appears to be a loss of chromosome 3. Patients with monosomy 3 follow a more aggressive clinical course than patients with a disomy chromosome 3 status. Gain in chromosome 8q is another cytogenetic marker for an adverse outcome (39,40).

Gene expression profiles classify patients into risk categories. Patients classified by this technique as having "class 2" tumors tend to follow an aggressive clinical course compared with those whose profiles indicate a class 1 profile (19). Additional subclasses are associated with prognosis (41). As mentioned above, there may be an association between a loss of BAP1 expression and class 2 gene expression profiling

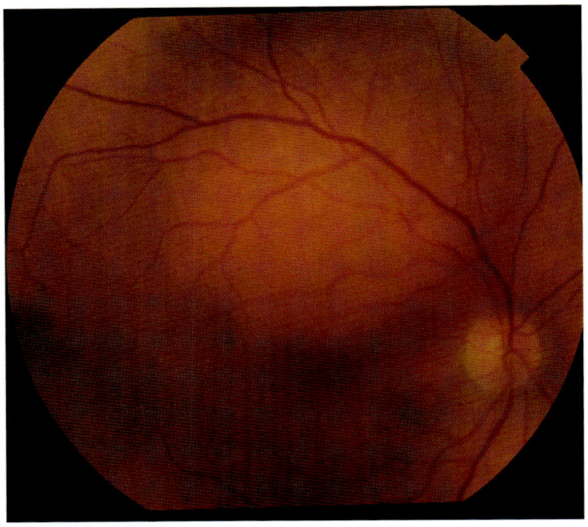

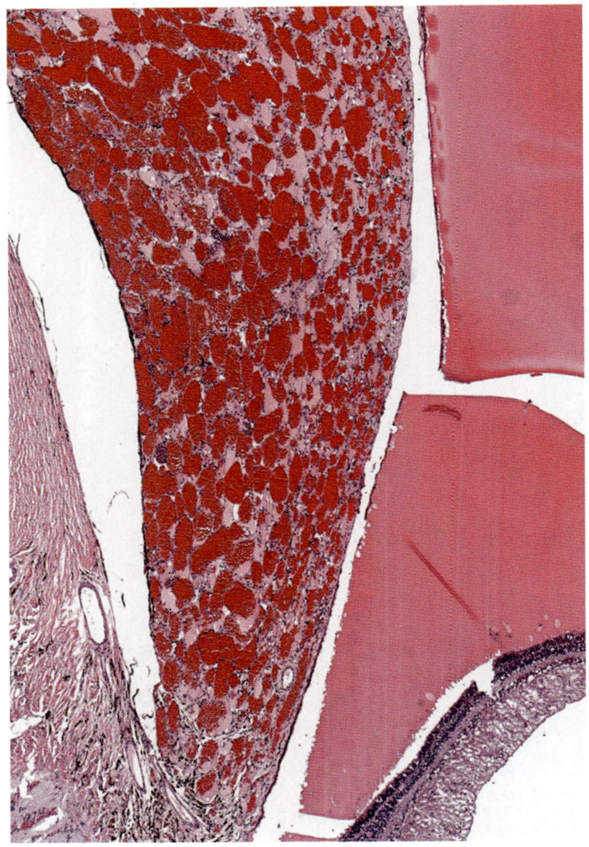

Figure 8-23

CIRCUMSCRIBED CHOROIDAL HEMANGIOMA

Above: Clinical appearance. (Courtesy of Drs. J. H. Francis and D. H. Abramson, New York, NY.)

Right: Circumscribed cavernous hemangioma of choroid. The separation of the lesion from the sclera is an artifact. There is a protein-rich, eosinophilic subretinal exudate.

(GEP) and it is possible that an absence of BAP1 expression by immunohistochemistry may emerge as a surrogate marker for detecting the class 2 GEP (42). Additionally, consideration of size (largest basal diameter) and other tumor characteristics, including GEP, may provide information more closely associated with outcome than GEP alone (43).

The expression of preferentially expressed antigen in melanoma (PRAME) is a marker of an adverse clinical outcome, and a comparison of gene expression profiling and PRAME expression is in progress (44). One recently published diagnostic algorithm first subclassifies uveal melanomas by chromosome 3 status and then subclassifies those tumors with disomy and monosomy status into prognostic subgroups by molecular markers (45).

OTHER UVEAL LESIONS

Choroidal hemangioma may be diffuse (as in the Sturge Weber syndrome) or circumscribed. Circumscribed choroidal hemangiomas are usually of the cavernous type (fig. 8-23). In the past, such lesions have been mistaken for uveal melanoma, leading to inappropriate enucleations. Contemporary imaging techniques have reduced the likelihood of incorrect clinical diagnoses and pathologists seldom encounter these lesions in enucleated eyes except as incidental findings (46).

Leiomyomas of the ciliary body may be histologically distinctive because the morphology suggests a neural origin (fig. 8-24). The smooth muscle in this area of the eye is presumed derived from neural crest, and diagnosis of mesectodermal leiomyoma for these benign tumors reflects the distinctive histogenesis (47). Immunohistochemistry may help establish the diagnosis. These tumors may express smooth muscle actin, desmin, h-caldesmon, CD56, and neuron-specific enolase; they do not express melanocytic, epithelial, or glial markers. It is possible to render an accurate diagnosis from material obtained through fine-needle aspiration biopsies (48).

The literature contains reports of rare peripheral nerve tumors including *ganglioneuroma* (49), *neurofibroma* (50), and *neurilemmoma* (51).

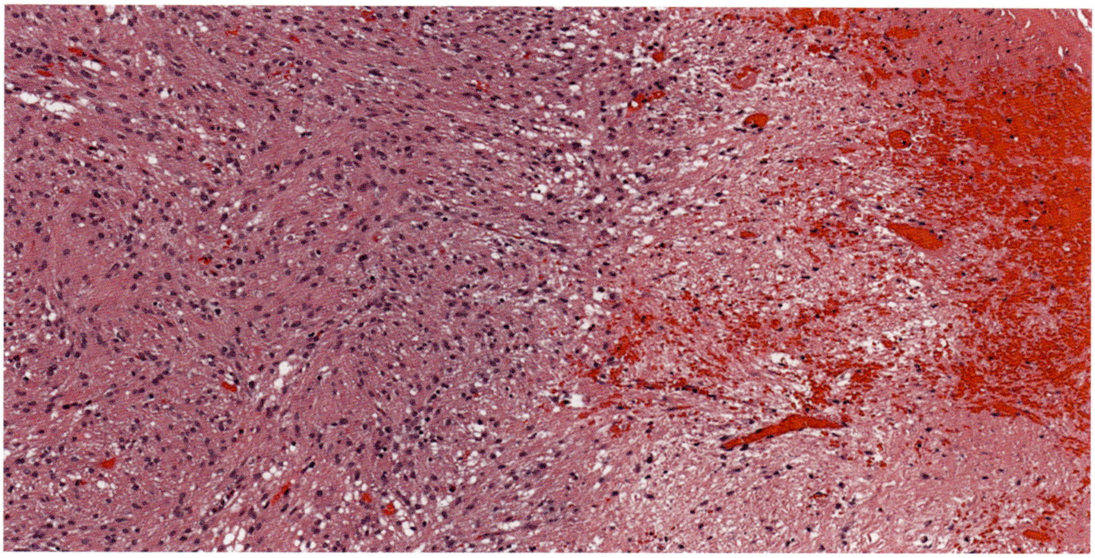

Figure 8-24

MESECTODERMAL LEIOMYOMA

Smooth muscle is identified at the left edge of the figure. From left to right, the lesional cells become more fibrillar as if looking at a neurilemmoma. This mixed morphology is characteristic of this lesion.

Xanthogranuloma may appear in the iris of young children. The lesion is typically very vascular and may result in elevation of intraocular pressure and spontaneous bleeding into the anterior chamber (hyphema). The histology is identical to its counterpart elsewhere. This condition seldom prompts an enucleation. Xanthogranulomas are accessible to fine-needle aspiration biopsy and treatment with topical corticosteroids is effective (52).

Although *choroidal osteoma* may have a place in the clinical differential diagnosis of choroidal melanoma, this rare lesion is unlikely to be seen histologically because imaging studies should lead to an accurate diagnosis. Histologically, bone is identified in the choroid (53). Bone, however, may form inside the eye, generally internal to Bruch membrane and therefore not in the choroid. Osseous metaplasia of the retinal pigment epithelium may explain bone in this location.

METASTASES TO THE UVEA

In one study from an eye bank, 13 percent of eyes from patients who died of cancer contained histologic foci of metastasis (54). A short life expectancy accompanies metastasis to the eye. If ocular metastases disturb vision, radiation therapy may provide a palliative measure to improve the quality of life in the terminal phases of the disease (55).

Metastases to the eye originate from nearly any source, although breast and lung cancers are the leading primary sites of tumors that disseminate to the eye. Tumors metastasize to any component of the uvea, including the iris, although the most common site is the choroid posteriorly where the microcirculation is exceptionally robust. Cancers may metastasize to ocular tissues other than the uvea, including the eyelid, conjunctiva, orbit, and retina.

Pathologists are likely to encounter tissue from tumors metastatic to the eye under two circumstances. If the eye is blind and painful, the eye may be removed. In cases where the primary tumor is unknown, a fine-needle aspiration biopsy or a partial resection (fig. 8-25) may identify the unknown primary or if vision is severely compromised and the primary tumor is unknown, a "diagnostic enucleation" may be performed.

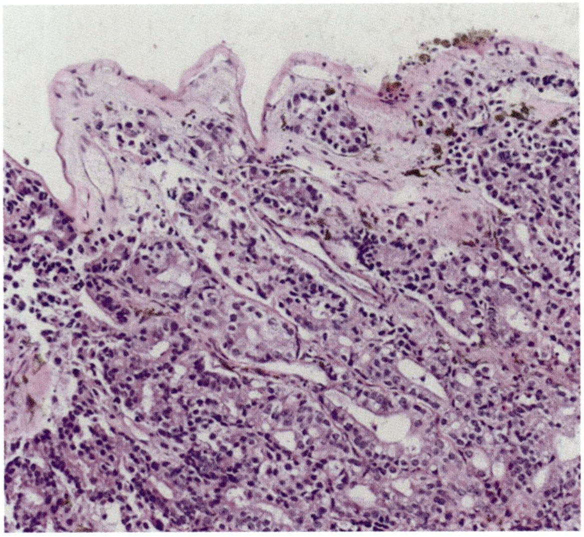

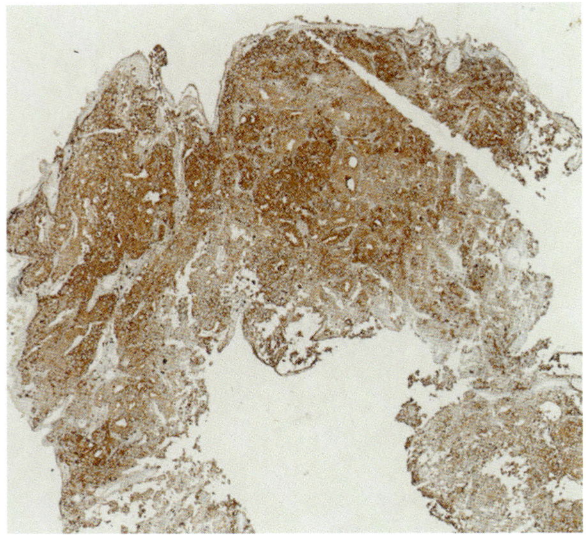

Figure 8-25

PROSTATIC CARCINOMA METASTATIC TO CHOROID

Left: The primary lesion in this patient was unknown and the choroidal lesion was biopsied. The undulating surface of the tissue is delimited by Bruch membrane and a small population of retinal pigment epithelial cells is identified at the top of the membrane, just to the right of center. The tumor is identified beneath Bruch membrane.

Right: The tissue is stained for the expression of prostatic acid phosphatase.

REFERENCES

1. Hawkins BS. The Collaborative Ocular Melanoma Study (COMS) randomized trial of pre-enucleation radiation of large choroidal melanoma: IV. Ten-year mortality findings and prognostic factors. COMS report number 24. Am J Ophthalmol 2004;138:936-51.
2. Damato B. The role of eyewall resection in uveal melanoma management. Int Ophthalmol Clin 2006;46:81-93.
3. Konstantinidis L, Damato B. Intraocular metastases—a review. Asia Pac J Ophthalmol (Phila) 2017;6:208-14.
4. Shields CL, Kaliki S, Livesey M, et al. Association of ocular and oculodermal melanocytosis with the rate of uveal melanoma metastasis: analysis of 7872 consecutive eyes. JAMA Ophthalmol 2013;131:993-1003.
5. Schwab C, Mayer C, Zalaudek I, et al. Iris freckles a potential biomarker for chronic sun damage. Invest Ophthalmol Vis Sci 2017;58:BIO174-9.
6. Shields CL, Kaliki S, Hutchinson A, et al. Iris nevus growth into melanoma: analysis of 1611 consecutive eyes: the ABCDEF guide. Ophthalmology 2013;120:766-72.
7. Demirci H, Shields CL, Shields JA, Eagle RC Jr, Honavar SG. Diffuse iris melanoma: a report of 25 cases. Ophthalmology 2002;109:1553-60.
8. Schwab C, Zalaudek I, Mayer C, et al. New insights into oculodermal nevogenesis and proposal for a new iris nevus classification. Br J Ophthalmol 2015;99:644-9.
9. Jakobiec FA, Silbert G. Are most iris "melanomas" really nevi? A clinicopathologic study of 189 lesions. Arch Ophthalmol 1981;99:2117-32.
10. Char DH. Surgical management of melanocytoma of the ciliary body with extrascleral extension. Comment On. Am J Ophthalmol 1994;118:404-5.
11. Chien JL, Sioufi K, Surakiatchanukul T, Shields JA, Shields CL. Choroidal nevus: a review of prevalence, features, genetics, risks, and outcomes. Curr Opin Ophthalmol 2017;28:228-37.

12. Juarez CP, Tso MO. An ultrastructural study of melanocytomas (magnocellular nevi) of the optic disk and uvea. Am J Ophthalmol 1980;90:48-62.
13. McLean IW, Foster WD, Zimmerman LE, Gamel JW. Modifications of Callender's classification of uveal melanoma at the Armed Forces Institute of Pathology. Am J Ophthalmol 1983;96:502-9.
14. Vader MJ, Madigan MC, Versluis M, et al. GNAQ and GNA11 mutations and downstream YAP activation in choroidal nevi. Br J Cancer 2017;117:884-7.
15. Mudhar HS, Doherty R, Salawu A, Sisley K, Rennie IG. Immunohistochemical and molecular pathology of ocular uveal melanocytoma: evidence for somatic GNAQ mutations. Br J Ophthalmol 2013;97:924-8.
16. Richetta A, Giustini S, Recupero SM, et al. Lisch nodules of the iris in neurofibromatosis type 1. J Eur Acad Dermatol Venereol 2004;18:342-4.
17. Scheie HG, Yanoff M. Iris nevus (Cogan-Reese) syndrome. A cause of unilateral glaucoma. Arch Ophthalmol 1975;93:963-70.
18. Gokhale R, Medina CA, Biscotti CV, Singh AD. Diagnostic fine-needle aspiration biopsy for iris melanoma. Asia Pac J Ophthalmol (Phila) 2015;4:89-91.
19. Harbour JW. A prognostic test to predict the risk of metastasis in uveal melanoma based on a 15-gene expression profile. Methods Mol Biol 2014;1102:427-40.
20. Klufas MA, Itty S, McCannel CA, Glasgow BJ, Moreno C, McCannel TA. Variable results for uveal melanoma-specific gene expression profile prognostic test in choroidal metastasis. JAMA Ophthalmol 2015;133:1073-6.
21. Singh AD, Kalyani P, Topham A. Estimating the risk of malignant transformation of a choroidal nevus. Ophthalmology 2005;112:1784-9.
22. Barr CC, Zimmerman LE, Curtin VT, Font RL. Bilateral diffuse melanocytic uveal tumors associated with systemic malignant neoplasms. A recently recognized syndrome. Arch Ophthalmol 1982;100:249-55.
23. Nielsen M, Dogrusoz M, Bleeker JC, et al. The genetic basis of uveal melanoma. J Fr Ophtalmol 2015;38:516-21.
24. Rimoldi D, Salvi S, Lienard D, et al. Lack of BRAF mutations in uveal melanoma. Cancer Res 2003;63:5712-5.
25. Singh AD, Rennie IG, Seregard S, Giblin M, McKenzie J. Sunlight exposure and pathogenesis of uveal melanoma. Surv Ophthalmol 2004;49:419-28.
26. de Lange MJ, Razzaq L, Versluis M, et al. Distribution of GNAQ and GNA11 mutation signatures in uveal melanoma points to a light dependent mutation mechanism. PLoS One. 2015;10:e0138002.
27. Singh AD, De Potter P, Fijal BA, Shields CL, Shields JA, Elston RC. Lifetime prevalence of uveal melanoma in white patients with oculo(dermal) melanocytosis. Ophthalmology 1998;105:195-8.
28. Abdel-Rahman MH, Pilarski R, Cebulla CM, et al. Germline BAP1 mutation predisposes to uveal melanoma, lung adenocarcinoma, meningioma, and other cancers. J Med Genet 2011;48:856-9.
29. Damato BE, Coupland SE. Differences in uveal melanomas between men and women from the British Isles. Eye (Lond) 2012;26:292-9.
30. Nichols EE, Richmond A, Daniels AB. Tumor characteristics, genetics, management, and the risk of metastasis in uveal melanoma. Semin Ophthalmol 2016;31:304-9.
31. Carvajal RD, Schwartz GK, Tezel T, Marr B, Francis JH, Nathan PD. Metastatic disease from uveal melanoma: treatment options and future prospects. Br J Ophthalmol 2017;101:38-44.
32. Kolandjian NA, Wei C, Patel SP, et al. Delayed systemic recurrence of uveal melanoma. Am J Clin Oncol 2013;36:443-9.
33. Stadigh A, Puska P, Vesti E, Ristimaki A, Turunen JA, Kivela TT. Ring melanoma of the anterior chamber angle as a mimicker of pigmentary glaucoma. Surv Ophthalmol 2017;62:670-6.
34. Yanoff M, Scheie HG. Melanomalytic glaucoma. Report of a case. Arch Ophthalmol 1970;84:471-3.
35. Collaborative Ocular Melanoma Study Group. Ten-year follow-up of fellow eyes of patients enrolled in collaborative ocular melanoma study randomized trials: COMS report no. 22. Ophthalmology 2004;111:966-76.
36. Prescher G, Bornfeld N, Hirche H, Horsthemke B, Jockel KH, Becher R. Prognostic implications of monosomy 3 in uveal melanoma. Lancet 1996;347:1222-5.
37. Sisley K, Rennie IG, Parsons MA, et al. Abnormalities of chromosomes 3 and 8 in posterior uveal melanoma correlate with prognosis. Genes Chromosomes Cancer 1997;19:22-8.
38. Onken MD, Worley L, Tuscan MD, Harbour JW. An accurate, clinically feasible multi-gene expression assay for predicting metastasis in uveal melanoma. J Mol Diagn 2010;12:461-8.
39. Werdich XQ, Jakobiec FA, Singh AD, Kim IK. A review of advanced genetic testing for clinical prognostication in uveal melanoma. Semin Ophthalmol 2013;28:361-71.
40. Dogrusoz M, Bagger M, van Duinen SG, et al. The prognostic value of AJCC staging in uveal melanoma is enhanced by adding chromosome 3 and 8q status. Invest Ophthalmol Vis Sci 2017;58:833-42.

41. Plasseraud KM, Wilkinson JK, Oelschlager KM, et al. Gene expression profiling in uveal melanoma: technical reliability and correlation of molecular class with pathologic characteristics. Diagn Pathol 2017;12:59.
42. Glasgow BJ, McCannel TA. Correlation of immunocytochemistry of BRCA1-associated protein 1 (BAP1) with other prognostic markers in uveal melanoma. Am J Ophthalmol 2018;189:122-6.
43. Dogrusoz M, Jager MJ, Damato B. Uveal melanoma treatment and prognostication. Asia Pac J Ophthalmol (Phila) 2017;6:186-96.
44. Reichstein D. New concepts in the molecular understanding of uveal melanoma. Curr Opin Ophthalmol 2017;28:219-27.
45. Robertson AG, Shih J, Yau C, et al. Integrative analysis identifies four molecular and clinical subsets in uveal melanoma. Cancer Cell 2018;33:151.
46. Singh AD, Kaiser PK, Sears JE. Choroidal hemangioma. Ophthalmol Clin North Am 2005;18:151-61.
47. Jakobiec FA, Font RL, Tso MO, Zimmerman LE. Mesectodermal leiomyoma of the ciliary body: a tumor of presumed neural crest origin. Cancer 1977;39:2102-13.
48. Kiratli H, Agin A, Onder S. Ciliary body mesectodermal leiomyoma diagnosed by fine needle aspiration biopsy. Ocul Oncol Pathol 2017;3:199-203.
49. Woog JJ, Albert DM, Craft J, Silberman N, Horns D. Choroidal ganglioneuroma in neurofibromatosis. Graefes Arch Clin Exp Ophthalmol 1983;220:25-31.
50. Burke JP, Leitch RJ, Talbot JF, Parsons MA. Choroidal neurofibromatosis with congenital iris ectropion and buphthalmos: relationship and significance. J Pediatr Ophthalmol Strabismus 1991;28:265-7.
51. Damato B, Damato EM, Konstantinidis L, Heimann H, Coupland SE. Choroidal schwannoma: a case series of five patients. Br J Ophthalmol 2014;98:1096-100.
52. Samara WA, Khoo CT, Say EA, et al. Juvenile xanthogranuloma involving the eye and ocular adnexa: tumor control, visual outcomes, and globe salvage in 30 patients. Ophthalmology 2015;122:2130-8.
53. Gass JD, Guerry RK, Jack RL, Harris G. Choroidal Osteoma. Arch Ophthalmol 1978;96:428-35.
54. Eliassi-Rad B, Albert DM, Green WR. Frequency of ocular metastases in patients dying of cancer in eye bank populations. Br J Ophthalmol 1996;80:125-8.
55. Cohen VM. Ocular metastases. Eye (Lond) 2013;27:137-41.

9 TUMORS OF THE NEUROSENSORY RETINA AND RETINAL PIGMENT EPITHELIUM

ANATOMY AND HISTOLOGY

Of the three coats of the eye (corneoscleral, uveal, and retinal), the retina is the inner-most and represents the neurosensory layer. The retina is located in the posterior portion of the eye and lies between the gelatinous vitreous body and the choroid, the vascular layer. The retina and ciliary body epithelia are formed by the ectodermal cells of the embryonic optic vesicle. The neuroepithelial layers arise from the neural crest and migrate anteriorly from the floor of the forebrain to form the optic vesicle and later the optic cup.

The retina is composed of 9 layers (fig. 9-1). The innermost layers encompass the ganglion cells, with their axons forming the optic nerve, and the retinal vasculature. The middle layer contains neurons (horizontal, amacrine, and bipolar), with synapses between ganglion cells, and also houses the body of the main glial cell of the retina, the Mueller cell. There is no fibroblastic stroma in the retina. The outer layers of the neurosensory retina contain the photoreceptors (cones and rods). The nonsensory layer is the retinal pigment epithelium (RPE), a hexagonal, cuboidal cell with apical microvilli that interact with the outer processes of the photoreceptors to maintain physiologic union and integrity between the RPE and the neurosensory retina (fig. 9-1, right). The basal portion of the RPE cell contains the nucleus and the pigment granules that are oval (football-shaped), unlike melanocytic pigment granules which are round (fig. 9-2). The basement membrane of the neuroepithelium forms part of Bruch membrane.

Bruch membrane is a five layer structure that in the inner portion is bound by the basement membrane of the RPE, at the center has two layers of collagen fibers alternating with an elastic fiber layer in the middle, and an outermost portion formed by the basement membrane of the choriocapillaris. This complex structure together with the RPE composes the outer portion of the blood-retinal barrier. The inner portion of the blood-retinal barrier is formed by the inner retinal vessels, which have tight junctions and are surrounded by pericytes.

The inner portion of the globe posteriorly is occupied by the vitreous humor (4 mL), a viscous gel-like material rich in hyaluronic acid. It is firmly attached to the retina, avascular, and mostly composed of water (more than 90 percent), collagen fibers, and rare cells (1).

The vascular choroid is composed of larger-caliber vessels at the outer portion toward the sclera and smaller vessels toward the retina, with a layer of capillaries facing Bruch membrane. The choroidal vessels are fenestrated in contrast with the retinal vessels which have tight junctions and are surrounded by pericytes. The stroma of the choroid contains mostly collagenous fibers, slender dendritic pigmented melanocytes more concentrated toward the suprachoroidal area, few fibrocytes, and rare lymphocytes and mast cells toward the far outer portion. Nerves and ganglion cells are also present. The areas where the bundles of medium-sized vessels and nerves enter and exit the eye through the sclera are called emissary channels and are sites of thinning of the sclera (fig. 9-3).

The anterior portion of the retina is limited by the ora serrata, which is the area where the neurosensory retina merges into the nonpigmented epithelium of the pars plana of the ciliary body and the RPE merges with the pigmented epithelium of the pars plana of the ciliary body (fig. 9-4). The ciliary body produces the aqueous humor which is secreted into the posterior chamber, the space between the iris and the lens, and travels to the anterior chamber where it filters out of the eye through the filtering system (trabecula, Schlemm canal, and aqueous veins) (fig. 9-5) (1).

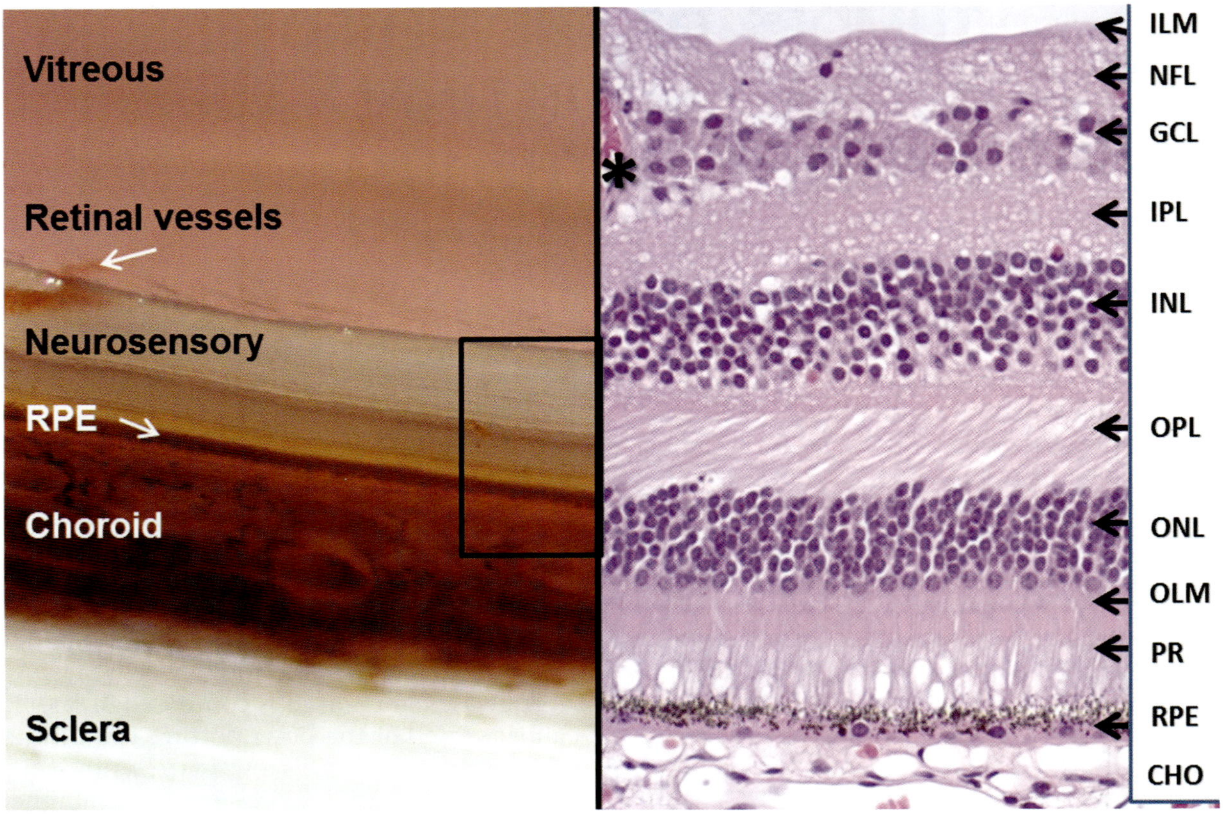

Figure 9-1

MACROSCOPIC ANATOMY AND HISTOLOGY OF THE RETINA

Left: Macroscopic view under stereoscopic dissecting microscope. Note the inner most portion the transparent translucent vitreous body, the white translucent neurosensory retina with the red-brown retinal vessels in the inner retinal layers, the pigmented retinal pigment epithelial (RPE) layer, the vascular choroid layer, and the most outer layer of the posterior pole, the sclera. The box indicates the approximate area of the microscopic view of the retina and choroid in right figure.

Right: Histologic section of the retina. There are nine neurosensory layers and the choroid. From the most inner layer of the retina, the inner limiting membrane (ILM), the nerve fiber layer (NFL), the ganglion cell layer (GCL), the inner plexiform layer (IPL), the outer plexiform layer (OPL), the outer nuclear layer (ONL), the outer limiting membrane (OLM), and the photoreceptor layer (PR). The non-sensory layer is the RPE. The inner layers of the retina contain the retinal vessels (*). The RPE rests on Bruch membrane, a structure that shares the RPE basement membrane with basement membranes from the choriocapillaris, a component of the choroid (CHO).

RETINOBLASTOMA AND RETINOCYTOMA

Definition and Classification. *Retinoblastoma* is a malignant tumor of the neural retina and is the most common intraocular malignancy in children, affecting 1 in 15,000 to 1 in 18,000 live births worldwide. Retinoblastoma may be congenital, sporadic, or hereditary, and presents predominantly in young children. The tumor may be unilateral, originating from a single site, or bilateral from multiple foci in one or both eyes.

Retinoblastoma is caused by a pathogenic variant in the *RB1* gene. This gene codes the retinoblastoma (RB) protein that controls the cell cycle. Children with the hereditary type of retinoblastoma have a susceptibility to developing other malignant tumors. Fewer than 2 percent of retinoblastomas are associated with *MYC* amplification and have different histopathologic findings (2).

The precise cellular origin of retinoblastoma is still unknown. The current understanding focuses on the histogenesis of retinoblastoma from a precursor neuronal stem-like cell. This cell may express markers different from those found in the neuronal/photoreceptor cells.

There is a form of the retinal tumor that is very well differentiated called *retinocytoma,* or *retinoma*

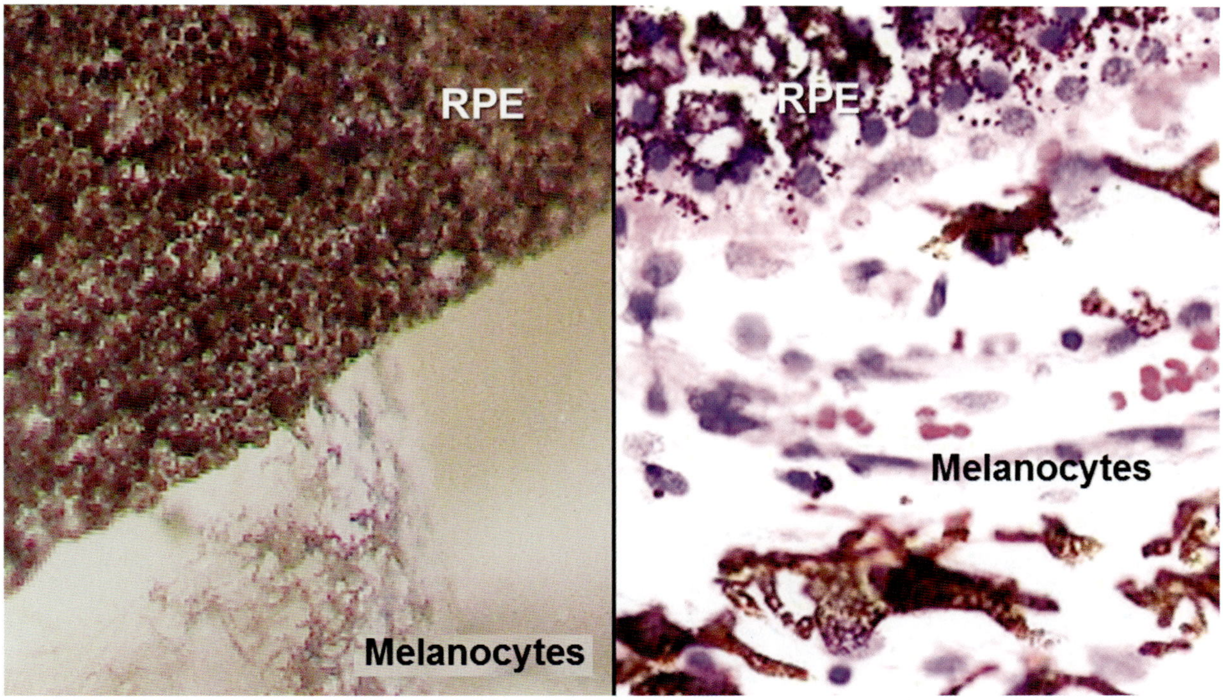

Figure 9-2

COMPARING CHOROIDAL DENDRITIC MELANOCYTES AND RETINAL PIGMENT EPITHELIAL CELLS (RPE)

Left: The stereoscopic microscope shows the cut edge of the sclera/choroid/RPE region. At the top left of the micrograph, the RPE monolayer is composed of hexagonal shaped cells. In the lower half of the image, the choroidal melanocytes with their dendritic appearance are admixed with the white fibrous stroma.

Right: A tangential section of the choroid (bottom) and the RPE (top) with choroidal vessels (at the center). The fine tan-brown, almost dust-like pigment granules in the dendritic choroidal melanocyte contrast with the well-formed, larger, round to oval (football-shaped), dark-brown pigment granules in the RPE cells.

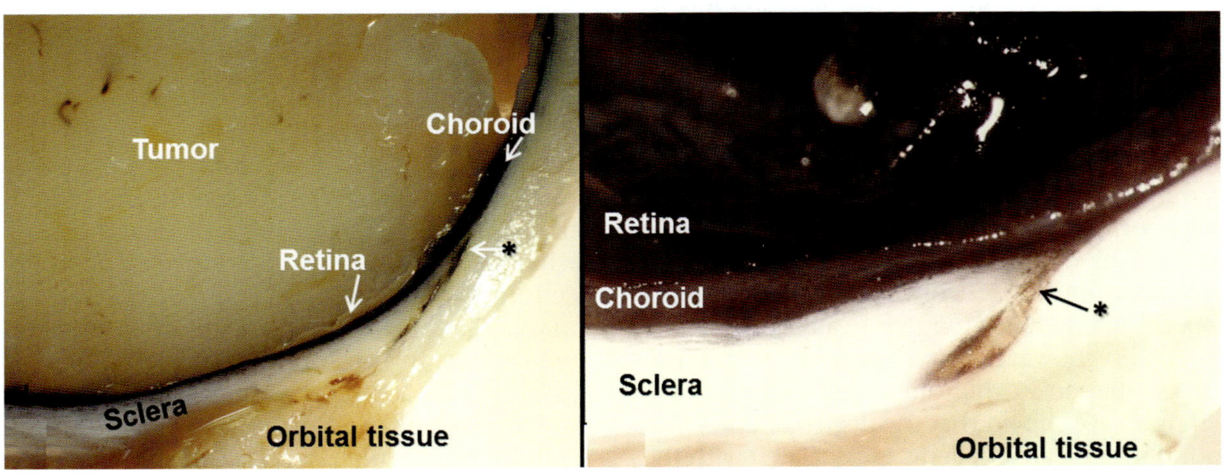

Figure 9-3

EMISSARY CHANNELS IN THE SCLERA

Left: This eye contains retinoblastoma.

Right: A section through an eye without intraocular tumor in which a scleral emissary channel (*) is seen. These channels are vascular/nerve bundles traversing the sclera to and from the extraocular tissues and choroid. The emissary channels are pathways through which tumors may exit the intraocular compartment into extraocular tissues.

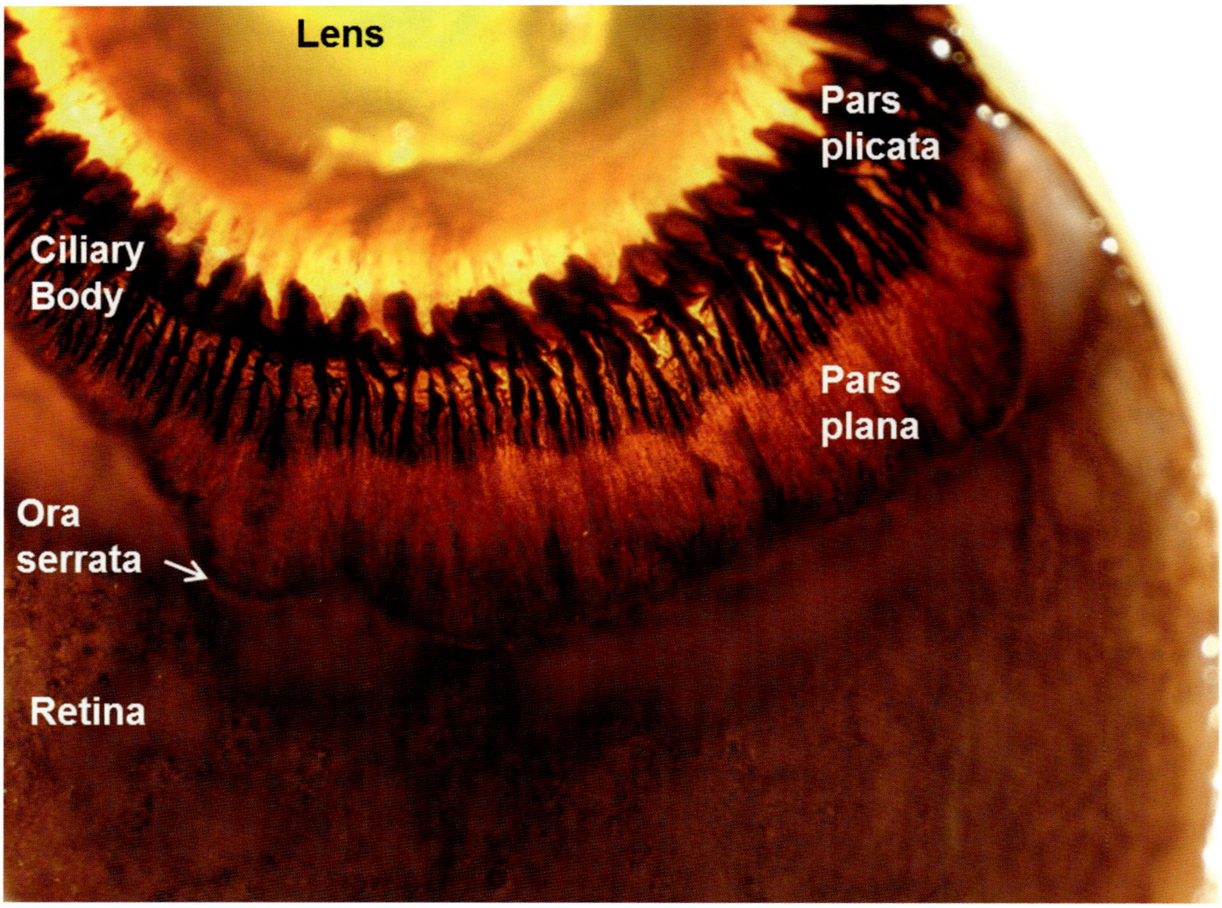

Figure 9-4

ANTERIOR SEGMENT STRUCTURES VISUALIZED FROM THE BACK OF EYE

The lens becomes yellow after fixation. The ciliary body is composed of the pars plicata (with folds of stroma covered by two layers of neuroepithelium) and the pars plana (flat portion composed by the two layers of neuroepithelium). The ora serrata (arrow) delineates the most anterior limit of the retina where it meets the pars plana of the ciliary body.

(3). Retinocytoma has the *RB1* mutation, similar to retinoblastoma. Patients that have been treated with chemotherapy show foci of retinocytoma-like tumor. This may represent a remaining well-differentiated tumor that is uncovered after the more poorly differentiated tumor is treated. There is the possibility, however, that chemotherapy induces differentiation in the tumor cells. Retrospective studies have shown that the degree of differentiation in untreated patients is associated with the age of the patient at enucleation (4).

Clinical Features. In most cases, children present before 3 years of age with a white reflex in the pupil (leukocoria). Other clinical presentations include strabismus and inflammatory or hemorrhagic intraocular events. Delay in the diagnosis of spontaneous necrotic retinoblastoma is usually secondary to the edema and inflammation induced by the necrotic ocular tissues, which may lead to a clinical diagnosis of orbital cellulitis. Older children, age 6 to 10 years, may present with an atypical pseudohypopyon (white cells in the anterior chamber, typically associated in a clinical setting with intraocular inflammation) and a diffuse infiltrate of tumor cells in the anterior portion of the retina, ciliary body, and anterior chamber. There is no large mass in the retina in this presentation.

The current clinical classification is based on response to treatment: chemotherapy, both systemic and local (5). There are two therapeutic goals: prevention of metastatic disease and,

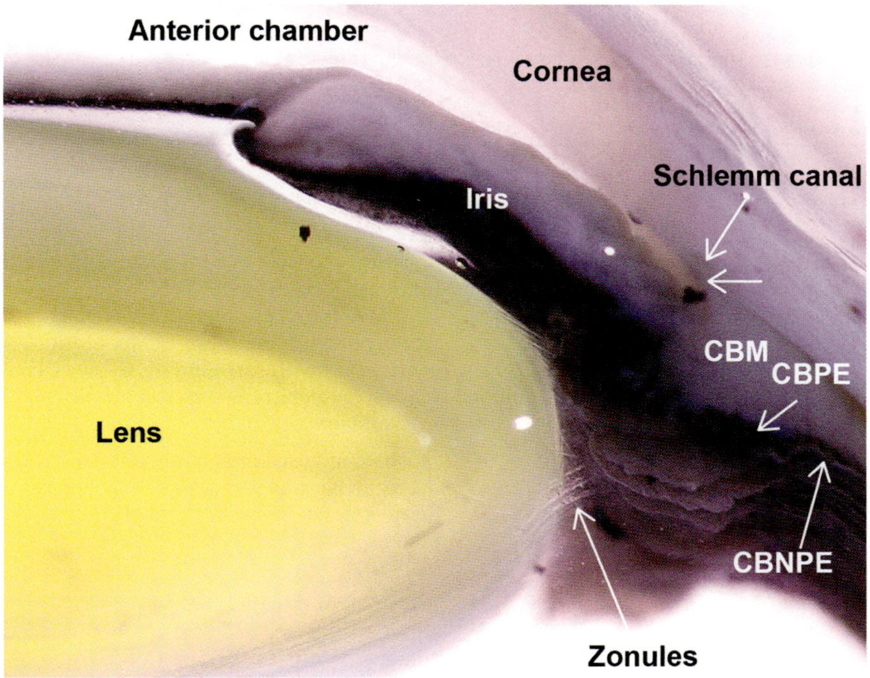

Figure 9-5

ANTERIOR SEGMENT AND ANGLE STRUCTURES OF EYE

The most anterior structure is the translucent avascular cornea. The anterior chamber is the space occupied by the aqueous humor between the cornea and the iris and lens. The angle of the anterior chamber contains the filtering system (trabecula [Trab]) and Schlemm canal) for the aqueous humor to exit the eye. Posterior to these structures is the ciliary body containing the ciliary muscle (CBM), the pigmented (CBPE) and nonpigmented (CBNPE) neuroepithelium responsible for the production of aqueous humor. The posterior chamber consists of the space between the iris and the ciliary body, zonules, and lens.

whenever possible, preservation of vision (ocular salvage). Clinical classifications may address both objectives.

These classifications divide the eyes into those containing small tumors, tumors with associated seeds and subretinal fluid, and large tumors with associated ocular findings. The latter are associated with the worst outcome for eye salvage and with a high probability of having high-risk disease resulting in metastasis and death (Appendix 8).

Until recently, the presence of tumor seeds carried an adverse prognosis for ocular salvage with most treatments. Lately, intraocular chemotherapy has been introduced as a treatment for vitreous seeds after or concurrent with other treatment modalities (6,7). Vitreous seeds are classified clinically into three groups based on their appearance and their different response to intravitreal chemotherapy (Table 9-1). Some studies show a significant difference between seed classes and required number of melphalan injections, cumulative/mean melphalan dose, and time to seed regression (6,7).

Retinoblastoma is one of the very few tumors that are diagnosed and treated without tissue-proven biopsy, thus, clinical features and imaging play a very important role in diagnosis and treatment. The presence of calcification in the tumor aids in the diagnosis of retinoblastoma, especially when there is an atypical presentation. A thorough evaluation is required to make the diagnosis and the examination is performed under anesthesia. Photographs obtained during the exam and ocular ultrasonography aid in confirming the diagnosis by locating the calcifications. Computerized tomography (CT) scan would be ideal for identifying calcifications, however, children under 1 year of age should not be exposed to radiation, especially if they carry the germline mutation. Magnetic resonance imaging (MRI) of the orbits and brain is performed to evaluate the status of the intraocular tumor and possible extraocular extension and invasion of the optic nerve. The brain MRI is indicated to exclude metastasis and the presence of pineoblastoma, a tumor that may develop in patients with *RB1* germline pathogenic variants (so-called *trilateral retinoblastoma*) (8).

Gross Findings. The gross findings reflect the tumor growth patterns and the association of the tumor to other ocular structures. Retinoblastomas arise from the retina and are gray, translucent masses with geographic areas of white to yellow necrosis and bright white calcifications with varying amounts of vessels (fig. 9-6).

Four growth patterns are identified on gross examination: endophytic growth, exophytic

Table 9-1

CLASSIFICATION AND HISTOPATHOLOGIC CORRELATION OF TUMOR SEEDS IN THE VITREOUS

Clinical Type/Class	Clinical Description[a]	Response to Intraocular Chemotherapy (Median)[b]	Microscopic Features[c]
Dust/type 1	Small granules of vitreous opacities, vitreous haze overlying tumor[d]	Time to regression, 2 to 3 weeks; no.[e] injections, 3; melphalan dose 20, µg	Scattered macrophages, necrotic cells, and individual viable tumor cells
Sphere/type 2	Spherically shaped opacities within vitreous; dust may be present around spheres; can be homogenously translucent with transparent or whitish center, or vice versa	Time to regression, 6 to 7 weeks; no. injections, 5; melphalan dose, 30 µg	Spherical clusters of viable cells; solid translucent: viable central cells indistinguishable from the remaining sphere; white/yellow center: necrotic/apoptotic central cells with surrounding viable cells; may have calcifications or surrounding macrophages
Cloud/type 3	Appear as a sheet or globule of dense collection of vitreous opacities and often with wispy edge; dust and spheres are sometimes also visible	Time to regression, 30 to 32 weeks; no. injections, 8; melphalan dose, 33 µg	Mostly composed of necrotic cellular debris, macrophages, and rare viable cells; cream-colored clouds have scattered red blood cells; there are open spheres of viable cells at the outer rim of the cloud
Post-Treatment Any type of seed can be present	Type 1, 2, or 3 seeds may remain; dust type with calcification or cream-colored cloud type predominate		Any or all of type 1, 2, or 3; red blood cells and macrophages in adjacent vitreous; changes consistent with pretreatment: calcification, pigment, regressed tumor, macrophages, rare viable tumor cell seeds

[a]Data from reference 6.
[b]Data from reference 7.
[c]Data from reference 9.
[d]Data from reference 8.
[e]No. = number of.

growth, diffuse growth, and necrotic tumor. The endophytic growth pattern is seen when the tumor grows from the retina into the vitreous cavity, the most inner ocular compartment. The endophytic tumor is usually dome-shaped and has the highest incidence of vitreous seeding. Vitreous tumor seeds are avascular aggregates of tumor cells that detach from the main tumor and grow in the vitreous. The endophytic tumor, or tumor seeds, invade the optic nerve head more frequently than other patterns of growth.

There are three basic types of vitreous seeds. Clinically described as type 1, "dust seeds" are seen grossly as small deposits of cells and material in the vitreous (fig. 9-7). Type 2A spheres are translucent and float in the vitreous; they alternate usually with type 2B spheres that show a solid white to cream-yellow center, often also with dust-type seeds (fig. 9-8). The type 3 cloud seed is a white to cream-yellow large deposit usually occupying the central anterior vitreous (fig. 9-9) (9).

The exophytic growth pattern is seen when the tumor grows from the retina into the subretinal space toward the choroid, the vascular coat of the posterior eye. The retina is usually detached and mostly replaced by tumor. Subretinal exudate accompanies most of the cases that may contain subretinal seeds. The seeds grow in a disk-like shape or as round to oval seeds and these may invade the RPE or the sub-RPE space. This type is associated with increased incidence of choroidal invasion (fig. 9-10)

Some retinoblastomas involve the retina diffusely. The retina is thickened, mostly anteriorly, and a small mass is possible in this area. The ciliary body and the posterior surface of the lens are covered by sheets of tumor cells. The tumor may cover the anterior surface of the iris or may have small nodules infiltrating the iris and the anterior chamber angle (fig. 9-11).

In the necrotic presentation, a large tan-brown to tan-white soft mass occupies the entire vitreous cavity. There is associated necrosis of retina, choroid, ciliary body, and iris, with the dispersion of pigment into the tumor and the necrotic tissues. The lens is cataractous in many cases; this is

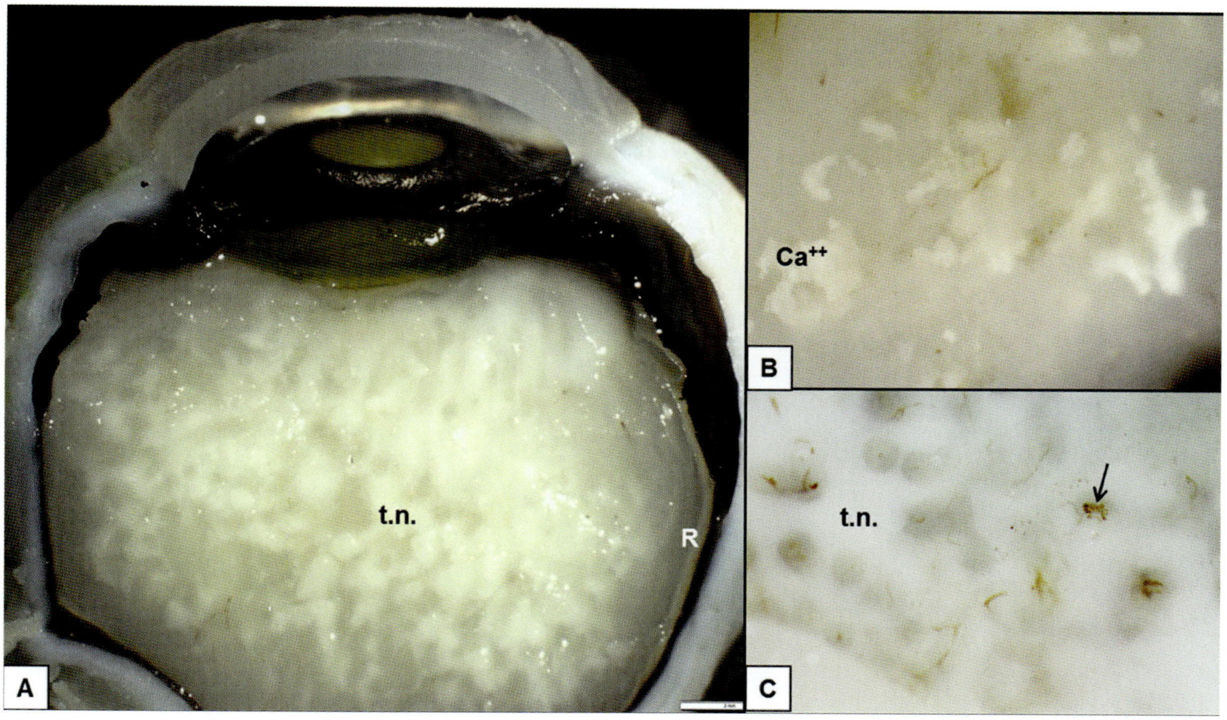

Figure 9-6

RETINOBLASTOMA

A: Scanning magnification view of an eye containing a retinoblastoma tumor that fills the vitreous cavity. The tumor is tan-gray, partially translucent, with geographic yellow-white areas corresponding to areas of tumor necrosis (t.n.).

B: Higher-power view shows the tumor with few vessels and bright white deposits of calcium (Ca++); about 80 percent of tumors show calcifications.

C: Higher magnification shows areas of tumor necrosis surrounding translucent tumor seen as tan-gray areas that surround tuffs of vessels (arrow).

usually not the case in other types of retinoblastoma. The cornea may be opaque with edema and the surface of the iris may be vascularized (iris neovascularization), with or without adhesions to the trabecular meshwork or cornea known as anterior synechia (fig. 9-12).

Microscopic Findings. Retinoblastoma is populated by retinoblasts, medium-sized to large cells with a high nuclear to cytoplasmic ratio, granular "salt and pepper" type chromatin, and scant delicate eosinophilic cytoplasm. Many mitotic figures and apoptotic cells may be identified (fig. 9-13A). At scanning magnification, large geographic areas of necrosis and foci of dystrophic tumor cell calcification may be identified (fig. 9-13B). The viable cells typically form a cuff around tumor vessels. The periphery of the cuff blends into areas of tumor necrosis, providing histologic evidence of the dependency on angiogenesis by retinoblastoma in the retina and vitreous (fig. 9-13C). The pattern of zones of viable tumor around blood vessels alternating with zones of necrosis is typically not identified in retinoblastomas that invade the choroid, suggesting a possible influence of the tumor microenvironment and degree of exposure to blood supply on behavior. Basophilic deposits in vessels and other ocular tissues and structures, such as the trabecular meshwork, the lens capsule, and subretinal exudate, are DNA fragments released from necrotic cells (Azzopardi phenomenon) (fig. 9-13D).

Retinoblastoma cells may be arranged in rosettes in varying amounts. The rosettes of retinoblastoma appear two-dimensionally as ringlets of cells that are only one cell layer deep (fig. 9-14, left). There are other conditions that produce retinal rosettes, such as the rosettes seen in the rare condition known as retinal dysplasia, but these rosettes are typically more than one cell layer thick (fig. 9-14, right).

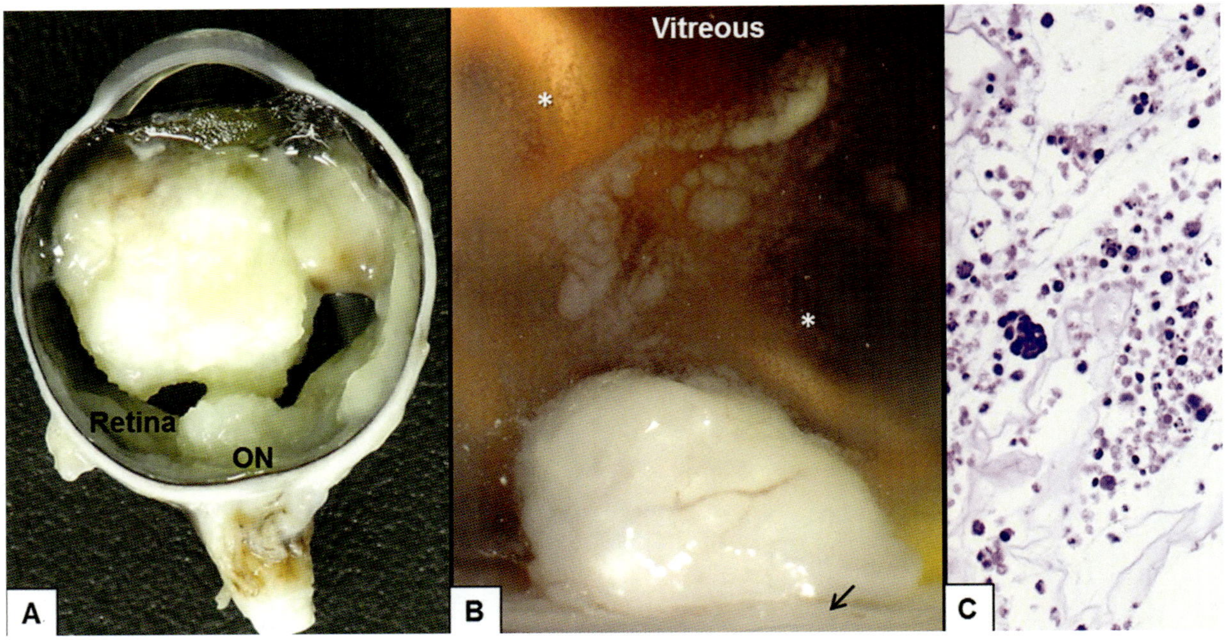

Figure 9-7

RETINOBLASTOMA: DUST-TYPE TUMOR SEEDS

A: Central pupil-optic nerve section of an eye with endophytic retinoblastoma that grows from the retina into the vitreous and seeds the head of the optic nerve with tumor.

B: High magnification of an endophytic tumor growing from the retina (arrow) into the vitreous and sprouting tumor seeds, many "dust-like," that give a ground-glass appearance to the vitreous (*).

C: Dust-type vitreous seeds are mostly necrotic cells, macrophages, and few clusters and single tumor cells.

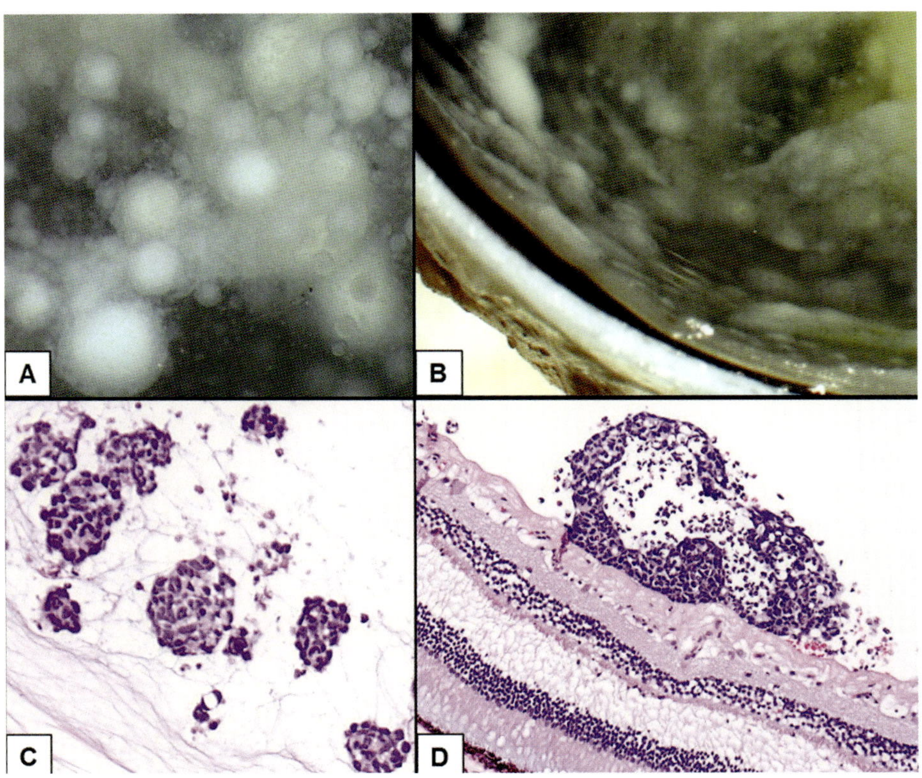

Figure 9-8

RETINOBLASTOMA: SPHERE-TYPE TUMOR SEEDS

A: Stereoscopic microscope at high magnification shows a cluster of sphere-type tumor seeds floating in the vitreous. Some seeds have a center that is white while others are translucent.

B: The seeds are on the surface of the retina.

C: Some seeds are translucent. The spheres are made of a solid proliferation of tumor cells. At the periphery they shed some single cells.

D: A large sphere with a central area of necrosis represents the white center seen grossly in B.

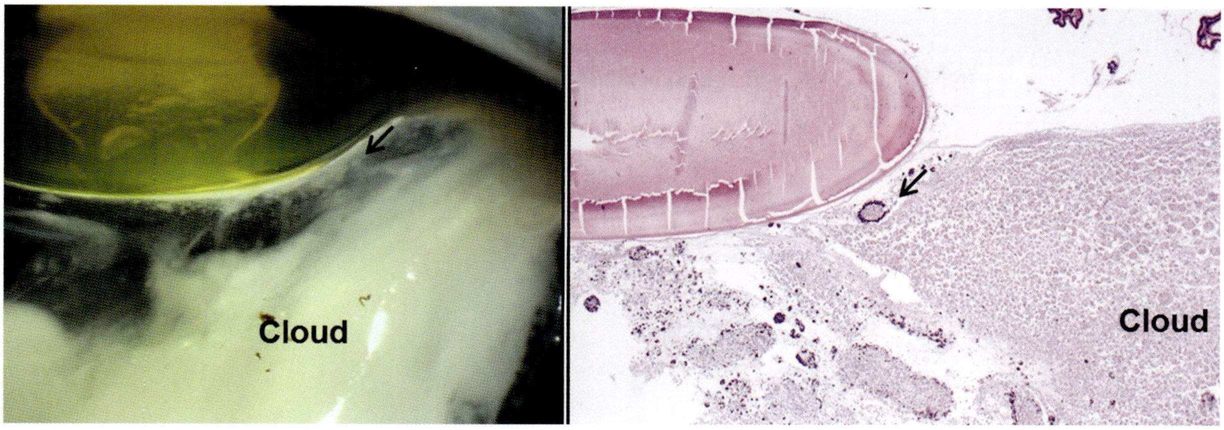

Figure 9-9

RETINOBLASTOMA: CLOUD-TYPE TUMOR SEEDS

Left: A cloud-type seed occupies most of the anterior vitreous behind the lens. Small sphere-type seeds are at the edge of the cloud (arrow).

Right: The cloud-type seed is made predominantly by necrotic cells and is lined or surrounded by sphere-type seeds (arrow).

Three types of rosettes are described in retinoblastoma. Two are unique to retinoblastoma and its analog, pineoblastoma: the Flexner-Wintersteiner rosettes and fleurettes. The Homer Wright rosette of retinoblastoma is identical morphologically to its counterpart in conditions such as neuroblastoma.

Flexner-Wintersteiner rosettes, the most typical type seen in retinoblastoma, are characterized by an empty center that appears to be lined by a membrane (fig. 9-15A). Homer Wright rosettes, like those seen in other tumors, are characterized by cytoplasmic extensions into the center of the structure (fig. 9-15B).

The third type of rosette, known as a fleurette, is characterized by the presence of well-differentiated cells with abundant eosinophilic cytoplasm that may be elongated or globular; the cells are attached to each other, resembling photoreceptors, a light microscopic observation confirmed by ultrastructural studies. In areas of tumor with abundant fluerettes, it may be difficult to detect mitotic figures or apoptotic bodies. If a tumor is composed exclusively of fleurettes, the lesion is designated as a retinocytoma or retinoma (fig. 9-15C). Some cases lack evidence of rosettes and these tumors are designated as poorly differentiated (fig. 9-15D).

Several types of seeding occur (Table 9-1). Rare single or small clusters of tumor cells, alternating with macrophages, and debris of necrotic cells, in the background of condensed vitreous defines the morphology of what is described clinically as dust-like seeds (fig. 9-7). Sphere-type seeds are tight groups of tumor cells and the seed itself is typically greater than 50 µm in diameter. The seeds corresponding to the translucent type are solid. Single cells detach at the rim of the sphere and appear to infiltrate into the adjacent vitreous (fig. 9-8). Those seeds corresponding to the white or yellow centers are characterized histologically as having a central area of necrosis. The cloud-type seeds are large deposits of necrotic cells with infiltrating macrophages and rimed by a few tumor spheres that partially open into the necrotic material (fig. 9-9) (9).

Exophytic retinoblastomas are tumors arising in the retina and predominantly extending into the subretinal space. Single or groups of tumor cells may form subretinal seeds within eosinophilic protein-rich subretinal exudate (fig. 9-10 C). The seeds may infiltrate beneath the retinal pigment epithelium but Bruch membrane remains intact in most cases (fig. 9-10D).

Choroidal invasion is defined by tumor deep to Bruch membrane. Massive choroidal invasion and invasion into the optic nerve are associated with an adverse outcome, as is a tumor that is detected outside the eye (extraocular extension).

There is a small group of retinoblastoma tumors that have different morphologic findings

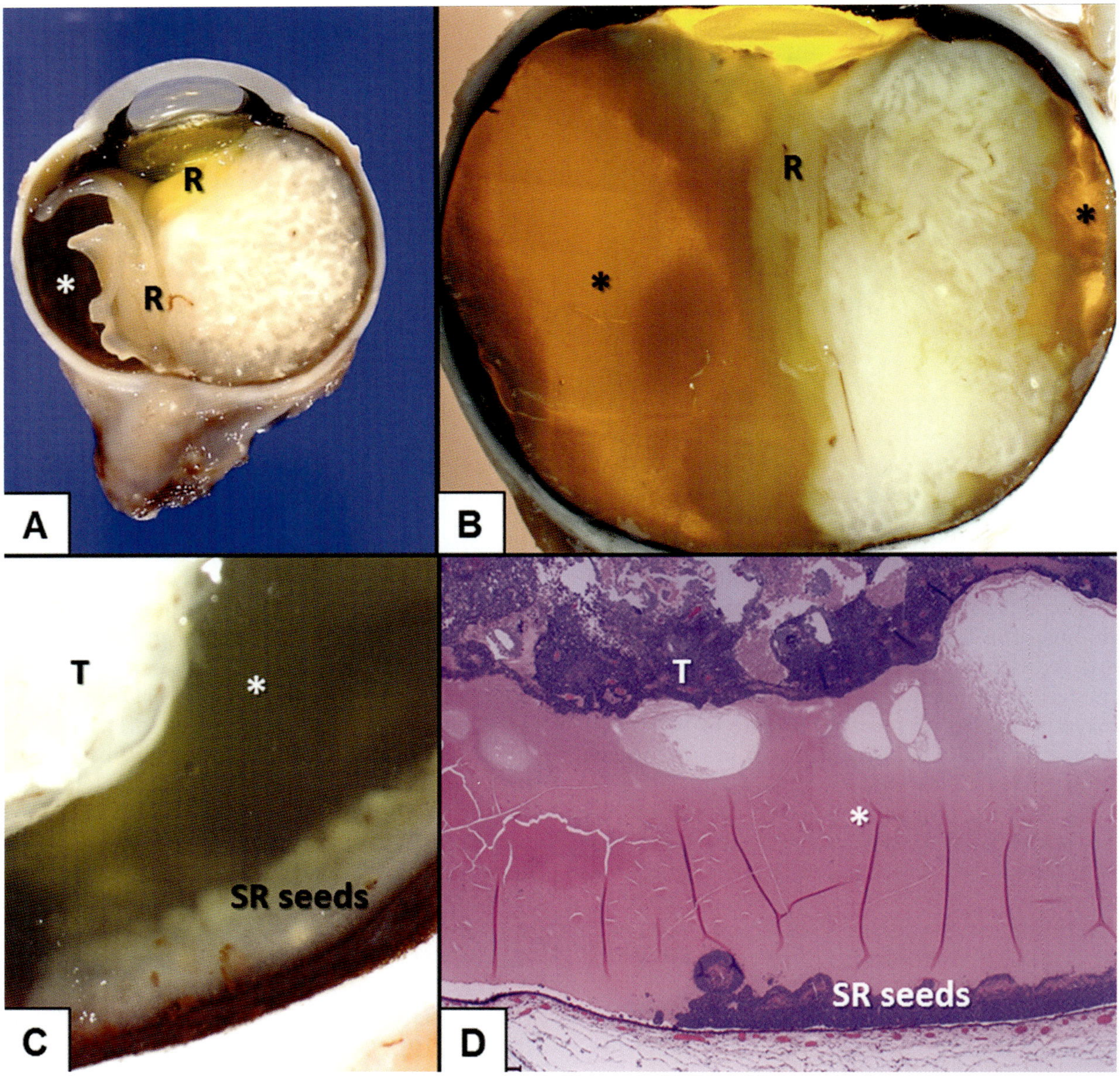

Figure 9-10

RETINOBLASTOMA: SUBRETINAL TUMOR SEEDS

A: Enucleated eye after removal of the superior calotte shows total retinal detachment (R) with subretinal fluid/exudate (*) and a retinoblastoma growing from the retina into the subretinal space (right).

B: Funnel-shaped retinal detachment (R), extensive proteinaceous exudate (*) that accumulates in the subretinal space, and tumor infiltrating the subretinal space are seen.

C: Higher magnification of an exophytic retinoblastoma (T) shows the solid layer or plaque of tumor seeds in the subretinal space over the retinal pigment epithelium covered by the subretinal exudate (*).

D: Vascularized tumor (T) overlies an eosinophilic proteinaceous exudate (*) that separates the tumor from the subretinal tumor seeds (SR seeds). These seeds are avascular and located over the RPE layer.

and are associated with *MYC* amplification instead of *RB1* mutation. Rosettes are not encountered in these tumors. The tumor cells are distinctive for round nuclei with prominent nucleoli.

Additional findings in some eyes with retinoblastoma include the presence of a neovascular membrane in the anterior surface of the iris, with or without anterior synechia (adhesions)

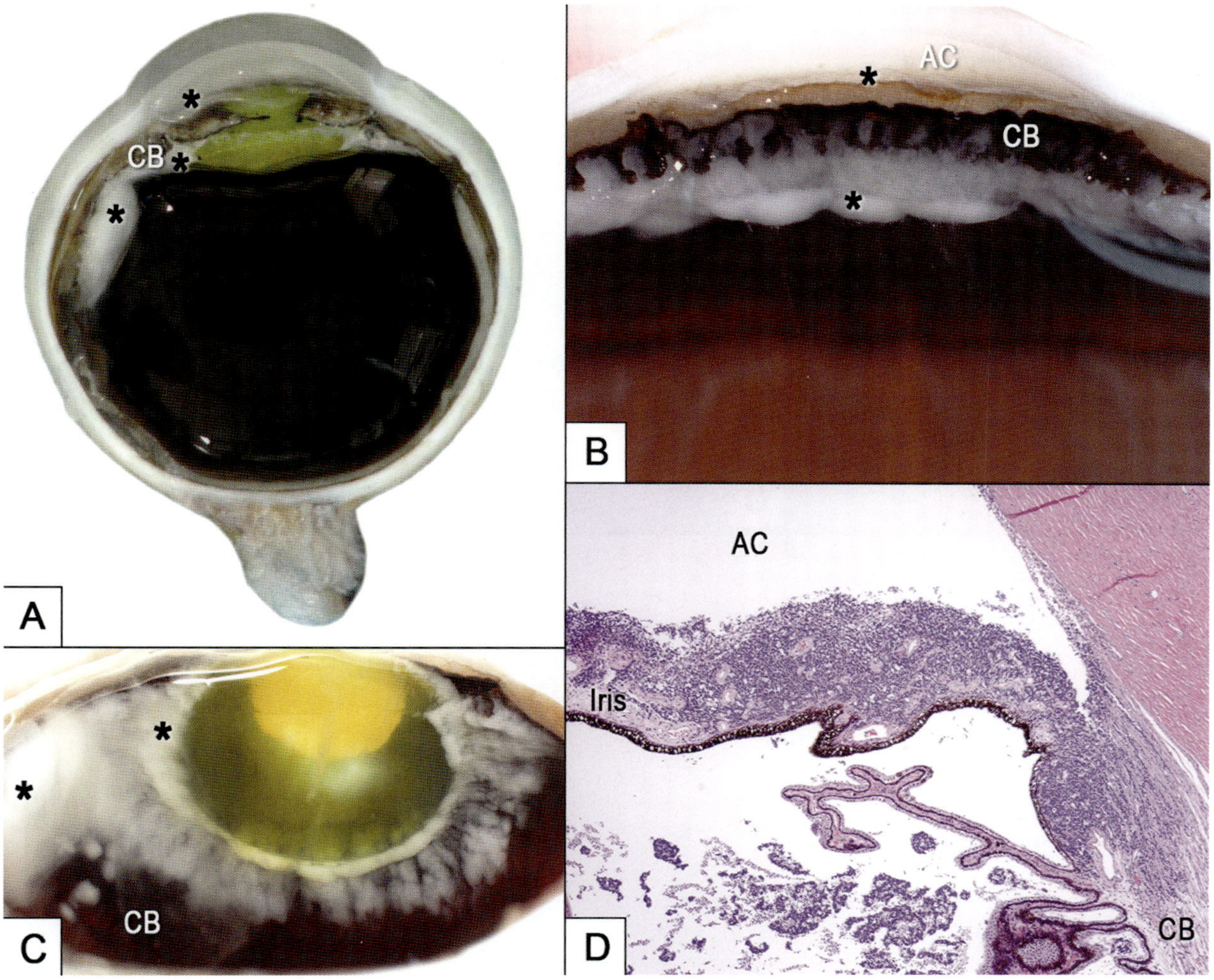

Figure 9-11

RETINOBLASTOMA: ANTERIOR DIFFUSE GROWTH PATTERN

A: A 6-year-old child has a small area of thickening of the peripheral retina by a fluffy white tumor (*). The tumor grows over the ciliary body (CB), extends into the posterior chamber (*), surrounds the lens, and grows into the anterior chamber (*).

B: Higher magnification of the cut superior calotte at the level of the peripheral ciliary body pars plicata (CB) shows the tumor (*) layering the surface of the ciliary body in the anterior chamber angle (AC).

C: Stereoscopic microscopic view of the ciliary body (CB) and posterior surface of lens when the eye is tilted down and seen from the back shows the white retinoblastoma covering these structures and surrounding the lens.

D: The tumor is seen in the posterior chamber in front of the ciliary body (CB) pars plicata and invading the iris into the trabecula at the anterior chamber (AC) angle and the anterior portion of the CB.

to the cornea, thereby obstructing access of aqueous humor to the trabecular meshwork (angle closure). With chronic angle closure, the lamina cribrosa of the optic nerve head is bowed posteriorly (neovascular glaucoma) (fig. 9-16, left). The finding of more than 90 percent necrosis of the tumor may be accompanied by necrosis of retina, RPE, choroid, ciliary body, and iris. The RPE, ciliary body, and iris epithelia detach and the pigment is dispersed in the tissues. Melanophages infiltrate the tumor and tissues. Congestion of vessels in the tumor and other tissues may be seen (fig. 9-12) (10).

Cytologic Findings. Retinoblastoma is one of the few tumors that are diagnosed and treated without tissue-proven biopsy. If contained within the eye, retinoblastoma has an excellent prognosis for cure, however, if the tumor is spread outside the eye there is a high recurrence rate. It is seldom necessary to obtain

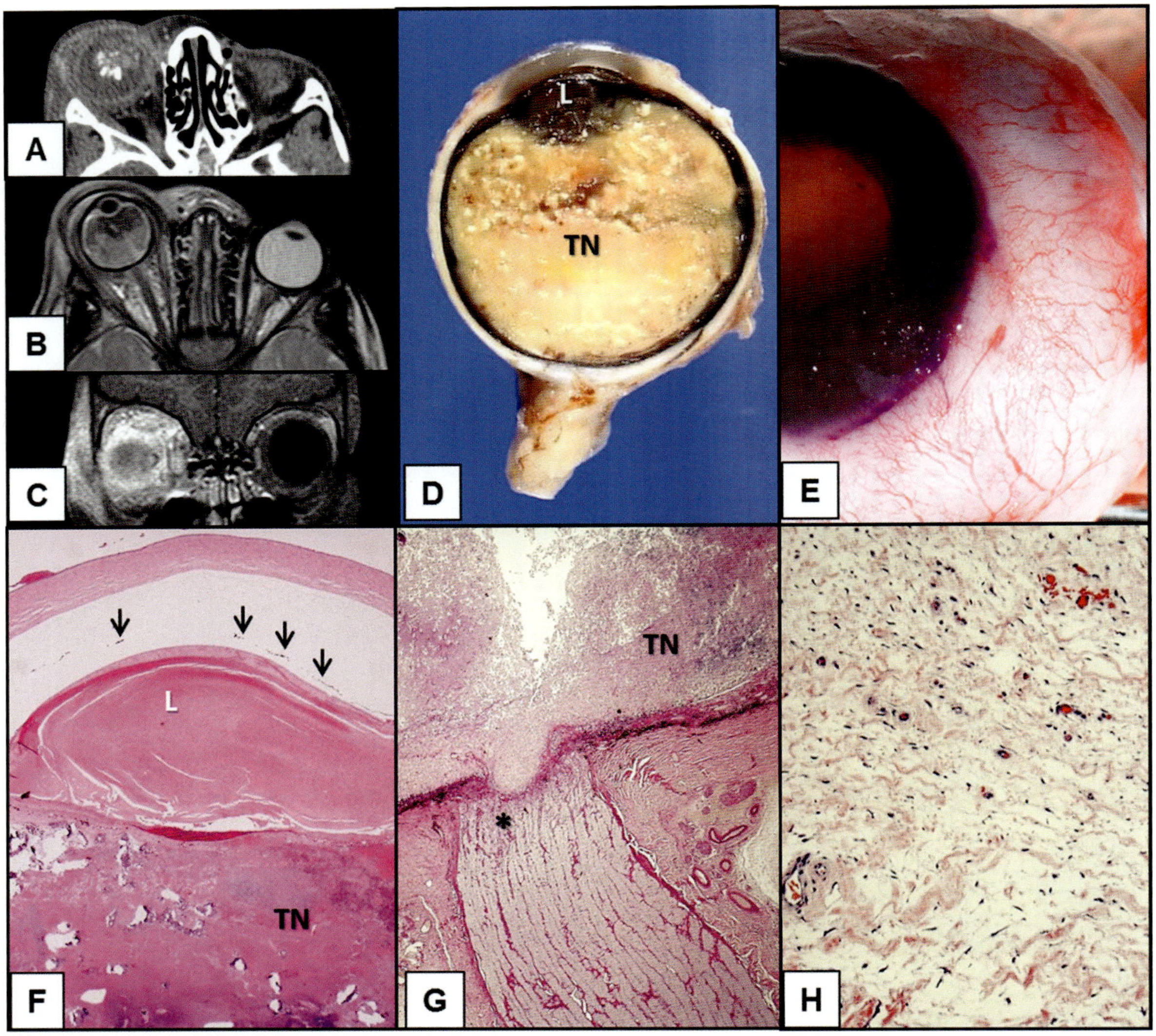

Figure 9-12

RETINOBLASTOMA: EXTENSIVELY NECROTIC (OVER 95 PERCENT)

A: CT image of a 2-year-old child who experienced eyelid edema and inflammation, accompanied by conjunctival edema (chemosis), pain, and proptosis. The cornea was opaque. The center of the proptotic eye contains bright white densities consistent with calcium.

B: T2-weighted magnetic resonance imaging (MRI) demonstrates irregular densities in the vitreous cavity of the right eye in contrast with the even water-type opacity seen in the vitreous in the left eye. There is also irregular opacification of the posterior orbital soft tissues. The optic nerve near the sclera also enhances.

C: T1-weighted MRI illustrates the enhancement of the periocular tissues.

D: Enucleated right eye after fixation in the same patient is remarkable for extensive tumor necrosis (TN), which is tan-yellow and soft alternating with bright white calcification in the anterior central vitreous cavity. The lens is dark yellow to brown, a reflection of a cataract. The lens is displaced forward and appears to touch the cornea.

E: The anterior portion of the eye shows the congested episcleral vessels, hazy cornea, and anterior chamber with dispersed pigment and blood.

F: The anterior portion of the eye with the cataractous lens (L) associated with areas of hemorrhage of the anterior portion of the necrotic tumor (TN). Tissue sectioning most likely displaced calcium from the tumor, resulting in what appears to be "empty spaces." Anterior to the lens there is only scattered pigment and membrane because most of the iris is necrotic (arrows).

G: The posterior portion of the eye including the optic nerve. Extensively necrotic tumor (TN) invades the optic nerve head focally and fills in the cup of the optic nerve head (cupping secondary to glaucoma). The lamina cribrosa is inflamed, seen as a hematoxyphilic area (*).

H: Extraocular fibrous adipose tissue with edema and inflammation, and with dilated blood vessels in the absence of tumor.

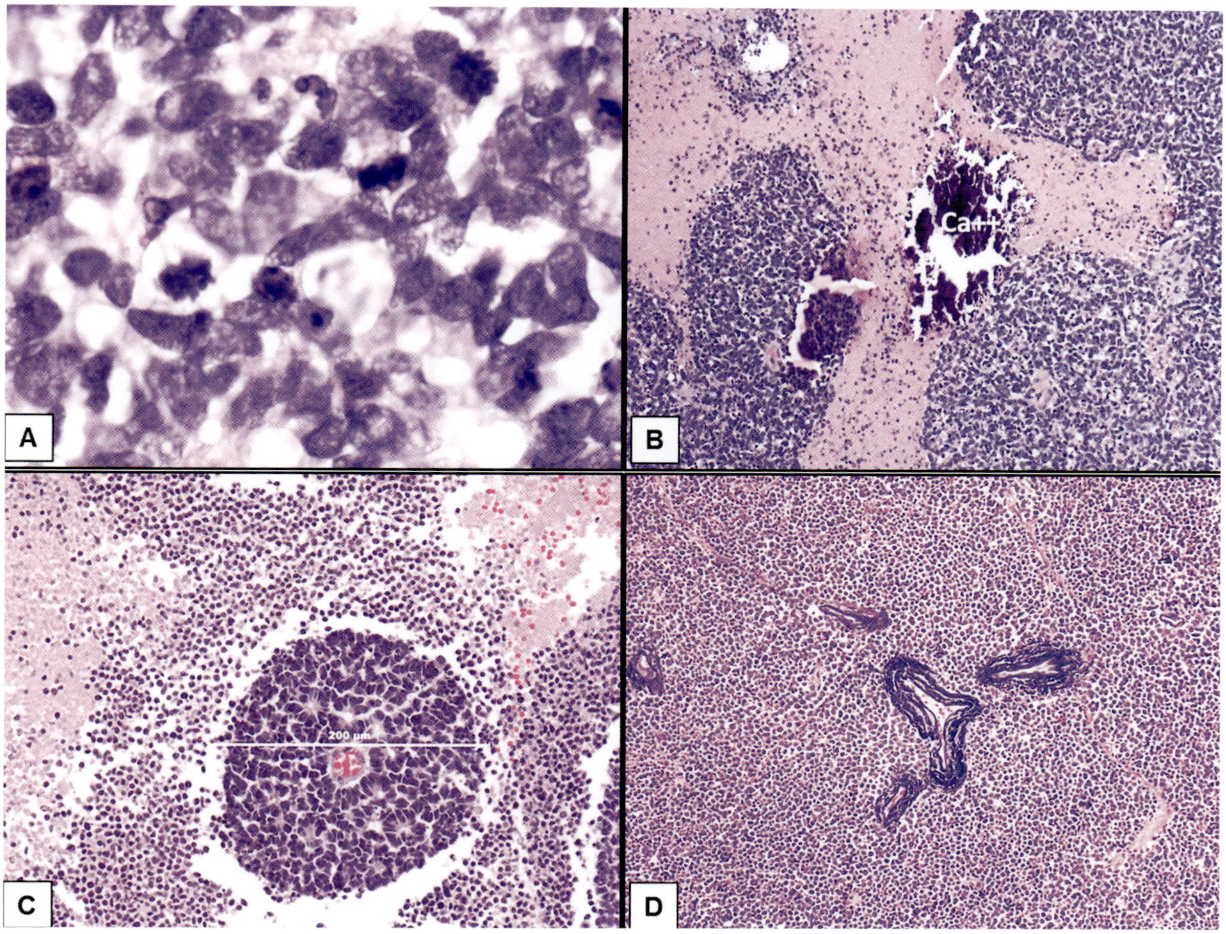

Figure 9-13

RETINOBLASTOMA

A: Medium to large cells with high nuclear to cytoplasmic ratio, granular "salt and pepper" type chromatin, and scant delicate eosinophilic cytoplasm (retinoblasts). Mitotic figures and apoptotic cells are seen.

B: Low-power view shows large geographic areas of necrosis and foci of dystrophic tumor cell calcification (Ca++).

C: Viable tumor cells form a cuff around tumor vessels. The periphery of the cuff blends into areas of tumor necrosis. This typically occurs over 100 µm from the tumor vessel (see measuring bar).

D: The basophilic deposits in vessels represent DNA fragments released from necrotic cells (Azzopardi phenomenon).

a biopsy from an eye with retinoblastoma but if such a procedure is necessary, then care must be taken to avoid the spread of the tumor outside the eye (11,12).

Cytology preparations are hypocellular because of the limited biopsy technique. The cells are medium to large, with a high nuclear to cytoplasmic (N:C) ratio. Mitoses and apoptotic cells are present. There is cell molding, and cell groups may form rosettes. If the tumor is sampled from the posterior pole, there is the possibility of sampling retina, and this may be a pitfall in interpretation. Retinal neurons may also have high N:C ratio and neuroendocrine type chromatin, however, they will lack apoptosis and mitosis (fig. 9-17).

Molecular Genetic Findings. *Genetic Susceptibility and Genetic Testing.* Mutations in the *RB1* gene are responsible for most cases of retinoblastoma. *RB1* is a tumor suppressor gene located on chromosome 13 (13). About 40 percent of cases are hereditary, the result of either a heterozygous de novo or an inherited germline *RB1* gene pathogenic variant (mutation). There is 50 percent chance that the offspring of individuals with retinoblastoma and a germline

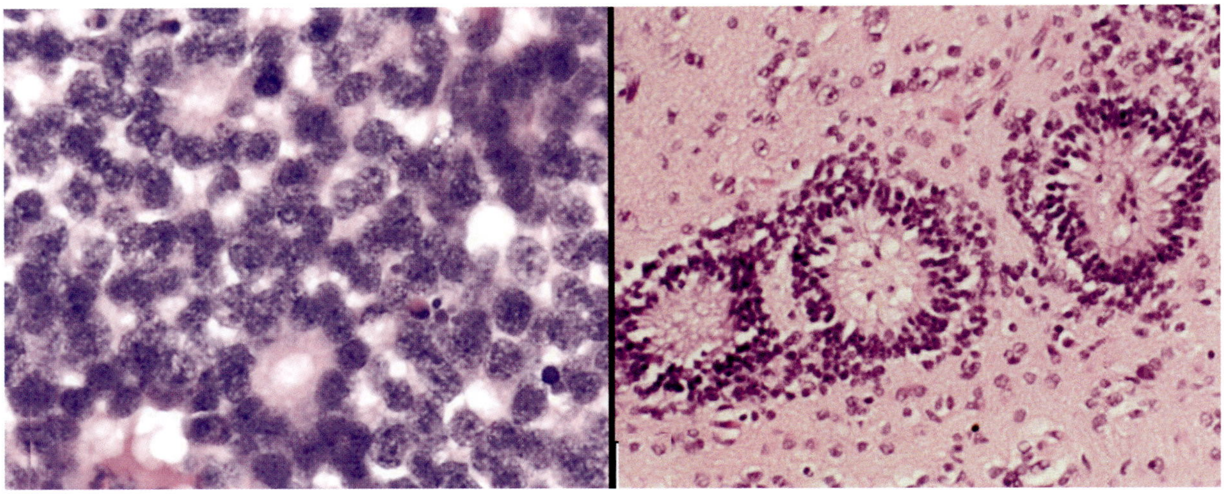

Figure 9-14

ROSETTES: RETINOBLASTOMA VS RETINAL DYSPLASIA

Left: The rosettes of retinoblastoma are arranged in ringlets that are only one cell layer deep.

Right: The rosettes of retinal dysplasia are more than one cell layer thick. In heterotopic brain in the retina, multiple layers of cells contribute to the rosettes.

mutation will acquire the pathogenic variant. The inheritance pattern is autosomal dominant.

Cancer-related morbidity and mortality are influenced by the germline status. Although most hereditary retinoblastomas are bilateral, unilateral retinoblastoma may be the first presentation within a family. In retinoma (retinocytoma) or retinoblastoma, both alleles have the pathogenic variant and additional mutations are required to develop retinoblastoma.

In 2 percent of patients with unilateral retinoblastoma and no detectable mutation in *RB1*, amplification of the *MYCN* oncogene is present (28-121 DNA copies). These children develop unilateral retinoblastoma, are diagnosed at a median age of 4.5 months, and have histopathologically distinct tumors with advanced features at diagnosis (2).

Genetic testing is indicated in children with retinoblastoma, especially when presenting as unilateral tumor without a previous family history. If tumor tissue is available, the pathogenic variants of the tumor are compared with peripheral (blood lymphocyte) DNA for the presence or absence of the same variants. If the pathogenic variant is only present in the tumor but not in peripheral cells, the tumor is most likely sporadic, although the possibility of germline mosaicism exists. If the patient (proband) carries the pathogenic variant in both the tumor and the lymphocytes, retinoblastoma is considered hereditary, and the causative mutation is specifically searched for in at-risk relatives. Proband and family members are followed according to their risk to identify tumors at the earliest stage, avoid complications of treatment and metastasis, and avoid unnecessary examination under anesthesia of children without the pathogenic variant.

Since the sensitivity of genetic testing may vary in different laboratories, the recommendation is that testing be performed at a certified laboratory with experience in retinoblastoma genetic testing. Post-test risk calculation by a genetics professional should take into account the estimated laboratory sensitivity to clarify the clinical risk category (population risk: low, intermediate, or high) (14).

Tumor Harvesting. Fresh tumor sampling should be done immediately after enucleation, by the surgeon or by the pathologist. There are two main techniques used to retrieve fresh tumor: 1) needle biopsy of the tumor through the sclera, or 2) opening a scleral window to retrieve tumor. The latter is preferred as there is a visualization of the tumor and the adjacent intraocular tissues. The optic nerve is the surgical margin to examine in retinoblastoma and should be sampled before manipulating and opening

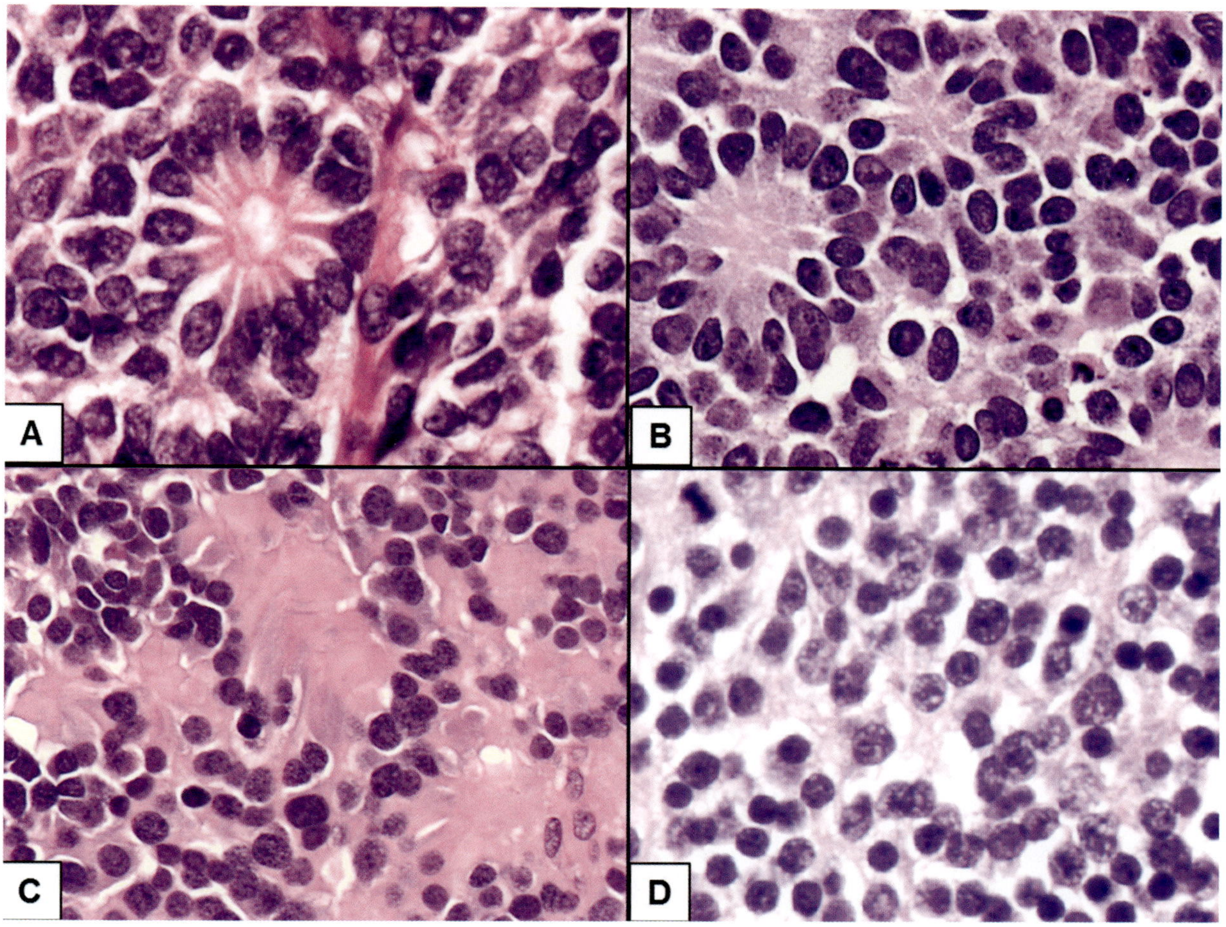

Figure 9-15

DIFFERENTIATION IN RETINOBLASTOMA

A: Flexner-Wintersteiner rosettes, typically seen in retinoblastoma, are composed of neoplastic cells with long cytoplasmic processes that converge upon a central lumen. The lumen appears to be bordered by basement membrane material, similar to the terminal bars of photoreceptors.

B: Homer Wright rosettes have cytoplasmic processes filling the center of the structure and these rosettes may be seen in other neuroblastic tumors.

C: Photoreceptor or neuronal differentiation is seen in very well-differentiated tumors (retinoma or retinocytoma). Mitoses and apoptosis are not seen in the well-differentiated fleurettes. Fleurettes are distinctive for well-formed photoreceptor-like structures with eosinophilic cytoplasm.

D: Poorly differentiated retinoblastoma lacks any rosettes and mitotic figures are present (upper left corner).

the eye for tumor harvesting (fig. 9-18A). A touch imprint can access tumor at the margin and possible tumor cells in meningeal spaces (fig. 9-18B–D). A cross-section of the margin is obtained and submitted for permanent sections.

After obtaining the optic nerve margin, the eye should be transilluminated to locate the tumor. If the eye does not transmit light because the entire posterior pole is filled with tumor, then the opening should be done either in the superior or inferior equator of the eye, horizontal plane (fig. 9-19). However, if the tumor is smaller and shows a shadow on transillumination, then the opening should be perpendicular to the tumor and away from the optic nerve or the anterior segment structures (fig. 9-19). This is to avoid including the opening into the pupil-optic (PO) nerve section to allow for evaluation of tumor invasion into the optic nerve and other structures (fig. 9-19) (15).

After tumor retrieval, the eye is placed in formalin for adequate fixation (48 hours). After

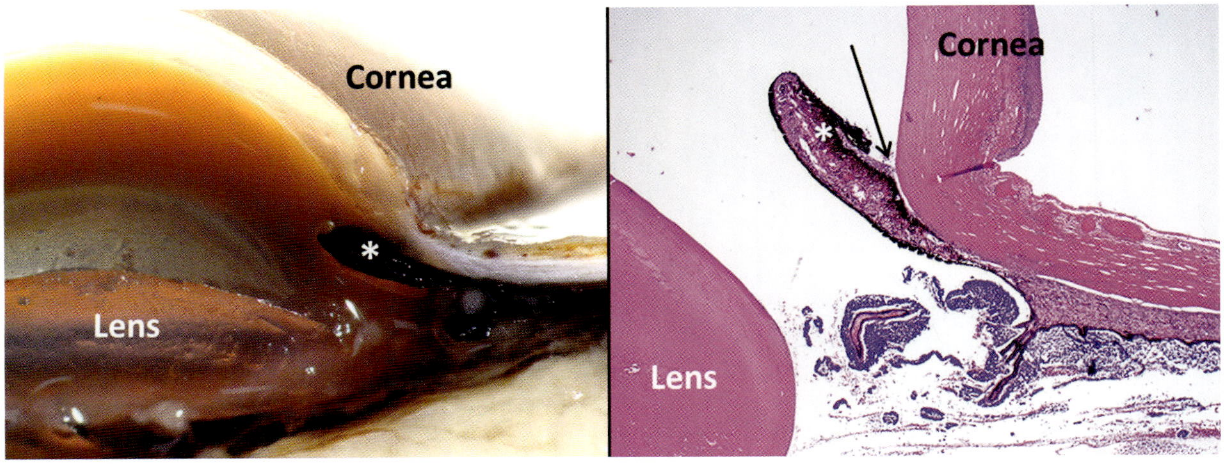

Figure 9-16

IRIS NEOVASCULARIZATION AND NEOVASCULAR GLAUCOMA IN RETINOBLASTOMA

Left: An eye with retinoblastoma (posterior to the lens) and anterior synechia, adhesions of the iris (*) to the cornea. A neovascular membrane, not visible grossly, contributed to the formation of the synechia.

Right: Histopathologic counterpart of the left figure. The neovascular membrane (arrow) pulls the iris toward the lens and secondarily closes the anterior chamber angle (blockage of the aqueous humor causes increase in the intraocular pressure and secondary glaucoma).

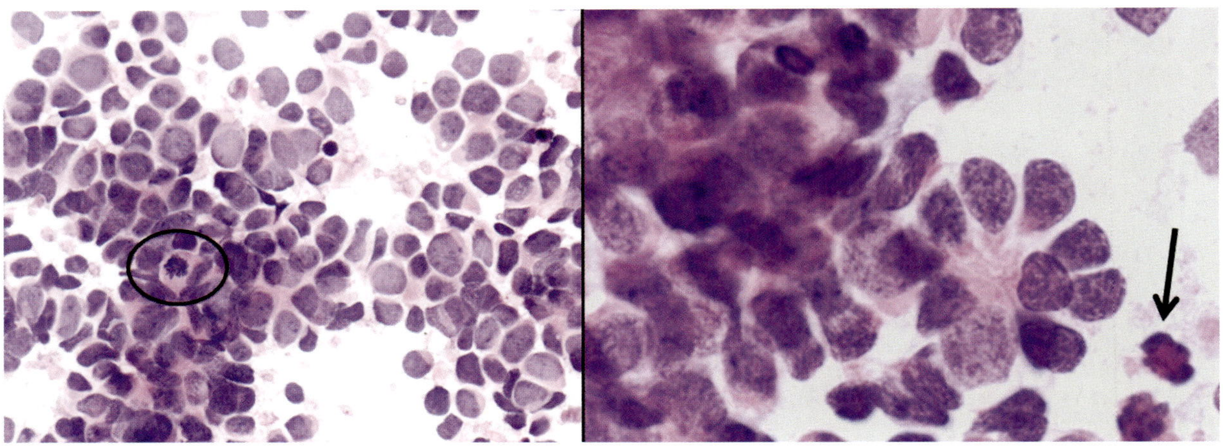

Figure 9-17

RETINOBLASTOMA: CYTOPATHOLOGIC FEATURES

Left: Touch imprint of retinoblastoma shows the presence of medium-sized cells with scant cytoplasm and hyperchromatic nuclei featuring salt and pepper type chromatin. A mitosis (circle) is easily found in these tumors.

Right: The detection of rosettes and apoptosis (arrow) differentiates tumor cells from possible nuclear layers of the retina that lack these features.

fixation, the eye is sectioned in three segments: the PO nerve section and two calottes (caps). One of the calottes should include the previous scleral window created to harvest the tumor. The calottes are then further sectioned in an anterior-posterior orientation ("bread loaf" pattern). The entire eye is submitted in four blocks: ON=optic nerve margin; PO=pupil-optic nerve section; and each calotte in each cassette in segments (fig. 9-20) (15).

The harvesting of fresh tissue through a scleral window introduces the possibility of artifactual "spread" of tumor cells into the choroid or outside the eye. The strategies for

Table 9-2

DIFFERENTIAL DIAGNOSIS OF RETINOBLASTOMA

	Retinoblastoma	Coats Disease	*Toxocara Canis* Endophthalmitis	Persistence of Fetal Vasculature (PFV)
Age	<2 years	4-10 years	6-11 years	At birth
Presentation	Unilateral/ bilateral	Unilateral	Unilateral	Unilateral/bilateral
Gender	Variable	Male (2:3)	Variable	Variable
Color	White/translucent	Yellow	White, gray	White
Retina	Tumor, retinal detachment	Exudate, retinal detachment, telangiectasias	Eosinophils, histiocytes in partial retinal detachment, and vitreous at the base	Retrolental membrane with funnel-shaped retinal detachment
Calcifications	Frequent	Infrequent	Infrequent	Infrequent
Other	US[a]/MRI/CT[a] scan = mass, calcification Genetics *RB1* pathogenetic variant or *MYC* amplification	Fluorescein angiogram = leaky vessels in retina; US/MRI/CT scan = no mass or calcification	Contact with puppies; negative ELISA excludes diagnosis	Other ocular anomalies: microphthalmia, abnormal ciliary body, iris shunt vessels

[a]US = ultrasound; MRI = magnetic resonance imaging; CT = computerized tomography; ELISA = enzyme-linked immunosorbent assay.

discriminating between artifactual spread and genuine choroidal invasion or extraocular extension are addressed below.

Immunohistochemical Findings. The diagnosis of retinoblastoma is made by morphology on routine hematoxylin and eosin (H&E)-stained slides. Retinoblastoma may need to be distinguished from inflammation, extramedullary hematopoiesis, or gliosis or in areas of complete necrosis. In these cases, immunohistochemistry to detect the expression of synaptophysin may be useful. Synaptophysin is positive in the cytoplasm of tumor cells, even those that are necrotic.

In rare cases, confirmation of retinoblastoma rather than another type of neuroendocrine tumor may be necessary, and immunohistochemistry for the retinoblastoma (RB) protein is used. Negative staining of tumor cells is consistent with the diagnosis of retinoblastoma in an otherwise synaptophysin-positive tumor (fig. 9-21A). Only about 85 percent of retinoblastomas have a true knockout production of the RB protein, the others have a dysfunctional RB protein, but the protein is produced and may be labeled by the RB antibody used (fig. 9-21B).

Differential Diagnosis. The clinical differential diagnosis includes several entities (Table 9-2), including Coats disease and medulloepithelioma; either may be seen in an eye removed for the presumptive clinical diagnosis of retinoblastoma (16). Coats disease mimics exophytic retinoblastoma because it presents in children 4 to 6 years of age, with serous retinal detachment secondary to telangiectatic vessels and aneurysms in the retinal vessels that leak. The exudate accumulates in the retina and subretinal space. Histologically, the subretinal space is characterized by a protein-rich exudate with cholesterol clefts and the presence of foamy cells that may contain melanin granules and therefore may represent retinal pigment epithelial cells that have consumed lipid. The "Coats triad" of exudate, cholesterolosis, and foam cells is not specific for Coats disease and may be seen in eyes removed for retinoblastoma. Retinal telangiectasis is the critical histologic marker of Coats disease (fig. 9-22).

Histopathologically, medulloepithelioma may resemble retinoblastoma and is the most common entity in the differential diagnosis. Medulloepithelioma may be benign or malignant, and the malignant component may be similar to retinoblastoma. In malignant medulloepithelioma there is residual medulloepithelioma in the lesion, with the typical neuroepithelium and extracellular basophilic material (see Medulloepithelioma below). RB protein is positive in the malignant component of medulloepithelioma but is negative in typical retinoblastoma.

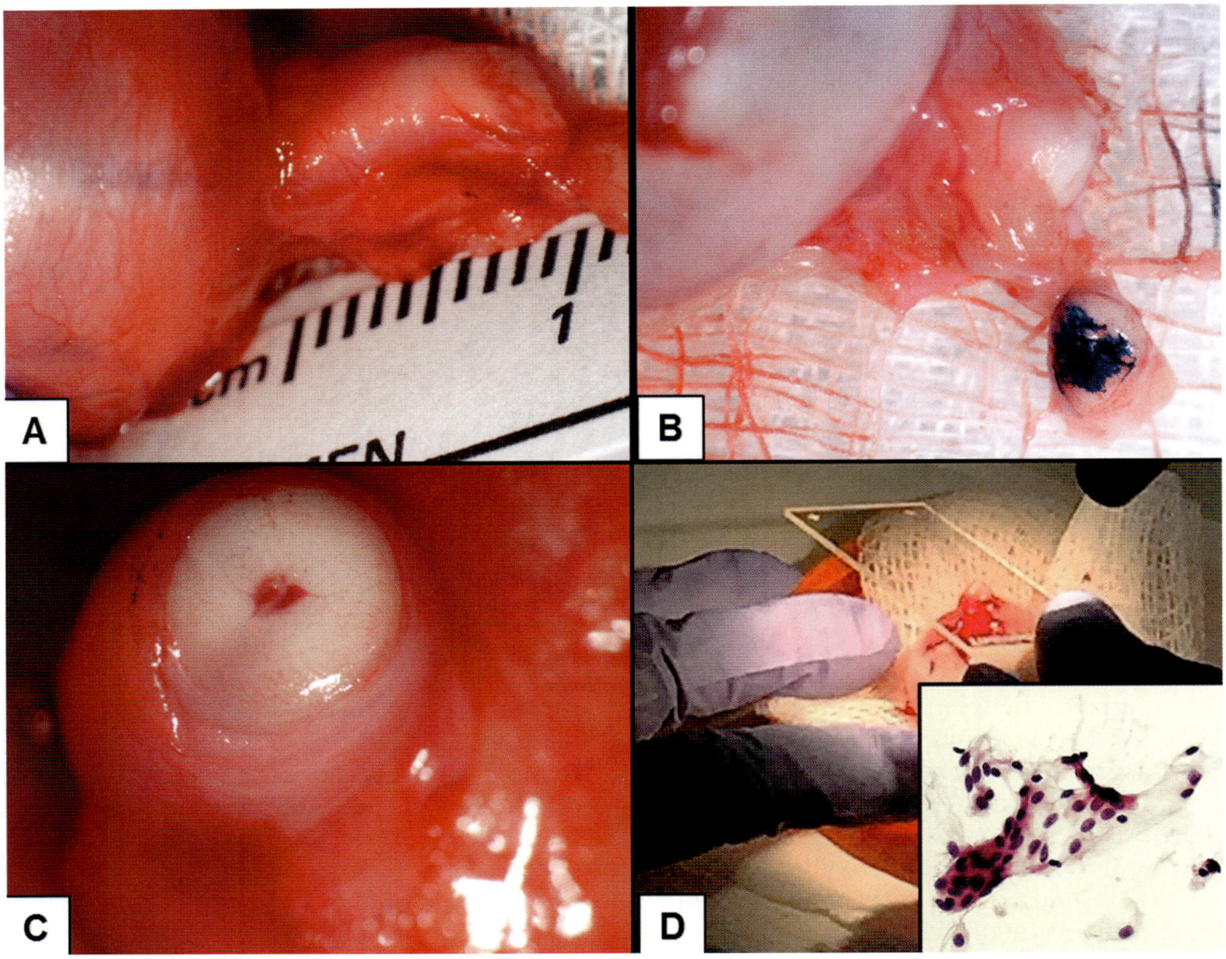

Figure 9-18

ENUCLEATED EYE WITH RETINOBLASTOMA FOR TUMOR RETRIEVAL

A: After receiving the eye and confirming laterality, the length of the optic nerve should be measured from the sclera to the tip of the margin (stretching the optic nerve distally from the eye). This measurement will appear in the pathology report as a staging parameter.

B: After inking the true margin of the optic nerve, a cross section is obtained about 2 mm from the distal nerve and submitted for histopathologic processing and examination.

C: Under the stereoscopic microscope, the cut section surface is evaluated for gross tumor. In this example, there are congested central vessels and a segmental area of atrophy (tan pink inferiorly).

D: Optionally, a touch imprint from the cut surface is used to evaluate the presence of tumor at this site. Insert shows the typical prominent meningothelial cells seen in children negative for tumor cytology.

Treatment. Retinoblastoma is treated by single, or more frequently, a combination of therapies that may include systemic chemotherapy (chemoreduction), intra-arterial chemotherapy, subconjunctival chemotherapy, plaque brachyradiotherapy, cryotherapy, photocoagulation, intravitreous chemotherapy, and primary enucleation (17,18). Oncologists are cautious about the use of therapies that may promote second cancers in children with germline *RB* mutations. Children with germline *RB* mutations are at life-long risk for the development of second malignancies regardless of the treatment received; the type of malignancies that develop, however, is associated with the type of treatment received. External radiation is seldom used to treat retinoblastomas because of a risk of osteosarcoma in the orbital bones in children with germline mutations.

Eyes seen in "secondary" enucleation may have tumors that are resistant or unresponsive

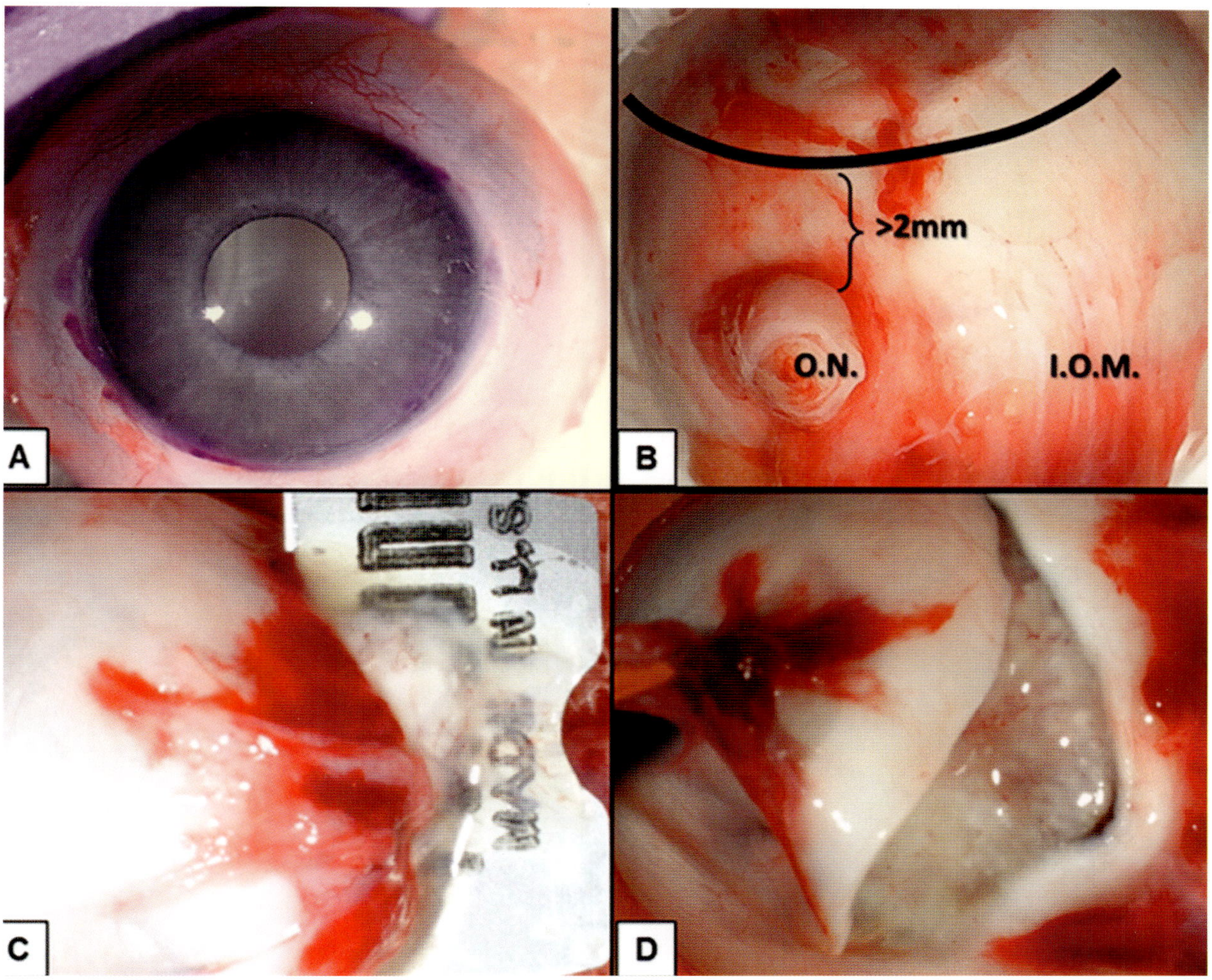

Figure 9-19

ENUCLEATED EYE FOR HARVESTING FRESH TUMOR BEFORE FIXATION

A: The white pupil is known as leukocoria.

B: View of the posterior surface of the eye. The inferior oblique muscle (IOM) inserts lateral (temporal) to the optic nerve. Therefore this is a right eye. If the eye is full of tumor, the process of transillumination will not be effective in identifying a well-delineated shadow cast by the tumor. In such cases, the preferred plane of section is the horizontal plane, as shown here. To retrieve fresh tumor before fixation a scleral window should be created in the same plane but away from the nerve to avoid tumor contamination (about 2 mm or more from the optic nerve in the equator of the eye).

C: With a sharp blade, steady, firm, left and right movements are performed on the sclera to penetrate it and the underlying choroid without pushing into the eye to avoid distorting and damaging the intraocular tissues. At the time of penetration of the blade into the eye, the friable tumor usually comes out over the opening. This is important when reviewing the histopathology as there will be tumor cells on the episcleral and orbital tissues in these areas.

D: The tumor is exposed by gently pulling the sclera superiorly to allow for tumor retrieval. If done correctly, the insertion of the superior oblique muscle serves as the place to grab and pull without disturbing the sclera, as shown here.

to treatment or have extensive complications of treatment. In these cases, the pathologic findings are associated with the regression of the tumor, the presence of resistant tumor, and the associated response of the ocular tissue to the treatment.

Eyes enucleated after radiotherapy or chemotherapy contain regression scars composed of glial tissue with foci of calcified retinoblastoma cells. Areas of photoreceptor/neuronal differentiation also are found, especially after chemotherapy (fig. 9-23A). In eyes that have been

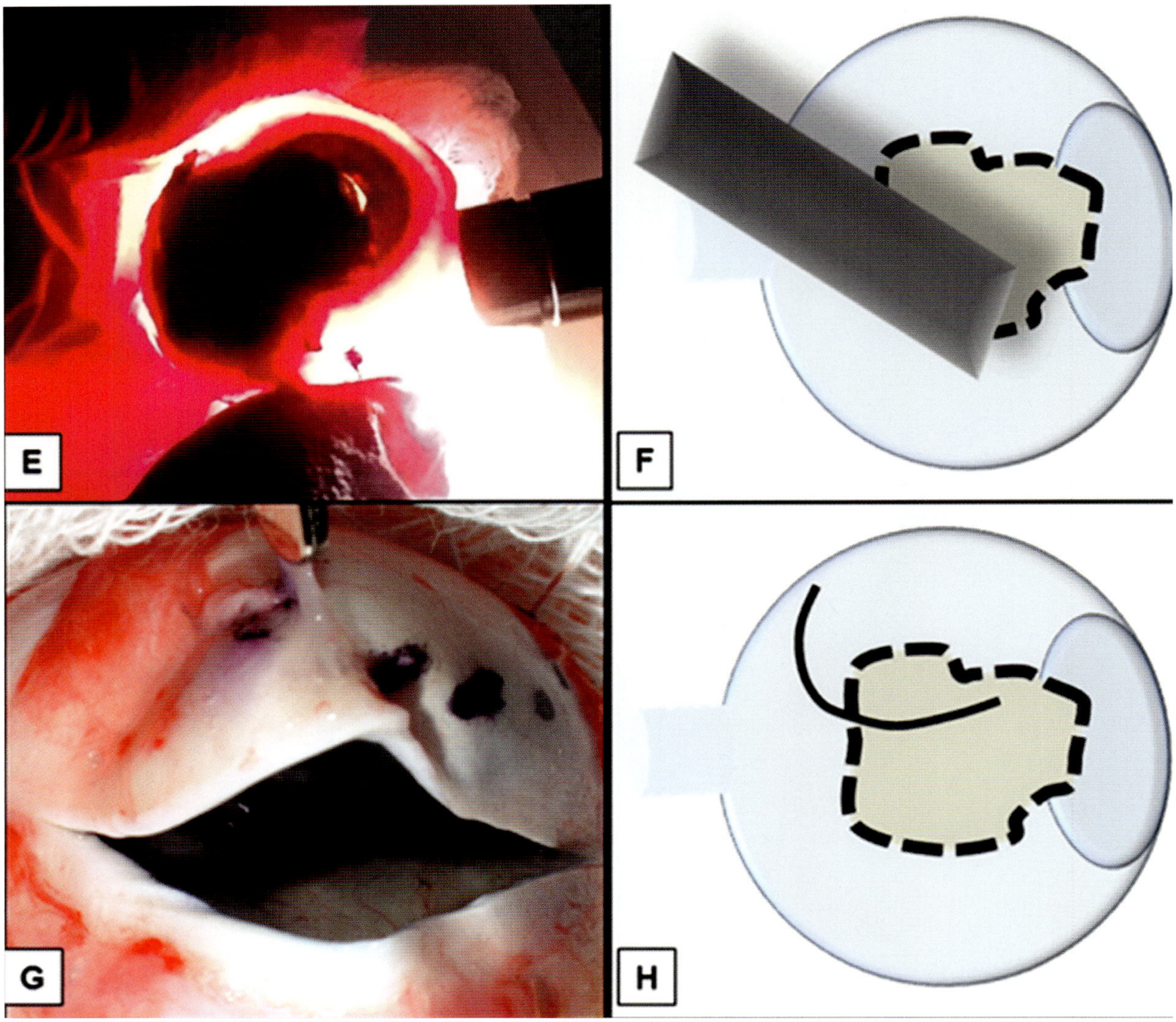

Figure 9-19, continued

E: Transillumination of the eye is performed in a dark room with a strong light source positioned in the cornea or in the equator of the eye and moving the eye to observe any shadows that are present. When a shadow is found, it is lined using a marker on the sclera. The shadow corresponds to tumor most of the time although it may be hemorrhage or retinal detachment.

F: To harvest cells in smaller tumors, a scleral window is opened at the edge of the shadow and not at the center (as shown in the diagram). The central larger portion is then examined in the pupil optic nerve section submitted for permanent sections.

G: The opening of the scleral-choroidal window is shown, leaving the edge of a previously drawn shadow with adequate amount for tumor harvesting.

H: The final location of the window in relationship with the nerve that leaves an undisturbed area to be included in the pupil-optic central section of the eye for permanent section.

treated with radiotherapy, neovascular membranes in the vitreous and necrotic retinal tissue may bleed and large areas of vitreous or subretinal hemorrhage may be present (fig. 9-23B).

After intra-arterial chemotherapy (IAC), histopathologic findings are mostly seen during the initial stages of the implementation of the IAC. These include thrombosed arteries, including the central retinal artery, and other intraocular and orbital vessels containing intraluminal particles of birefringent foreign material, surrounded by foreign body giant cells (fig. 9-23C). The intravascular material may be partially calcified (19). Other findings include focal areas of atrophic choroid with associated atrophic outer retinal layers, and focal arterial (orbital, retinal, or choroidal)

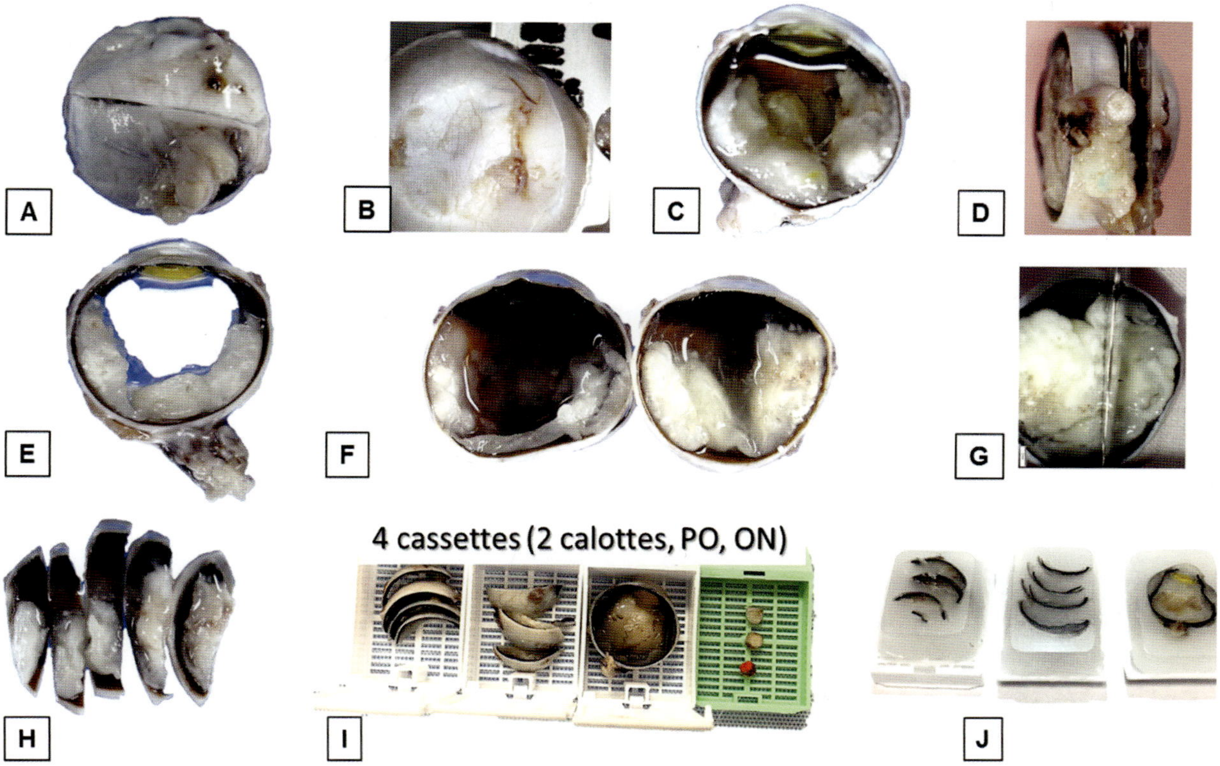

Figure 9-20

DISSECTION AND SAMPLING OF THE EYE AFTER FIXATION

A: A previously opened eye (superior scleral window) after fixation.

B: To obtain a central pupil-optic nerve (PO) section, a new cut is made about 0.5 mm superior to the optic nerve but below the previous cut to include the scleral window in the newly created calotte (cap). In those eyes that have not been opened previously, the cut is made 0.5 mm superior to the optic nerve in the plane of section selected after transillumination. If the eye is full of tumor and there is no transillumination, a horizontal section is preferred. The section aims to cut from the posterior pole to the cornea, entering the limbus at about 2 to 3 mm into the cornea.

C: An eye after removing the first calotte (in this case superior). This view allows for identification of the tumor and the association with other intraocular structures. In this eye, an endophytic tumor replaces the retina and covers the head of the optic nerve.

D: To complete the sections of the eye when obtaining the calottes, the type of sectioning movement changes from a lateral right to left seesaw movement to a guillotine fast and precise movement. At the time the blade passes the equator of the eye, about 5 mm from the limbus, the eye with the blade is turned so the corneal surface is on the cutting board surface and the optic nerve faces the ceiling. The blade is the held with the two hands firmly and a guillotine-like precise cut is made to the edge of the lens and cornea without dragging the retina or dislocating the lens.

E: PO sections without the two calottes. The optic nerve has not been incised nor has the pupil. The true PO section will be obtained by the histotechnologist at time of microtomy of this segment. This is to preclude cutting into important optic nerve pathology (invasion) and not having the area seen in the histologic sections.

F: Two calottes each have different pathologic findings and amounts of tumor, which should be described in the macroscopic description to correlate later with histologic findings and staging, especially noting choroid, anterior segment, and extraocular extension of the tumor.

G. Calottes are further sectioned in an anterior-posterior direction and in a bread loaf manner to obtain more choroidal and scleral surface to examine histologically.

H: A calotte after sectioning.

I: After macroscopic dissection, the entire eye is included in four cassettes with the calottes sectioned. Tall cassettes are required to hold the thicker sections of the eye (long processing similar to that used for breast mastectomy specimens). The optic nerve fits in the regular cassettes and if a long nerve is received, the distal 5 to 10 mm are submitted in cross sections as shown here.

J: The paraffin blocks of the eye show the on-edge embedding of the calottes and the complete PO section.

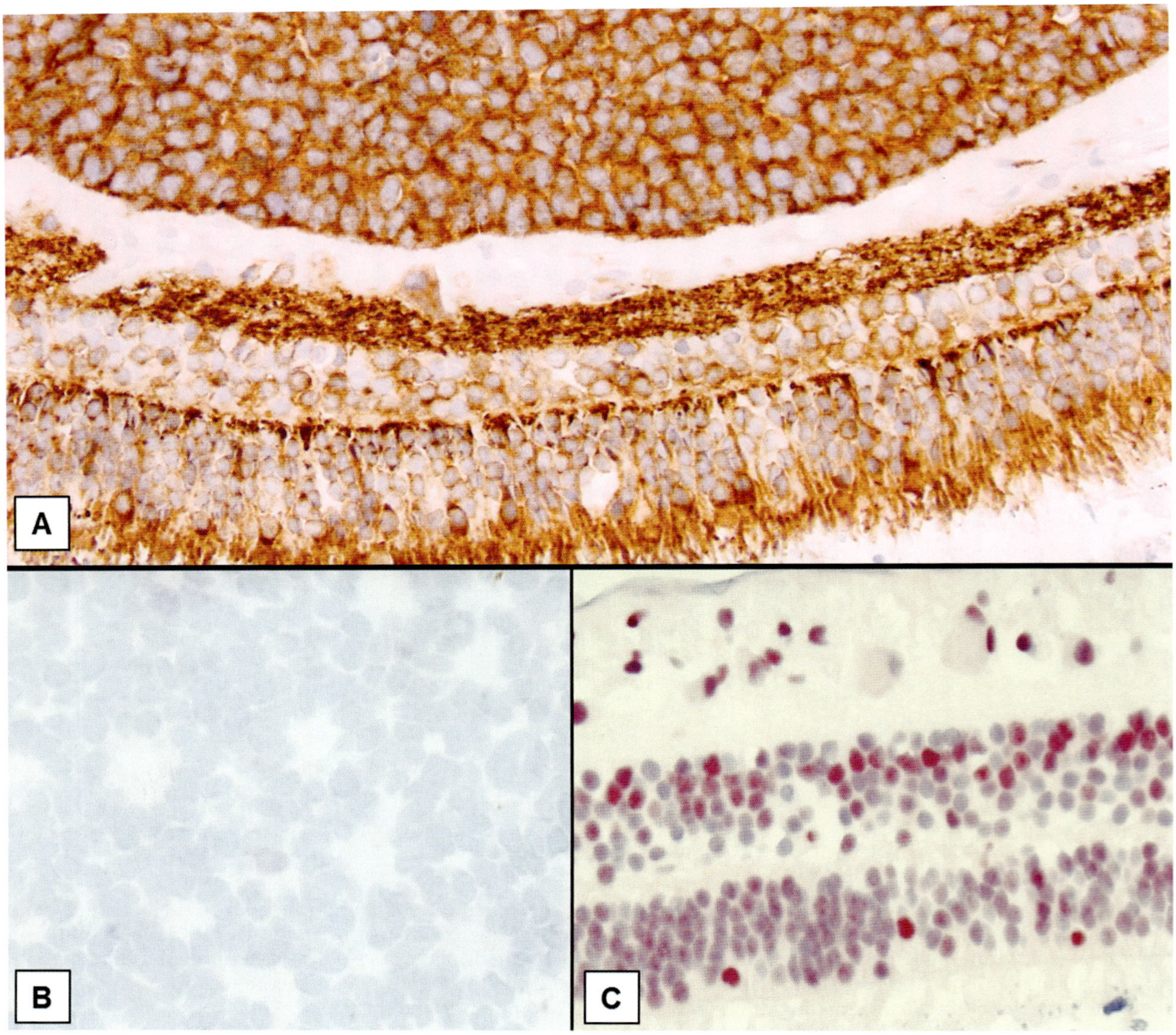

Figure 9-21

RETINOBLASTOMA: IMMUNOHISTOCHEMISTRY

A: Synaptophysin antibody is an excellent marker for retinoblastoma because it stains the cytoplasm of the tumor cells consistently (compare staining in the tumor at the top with staining in the underlying retina). Synaptophysin highlights the synaptic regions of the retina and the outer segments of rod-type photoreceptors.

B: Retinoblastoma protein is negative. The *RB* gene is mutated and not producing the protein.

C: The retina in a patient with sporadic retinoblastoma shows positive nuclear staining with RB protein antibody.

subendothelial smooth muscle proliferations (fig. 9-23D) (20). The residual tumors may have extensive photoreceptor/neuronal differentiation and zones of gliosis with calcification. These findings are similar to those seen in eyes treated with intravenous chemotherapy. In eyes that have undergone local therapy, including laser photocoagulation, cryotherapy, or plaque radiation, large areas of chorioretinal scarring with retinal pigment epithelium dispersion and proliferation into the retina may be seen.

After intravitreous chemotherapy, the vitreous seeds may be persistent. Most are of dust or cloud type in a background of vitreous bands with calcified seeds.

Metastasis and Recurrence. The untreated eye fills with retinoblastoma and the normal ocular tissue structures are destroyed. The most

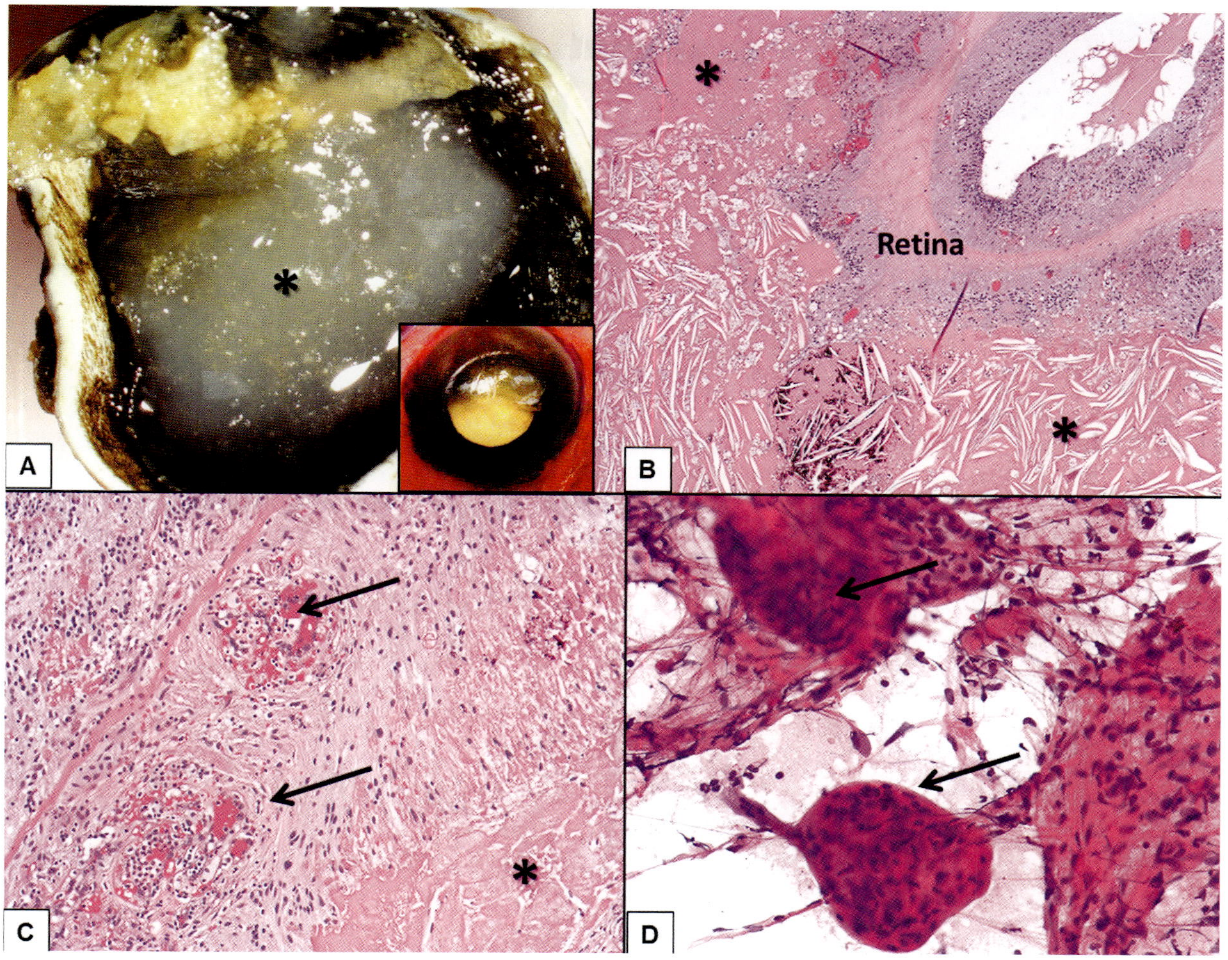

Figure 9-22

COATS DISEASE

A: A sectioned eye with Coats disease features an extensive subretinal proteinaceous exudate (*) with small shiny cholesterol crystals and tan-yellow material (hemorrhage) in the most anterior portion of the exudate. The insert shows the leukocoria/ xanthocoria in this eye.

B: Corresponding histopathologic section of the eye seen in A with detached retina and eosinophilic proteinaceous exudate (*) containing foamy (lipid) histiocytes and many cholesterol crystals (spaces left by the crystals after dissolved during processing).

C: Aneurysmatic leaking vessels (arrows) in the retina are associated with proteinaceous exudate (*) and hemorrhage.

D: Touch imprint of the unfixed retina shows the aneurysms (arrows).

important route of spread is through the optic nerve. The tumor may spread directly along the nerve fiber bundles into the optic chiasm and brain parenchyma and may infiltrate through the pia into the subarachnoid space. Retinoblastoma cells in the subarachnoid space may reach the cerebrospinal fluid (CSF) and spread into the brain and spine.

The second major route of spread is through the massive involvement of the choroid into the orbit via either scleral canals (areas within the sclera where ciliary vessels, nerves, and vortex veins enter or exit the eye) or by direct extension through the sclera (21). Extraocular extension generally occurs within 6 months if intraocular tumors are left untreated. Extraocular extension dramatically increases the chances of hematogenous and lymphatic spread.

Metastasis can occur by hematogenous dissemination secondary to orbital and bone

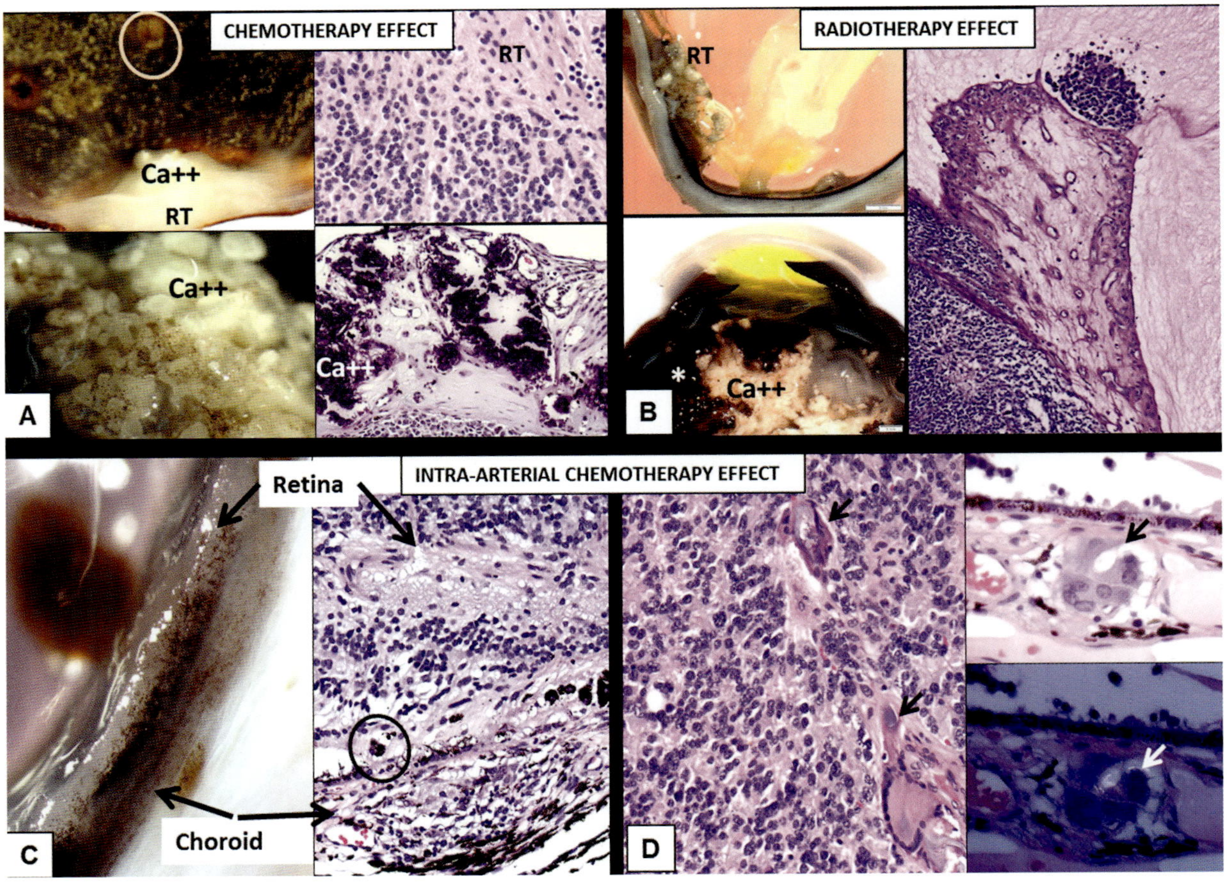

Figure 9-23

POST-TREATMENT EFFECT IN EYES WITH RETINOBLASTOMA

A: *Top left:* Systemic chemotherapy produces regression of the tan-gray lesion (RT) with large chunky calcifications (Ca++). In this eye there are also chorioretinal scars (circle) left by local treatment (laser or cryotherapy). *A, top right:* histopathology of the regressed tumor (RT) shows a retinocytoma-like lesion with neuronal or photoreceptor differentiation and an absence of mitoses or apoptosis. *A, bottom left:* chunky large calcifications (Ca++) in the regressed tumor. *A, bottom right:* histopathology of the same area with large chunky calcifications (Ca++) in a gliotic tissue.

B: Radiotherapy effect in an eye with retinoblastoma. *B, top left:* the pupil-optic nerve section of this eye with a regressed tumor (RT) features calcifications and a funnel-shaped, totally detached retina at the center. *B, bottom left:* postradiation treatment shows areas of vitreous, retinal, and subretinal hemorrhages (*) with a detached retina and extensively calcified tumor (Ca++). *B, right panel:* epiretinal and tumor neovascular membrane.

C: Effects after intra-arterial chemotherapy. *C, left panel:* gross photograph of a chorioretinal scar with RPE migration into the retina. *C, right panel:* histologic section of the same area with the RPE migration encircled. In eyes that have undergone local therapy, including laser photocoagulation, cryotherapy, or plaque radiation, large areas of chorioretinal scarring with RPE dispersion and proliferation into the retina are seen. Retinal atrophy and gliotic attachment to the RPE layer.

D: A mostly regressed tumor with several multinucleated foreign body type cells (arrows) with fiber-like blue-gray material in the cytoplasm. *D, top right:* high-power view of a foreign body-type giant cell in the choroid. *D, bottom right:* polarization discloses birefringent polarizable foreign material (arrow).

invasion or when lymphatic invasion reaches the lymph nodes. Widespread metastases are found in lung, bone, and brain. Lymphatic dissemination occurs in tumors that spread anteriorly into the conjunctiva and eyelids or extend into extraocular tissues. Lymphatic vessels and lymphoid tissue are absent in the orbit and intraocular tissues. In the ocular region, only conjunctiva and skin have lymphatic channels. Tumors must first reach these areas to permeate the lymphatic vessels and then spread into regional lymph nodes (fig. 9-24).

Retinoblastoma metastases are less differentiated than intraocular tumors. Fleurettes have

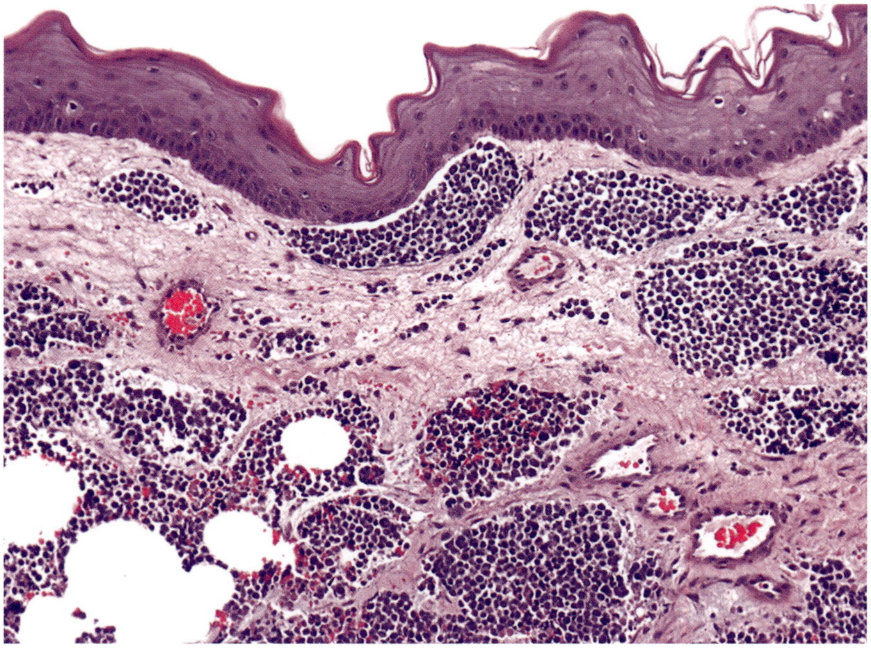

Figure 9-24

EXTRAOCULAR EXTENSION OF RETINOBLASTOMA

Extension of retinoblastoma in the stroma of conjunctiva and some intralymphatic tumor. Empty spaces represent silicon oil filtered from the intraocular space into the stroma (silicon oil was used during vitrectomy in an unrecognized retinoblastoma patient, which caused a iatrogenic extraocular extension and systemic metastasis).

never been described in metastatic tumors and Flexner-Wintersteiner rosettes are rarely encountered. Recurrences and metastasis usually occur within the first 2 years of diagnosis of retinoblastoma, while second primary malignancies occur more than a decade after diagnosis. When well-differentiated extraocular tumors appear outside of the orbit, a diagnosis of a primary primitive neuroectodermal tumor (PNET) must be considered. In patients with germline mutations who develop second primary malignancies, the histopathologic features and time of presentation of the tumor are helpful features for differentiation from new primary tumors.

Trilateral Retinoblastoma. Primary retinoblastomas of the pineal and parasellar sites have been called trilateral retinoblastoma (pinealoblastoma) and usually present as single tumors. Trilateral retinoblastoma is a well-recognized, although rare, syndrome. Most reported cases involve patients with a family history of retinoblastoma and the disease is usually fatal. There are recent observations that newer methods of chemoreduction can minimize the risk of developing trilateral retinoblastoma (22).

These tumors can appear several years after successful treatment of intraocular retinoblastoma. They are usually more differentiated than the primary tumor and may contain numerous rosettes, fleurettes, and individual cells with photoreceptor differentiation. The presentation of trilateral retinoblastoma contrasts with metastatic retinoblastoma since the latter presents as multiple, undifferentiated tumors within the first 2 years of initial treatment.

Prognosis. Metastatic retinoblastoma is still associated with a poor prognosis, especially when recurrence occurs in the central nervous system or is associated with a trilateral presentation. Most clinical findings are not useful in predicting the occurrence of metastasis in children with retinoblastoma; histopathologic findings are better estimates for predicting such risk.

The most important prognostic indicators for the development of metastasis are: significant "massive" choroidal invasion, invasion into the optic nerve, presence of tumor at the optic nerve resection margin, and extraocular extension of tumor into the orbit. Each of these characteristics is explained below.

Tumor Invasion into the Choroid. Massive choroidal invasion by tumor increases the possibility of hematogenous spread, either via the choroidal vessels or, more frequently, by extension through the sclera into the orbital tissues. Massive choroidal invasion is defined as invasion into the choroid that expands the choroid 3 mm or more in diameter to reach the sclera (fig. 9-25A,B). Focal invasion is any tumor in the choroid that is less than 3 mm (fig. 9-25C). If there are multiple

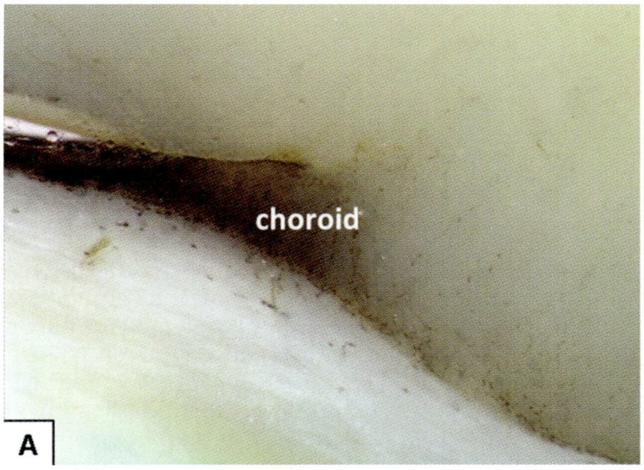

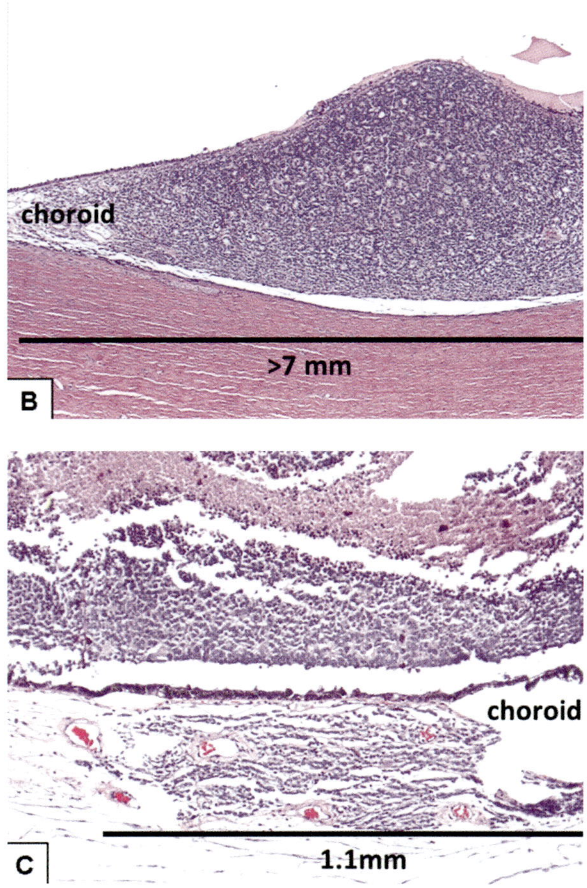

Figure 9-25

POSTERIOR UVEAL (CHOROIDAL) INVASION BY RETINOBLASTOMA

A: Cut section of an eye with massive choroidal invasion (over 3 mm) by retinoblastoma. Note the dissolution of the RPE and the replacement of the choroid by tumor in the right portion of the figure. Note the absence of necrosis.

B: Histopathologic correlation of the invasion of the choroid seen in A. Dissection and replacement are present at the edge of invasion by tumor, which has some rosettes. It measures more than 7 mm in maximum diameter.

C: Small focus (less than 3 mm) of tumor invasion in the choroid. The infiltrating tumor cells distend and replace the choroidal spaces and surround larger vessels with finger-like projections at the edges of the infiltrate. These are characteristic features of true choroidal invasion.

separated small foci of choroidal invasion, the measurements should be added to have the maximum amount of invasion (15).

The significance of choroidal involvement and its effect on survival outcome remains controversial. In the literature, some degree of choroidal invasion is reported in 12 to 62 percent of eyes enucleated for retinoblastoma. There is a wide range of reported mortality rates, from 11 to 81 percent, in eyes with massive choroidal invasion (23). Most authors recognize that massive choroidal invasion is associated with a worse prognosis when associated with optic nerve invasion posterior to the lamina cribrosa (retrolaminar invasion) (24). Given the uncertainty of the significance of massive choroidal invasion as an independent risk factor for metastasis, choroidal invasion should be recorded when detected and a measurement of the choroidal involvement provided. In this way, oncologists may decide whether to administer adjuvant chemotherapy to these patients while data on the risk for metastasis after choroidal invasion continue to accumulate.

When eyes are opened to harvest fresh tissue for assessment of retinoblastoma status, it is possible (even likely) that friable retinoblastoma tissue will be introduced into the choroid. In such cases, fragments of the tumor, disassociated within the collagenous framework of the choroid, are identified. Small fragments of retinoblastoma tissue may also be identified outside the eye, without evidence of choroidal invasion or optic nerve invasion, and such small "extraocular" fragments of tumor tissue are also likely to be artefactual (fig. 9-26). The presence of fragments of tumor tissue in the choroid or outside the eye should be recorded as likely artifactual so that a secondary review of slides, perhaps years after the enucleation, reflects the assessment of the eye at the time of slide interpretation.

Tumor Invasion into the Optic Nerve. Tumor invasion into the neurosensory tissue anterior to the lamina cribrosa (the "lamina retinalis") and

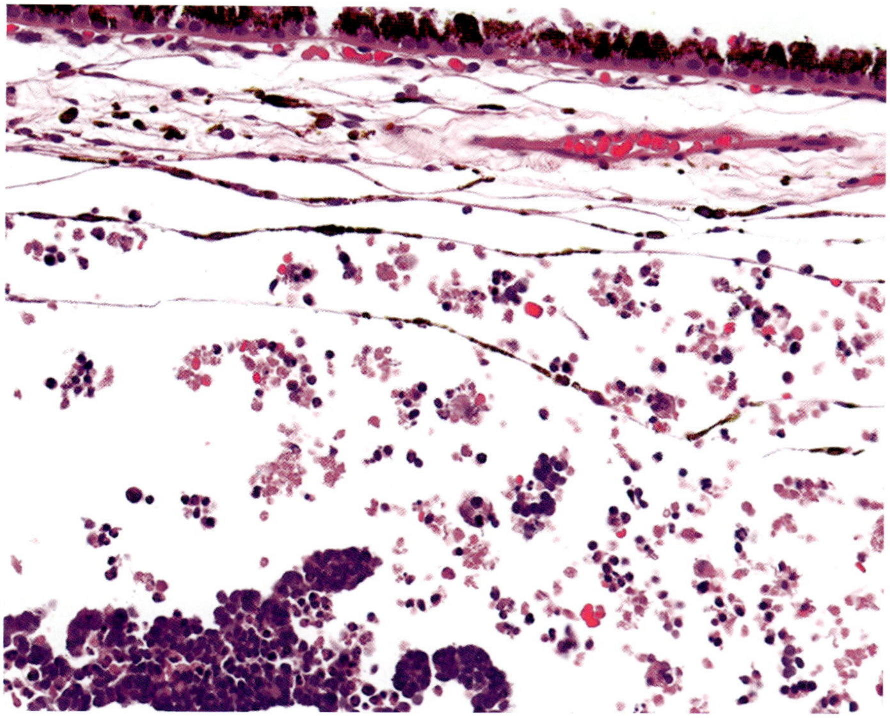

Figure 9-26

ARTIFACTS OF FRESH TUMOR RETRIEVAL IN THE UVEA (CHOROID) MIMICKING INVASION BY RETINOBLASTOMA

The choroidal spaces are open and the tumor with abundant necrotic cells, floats in the center without pushing borders. The large focus of tumor is surrounded by necrotic cells while the reverse happens in true invasion in which the necrosis is at the center.

into the lamina cribrosa is associated with a low mortality rate (10 and 29 percent, respectively). However, when a tumor is identified posterior to the lamina cribrosa, the associated mortality rate increases to 42 percent and continues to increase with deeper degrees of invasion. If a tumor is identified at the transected surgical margin, the associated mortality rate may be as high as 80 percent (fig. 9-27A). The importance of obtaining a long segment of the optic nerve at the time of enucleation is emphasized by these data (25–27).

In recent studies, the depth of tumor invasion into the optic nerve (more than 1.5 mm) has become important in predicting metastasis (unpublished data). To measure the invasion of the optic nerve there are two approaches. Measurement from the inner limiting membrane to the deepest point of invasion into the nerve tissue is preferred if there is a visible limiting membrane (fig. 9-27B). If the tumor has obliterated the optic nerve limiting membrane, then the measurement should be taken from the level of Bruch membrane to the deepest point of tumor invasion into the optic nerve tissue (fig. 9-27C). Tumor in the meningeal layers is also associated with central nervous system (CNS) recurrence and should be reported in the histopathologic interpretation.

Scleral and Extrascleral Extension. The tumor may infiltrate the sclera either by direct extension from massive choroidal invasion or by tracking the emissary channels (fig. 9-28). Deeper levels of the block should be obtained in these cases to identify extrascleral extension, if present. Extrascleral extension is a risk factor for orbital recurrence after enucleation and for metastasis, and this tumor characteristic should be recorded in the pathology report (28).

Tumor at Surgical Margins and in the Orbit. Retinoblastoma tumor cells are left in the orbit when the tumor is present at the cut end of the optic nerve (fig. 9-29). After the orbital soft tissues are invaded, the tumor may spread directly into the orbital bones, through the sinuses into the nasopharynx, or via the various foramina and bony openings into the cranium. In cases of extraocular extension of a tumor before enucleation, the soft tissues of the orbit represent additional surgical margins. Sometimes pre-enucleation chemotherapy is prescribed if extraocular extension is suspected. The objective is to treat the extraocular tumor and to reduce the tumor burden before enucleation. In these cases, evaluation of orbital soft tissue margins to access the presence of a residual tumor is of prognostic importance. Extraocular extension dramatically increases the possibility of

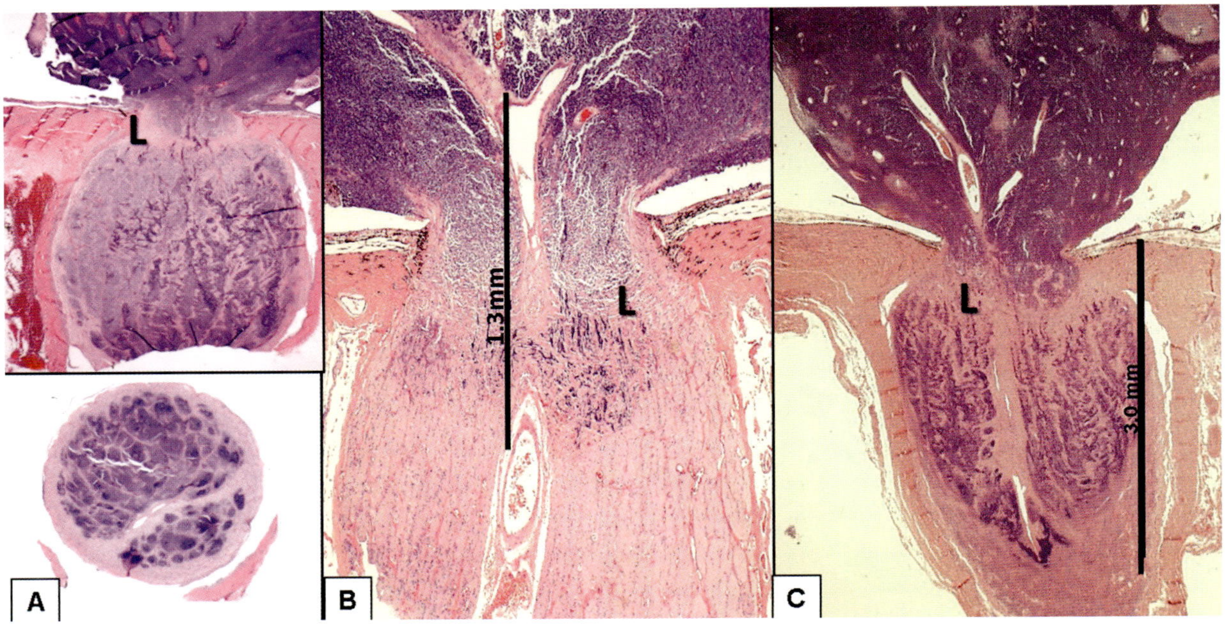

Figure 9-27

OPTIC NERVE INVASION BY RETINOBLASTOMA

A: *Top:* There is extensive invasion into the optic nerve to the resection margin. The nerve is expanded toward the meninges by the tumor. There is extensive tumor posterior to the lamina cribrosa (L). *A, bottom:* cross section of the optic nerve approaching the margin shows the tumor following the nerve bundles.

B: Optic nerve with postlaminar (L) invasion by tumor that measures 1.3 mm from the inner limiting membrane of the nerve head to the deepest invasion site.

C: Tumor invades the optic nerve in the postlaminar (L) region and obliterates the head of the nerve and the lamina cribrosa. The tumor measures 3 mm from the Bruch membrane to the deepest point of invasion.

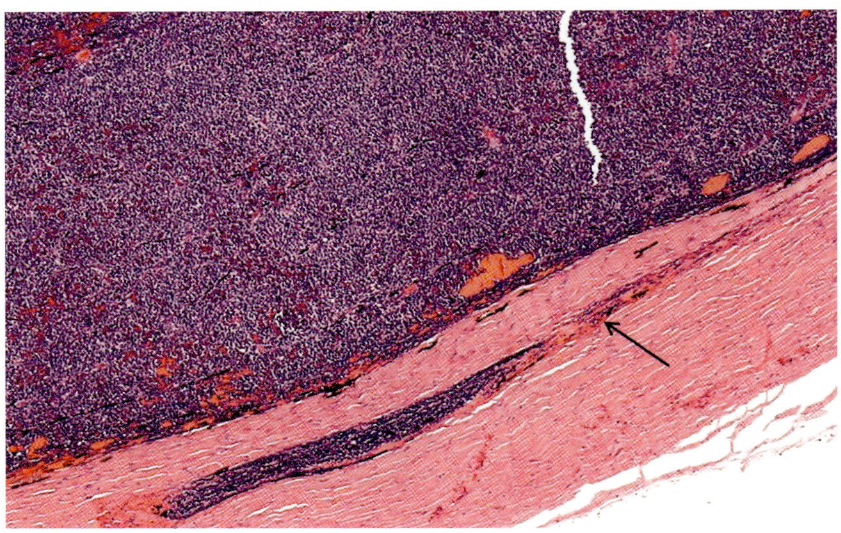

Figure 9-28

INTRASCLERAL INVASION BY RETINOBLASTOMA

The tumor tracks the emissary channels where vessels and nerves enter and exit the eye, through the sclera (arrow). In this case, the tumor is only intrascleral as serial sections showed no episcleral invasion.

hematogenous dissemination and may provide tumor access to conjunctival lymphatic vessels and subsequent lymph node metastasis.

Anterior Segment Involvement. There is limited information on the significance of the extension of retinoblastoma into the iris, ciliary body, anterior chamber, trabecular meshwork, and cornea (fig. 9-30). Anterior segment involvement is frequently associated with other high-risk features and the value of this finding as an independent variable in predicting metastasis is uncertain. At this time, if the anterior segment is involved by tumor, it should be reported. Iris neovascularization, if present, should also be documented, if for no other reason than to confirm the clinical detection of this finding. The role of high intraocular pressure in facilitating tumor invasion into the optic nerve is unclear.

Minimal Disease Detection (Liquid Biopsy). Currently, patients with systemic metastasis can be cured with intensive therapies if there is no CSF involvement. Tumor recurrence in the bone marrow shows a rapid response and high treatment success rate with the use of intensive chemotherapy. The challenge remains for those children without overt CNS dissemination but who have small tumor seeds in the brain or CSF because when they develop the overt CNS relapse the disease fails to respond to any kind of treatment. Recent reports have published the use of "liquid biopsy" for assessment of recurrence of disease at extraocular sites using blood and CSF (29).

Staging. The 8th edition of the American Joint Commission for Cancer (AJCC) staging of retinoblastoma describes the features that define the extent of intraocular tumor used to predict ocular salvage (previously called "group") and the risk of extraocular extension, T stage, for each eye. Retinoblastoma staging also includes the standard node (N) and metastasis (M) fields. In bilateral retinoblastoma, each eye may have a different extent of intraocular tumor, scored separately, but for overall staging, the T stage of the worst eye is used.

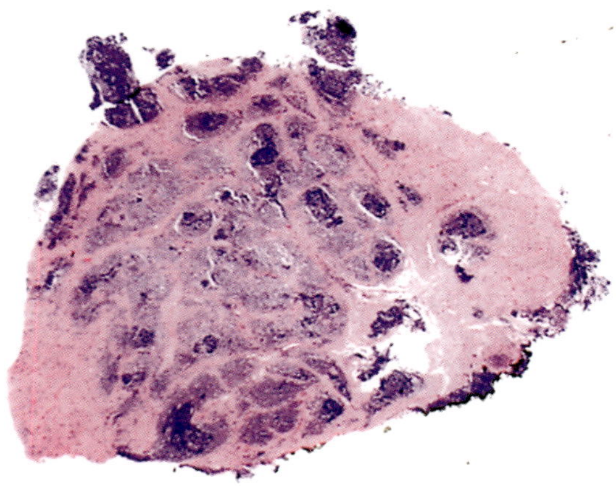

Figure 9-29

OPTIC NERVE MARGIN INVOLVED BY RETINOBLASTOMA

Histologic section at the level of the optic nerve margin shows tumor cells, some of which are crushed.

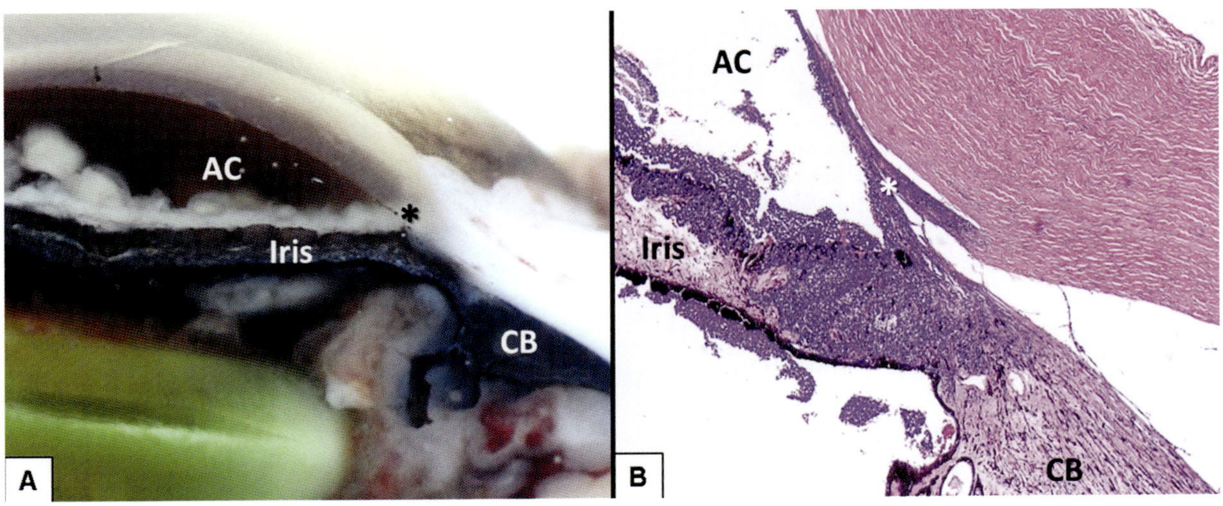

Figure 9-30

ANTERIOR CHAMBER INVASION BY RETINOBLASTOMA

A: Cut surface of the eye with retinoblastoma that has invaded the anterior chamber (AC), angle (*) posterior chamber (between iris and ciliary body [CB]), and anterior vitreous.

B: Histologic correlation of the macroscopic picture shows retinoblastoma tumor infiltrating the root and surface of the iris, the angle (*) structures including the trabecular meshwork, and the anterior chamber (AC).

Table 9-3

PATHOLOGIC FEATURES FOR STAGING RETINOBLASTOMA[a]

Categories: Tumor, Lymph Node, Metastasis, Hereditary Trait	Description of Findings	Stage Grouping
Tumor: confined to the eye	pT1. Without invasion or with minimal <3 mm invasion into the choroid or focal prelaminar/laminar optic nerve head invasion pT2. a. Concomitant invasion of choroid (<3 mm) and prelaminar/laminar optic nerve b. Anterior chamber structures, single or concomitant with each other pT3. Significant local invasion a. Choroidal invasion >3 mm in one foci or multiple foci that when adding the sizes measure >3 mm in maximum diameter b. Postlaminar optic nerve invasion (PLONI), not reaching the cut margin c. Any partial invasion of the sclera without extrascleral extension	I: pT (1,2,3), pN(O), M(0), H(any) II: pT (4), pN(O), M(0), H(any) III: pT (any), pN(1), M(0), H(any) IV: pT (any), pN(any), M(1), H(any)
Tumor: Extraocular tumor	pT4. Tumor at transected margin of optic nerve or in meninges, extraocular extension into episclera, conjunctiva, extraocular muscles or bone	
Lymph node status	pN0. Without histopathologic metastasis to lymph nodes pN1. histopathologic confirmation of metastasis to lymph nodes (prelaminar)	
Distant Metastasis	pM0. No histopathologic proof of distant metastasis pM1a. Histopathologic evidence of systemic metastasis (bone marrow, liver, soft tissue, etc) pM1b. Histopathologic/cytologic diagnosis of CNS metastasis	
Hereditary Trail	HX. Unknown status of constitutional *RB1* gene mutation H0. No constitutional *RB1* gene mutation H1. Bilateral or trilateral (intracranial primitive neuroectodermal tumor), family history of retinoblastoma, or constitutional *RB1* gene mutation	

[a]Data from reference 30.

Evidence-based data have supported the changes in the 8th edition, derived from published literature and from the retrospective survey conducted by the Ophthalmic Oncology Task Force (OOTF). In this new classification, the clinical staging (cT) of the tumor is based on size and associated ocular changes that predict failure to treatment and suggest tumor progression to possible high-risk features (see Appendix 8).

Pathologic staging takes into account the degree of invasion into the ocular structures. pT1 shows no invasion into the optic nerve, choroid, or sclera. pT2 shows local invasion into the above-mentioned structures, and pT3 has significant local choroidal invasion (over 3 mm), or retrolaminar optic nerve invasion, or partial thickness involvement of inner sclera. In pT4 there is histopathologic evidence of extraocular tumor (Table 9-3).

For metastatic staging, the groups are divided into pM1a (pathologic evidence of tumor at distant metastasis [not CNS]) versus pM1b (CSF or CNS parenchyma). In the 8th edition of the AJCC there is a new definition, heritable trait, denoted by the capital letter "H" (see chapter 1). The germline predisposition significantly affects patient outcomes, including cancer-related morbidity and mortality (Table 9-3). The combination of findings defines the final stage group. The AJCC classification only applies to unilateral or bilateral ocular tumors. Patients with trilateral retinoblastoma are staged by systems for brain and spinal cord tumors (30).

GLIAL TUMORS AND TUMOR-LIKE CONDITIONS

Astrocytoma

Definition. *Astrocytomas* are low-grade neoplasms arising from the retina or optic nerve head. Retinal astrocytomas are usually seen in children and young adults. These tumors are associated with tuberous sclerosis complex (TSC) or neurofibromatosis.

Clinical Features. Retinal astrocytomas are present in up to half of the patients with TSC. These lesions arise from the inner retina, usually as well-defined, elevated nodules within the vascular arcades (the boundaries outlining the macula), adjacent to the optic nerve, or less frequently, in the retinal periphery. The tumors may present with vitreous seeds. Patients with TSC may have multiple lesions in one eye or bilateral lesions.

Optical coherence tomography (OCT) of the retina demonstrates a gradual transition between the normal retina and the lesion. Features include hyper-reflectivity with retinal disorganization, "moth-eaten" spaces, and posterior shadowing (31).

Gross Findings. Retinal astrocytomas are well-defined gray-pink lesions in the retina.

Microscopic Findings. The histopathology is characterized by two types of cells: fascicles of cytologically bland spindle-shaped astrocytes that are positive for glial fibrillary acidic protein (GFAP) (fig. 9-31), and large cells with abundant eosinophilic cytoplasm, round nuclei, and prominent nucleoli. The latter cells resemble the subependymal giant cell astrocytoma (SEGA) found in the brain, and express both glial and neuronal markers. Tumors of this type are benign but may grow to fill the globe posteriorly, thus, the term *aggressive retinal astrocytoma.* Features typical of pilocytic astrocytoma such as Rosenthal fibers are usually absent, however, they sometimes are seen in the rare optic nerve pilocytic astrocytoma that extends into the retina. Calcification and necrosis are histologic features often seen; mitoses are characteristically absent.

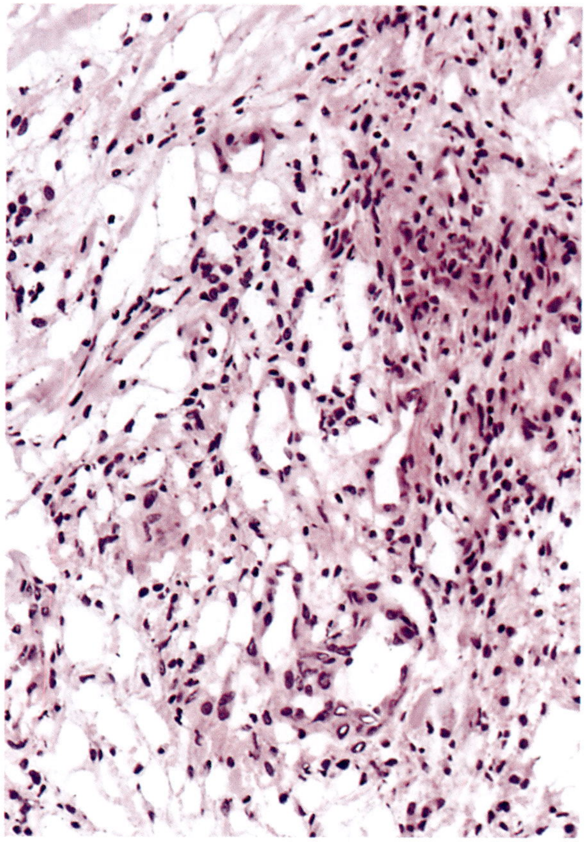

Figure 9-31

ASTROCYTOMA OF THE RETINA

There are fascicles of cytologically bland spindle-shaped astrocytes in a loose background. The cells have abundant eosinophilic cytoplasm and no atypia. This area was seen in the inner layers of the retina (nerve fiber layer expanded by the lesion).

Treatment and Prognosis. Astrocytomas may threaten vision, depending on their location and size, but these lesions are not clinically aggressive or fatal. Zones of necrosis may develop in retinal astrocytomas, especially those with the large SEGA type cells, accompanied by iris neovascularization and possibly glaucoma, as with other necrotic intraocular lesions. Lesions that grow may be treated with Verteporfin photodynamic therapy (32).

Retinal Gliosis

Definition. Retinal astrocytes, Mueller cells, participate in *gliosis* that forms benign mass lesions within the eye. These lesions occupy less than 25 percent of the posterior ocular compartment and typically feature prominent vascularity. They tend to develop in a background of retinal disease or previous injury. Reactive gliotic mass lesions that fill 25 to 50 percent of the posterior compartment are designated *submassive retinal gliosis*, reserving the term *massive gliosis* for reactive lesions that occupy more than half the space of the posterior compartment. Synonyms include *reactive retinal gliosis, retinal angiomatous mass, presumed acquired hemangioma, reactive retinal astrocytic tumor,* and *vasoproliferative retinal tumor* (33).

Clinical Features. Retinal gliosis occurs in adults without gender predilection. It is usually associated with other ocular conditions (Table 9-4). The typical location of this lesion is the inferotemporal pre-equatorial retina. The lesion may present with decreased vision or as an incidental finding during a dilated fundus examination. Subretinal exudates accompany the lesion and cause shallow nonbullous retinal detachment. RPE hyperplasia is seen at the base of the lesion. These features are associated with decreased vision. The vascular component of the lesion leaks fluorescein on fluorescein angiography.

Gross Findings. Focal reactive gliosis is a pink to yellow to orange nodule that has a vascular appearance. It may be adjacent to flat retinal detachment with subretinal exudate. In cases of massive reactive gliosis, the entire posterior compartment of the eye may be filled by a yellow to orange rubbery mass.

Microscopic Findings. The lesion is composed of spindle-shaped fibrous astrocytes that may contain Rosenthal fibers, admixed capillaries, and focally sclerotic vessels. Proteinaceous exudates in the retina and a subretinal space contain cholesterol clefts and giant cell reaction. The base of the lesion shows RPE hyperplasia. There may be fibrous or osseous metaplasia with an associated lymphocytic infiltrate in the choroid. Immunohistochemical stains are positive for GFAP, vimentin, and S-100 protein in astrocytic cells; mitotic figures are absent or rare, and the prominent blood vessels are periodic acid–Schiff (PAS) positive.

Table 9-4

OCULAR CONDITIONS WITH RETINAL GLIOSIS SIMULATING NEOPLASM

Congenital/Hereditary
Retinitis pigmentosa
Coats disease
Aniridia
Familial exudative vitreoretinopathy
Congenital hypertrophy of retinal pigment epithelium (CHRPE)
Coloboma of retina and choroid
Inflammatory/Infectious
Toxoplasma retinitis
Pars planitis (intermediate uveitis)
Histoplasmosis
Idiopathic choroiditis
Idiopathic peripheral retinal vasculitis
Other
Retinopathy of prematurity (ROP)
Traumatic chorioretinopathy
History of retinal detachment repair

Differential Diagnosis. The principal entities in the differential diagnosis are retinal astrocytoma and hemangioblastoma. The associated tumor vasculature is distinctive. The presence of engorged feeding vessels characterizes hemangioblastomas. The stromal cells that are so distinctive of hemangioblastoma elsewhere are not seen in reactive gliosis.

Treatment and Prognosis. Patients have a long course requiring multiple local therapies that may culminate in enucleation. The prognosis for vision and ocular salvage for the various forms of reactive retinal gliosis that forms masses is poor.

VASCULAR TUMORS

Hemangioblastoma

Definition. *Retinal hemangioblastoma* is similar histologically to the lesion described in the CNS and discussed in detail elsewhere in this series. The development of retinal hemangioblastomas commonly accompanies a pathogenic variant in the *VHL* gene. (34).

Clinical Features. Sporadic tumors are generally solitary and unilateral, whereas those occurring in patients with *VHL* mutation are frequently multiple and bilateral. The median age at diagnosis in sporadic cases is 36 years and in *VHL* cases, 17 years. There is no gender predilection.

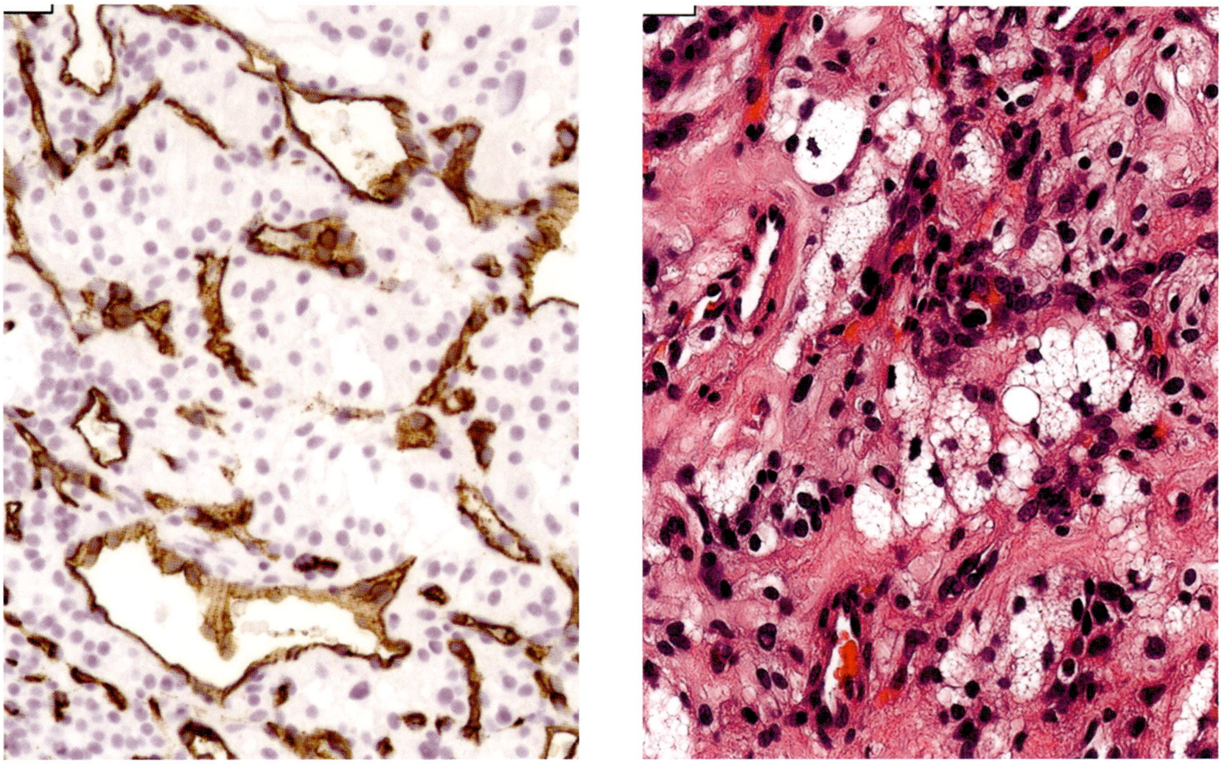

Figure 9-32

HEMANGIOBLASTOMA OF THE RETINA

Left: Capillary sized vessels depicted here by CD31 stain are surrounded by bland-appearing stromal cells.

Right: Stromal cells are distinctive for vacuolated cytoplasm and small nuclei alternating with other cells without the vacuoles, all surrounding vessels.

Approximately 50 percent of *VHL* affected patients develop retinal hemangioblastoma (34).

Funduscopic examination shows that small hemangioblastomas are flat or slightly raised clumps of capillary vessels. Prominently dilated vessels may be associated with white gliotic tissue at the surface of the lesion. The lesion appears more frequently in the temporal retinal quadrants but may have a juxtapapillary location or develop on the optic nerve. When the lesions are in the juxtapapillary region, they are flatter and associated with serous retinal detachment. Lipid exudates may surround larger tumors.

The dense capillary plexus serves as a shunt, so the feeding artery and the draining vein causing dilatation of the vessel lead to a characteristic clinical appearance. The leaking vessels may produce an exudative retinal detachment, with the gliotic vitreous adhesion inducing traction and further detachment of the retina. In the most advanced cases, the size of the lesion and the extensive exudative retinal detachment lead to phthisis bulbi, atrophy of the eye coupled with internal disorganization.

Gross Findings. Orange to pink nodules protrude from the retina into the vitreous or in the optic nerve. The nodules are associated with a feeding engorged retinal artery and surrounded by yellow retinal exudates. The pathologist may receive an en-block excisional specimen or, more commonly, an enucleation specimen. In the latter, the eye may show extensive retinal detachment with exudates associated with tumor.

Microscopic Findings. Hemangioblastomas are composed of capillary-sized vascular channels surrounding stromal cells (fig. 9-32, left). The stromal cells are vacuolated. A small cluster of angiomesenchymal cells, tumorlet cells (fig. 9-32, right), are present in both ocular and cerebellar hemangioblastomas.

The stromal cells and the tumorlet cells may express CD133, a marker present on hematopoietic, endothelial, and neural progenitor cells. Stromal cells may carry the genetic variant of the *VHL* gene, suggesting that these cells are genuinely neoplastic. Clusterin may function as a tumor suppressor in the *VHL* gene product (pVHL) because of markedly decreased expression in the tumor stromal cells, in contrast with retinal and optic nerve tissues (35).

Differential Diagnosis. For small, incidentally found lesions, the differential diagnosis is with intraretinal microaneurysms. These lack the stromal cells. For the more typical medium to large lesions, the principal differential diagnosis is with reactive retinal astrocytic proliferation (sometimes called a vasoproliferative tumor) and retinal astrocytoma, but these lesions do not have the prominent feeding artery and draining vein characteristic of hemangioblastoma. Coats disease is mimicked in those tumors with exudative retinal detachments, and the presence of the stromal cells typical of hemangioblastoma should differentiate these entities.

Treatment and Prognosis. Laser photocoagulation or cryotherapy adequately treats small to medium-sized retinal hemangioblastomas that do not involve the optic nerve. For more massive tumors, surgical en-block resection has produced excellent visual outcomes. Patients with tumors involving the optic nerve have a guarded prognosis and may be treated with anti-VEGF agents.

Overall morbidity and mortality in patients with the *VHL* germline mutation depend on the specific genotypes and phenotypes. The types of pathogenic variants determine the association with other tumors that the patients may develop (CNS hemangioblastoma, pheochromocytoma, renal cell carcinoma, pancreatic cysts, and tumors). The development of CNS tumors and renal cell carcinoma impacts morbidity and mortality. Patients with complete deletions of the *VHL* gene are much less likely to develop retinal hemangioblastomas (9 percent), than patients with partial deletions, simple missense, and nonsense mutations (45 percent). About 88 percent of patients with *VHL* retain 20/20 or better vision in at least one eye. Tumors that grow to a large size during childhood are associated with the worst prognosis for ocular salvage and these eyes may require enucleation because they are blind and painful.

Retinal Cavernous Hemangioma

Definition. *Retinal cavernous hemangiomas* are hamartomatous vascular lesions in the retina. Most lesions arise sporadically, but this condition is familial in about 20 percent of patients with pathogenic variants in the *KRIT1/CCM1* gene (36).

Clinical Features. Patients are generally asymptomatic, and routine eye examinations detect the lesions. The most common location of a cavernous hemangioma is the temporal quadrant and retinal periphery (75 percent of patients). When the hamartoma is present in the posterior retina and involves the macula, vision is affected (36).

The lesion arises from the inner retinal circulation. It is formed by engorged, dilated vessels that may involve the full thickness of the retina. The vessels are distinctive, with multiple thin-walled saccular aneurysms on the surface of the retina. White gliotic tissue surrounds the aneurysms. There may be retinal or vitreous hemorrhage. Fluorescein angiography demonstrates a characteristic low flow rate in the vessels. Skin and CNS involvement (14 percent of patients have CNS involvement) may accompany the ocular findings.

Gross Findings. These are usually incidental findings. Dark red to brown aneurysmal vessels occupy most of the retinal periphery, with vascular protrusion into the vitreous. Areas of hemorrhage may be present in the overlying vitreous.

Microscopic Findings. Clusters of dilated, thin-walled vascular spaces are lined by endothelial cells. They dissect the layers of the retina, replacing the full thickness (36). Similar to other cavernous hemangiomas, organized thrombus may be identified. The adjacent retina shows areas of gliosis.

Differential Diagnosis. The histopathologic differential diagnosis is mainly with Coats disease and hemangioblastoma. In cavernous hemangioma, there are no associated exudates in the retina or exudative retinal detachment, as seen in Coats disease. Dilated feeder vessels or draining vessels are not seen in cavernous hemangiomas as they are in hemangioblastomas.

Treatment and Prognosis. The treatment is usually by observation since the vitreous hemorrhages generally are focal and clear spontaneously. The lesions in the macular area may affect vision permanently. Patients with lesions in other locations have an excellent prognosis because cavernous hemangiomas do not progress. Patients should undergo evaluation for CNS hemangiomas.

MEDULLOEPITHELIOMA

Definition. *Medulloepithelioma* is a primary intraocular tumor characterized by cords of multilayered neuroepithelial cells of the embryonal type that resemble the medullary epithelium of the optic cup.

Clinical Features. The majority of medulloepithelioma tumors arise from nonpigmented epithelium of the ciliary body; some arise from the head of the optic nerve or the retina. The age at presentation of medulloepithelioma is frequently between 2 and 4 years (range, less than 1 to 66 years), and 63 percent of affected patients are girls or women. Patients present with poor vision or blindness, followed by the detection of a white pupil (leukocoria), strabismus, or red painful eye (37).

By ophthalmologic examination, one sees a gray, white-yellow, or pink ciliary body mass, which in 61 percent of patients includes cysts in the tumor or adjacent to it. Iris neovascularization, secondary glaucoma, or a neoplastic cyclitic membrane is present in about half of the patients. Lens defect or coloboma caused by segmental loss of zonular fibers is present in 20 percent.

The diagnosis is often delayed because patients present with neovascular glaucoma and are treated often with shunting surgical procedures (38). The tumors may track the tubing into the subconjunctival reservoir, especially if they have a malignant component and become extraocular.

Some patients with ciliary body medulloepithelioma have the *DICER1 syndrome*; 5 percent have a history of pleuropulmonary blastoma (PPB) and less than 1 percent of these patients develop ciliary body medulloepithelioma (39). The DICER1 syndrome is a tumor-predisposition disorder that tends to affect children and young adults with a mixture of rare and common tumors that include pleuropulmonary blastoma, cystic nephroma, Sertoli-Leydig cell tumors, thyroid cancer, and multinodular goiter.

Gross Findings. Medulloepithelioma is a fleshy, white-gray to yellow-pink mass with intratumoral cysts located on the ciliary body or between the ciliary body and the lens. In some cases, the tumor extends to the iris and anterior chamber as translucent membranous or cystic masses. The mean tumor size is 11 mm (range, 2 to 30 mm). Cataract may be associated, and a cyclitic membrane is present frequently (fig. 9-33A).

Microscopic Findings. The neoplastic epithelium resembles embryonic medullary epithelium. It forms anastomosing cords, bands, and ribbons of polarized neuroepithelial cells that are surrounded by relatively acellular myxoid stroma rich in hyaluronidase-sensitive mucopolysaccharide that stains with Alcian blue (fig. 9-33B,C). Frequently, the tumor has a diktyoma-like pattern that is composed of a net-like disposition of thin interlacing strands of neuroepithelium. In about 30 percent of cases, neoplastic cells contain pigmentation. The tumors may have cystic spaces.

Some tumors contain heterologous tissue components. The most frequently encountered is cartilage, but bone, muscle, and brain tissue may also be present (fig. 9-33D).

In 65 to 80 percent of tumors, there is a histologically malignant component, with poorly differentiated neuroblastic cells resembling retinoblastoma. This variant of medulloepithelioma may contain Homer Wright or Flexner-Wintersteiner rosettes. The heterologous component may also show sarcomatous transformation, with marked atypia, pleomorphism, increased mitotic activity, and necrosis (fig. 9-33E). The most important criteria for malignancy is extraocular extension because this is a critical finding correlated with tumor-associated mortality (fig. 9-33F) (40).

Immunohistochemical Findings. The neuroepithelial component of medulloepithelioma is immunoreactive for vimentin, neuron-specific enolase, pancytokeratins, and CK18, but negative for CK7, CK20, and epithelial membrane antigen. Other antibodies are expressed inconsistently: chromogranin, synaptophysin, GFAP, S-100 protein, and HMB-45 (41). LIN28A protein is detected in almost all medulloepitheliomas, and the intensity of staining correlates with aggressive behavior (invasion into ocular tissues).

Molecular Genetic Findings. Most medulloepitheliomas are sporadic. Rarely, medulloepithelioma is a component of the DICER1

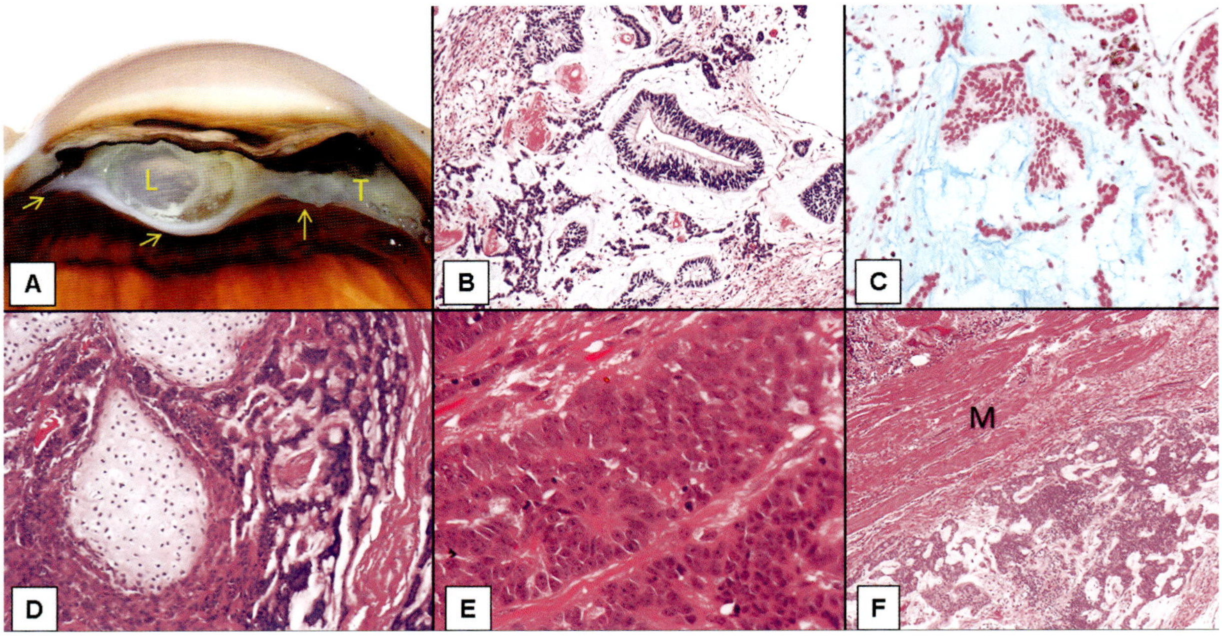

Figure 9-33

MEDULLOEPITHELIOMA

A: A small medulloepithelioma (T) is located at the ciliary body. Cyclitic neoplastic and gliotic membrane (arrows) extends from the tumor to the opposite ciliary body, passing behind the lens (L). The lens is cataractous.

B: The tumor contains tubular structures that resemble embryonic medullary epithelium with polarized neuroepithelial cells (at the center). There are surrounding anastomosing cords, bands, and ribbons of tumor cells and relatively acellular myxoid stroma.

C: The stroma is composed of hyaluronidase-sensitive mucopolysaccharide-rich stroma.

D: There are heterologous components in this tumor, like this focus of cartilage.

E: Malignant foci of retinoblastoma-like tumor with mitoses and rosettes are found in tumors with malignant histopathologic features.

F: Medulloepithelioma in an extraocular location infiltrates the striated extraocular muscle (M).

pathogenic variant syndrome. DICER1 is an endonuclease critical for the production of miRNA, which helps regulate gene expression in embryologic and early developmental stages (1). Medulloepitheliomas are associated with miRNA dysregulation (significantly upregulated and downregulated miRNAs). The miRNA dysregulation is similar to that seen in glioblastomas and retinoblastomas. Recurrent somatic *DICER1* mutations may accompany some cases of sporadic intraocular medulloepithelioma without pleuropulmonary blastoma syndrome. Recurrent aberrations in chromosomes 1p, 4, 8, and 16p have been identified in medulloepithelioma (39).

Differential Diagnosis. Clinically and histopathologically, retinoblastoma is in the differential diagnosis, because medulloepithelioma may contain heterologous components, and one of the components may resemble retinoblastoma. In these cases, LIN28A immunostaining is specific for medulloepithelioma since retinoblastoma is negative. RB protein expression is negative in retinoblastoma and positive in medulloepithelioma.

The condition is known as "persistent fetal vasculature" also needs to be distinguished from medulloepithelioma. Almost all medulloepitheliomas have an accompanying neoplastic cyclitic membrane that may be vascularized. Neoplastic proliferation of neuroepithelium is absent in persistent fetal vasculature.

Treatment and Prognosis. Medulloepithelioma is treated by enucleation to prevent any extraocular extension of tumor. Only a few small tumors have been resected en block with variable degrees of success. Most tumors treated locally tend to recur, and when this happens, enucleation is the preferred therapeutic strategy.

Survival after treatment of medulloepithelioma is excellent if there is no extraocular extension. When tumor extends into the extraocular tissue, such as the conjunctiva (where these tumors have access to lymphatics), orbit, and intracranial compartment, these cases have had a fatal outcome.

Histopathology (the basis for the current classification) and molecular findings have no demonstrated association with prognosis. Some have advocated for the use of a grading scheme (40): grade I tumors (current World Health Organization [WHO] benign designation), grade II tumors (tumor progression designated when pleomorphism, increased mitotic activity, or local invasion is detected), and grade III tumors (tumor transformation to metastatic potential when there is extrascleral extension or metastasis). Metastatic disease only occurs in cases with extraocular extension.

The pathology report should state whether the medulloepithelioma has a heterologous or malignant component (which type) and whether the tumor has invaded ocular tissues with extraocular extension and the location of the extension (conjunctiva versus orbit versus optic nerve cut end). The report should include mention of the status of margins.

COMBINED HAMARTOMA OF THE RETINA AND RETINAL PIGMENT EPITHELIUM (JUXTAPAPILLARY HAMARTOMA), ADENOMA, AND ADENOCARCINOMA OF THE RETINAL PIGMENT EPITHELIUM

Combined Hamartoma of the Retina and Retinal Pigment Epithelium

Combined hamartoma of the retina and RPE (CHR-RPE) contains disorganized retinal elements and proliferation of RPE. In simple hamartoma of the RPE, only the RPE forms the lesion.

Hamartomas of the retina and RPE are rare lesions. Patients range in age from infants to adolescents (up to 15 years of age). Although most lesions are unilateral and sporadic, CHR-RPE occurs in neurofibromatosis types 1 and 2, Gorlin syndrome, and brachio-oculo-facial syndrome (42).

CHR-RPE lesions have a variable amount of pigment, telangiectatic vessels, and white glial tissue with epiretinal membrane formation. This tissue may involve the adjacent retinal vessels and lead to progressive decline in vision. The lesions may be slightly elevated, and can become thicker over time. Simple hamartomas of the RPE are well circumscribed, mostly flat, pigmented lesions with prominent involvement of the epiretinal area; they are located in the posterior pole. Simple RPE hamartomas have fewer or no vascular or glial components.

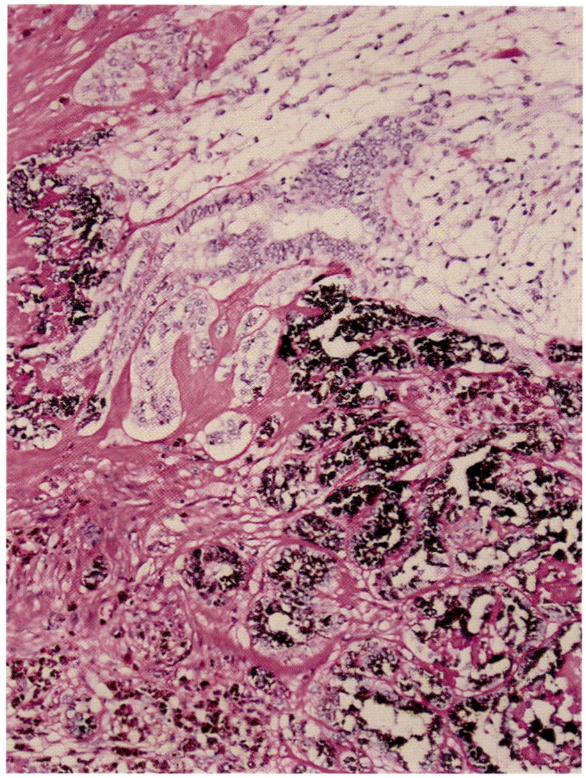

Figure 9-34

COMBINED HAMARTOMA OF THE RETINA AND RETINAL PIGMENT EPITHELIUM (CHR-RPE)

Glial and retinal tissue and single cells, cords, and tubules of retinal pigment epithelium (RPE) form this mass located adjacent to the optic nerve head. The cords and nest of RPE are surrounded by PAS-positive membranes.

The dilated vessels of CHR-RPE are admixed with glial tissue and single cells, cords, and tubules of RPE (fig. 9-34). An associated glial epiretinal membrane with pigmented cells is present in most cases.

The differential diagnosis of CHR-RPE is related to the age of the patient at presentation and the amount of RPE component. In children with a hypopigmented lesion that presents with leukocoria, retinoblastoma is the main entity in the differential diagnosis. In adults, the differential diagnosis of pigmented CHR-RPE includes uveal melanoma and melanocytoma. Clinically and ultrasonographically, it is not difficult to distinguish CHR-RPE from retinoblastoma or uveal melanoma. Although much less commonly encountered, adenoma of the RPE, persistent fetal vasculature, and morning glory disc malformations are often difficult to differentiate (43).

Adenoma of the Retinal Pigment Epithelium

Adenomas of the RPE are acquired tumors derived from the RPE, usually diagnosed during adulthood at a mean age of 50 years. Patients often have a history of inflammation or injury, and because of this association, RPE adenomas are thought to arise from prior reactive RPE hyperplasia. The tumors usually occur in the periphery of the retina but can be found in the peripapillary regions as well.

Adenomas of the RPE present as a unilateral, single, round to oval, elevated, usually pigmented mass. This neoplasm features slow growth, with eventual invasion into the retina, where it develops a retinal feeding artery and draining vein. Other associated changes are lipid exudates in the overlying retina or adjacent to the tumor (44).

Tumors consist of polygonal cells with variable pigmentation that forms cords, tubules, and glands. These are surrounded by a basement membrane and admixed with fibrous appearing stroma. PAS highlights the basement membranes. Melanocytic (S-100 protein and HMB-45 especially) and epithelial markers (low-weight keratins) are expressed. The entities in the differential diagnosis are adenocarcinoma of the RPE, choroidal melanoma, congenital hypertrophy of the RPE, combined hamartoma of the RPE, and hyperplasia of the RPE (45).

Treatment depends on specific findings and associated retinal changes. If the tumors are asymptomatic and small, observation alone is warranted. For more extensive lesions in the retina that are anterior to the equator and that enlarge, the lesion may be locally resected. Exudative retinal detachment may complicate this lesion and treatment directed at the retinal detachment may be required. Malignant transformation is rare.

Adenocarcinoma of the Retinal Pigment Epithelium

General Features. *Adenocarcinomas of the RPE* appear in middle-aged adults. The characteristics are similar to those of adenomas of the RPE and are also associated with previous trauma, inflammation, and reactive retinal lesions. This overlap makes differentiation between adenoma and adenocarcinoma of the RPE clinically challenging. They occur in the same location, anterior to the choroid and in the RPE layer. The dome-shaped pigmented lesion with prominent vessels and associated exudative retinal detachment is common to both. By ultrasonography, the tumors show high internal reflectivity and acoustic solidity. However, the RPE tumors lack the prominent internal vascularity found in the majority of melanomas.

Microscopic Findings. Adenocarcinomas of the RPE are composed of vacuolated, mostly hexagonal cells with varying amounts of pigment. The pigment granules differ from those of choroidal melanocytes because the RPE contains elongated "football"-shaped granules (see fig. 9-2). RPE tumors located in the peripapillary posterior regions feature a distinctive arrangement of strands of tumor cells associated with prominent PAS-positive connective tissue (fig. 9-35A). Other tumors show neoplastic cells in nests, tubules, and glandular configurations.

The criteria for malignancy include nuclear atypia, cellular pleomorphism, rare mitotic activity, and choroidal and retinovitreal invasion. Nuclear atypia is identified by the presence of prominent nucleoli (fig. 9-35B) (46).

The neoplastic RPE cells are immunoreactive for the melanocytic markers S-100 protein, Melan A (Mart- 1), MITF, and SOX10 and epithelial markers of low molecular weight including CK7, AE1/3, CAM5.2, and epithelial membrane antigen. They also stain for vimentin (45,46).

Differential Diagnosis. The most common entity in the differential diagnosis is melanoma, especially in adenocarcinomas with prominent nucleoli which may be confused with epithelioid-type melanoma cells. If the lesions are partially pigmented and atypical, they may be confused with adenocarcinomas of the nonpigmented epithelium of the ciliary body, lesions that may be more infiltrating and have

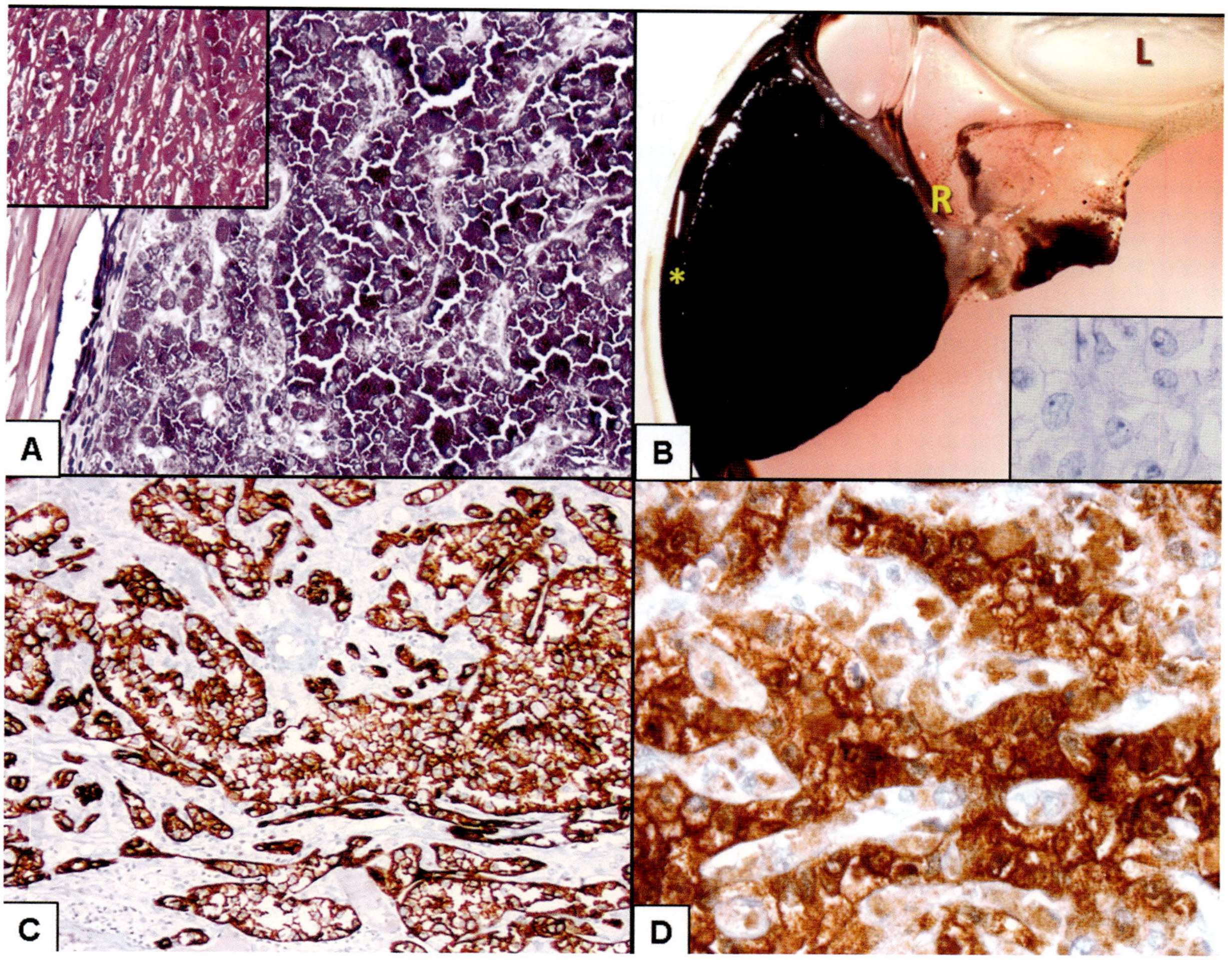

Figure 9-35

ADENOCARCINOMA OF THE RETINAL PIGMENT EPITHELIUM

A: Bleached section shows nests and tubules of neoplastic RPE invading the choroid (the sclera is at the left). The insert shows the PAS-rich stroma between the tumor cells.

B: The pupil-optic nerve section with a highly pigmented RPE tumor invading the overlying retina (R) but not involving the choroid (*). The vitreous contains pigmented cells between the tumor and the lens (L). The insert shows a bleached section with prominent nucleoli of the RPE cells.

C: Pancytokeratin stains the strands, nests, and glandular structures.

D: S-100 protein stains the tumor cells.

extraocular extension (47). The PAS-positive septa/connective tissue, the linear and tubular arrangement of the cells, and the location of the tumor on the Bruch membrane and not in the choroid support the diagnosis of RPE tumor. However, adenocarcinomas of the non-pigmented epithelium of the ciliary body may also have PAS-positive material and a similar immunoreactive profile. In these cases, a tumor confined to the ciliary body area that is mostly nonpigmented helps in differentiating both entities. In cases of complete replacement of the intraocular structures, as happens in blind eyes, this differentiation may be impossible (47). Immunohistochemistry shows that not only are the melanocytic markers positive but the epithelial markers are strongly positive in most cells (fig. 9-35C,D).

Treatment and Prognosis. Visual complications can arise in large, rapidly growing tumors or those aggressively invading the retina and vitreous. For tumors contained in the eye, without extrascleral extension, prognosis is excellent. Metastatic RPE adenocarcinoma has been documented in tumors that extend widely into the stroma, especially in blind eyes that develop the tumor without being noticed and come to clinical attention only when accompanied by proptosis (extraocular extension) (47,48).

SECONDARY AND METASTATIC TUMORS

General Features. *Secondary tumors* invade from the adjacent ocular structures into the retina. They also occur as metastases into the vitreous and retina.

The most frequent tumors to secondarily invade the retina arise in the uveal tract, and of these, uveal melanoma is the most frequent. Uveal melanoma may break through Bruch membrane and invade the subretinal space, forming the typical mushroom configuration, with a neck at the site of rupture and the outgrowth into the subretinal space. These tumors may further invade the retina and rarely form vitreous seeds.

Tumors from the optic nerve may invade into the adjacent retina, a finding that is usually seen in melanocytomas and hemangioblastomas. Gliomas of the optic nerve also involve the retina and vitreous.

Vitreoretinal metastases are rare, occurring less frequently than those to the choroid or orbit, although concomitant choroidal and retinal metastases are seen. Few cases are reported in the literature, and these involve carcinomas and melanoma. Tumors metastasizing to the retina include primary small cell carcinoma of the lung, esophageal carcinoma, liver carcinoma (cholangiocarcinoma), and cutaneous melanoma (49).

The retina and vitreous may be involved by lymphoma. The clinical history together with immunohistochemical and molecular findings is essential for reaching the diagnosis.

Clinical Features. Patients with metastatic disease present with a rapid onset of loss of visual acuity, floaters, and metamorphopsia. Direct visualization of the fundus and imaging techniques usually reveal a white-cream thickening of the retina. In cases of pigmented metastasis, pigmentation is mottled or solid pigment is seen in the thickened retina.

Vitreoretinal metastases of carcinoma or amelanotic melanoma may masquerade as inflammatory disease, especially simulating parasitic or fungal infections, acute retinal necrosis (herpetic, viral retinitis), or primary intraocular lymphoma. These amelanotic metastases can present with retinal vascular thrombosis (branch retinal artery occlusion or branch vein thrombosis) with associated intraretinal hemorrhage (49).

Microscopic Findings. Metastatic tumors presumably arise at the retina through its inner circulation. Thus, they first infiltrate retina. Many lesions break through the inner limiting membrane and invade into the vitreous. In these cases, the cells track the surface of the retina, forming a neoplastic epiretinal membrane. Floating tumor seeds in the vitreous are also seen. If the tumor expands the retina, the neoplastic cells may grow into the subretinal space, forming subretinal tumors that may mimic melanoma tumors from the choroid. Some tumors track the retina toward the optic nerve head and invade the nerve.

Prognosis. The prognosis is usually poor since many patients also present with either disseminated metastases or metastasis to the brain. Prognosis also depends on the type of tumor (50).

REFERENCES

1. Hogan MJ, Alvarado JA, Weddell JE. The Limbus. In: Histology of the human eye: an atlas and textbook. Philadelphia: W. B. Saunders; 1971.
2. Rushlow DE, Mol BM, Kennett JY, et al. Characterisation of retinoblastomas without RB1 mutations: genomic, gene expression, and clinical studies. Lancet Oncol 2013;14:327-34.
3. Dimaras H, Khetan V, Halliday W, et al. Loss of RB1 induces non-proliferative retinoma: increasing genomic instability correlates with progression to retinoblastoma. Hum Mol Genet 2008;17:1363-72.
4. Eagle RC Jr. High-risk features and tumor differentiation in retinoblastoma: a retrospective histopathologic study. Arch Pathol Lab Med 2009;133:1203-9.
5. Murphree AL. Intraocular retinoblastoma: the case for a new group classification. Ophthalmol Clin N Am 2005;18:41-53.

5a. Shields CL, Shields JA. Basic understanding of current classification and management of retinoblastoma. Curr Opin Ophthalmol 2006;17:228-34.

6. Munier FL. Classification and management of seeds in retinoblastoma. Ellsworth Lecture Ghent August 24, 2013. Ophthalmic Genet 2014;35:193–207.
7. Francis JH, Abramson DH, Gaillard MC, et al. The classification of vitreous seeds in retinoblastoma and response to intravitreal melphalan. Ophthalmology 2015;122:1173-9.
8. Blach LE, McCormick B, Abramson DH, et al. Trilateral retinoblastoma—incidence and outcome: a decade of experience. Int J Radiat Oncol Biol Phys 1994;29:729-33.
9. Amram AL, Rico G, Kim JW, et al. Vitreous seeds in retinoblastoma: clinicopathologic classification and correlation. Ophthalmology 2017;124:1540-7.
10. Chong EM, Coffee RE, Chintagumpala M, Hurwitz RL, Hurwitz MY, Chévez-Barrios P. Extensively necrotic retinoblastoma is associated with high-risk prognostic factors. Arch Pathol Lab Med 2006;130:1669-72.
11. Shields JA, Shields CL, Ehya H, Eagle RC Jr, De Potter P. Fine-needle aspiration biopsy of suspected intraocular tumors. The 1992 Urwick Lecture. Ophthalmology 1993;100:1677-84.

12 Chawla B, Tomar A, Sen S, Bajaj MS, Kashyap S. Intraocular fine needle aspiration cytology as a diagnostic modality for retinoblastoma. Int J Ophthalmol 2016;9:1233-5.

13. Squire J, Gallie BL, Phillips RA. A detailed analysis of chromosomal changes in heritable and non-heritable retinoblastoma. Hum Genet 1985;70:291-301.
14. Skalet AH, Gombos DS, Gallie BL, et al. Screening children at risk for retinoblastoma: consensus report from the American Association of Ophthalmic Oncologists and Pathologists. Ophthalmology 2018;125:453-8.
15. Sastre X, Chantada GL, Doz F, et al. Proceedings of the consensus meetings from the International Retinoblastoma Staging Working Group on the pathology guidelines for the examination of enucleated eyes and evaluation of prognostic risk factors in retinoblastoma. Arch Pathol Lab Med 2009;133:1199-202.
16. Shields JA, Shields CL, Parsons HM. Differential diagnosis of retinoblastoma. Retina 1991;11:232-43.
17. Gombos DS, Chévez-Barrios P. Current treatment and management of retinoblastoma. Curr Oncol Rep 2007;9:453-8.
18. Funes S, Sampor C, Villasante F, et al. Feasibility and results of an intraarterial chemotherapy program for the conservative treatment of retinoblastoma in Argentina. Pediatr Blood Cancer 2018;65:e27086.
19. Eagle RC Jr, Shields CL, Bianciotto C, Jabbour P, Shields JA. Histopathologic observations after intra-arterial chemotherapy for retinoblastoma. Arch Ophthalmol 2011;129:1416-21.
20. Bohm KJ, Gobin YP, Francis JH, et al. Choroidal infarction following ophthalmic artery chemotherapy. Int J Retina Vitreous 2018;4:16.
21. Chantada GL, Dunkel IJ, Antoneli CB, et al. Risk factors for extraocular relapse following enucleation after failure of chemoreduction in retinoblastoma. Pediatr Blood Cancer 2007;49:256-60.
22. Shields CL, Meadows AT, Shields JA, Carvalho C, Smith A. Chemoreduction for retinoblastoma may prevent intracranial neuroblastic malignancy (trilateral retinoblastoma. Arch Ophthalmol 2001;119:1269-72.
23. Shields CL, Shields JA, Baez KA, Cater J, De Potter PV. Choroidal invasion of retinoblastoma: metastatic potential and clinical risk factors. Br J Ophthalmol 1993;77:544-8.
24. Bosaleh A, Sampor C, Solernou V, et al. Outcome of children with retinoblastoma and isolated choroidal invasion. Arch Ophthalmol 2012;130:724-9.

25. Chévez-Barrios P, Eagle R, Marback EF. Histopathologic features and prognostic factors. In: Singh AD, Murphree AL, Damato BE, eds. Clinical ophthalmic oncology: retinoblastoma, 2nd ed. Hiedelberg: Springer; 2015:167.
26. Chantada GL, Fandiño AC, Guitter MR, et al. Results of a prospective study for the treatment of unilateral retinoblastoma. Pediatr Blood Cancer 2010;55:60-6.
27. Chantada GL, Guitter MR, Fandiño AC, et al. Treatment results in patients with retinoblastoma and invasion to the cut end of the optic nerve. Pediatr Blood Cancer 2009;52:218-22.
28. Cuenca A, Giron F, Castro D, et al. Microscopic scleral invasion in retinoblastoma: clinicopathological features and outcome. Arch Ophthalmol 2009;127:1006-10.
29. Torbidoni AV, Laurent VE, Sampor C, et al. Association of cone-rod homeobox transcription factor messenger RNA with pediatric metastatic retinoblastoma. JAMA Ophthalmol 2015;133:805-12.
30. Mallipatna A, Gallie B, Chévez-Barrios P, et al. Retinoblastoma. In: Amin MB, Edge SB, Greene FL, et al., eds. AJCC Cancer Staging Manual, 8th ed. New York: Springer; 2017:819-31.
31. Ulbright TM, Fulling KH, Helveston EM, Astrocytic tumors of the retina. Differentiation of sporadic tumors from phakomatosis-associated tumors. Arch Pathol Lab Med 1984;108:160-3.
32. Eskelin S, Tommila P, Palosaari T, Kivelä T. Photodynamic therapy with verteporfin to induce regression of aggressive retinal astrocytomas. Acta ophthalmol 2008;86:794-9.
33. Poole Perry LJ, Jakobiec FA, Zakka FR, et al. Reactive retinal astrocytic tumors (so-called vasoproliferative tumors): histopathologic, immunohistochemical, and genetic studies of four cases.. Am J Ophthalmol 2013;155:593-608.
34. Chew EY. Ocular manifestations of von Hippel-Lindau disease: clinical and genetic investigations. Trans Am Ophthalmol Soc 2005;103:495-511.
35. Zhou M, Shen D, Head JE, et al. Ocular clusterin expression in von Hippel-Lindau disease. Mol Vis 2007;13:2129-36.
36. Wang W, Chen L. Cavernous hemangioma of the retina: a comprehensive review of the literature (1934-2015). Retina 2017;37:611-21.
37. Saunders T, Margo CE. Intraocular medulloepithelioma. Arch Pathol Lab Med 2012;136:212-6.
38. Pastora-Salvador N, Abelairas-Gómez J, Peralta-Calvo J, et al. Atypical clinical presentation and long term survival in a patient with optic nerve medulloepithelioma: a case report. J Med Case Rep 2012;6:123.
39. Sahm F, Jakobiec FA, Meyer J, et al. Somatic mutations of DICER1 and KMT2D are frequent in intraocular medulloepitheliomas. Genes Chromosomes Cancer 2016;55:418-27.
40. Verdijk RM. On the classification and grading of medulloepithelioma of the eye. Ocul Oncol Pathol 2016;2:190-3.
41. Jakobiec FA, Kool M, Stagner AM, et al. Intraocular medulloepitheliomas and embryonal tumors with multilayered rosettes of the brain: comparative roles of LIN28A and C19MC. Am J Ophthalmol 2015;159:1065-74.
42. Schachat AP, Shields JA, Fine SL, et al. Combined hamartomas of the retina and retinal pigment epithelium. Ophthalmology 1984;91:1609-15.
43. Font RL, Moura RA, Shetlar DJ, Martinez JA, McPherson AR. Combined hamartoma of sensory retina and retinal pigment epithelium. Retina 1989;9:302-11.
44. Shields JA, Shields CL, Gündüz K, Eagle RC Jr. Neoplasms of the retinal pigment epithelium: the 1998 Albert Ruedemann, Sr, memorial lecture, Part 2. Arch Ophthalmol 1999;117:601-8.
45. Shields JA, Eagle RC, Dutton J, Ehya H, Shields CL. Adenocarcinoma of the retinal pigment epithelium: clinicopathologic correlation with paradoxical immunohistochemical findings. JAMA Ophthalmol 2014;132:1249-52.
46. Fan JT, Robertson DM, Campbell RJ. Clinicopathologic correlation of a case of adenocarcinoma of the retinal pigment epithelium. Am J Ophthalmol 1995;119:243-5.
47. Shields JA, Eagle RC Jr, Shields CL, De Potter P. Acquired neoplasms of the nonpigmented ciliary epithelium (adenoma and adenocarcinoma). Ophthalmology 1996;103:2007-16.
48. Klein KA, Lally DR, Taney LS, Laver NV, Duker JS. Retinal pigment epithelial adenocarcinoma presenting as an amelanotic mass. Ophthalmic Surg Lasers Imaging Retina 2015;46:369-72.
49. Shields CL, McMahon JF, Atalay HT, Hasanreisoglu M, Shields JA. Retinal metastasis from systemic cancer in 8 cases. JAMA Ophthalmol 2014;132:1303-8.
50. Klufas MA, McCannel CA, McCannel TA. Simultaneous retinal and choroidal metastases in lung adenocarcinoma. JAMA Ophthalmol 2016;134:e155368.

10 TUMORS OF THE OPTIC NERVE AND OPTIC NERVE HEAD

ANATOMY AND HISTOLOGY

Embryologically, the optic nerve is derived from diencephalon, forming the optic stalk that connects to the optic vesicle (or the optic cup). The neurosensory retina develops from the inner layer of this cup. The axons of the optic nerve originate from the retinal ganglion cells. At birth, there are approximately 1.2 million axons in the optic nerve.

The optic nerve has an intraocular component, the optic nerve head (0.7 to 1.0 mm in length), an intraorbital portion (25 to 30 mm), an intracanalicular component inside the bone between the orbit and intracranial space (4 to 10 mm), and an intracranial portion that extends to the chiasm (about 10 mm). The central optic nerve vessels are branches from the ophthalmic artery (and vein), which pierce the meninges 10 to 15 mm behind the sclera. These vessels take a central position and become the central retinal vessels upon entering the eye (fig. 10-1).

Histologically, axons (nerve fibers), glial cells, and vessels contribute to the architecture of the optic nerve. The intraocular component of the nerve contains astrocytes and microglia. Oligodendrocytes are present in the extraocular component of the nerve. Although the intraocular component of the optic nerve is not typically myelinated, the postlaminar optic nerve is myelinated by oligodendrocytes. The optic nerve is a tract of the central nervous system and meninges (pia, arachnoid with meningothelial cells, and dura) envelop the nerve. The pia of the optic nerve contributes fibrous septa that subdivide the nerve into fascicles (figs. 10-2, 10-3) (1).

OPTIC NERVE SHEATH MENINGIOMA

Definition and Epidemiology. *Meningioma* develops from meningothelial cells. Meningiomas involving the optic nerve and orbit are more frequently secondary from those arising intracranially (from the sphenoid wing). Intraorbital meningiomas arise from the optic nerve sheath (8 percent from the intracanalicular portion and 92 percent from the intraorbital portion), but some are believed to arise from a precursor cell in the orbital tissues.

The etiology of optic nerve meningiomas is unknown. The literature reports an association between a history of radiation exposure or neurofibromatosis type 2 (NF2) with the development of optic nerve meningioma. Most optic nerve meningiomas are unilateral, and when bilateral, occur in children with NF2 (2).

Clinical Features. Similarly to other meningiomas, those arising from the optic nerve tend to develop in the fifth decade of life (average, 40 years of age). There is overexpression of progesterone and estrogen receptors in meningiomas, which may explain the association with an increased incidence in women (3).

Primary optic nerve meningiomas arise from the arachnoid, and encircle and compress the nerve. This compression may compromise the optic nerve circulation. Thus, patients may present with optociliary shunt vessels, optic nerve atrophy, and progressive loss of vision (Hoyt-Spencer triad). More often, patients complain of progressive vision loss, with alterations in color vision and field defects that develop over time, sometimes more than 12 months. The mass tumor effect may produce proptosis and alteration of ocular movements (strabismus) (2).

Imaging (computerized tomography [CT] scans and magnetic resonance imaging [MRI]) show the lesion usually in the retrobulbar intraconal space, compressing nerve and expanding into the soft tissues. The effect of the atrophic optic nerve and the meningeal tumor by imaging gives the impression of "railroad tracks."

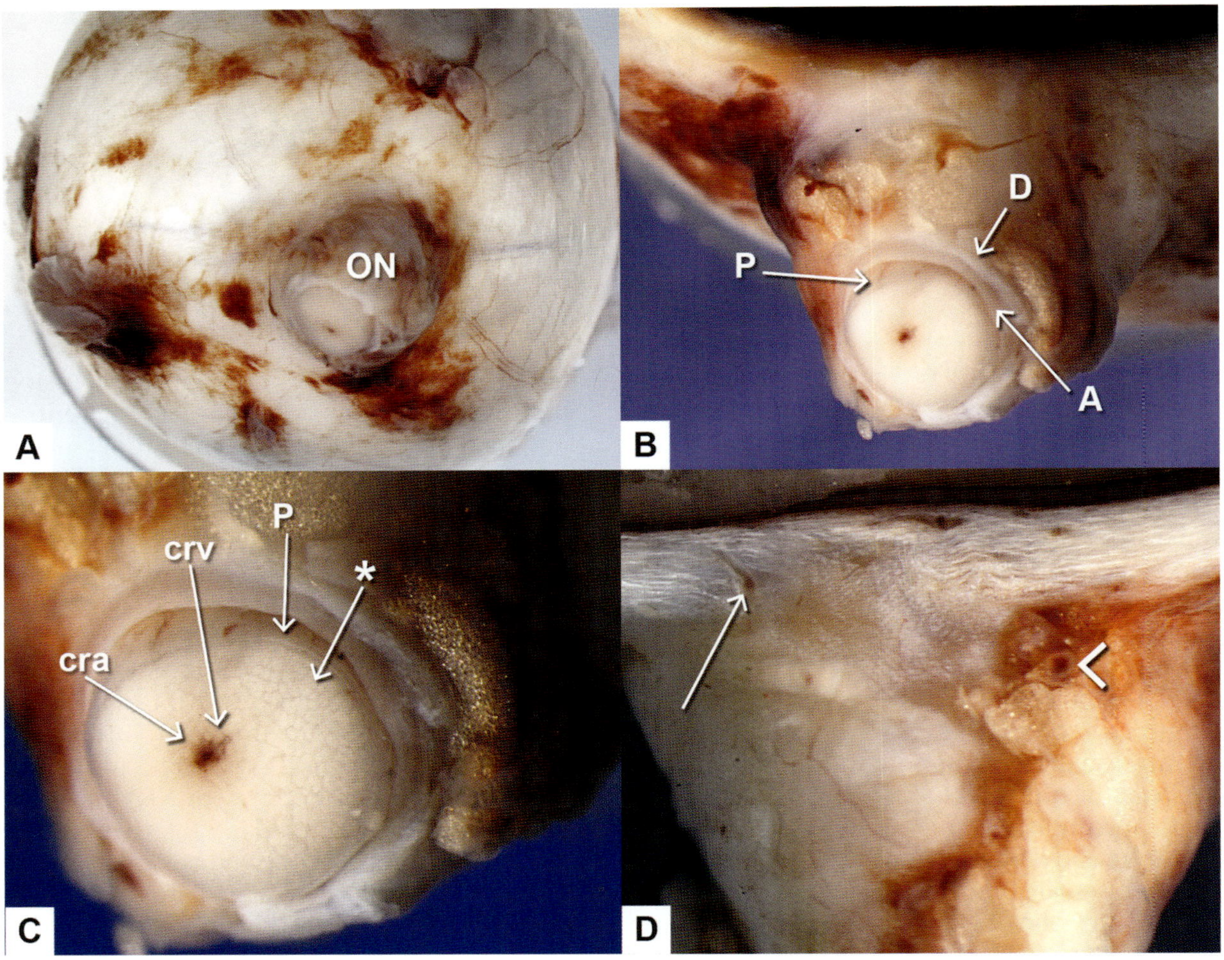

Figure 10-1

ANATOMY OF OPTIC NERVE

A: The optic nerve, located in the posterior pole, is covered by meninges.

B: Cross section of meningeal coats: the outer coat is the dura mater (D), which is partially covered by adipose tissue; the middle coat is the arachnoid (A); and the inner coat is the pia mater (P).

C: Closer view of the cut section shows the central vessels: the central retinal artery (cra) and the central retinal vein (crv). The pia mater (P) extends collagenous tissue into the optic nerve to form the fibrous septa (*), which separate the nerve into axonal bundles and carry the vessels of the nerve.

D: The cut section of the sclera adjacent to the nerve shows the rich vasculature that feeds the optic nerve head. The arrow shows the short ciliary vessel piercing the sclera and reaching the choroid, and the arrowhead shows the artery of the circle of Singh that gives branches to the optic nerve at the levels of the lamina cribrosa and the meninges. Both arterial branches arise from the ophthalmic artery.

Gross Findings. The orbital portion of the optic nerve may show a plaque-like thickened mass replacing leptomeninges and compressing the optic nerve, which appears atrophic. The tumor extends in nodules into the orbital soft tissues if it has extended from the arachnoid to the other meningeal sheaths and into the adjacent tissues. The extension of the tumor depends on tumor evolution and time of surgical resection.

Primary meningiomas of the optic nerve may be entirely intradural or may extend extradurally. By contrast, primary optic nerve gliomas are always intradural. In the context of primary optic nerve glioma that develops in neurofibromatosis, meningeal hyperplasia may be histologically indistinguishable from meningioma. A shallow incisional biopsy from an enlarged optic nerve may not show meningioma or

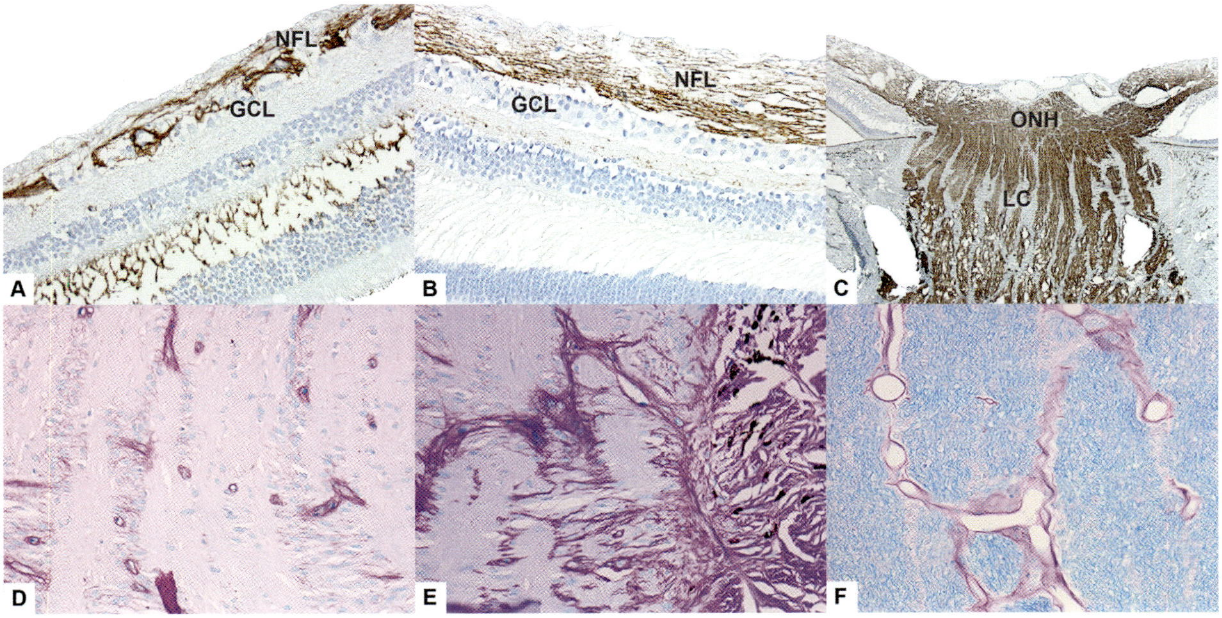

Figure 10-2

HISTOLOGY OF THE OPTIC NERVE

A: Axons of the ganglion cells located in the ganglion cell layer (GCL) of the retina continue into the optic nerve. In the GCL, there are also astrocytes surrounding vessels and extending into the nerve fiber layer (NFL), seen here with the glial fibrillary acid protein (GFAP) immunostain.

B: Axons of the ganglion cells (GCL), highlighted by neurofilament immunostain, constitute the nerve fiber layer (NFL). The NFL is parallel to the internal limiting membrane of the retina and the deeper retinal layers orient themselves perpendicular to the NFL.

C: Low-power view of the optic nerve highlights the retinal nerve fiber layer, the optic nerve head (ONH), and axons passing through the lamina cribrosa (LC) to form the orbital portion of the optic nerve.

D: The nerve fibers of the optic nerve head are negative for myelin. The periodic acid–Schiff (PAS) stain highlights the vessels.

E: The lamina cribrosa area is nonmyelinated, and PAS stain highlights the rich vasculature and fibrous tissue present in this area. To the right are a few melanocytes in the deep choroid/sclera at the edge of the optic nerve head/lamina cribrosa.

F: Post-lamina cribrosa optic nerve shows light blue staining of the myelin that surrounds the nerve fibers/axons located between the fibrovascular septa, highlighted by the PAS stain.

meningeal hyperplasia in the case of glioma in the context of neurofibromatosis (especially if neurofibromatosis history is not known).

The distinction between primary optic nerve glioma and primary optic nerve meningioma is essential because, in children, gliomas tend to grow slowly while meningiomas tend to grow rapidly, with possible extension into the intracranial cavity. It is crucial for the pathologist to know whether the biopsy was within the nerve (intradural) or extradural. Tissue that is extradural and that shows the histologic features of meningioma is a meningioma (4).

Microscopic Findings. Although all histologic subtypes of meningiomas may be present in the optic nerve, most are meningothelial or transitional types (World Health Organization [WHO] grade I). Meningothelial meningiomas feature whorls of medium-sized epithelioid cells, oval to round nuclei with occasional intranuclear pseudoinclusions or clearing, and abundant eosinophilic cytoplasm. The cells have indistinct borders, and fibrous septa separate aggregates of cells (fig. 10-4). The transitional type features a mixture of epithelioid meningothelial cells in whorls and fibrous elongated meningothelial type cells with psammoma bodies. These tumors are well vascularized and may contain microcystic changes.

The transitional type may invade the meninges and extend beyond the optic nerve (fig. 10-5). This type may show atypical features such as invasion of bone or increased mitotic activity (WHO grade II).

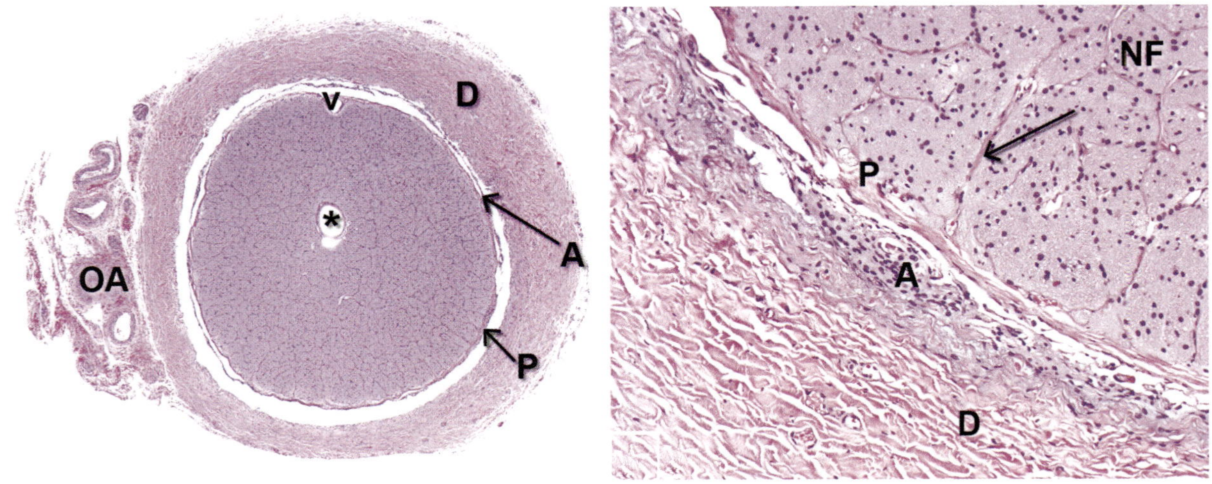

Figure 10-3

HISTOLOGY OF POST-LAMINAR OPTIC NERVE

Left: Cross section of the post-laminar optic nerve at low power shows the central retinal artery (*) in the center of the nerve. The central retinal vein is adjacent to the central retinal artery, and also seen in the periphery (v). The meninges cover the optic nerve with the pia matter (P) in the inner position, then the arachnoid (A), and the outermost dura mater (D). The ophthalmic artery (OA) is on one side of the nerve.

Right: The dura (D) is remarkable for its dense collagenous tissue and small vessels. The arachnoid (A) is in the middle and contains the meningothelial cells and vessels. The pia is a collagenous fibrous layer attached to the neural portion of the optic nerve and provides collagenous strands to form the collagenous fibers of the septa (arrow). The optic nerve septa divide the nerve fibers (NF) into bundles and contain the small vessels that nourish the optic nerve. The nerve fiber (NF) component of the optic nerve alternates with astrocytes, oligodendrocytes (only in the post-laminar optic nerve), and some microglia seen here as nuclei.

Molecular Genetic Findings. Primary optic nerve meningiomas have alterations in chromosomes 17 and 19, in contrast with meningiomas arising from intracranial locations that harbor anomalies in chromosome 22. Most optic nerve sheath meningiomas are sporadic. In a small number of cases, patients are positive for NF2 (2 to 8 percent) (5).

Differential Diagnosis. The principal histologic entities in the differential diagnosis are other primary tumors arising from meninges such as hemangiopericytoma or melanocytoma, and conditions such as metastatic tumors from such sites as the breast or lung; secondary infiltrating tumors such as lymphoma, leukemia, and melanoma; intracranial meningioma; and inflammatory or reactive lesions such as idiopathic orbital inflammation associated with IgG4 disease or leptomeningeal gliosis. Radiologic, clinical, and histopathologic findings should be analyzed together to avoid misinterpretation.

Prognosis. The visual prognosis is usually dismal in patients with meningiomas because of nerve compression and vascular compromise. However, because the majority of the optic nerve sheath meningiomas are WHO grade I, patients have an excellent prognosis for survival. Only rarely does an atypical meningioma invade intracranially and compromise the otherwise excellent prognosis.

OPTIC NERVE GLIOMA

Definition. *Optic nerve gliomas*, or *astrocytomas*, develop from intraneural astrocytes and arise in any portion of the optic nerve. Although the most common type of astrocytoma seen in the intraorbital portion of the optic nerve is the *low-grade pilocytic astrocytoma*, other types of astrocytoma are also seen. The rare *glioblastomas* that occur in the optic nerve have an unidentified cell of origin although some experts contend that these may arise from supporting astroglia.

Clinical Features. Children are most often affected. About 90 percent of tumors develop within the first two decades of life, and the peak age is about 7 years. There is a slight predilection for females (3 to 2). Gliomas of the optic

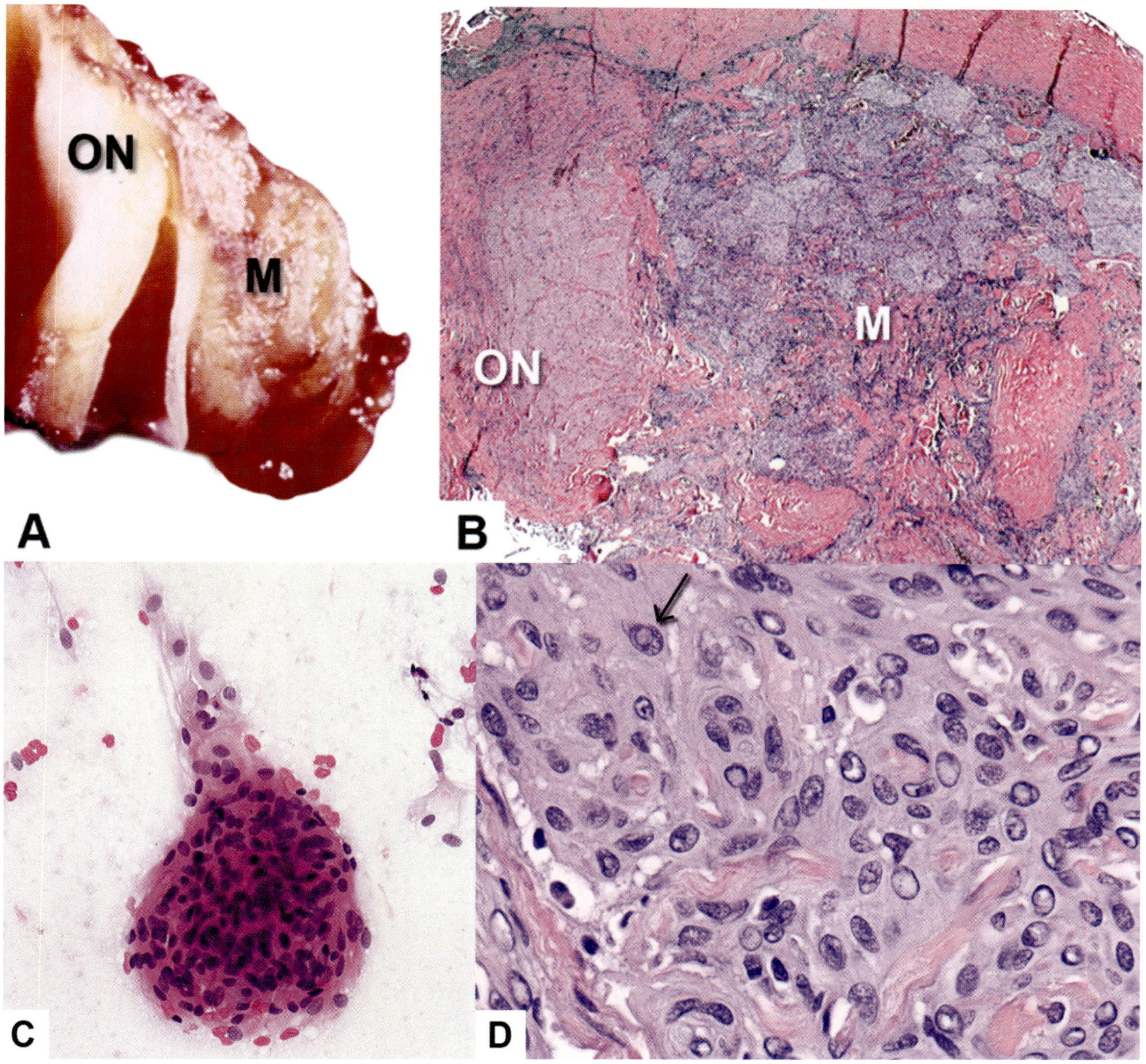

Figure 10-4

MENINGIOMA OF THE OPTIC NERVE

A: Macroscopic photograph of an en bloc excision of an optic nerve meningioma. The tumor (M) arises from the meninges adjacent to the optic nerve (ON).

B: Low-power view of a meningioma (M) infiltrating the meninges of the optic nerve (ON).

C: Cytologic touch imprint of meningioma shows groups of epithelioid cells with abundant eosinophilic cytoplasm, indistinct cell borders, and bland oval nuclei.

D: Histologic section of a meningothelial-type meningioma composed of medium to large epithelioid cells with abundant cytoplasm and indistinct cell borders. Oval nuclei have occasional intranuclear pseudoinclusions (arrow).

nerve represent two thirds of the primary optic nerve tumors.

The clinical presentation depends on the age of the patient and site of the tumor when diagnosed. Visual impairment, optic nerve atrophy or swelling, and an orbital tumor mass effect causing proptosis are the usual manifestations (6).

Molecular Genetic Findings. Seventy percent of optic nerve gliomas develop in patients with NF1, and approximately 20 percent of NF1 patients have bilateral gliomas (7). Allelic

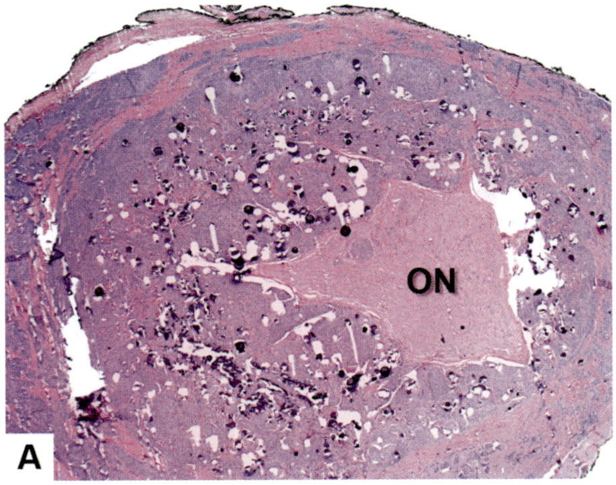

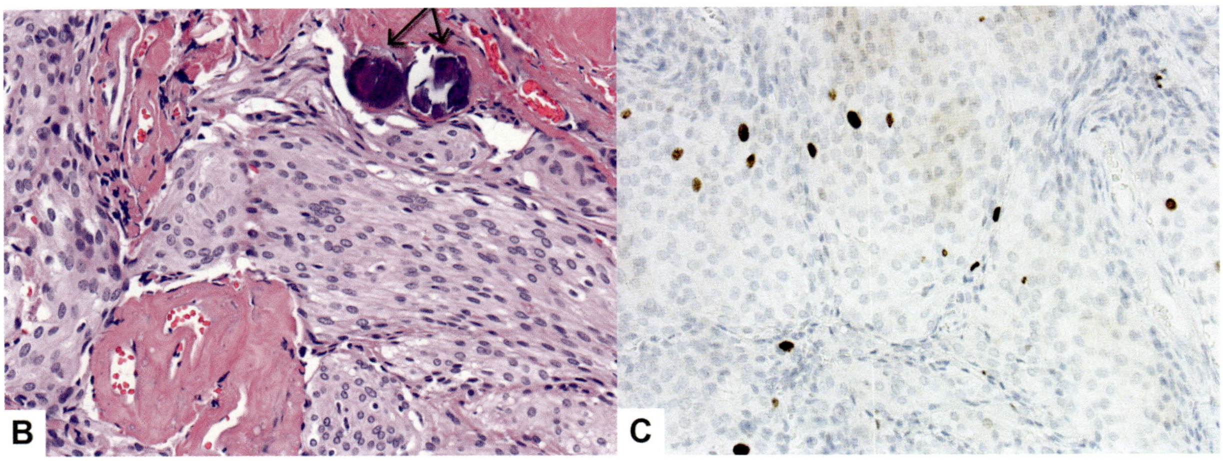

Figure 10-5

PSAMMOMATOUS (FIBROUS) MENINGIOMA

A: The tumor is encroaching on and focally invading into the optic nerve (ON). It has breached the dura and entered the orbit. The tumor features many basophilic calcifications (psammoma bodies).

B: The meningothelial cells dissect the tissue and are associated with psammoma bodies (arrows).

C: The proliferation rate, as seen here with MIB-1 (Ki-67) immunohistochemistry, shows no significant increase.

losses and deletions in the *NF1* gene on the long arm of chromosome 17 and upregulation of the *NF1* gene transcripts appear in pilocytic astrocytomas of the optic nerve. (8). Pilocytic astrocytomas in the optic nerve generally lack the mutations seen in aggressive gliomas of other sites, including mutations in *TP53* gene and *IDH* mutations. Some of these low-grade astrocytomas show activation of the MAPK1 and CKN2A pathways, suggesting a possible target for treatment.

Differential Diagnosis. The clinical differential diagnosis of astrocytomas of the optic nerve includes optic nerve gliosis, piloid gliosis, secondary involvement of the meninges by astrocytoma mimicking meningioma, reactive meningothelial cells in pilocytic astrocytomas mimicking meningioma, anaplastic pilocytic astrocytomas mimicking glioblastoma, necrotic and reactive astrocytes in viral and inflammatory neuritis, and metastatic and secondary tumors mimicking glioblastoma. (These are detailed in reference 5a).

Prognosis. The most important prognostic factor in optic nerve gliomas is the type of tumor, and pilocytic astrocytomas, which grow slowly, carry a good prognosis. Most remain stable for long periods and even regress. Others may progress after a long time of stable growth and may diffusely infiltrate the chiasm and cerebral parenchyma, with consequent morbidity and possible mortality. High-grade gliomas/glioblastomas have the same dismal prognosis as other glioblastomas in the CNS despite treatment (9).

Pilocytic Astrocytoma

Pilocytic astrocytomas arise from the astrocytes within the parenchyma of the nerve. The shape of the optic nerve expands in a fusiform configuration due to tumor. Pilocytic astrocytomas frequently have a biphasic pattern, with

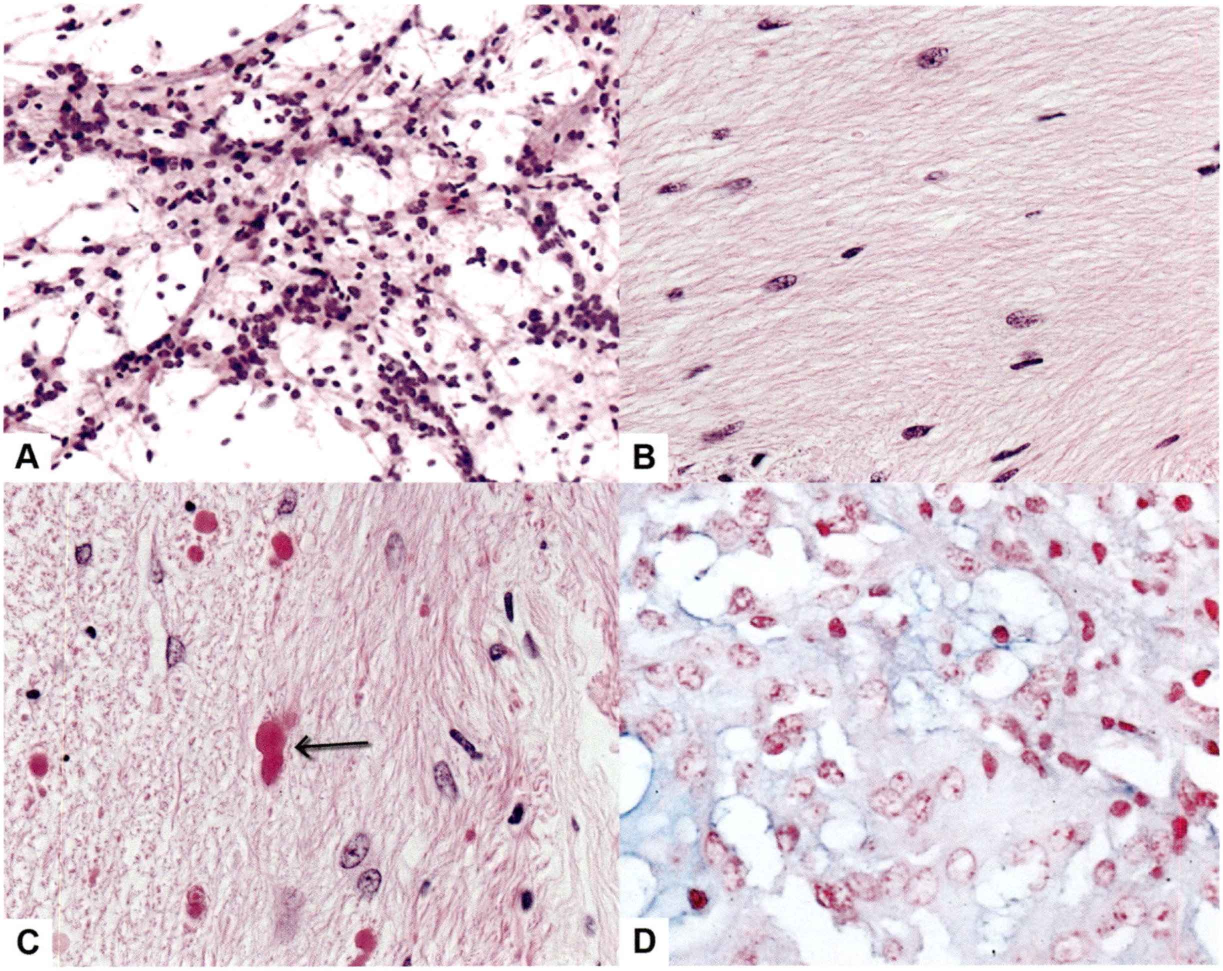

Figure 10-6

PILOCYTIC ASTROCYTOMA

A: Cytologic touch imprint preparation shows the delicate fibrillary cytoplasm of the neoplastic astrocyte and the bland nuclear features.

B: Delicate spindle glial cells with oval nuclei and abundant fibrillary cytoplasm are seen in this pilocytic astrocytoma.

C: Dense eosinophilic beaded deposits (Rosenthal fibers [arrow]) and eosinophilic granular bodies in the background are typical of these tumors.

D: Some areas of pilocytic astrocytomas contain microcystic spaces. The mucin is highlighted here with an Alcian blue stain.

alternating loose clear cells of bipolar astrocytes with microcystic spaces occupied by mucin and denser areas composed of bipolar astrocytes compacted against vessels. The cells are bland spindle cells with delicate cytoplasm, oval nuclei, Rosenthal fibers, and protein droplets. Eosinophilic round deposits may be present (fig.10-6).

Verhoeff described three types of optic nerve pilocytic astrocytoma (1). One type, a finely reticulated type, resembles the native neuroglia of the optic nerve, with small oval or round nuclei in the background of delicate syncytial cytoplasm. The round nuclei may mimic those in the oligodendrocytes (oligodendroglioma is extremely rare in the optic nerve), and the diffuse expansion of the stroma may be difficult to differentiate from diffuse astrocytoma (2). Another type is the coarsely reticulated type featuring the more typical biphasic pattern and frequently containing Rosenthal fibers, protein droplets, and eosinophilic deposits. The vasculature varies in concentration and may feature distinctive hyalinized blood vessels (5). A coarsely fibrillated type, or spindle cell

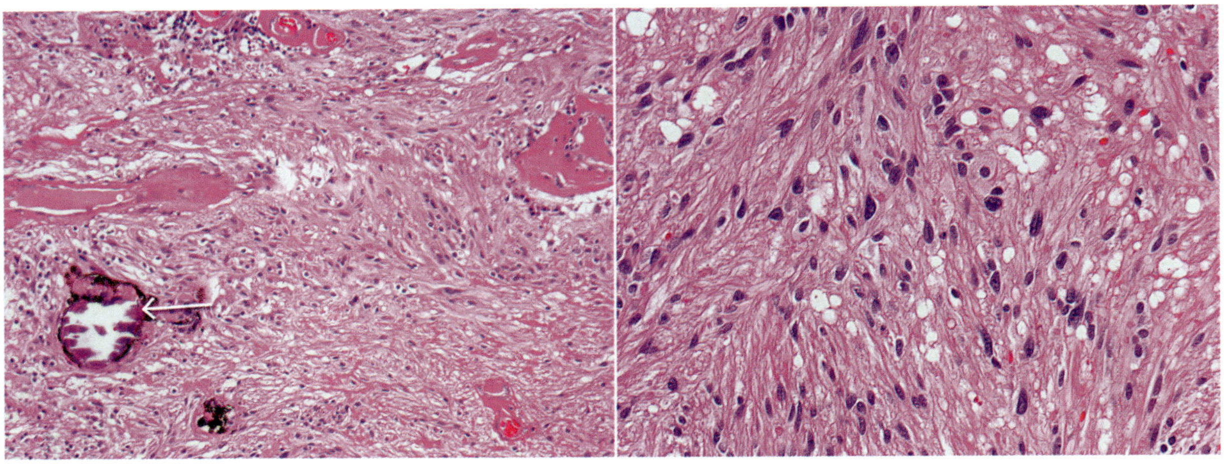

Figure 10-7

ASTROCYTIC HARMATOMA

Left: The proliferation of bland astrocytes/glial cells is associated with dilated vessels and focal calcification (arrow). Right: Detailed view of the proliferation of spindle glial cells with occasional vacuolization of the cytoplasm.

pattern, features well-formed bundles of coarse neuroglial fibrils and spindle cells.

Pilocytic astrocytomas stain diffusely with glial fibrillary acidic protein (GFAP).

Other Gliomas

Glioblastoma (WHO grade IV) may rarely develop in the optic nerve. It has the same morphologic and cytologic features (mitoses, gemistocytes, oligodendroglial-like cells) as do glioblastomas in other parts of the brain, with microvascular proliferation and necrosis. These tumors may also feature *IDH* mutated or wild type, and intact or loss of *ATRX*.

Anaplastic astrocytoma may be seen in the optic nerve as a primary tumor. It is a WHO grade III tumor that is cellular, mitotically active, and diffusely infiltrating.

ASTROCYTIC HAMARTOMA

Astrocytic hamartomas arise most commonly from the retina but can also develop from the optic nerve head (see chapter 9). There is a predisposition for these tumors to arise in the context of NF2 or tuberous sclerosis.

The key funduscopic finding is a discoid plaque-like mass that protrudes into the vitreous. Histopathologically, these lesions are formed by compact, bland, spindle-shaped astrocytes that may contain areas of cystic degeneration and calcification (fig. 10-7). The cells stain with GFAP.

These lesions are benign, and although they may compromise vision, they do not become malignant. Occasionally, calcifications are large and prominent, and some may refer to the lesion as "giant calcified drusen."

MELANOCYTOMA

Definition and Clinical Features. *Melanocytoma* is a benign melanocytic congenital tumor located in the anterior portion of the optic nerve and disc; it may also develop in the uveal tract. Melanocytoma is generally unilateral, slightly more frequent in women than men (1.5 to 1.0 ratio), and the diagnosis established in the fifth decade of life. The literature refers to rare cases of bilateral involvement in children.

Clinical Features. Melanocytomas are usually located in the optic nerve or adjacent choroid. More than 85 percent of patients are asymptomatic. The remaining patients have symptoms associated with abnormalities of the visual fields and rarely (less than 10 percent of patients with symptoms) total visual loss. Visual loss or visual field defects are not typical of malignant transformation.

Gross Findings. Melanocytoma is a dark-brown nodule on the optic nerve head that protrudes into the vitreous either centrally or

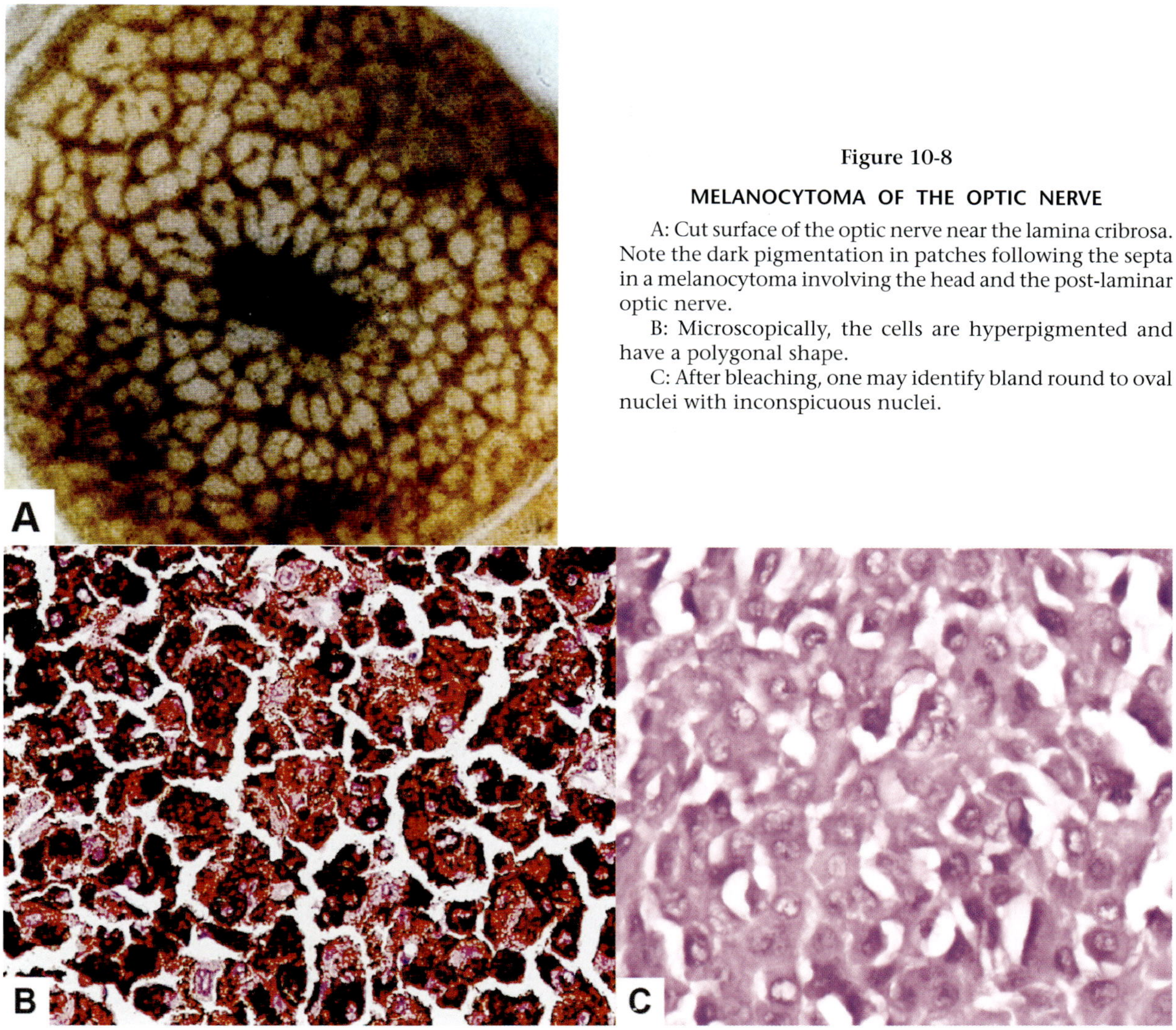

Figure 10-8

MELANOCYTOMA OF THE OPTIC NERVE

A: Cut surface of the optic nerve near the lamina cribrosa. Note the dark pigmentation in patches following the septa in a melanocytoma involving the head and the post-laminar optic nerve.

B: Microscopically, the cells are hyperpigmented and have a polygonal shape.

C: After bleaching, one may identify bland round to oval nuclei with inconspicuous nuclei.

on the edge of the optic nerve disc. The tumor may extend into the retrolaminar region but close to the lamina cribrosa. Cross section shows the infiltration between the septa (fig. 10-8A).

Microscopic Findings. Heavily pigmented, round to polygonal cells populate these lesions. After bleaching to remove dense melanin, it is possible to visualize round to slightly oval nuclei with occasional small nucleoli. No cytologic atypia or mitoses are detected (fig. 10-8B,C). Areas of necrosis may be present and may contain melanophages and a lymphocytic infiltrate. Melanophages are present adjacent to the tumor into the vitreous and on the retina.

Molecular Genetic Findings. Melanocytomas are believed to arise from melanocytes similar to those found in the uveal tract and other central nervous system (CNS) locations derived from the neural crest. There are few molecular or genetic studies on melanocytomas. Investigators detected *BAP1* mutations in those cases that transformed into melanomas. One case featured the *GNAQ* mutation (*GNAQ* mutations are usually seen in uveal melanomas as initiating events). A somatic *BAP1* mutation accompanied one of the two cases of *BAP1*-positive melanocytoma. Patients with *BAP1* tumor predisposition syndrome may develop melanocytoma, but studies have not confirmed this association (10). Gene expression profiling in three optic nerve melanocytomas showed all cases with class 1A signature (low risk of metastasis in uveal melanomas) (11), supporting the benign course of the majority of melanocytomas.

Prognosis. In a large series (12), malignant transformation was seen in fewer than 2 percent of all melanocytomas studied. Tumors are usually stable in size, and some may have focal infarcts. There is no association between infarction in melanocytoma and malignant transformation. Malignant transformation may occur many years after initial diagnosis (12,13).

PRIMARY MELANOMA

Primary melanomas are sporadic tumors that may arise from the optic disc or nerve. The tumors may also arise from melanocytes in the meninges or septa. Predisposing factors include nevus of Ota and melanosis oculi. Most of the melanomas seen in the optic nerve arise in the uveal tissue and secondarily invade the nerve or meninges, or they are cutaneous metastatic melanomas.

HEMANGIOMA, HEMANGIOPERICYTOMA/ SOLITARY FIBROUS TUMOR, HEMANGIOBLASTOMA

Hemangiomas, capillary or cavernous, may arise from the optic nerve, but they are rare lesions. *Hemangiopericytoma,* or currently called *solitary fibrous tumor*, arises from the meninges (the dura) and is a fibroblastic tumor with prominent hemangiopericytoma-like vessels. This tumor is rare, but more frequent in women (20 to 50 years of age). The lesion, located behind the eye within the muscle cone, exhibits slow and painless growth and may cause vision loss and exophthalmos.

Microscopically, solitary fibrous tumors are those similar those seen elsewhere. They are formed by oval cells with bland nuclei. Mitoses vary but are usually scant. The stroma is vascular and forms the staghorn pattern. Intercellular collagen and areas of necrosis may be seen.

Immunohistochemistry is similar to that seen in other sites. The *NAB2-STAT6* fusion involves the 12q13 gene locus, resulting in STAT6-positive staining. CD34, BCL2, and CD99 are variably positive by immunohistochemistry and seen in over 85 percent of cases.

The differential diagnosis includes meningioma, hemangioblastoma, and sarcoma. Patient prognosis depends on the high-risk features present in the tumor (more than 5 cm in size, hypercellularity, necrosis, increased mitotic activity, cellular atypia, and hemorrhage) and positive margins.

Hemangioblastomas develop in the head of the optic nerve, but they may involve the intraorbital and intracanalicular portions of the nerve. They are red-orange lesions that are vascular. They present in the setting of von Hippel-Lindau disease, possibly in association with retinal hemangioblastoma. The histopathologic features are identical to those in the retina. Tumor infiltrates the nerve between the axons (fig. 10-9) (14,15).

SECONDARY AND METASTATIC TUMORS

Definition. *Secondary tumors* of the optic nerve are tumors that invade the optic nerve from adjacent tissue (choroid, retina, sclera, orbit, brain) or are metastatic to the nerve.

General Features. Secondary involvement from ocular tissues is more frequent than from intracranial lesions. The least common occurrence is metastasis (about 1 to 4 percent of ocular metastasis). Metastasis is bilateral in a fifth of the patients, and more than 70 percent are contiguous with peripapillary choroid (16). Occasionally, juxtapapillary melanomas of the uvea infiltrate the meninges without spreading diffusely. Retinoblastoma is the most common secondary tumor involving the nerve in children.

Secondary involvement by lymphoma/leukemia is more frequent than clinically noted, and tumor infiltrates in the nerve or meninges are identified in many autopsy eyes of patients with leukemia (17,18). Primary lymphoma of the optic nerve is extremely rare with only a few cases reported in the literature. Most frequently, there is invasion of the optic nerve, retina, and CNS of the ocular-CNS primary nongerminal type, diffuse large B-cell lymphoma (16,19).

Clinical Features. The most common presentations of secondary tumors involving the optic nerve are visual loss and visual field defects. Ophthalmoscopic findings include swelling of the optic nerve head with flame-type hemorrhages, nodular lesions, or diffuse infiltrates. Fluorescein angiograms demonstrate areas of hyperfluorescence or hypofluorescence. Ultrasonography may not be helpful because of the location of the lesion; MRI shows the irregularity and thickening of the optic nerve. Lymphomatous involvement in immunocompromised patients may accompany infections (Epstein-Barr virus [EBV], human T-cell lymphoma virus

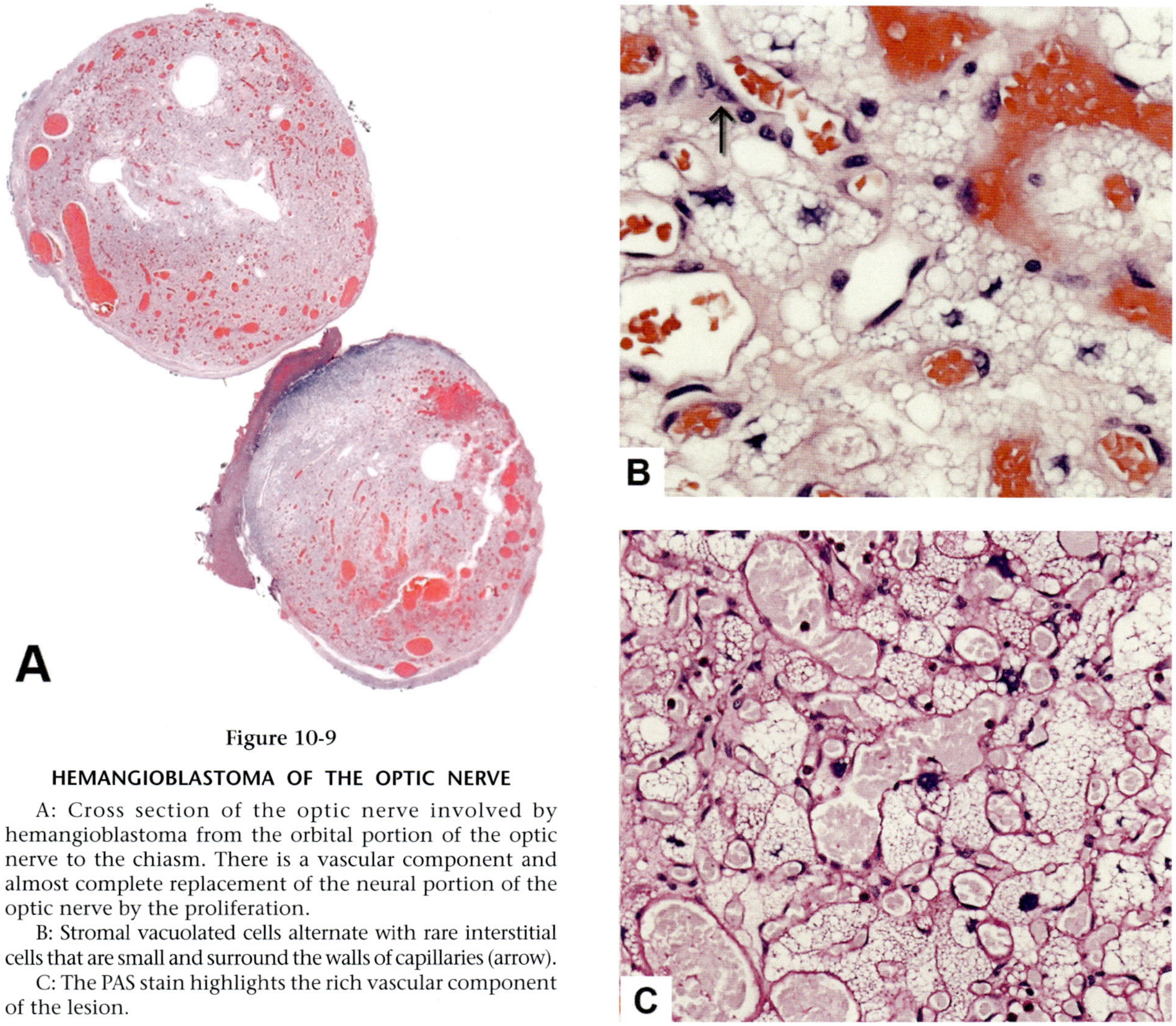

Figure 10-9

HEMANGIOBLASTOMA OF THE OPTIC NERVE

A: Cross section of the optic nerve involved by hemangioblastoma from the orbital portion of the optic nerve to the chiasm. There is a vascular component and almost complete replacement of the neural portion of the optic nerve by the proliferation.

B: Stromal vacuolated cells alternate with rare interstitial cells that are small and surround the walls of capillaries (arrow).

C: The PAS stain highlights the rich vascular component of the lesion.

[HTLV1], and human immunodeficiency virus [HIV]) and autoimmune diseases.

Microscopic Findings. The histopathologic findings are consistent with the type of tumor involving the optic nerve. Uveal melanomas that are peripapillary may invade the edge between the optic nerve and the choroid and reach the meninges. They usually do not diffusely infiltrate the meninges but only form a localized nodule (fig. 10-10). Neuroinvasive melanomas (usually of epithelioid type) are rare, and they invade the nerve causing necrosis, many associated with secondary neoplastic glaucoma (tumor cells invade the angle) (fig. 10-11) (20).

Retinoblastoma in children may invade the optic nerve. Assignment of prognosis for metastasis hinges on the amount and location of the invasion. Post-laminar optic nerve invasion (PLONI) by retinoblastoma has shown a proportional increase in probability for CNS involvement with the amount and distance from the head of the optic nerve into the post-lamina cribrosa nerve tissue. Tumors in the meninges also have a high risk of CNS recurrence because the meningeal space of the optic nerve contains cerebrospinal fluid (CSF) that connects with the CSF and meninges in the brain. Tumor at the cut margin of the optic nerve has more than 80 percent risk of local recurrence in the orbit (fig. 10-12) (21).

Lymphoma and leukemia secondarily involve the optic nerve more frequently than primary tumors, and the findings reflect the type of lesion and location. Low-grade tumors

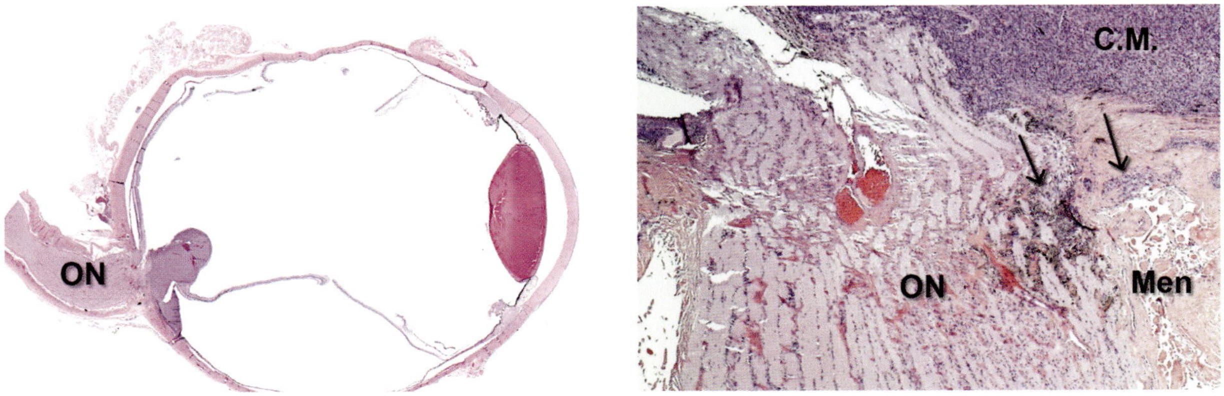

Figure 10-10

CHOROIDAL MELANOMA INVADING THE OPTIC NERVE

Left: An enucleated eye with a choroidal melanoma predominantly involving the peripapillary choroid and the optic nerve head. The post-laminar optic nerve (ON) is free of tumor involvement.

Right: View of the edge of the optic nerve (ON) and the choroidal melanoma (CM) shows invasion of the partially pigmented tumor (arrow) into the optic nerve at the edge with the meninges (Men) and the tumor in the sclera involving the meninges focally (arrows).

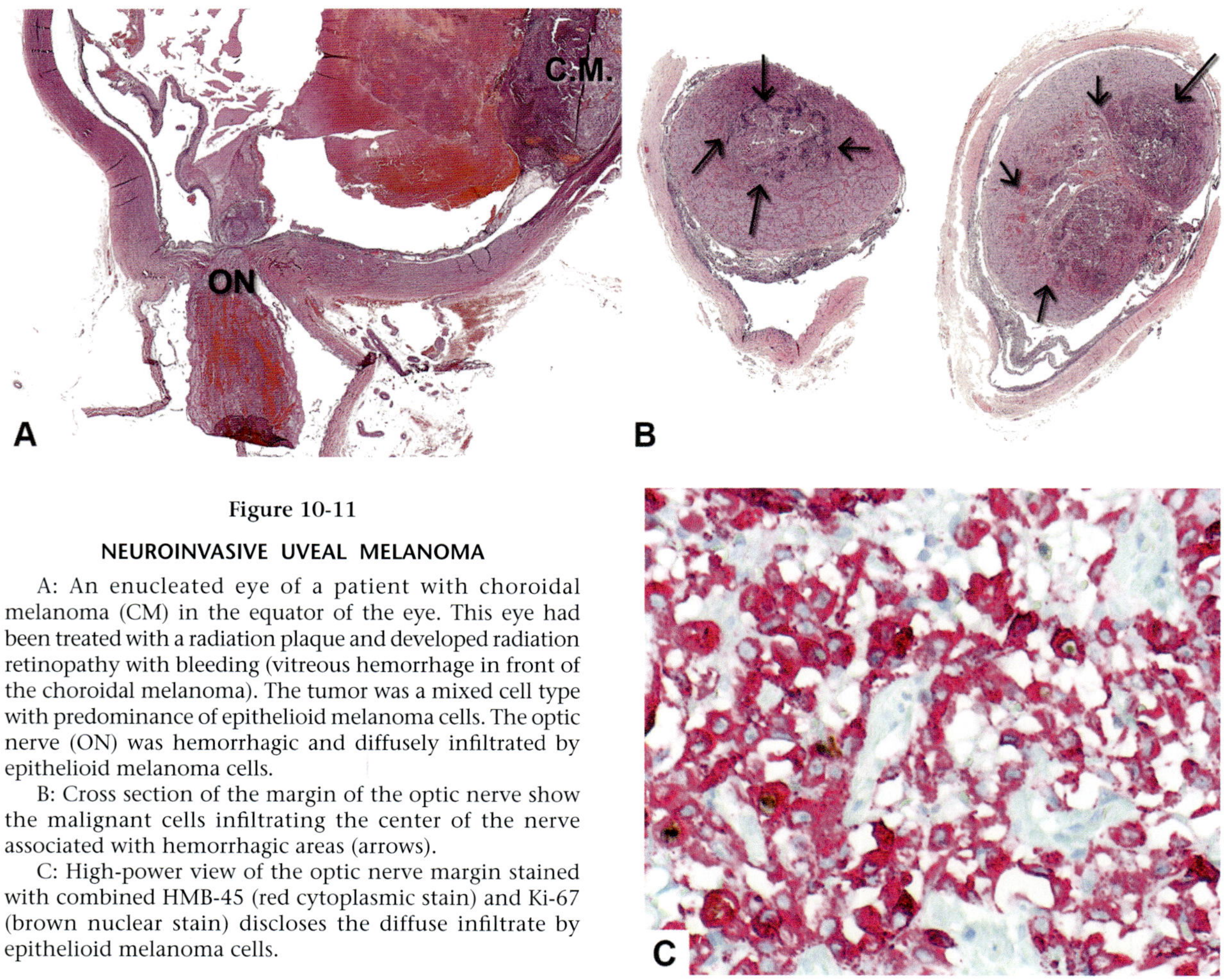

Figure 10-11

NEUROINVASIVE UVEAL MELANOMA

A: An enucleated eye of a patient with choroidal melanoma (CM) in the equator of the eye. This eye had been treated with a radiation plaque and developed radiation retinopathy with bleeding (vitreous hemorrhage in front of the choroidal melanoma). The tumor was a mixed cell type with predominance of epithelioid melanoma cells. The optic nerve (ON) was hemorrhagic and diffusely infiltrated by epithelioid melanoma cells.

B: Cross section of the margin of the optic nerve show the malignant cells infiltrating the center of the nerve associated with hemorrhagic areas (arrows).

C: High-power view of the optic nerve margin stained with combined HMB-45 (red cytoplasmic stain) and Ki-67 (brown nuclear stain) discloses the diffuse infiltrate by epithelioid melanoma cells.

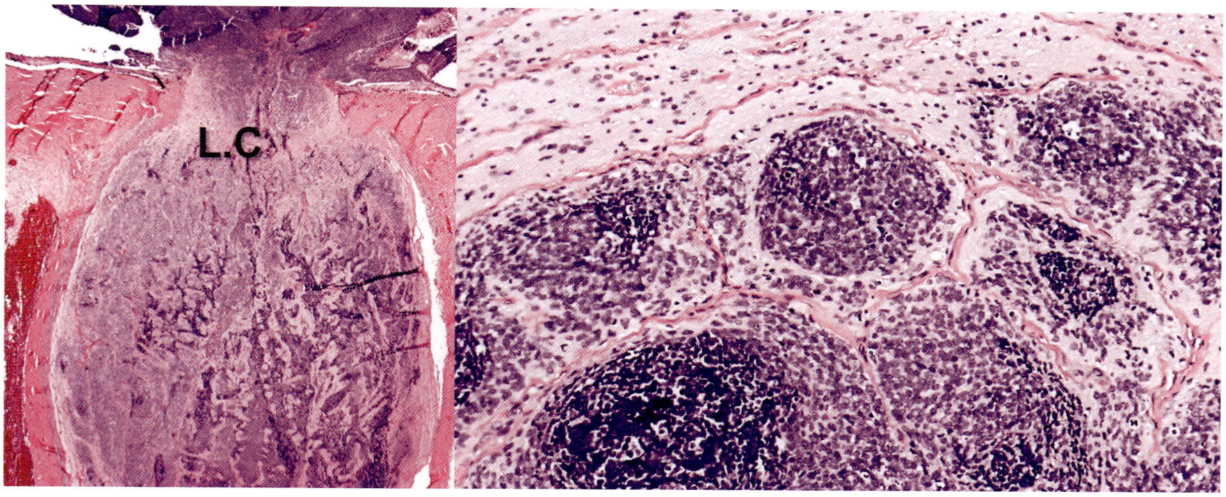

Figure 10-12

RETINOBLASTOMA WITH POST-LAMINAR OPTIC NERVE INVASION

Left: The optic nerve is diffusely infiltrated by retinoblastoma in all the segments: optic nerve head or prelaminar, laminar (LC) and post-laminar nerve. The nerve expands toward the meninges in the post-laminar region secondary to tumor infiltration.

Right: Cross section of the optic nerve margin. The infiltrate of retinoblastoma cells dissects the nerve fibers between the septa. This patient had a fatal recurrence in the central nervous system despite intensive treatment.

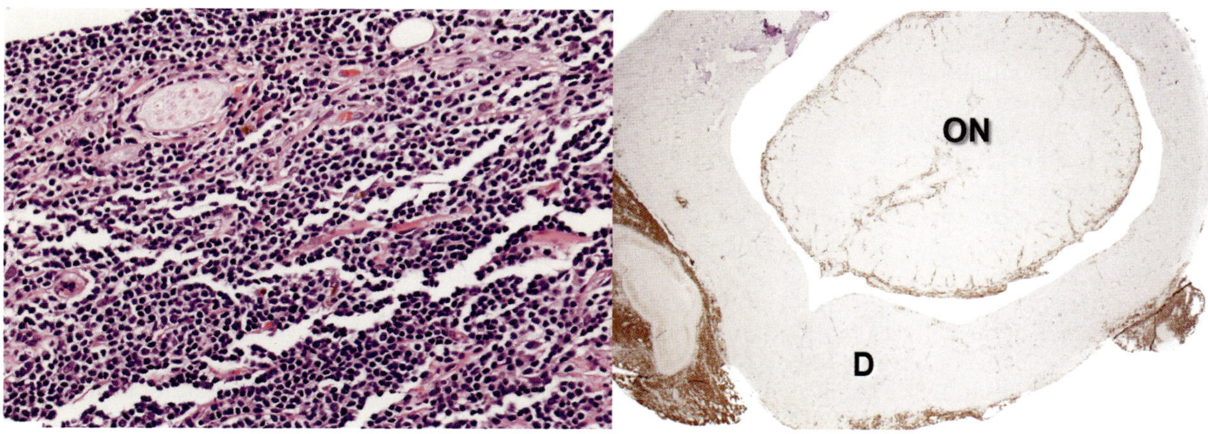

Figure 10-13

LOW-GRADE EXTRANODAL MARGINAL ZONE (ENMZ) B-CELL LYMPHOMA INVOLVING THE OPTIC NERVE

Left: Diffuse infiltrate of small B cells in the dura and arachnoid of the optic nerve in a patient with ENMZ lymphoma.

Right: Cross section of the optic nerve shows the B-cell infiltrate stained with CD20 in the meninges, including the dura mater (D) and into the optic nerve (ON) neural tissue through the fibrous septa.

(extranodal marginal zone B-cell lymphomas, chronic lymphocytic lymphoma/leukemia) infiltrate the meninges without forming a mass (fig. 10-13). These tumors may be associated with amyloid deposition and occlude the central artery secondary to the arterial amyloidosis.

Metastatic breast and lung carcinomas are the most commonly found secondary tumors in the optic nerve, but the literature cites reports of prostate metastases, with 20 percent of tumors from an unknown primary. Immunohistochemistry and clinical history are vital for diagnosing metastatic carcinoma (18).

REFERENCES

1. Hogan MJ, Alvarado JA, Weddell JE. Histology of the human eye: an atlas and textbook. Philadelphia: Saunders; 1971:523-606.
2. Shapey J, Sabin HI, Danesh-Meyer HV, Kaye AH. Diagnosis and management of optic nerve sheath meningiomas. J Clin Neurosci 2013;20:1045-56.
3. Wan WL, Geller JL, Feldon SE, Sadun AA. Visual loss caused by rapid progressive intracranial meningiomas during pregnancy. Ophthalmology 1990;97:18-21.
4. Stern J, Jakobiec FA, Housepian EM. The architecture of optic nerve gliomas with and without neurofibromatosis. Arch Ophthalmol 1980;8:505-11.
5. Bosch MM, Wichmann WW, Boltshauser E, Landau K. Optic nerve sheath meningiomas in patients with neurofibromatosis type 2. Arch Ophthalmol 2006;124:379-85.
6. Robert-Boire V, Rosca L, Samson Y, Ospina LH, Perreault S. Clinical Presentation and outcome of patients with optic pathway glioma. Pediatr Neurol 2017;75:55-60.
7. Toonen JA, Anastasaki C, Smithson LJ, et al. NF1 germline mutation differentially dictates optic glioma formation and growth in neurofibromatosis-1. Hum Mol Genet 2016;25:1703-13.
8. Pan Y, Bush EC, Toonen JA, et al. Whole tumor RNA-sequencing and deconvolution reveal a clinically-prognostic PTEN/PI3K-regulated glioma transcriptional signature. Oncotarget 2017;8:52474-87.
9. Wong JY, Uhl V, Wara Sheline GE. Optic gliomas. A reanalysis of the University of California, San Francisco experience. Cancer 1987;60:1847-55.
10. Francis JH, Wiesner T, Milman T, et al. Investigation of Somatic GNAQ, GNA11, BAP1 and SF3B1 mutations in ophthalmic melanocytomas. Ocul Oncol Pathol 2016;2:171-7.
11. de Alba MA, Villegas VM, Gold AS, et al. Clinical findings and genetic expression profiling of three pigmented lesions of the optic nerve. Case Rep Ophthalmol Med 2015;2015:590659.
12. Shukla SY, Shields JA, Eagle RC, Shields CL. Transformation of optic disc melanocytoma into melanoma over 33 years. Arch Ophthalmol 2012;130:1344-7.
13. Shields JA, Shields CL, Eagle RC Jr. Melanocytoma (hyperpigmented magnocellular nevus) of the uveal tract: the 34th G. Victor Simpson lecture. Retina 2007;27:730-9.
14. Staub BN, Livingston AD, Chévez-Barrios P, Baskin DS. Hemangioblastoma of the optic nerve producing bilateral optic tract edema in a patient with von Hippel-Lindau disease. Surg Neurol Int 2014;5:33.
15. Zhou M, Shen D, Head JE, et al. Ocular clusterin expression in von Hippel-Lindau disease. Mol Vis 2007;13:2129-36.
16. Shields JA, Shields CL, Singh AD. Metastatic neoplasms in the optic disc: the 1999 Bjerrum Lecture: part 2. Arch Ophthalmol 2000;118:217-24.
17. Kincaid MC, Green WR. Ocular and orbital involvement in leukemia. Surv Ophthalmol 1983;27:211-32.
18. Font RL, Ferry AP. Carcinoma metastatic to the eye and orbit III. A clinicopathologic study of 28 cases metastatic to the orbit. Cancer 1976;38:1326-35.
19. Nelson C, Hertzberg B, Klintworth G. A histopathologic study of 716 unselected eyes in patients with cancer at the time of death. Am J Ophthalmol 1983;95:788-93.
20. Gorham JP, Szalai E, Wells JR, Grossniklaus HE. Retinoinvasive uveal melanoma: report of 2 cases and review of the literature. Ocul Oncol Pathol 2017;3:292-5.
21. Chantada GL, Casco F, Fandiño AC, et al. Outcome of patients with retinoblastoma and postlaminar optic nerve invasion. Ophthalmology 2007;114:2083-9.

11 INTRAOCULAR HEMATOLYMPHOID NEOPLASMS

Ophthalmic pathologists divide intraocular hematolymphoid neoplasias into primary and secondary types. These are further subdivided by the location of the predominant intraocular tissue involved and the cellular type.

Intraocular lymphomas are rare; the exact incidence is unknown because of the difficulty in obtaining diagnostic material and in recognizing the diagnosis. The most common lymphoma is the primary vitreoretinal lymphoma, followed by primary choroidal lymphoma, ciliary body lymphoma, and least frequently, iris lymphoma (fig. 11-1).

Secondary intraocular involvement by systemic or nodal lymphomas may be present in the vitreous, retina, and especially, uvea. It is crucial to adequately evaluate patients with intraocular lymphoma for possible extraocular primary lymphoma metastasis to the eye (1,2).

PRIMARY VITREORETINAL LYMPHOMA

Definition and Epidemiology. *Primary vitreoretinal lymphoma* (PVRL) consists of large atypical lymphoma cells that infiltrate the vitreous, retina, subretinal pigment epithelium, and optic nerve. The central nervous system (CNS) is frequently involved, either synchronous or asynchronous with the visual presentation, but lymph nodes, bone marrow, or other visceral organs are not involved.

Most PVRLs are of B-cell lineage, with rare cases of large T- or natural killer (NK)-cell lymphomas (1–3). PVRL is an aggressive non-Hodgkin lymphoma, a diffuse large B-cell lymphoma (DLBCL) as classified by the World Health Organization (WHO) lymphoma classification, with an immunophenotype of nongerminal type.

PVRL has no gender predilection and presents predominantly in patients over 50 years of age (average, 63 years). PVRL may develop in younger patients with immunosuppressive conditions, especially in human immunodeficiency virus (HIV)-positive patients. There is also an association with Epstein-Barr virus (EBV).

PVRL is a rare disease and challenging to diagnose. There is no central tumor registry for this disorder and the incidence is unknown. There has been a steady increase in the number of cases since the 1980s when vitrectomies started to be a routine procedure, making possible the diagnosis of PVRL (2017 calculated incidence of 0.46 per 100,000 people/years). In recent years, most cases are in immunocompetent patients without a known cause, risk factor, or association.

Clinical Features. The clinical presentation of PVRL varies significantly, and this is an essential reason for a delay in diagnosis. PVRL has an insidious presentation and may mimic different ocular diseases, hence, it is often referred to as a masquerading syndrome. The usual time

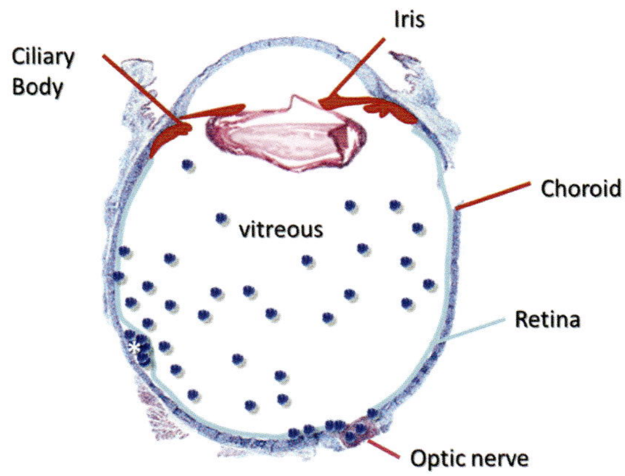

Figure 11-1

INTRAOCULAR SITES OF HEMATOLYMPHOID LESIONS

The blue cells represent the most common sites of involvement: the vitreoretinal (and optic nerve) sites. Lymphoma of retina may invade the subretinal space, forming pockets of lymphoma cells (*).

from initial symptoms to diagnosis is about 6 months (up to 2 to 3 years). It is bilateral in 70 to 90 percent of cases although because of the uneven lymphomatous involvement of the intraocular structures, it may present as unilateral disease when first diagnosed. The symptoms are nonspecific, and most patients present with visual alterations (blurry vision [50 percent], loss of visual acuity [30 percent], and floaters [20 percent]) (1,2).

An initial finding may be the presence of lymphoma cells in a perivascular retinal location, mimicking vasculitis with perivascular sheathing. The lymphoma cells accumulate in the subretinal space, specifically in and under the retinal pigment epithelium (RPE), giving an uneven appearance and usually interpreted as drusen or RPE disease. As the disease progresses, the cells further accumulate in the retina and may replace it (full thickness), causing segmental areas of necrosis that render a creamy white appearance to the affected retina. The lymphoma cells remain viable near the choroid (under RPE and subretinal space), but as they permeate the inner limiting membrane of the retina into the vitreous (avascular structure), they become necrotic. These create clumps of necrotic cells that are associated with the macrophages cleaning the cellular debris and giving the impression clinically of granulomatous vitritis.

Although the lymphoma cells accumulate in the sub-RPE area, Bruch membrane prevents infiltration into the choroid. It is unusual to have very large infiltrates of PVRL in the choroid. The cellular infiltrate may involve the optic nerve head and optic nerve, giving the appearance of an edematous nerve head (papilledema). Less frequently, the cells in the vitreous migrate and accumulate in the anterior segment, especially in the anterior chamber, attaching to the endothelial surface and mimicking the keratic precipitates seen in granulomatous iritis/uveitis. They also may cover and infiltrate the trabecular meshwork and cause secondary glaucoma, a rare presentation. Secondary tissue damage (necrosis, leaking vessels, and infarction) causes hemorrhage and neovascularization, especially in the iris, resulting in hyphema or secondary neovascular glaucoma (2).

Imaging, such as fluorescein angiography, highlights the leaking nature of the affected vessels with an accumulation of fluorescein in the subretinal space. Ocular coherence tomography (OCT) shows the extent of retinal involvement but only in initial stages of the disease. When the majority of cells are in the retina, as in later stages, the vitreous opacities and the thickness of the retinal infiltrate prevent an accurate evaluation of the retinal disease.

CNS disease may present with different neurologic signs and symptoms, depending on the specific CNS area involved. Because of the frequent association of ocular lymphoma with CNS lymphomatous involvement, patients with apparent uveitis and neurologic symptoms should be worked up to exclude PVRL (1,2).

Etiology and Molecular Biology. The WHO lymphoma classification has a unique category for PVRL (and associated CNS lymphomas) that is part of the DLBCL category because of the unique biological features. PVRL cells express B-cell markers (CD20, PAX5, CD79a) and many also express BCL2, BCL6, MUM1, and monotypically heavy or light chains, but they lack expression of CD10 or plasma cell markers. These correspond to the activated B-cell (ABC) subtype of the DLBCLs and specifically of a late stage of maturation of nongerminal B-cell type (1,2,4).

Only a few papers addressing the utility of molecular genetic studies are available. These tests should be ordered with caution. Many intraocular biopsy samples contain tiny amounts of lymphoma cells and inflammatory cells that may contribute to uneven results.

Some authorities extend information known about primary CNS lymphomas and apply these data to PVRL. Recent publications highlight three critical genes with pathogenic variants. The most frequent gene association found (about 70 percent of the PVRLs) is in the gene encoding myeloid differentiation factor 88 (*MYD88*), which is a member of the toll-like receptor pathway and plays a central role in the innate and adaptive immune response (5). Also recently associated with PVRL is a member of the B-cell receptor signaling pathway, *CD79B* (CD79b Molecule), a protein-encoding gene. The known pathways related to *CD79B* are hematopoietic stem cell differentiation, lineage-specific (B cell) markers, and Fc-GammaR pathways. The last group is transcription factors from the nuclear factor kappa-light chain-enhancer of activated B cells (NF-κB) family, which binds

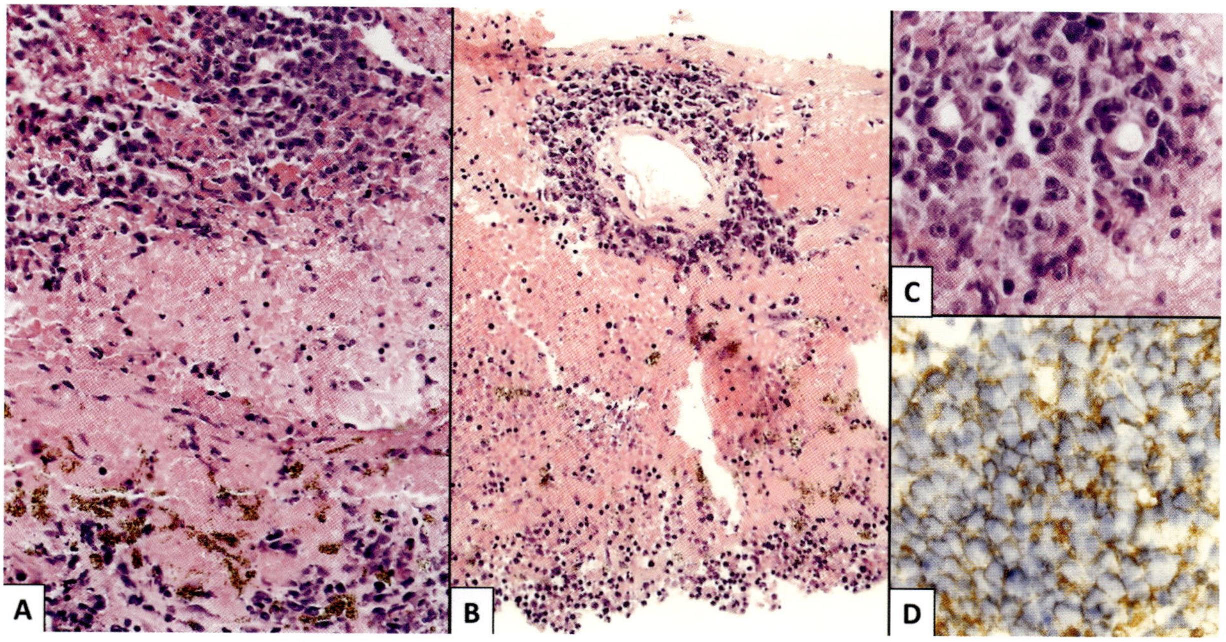

Figure 11-2

PRIMARY VITREORETINAL LYMPHOMA

A: The retina is infiltrated by large lymphoma cells with extensive necrosis of outer layers (bottom) and retinal pigment epithelium (pigmented cells).

B: Mostly necrotic retina with perivascular viable large lymphoma cells.

C: High-power view of the perivascular infiltrate shows large cells with irregular nuclear contours and occasional nucleoli with some apoptotic cells.

D: Same area as seen in C with immunohistochemistry using CD20 antibody. The collective findings are those seen in diffuse large B-cell lymphoma, primary vitreoretinal lymphoma (PVRL).

to DNA in the promoter regions of genes regulating cellular processes. NF-κB is interesting because in human cancer it has been linked with initiation, development, metastasis, and resistance to treatment. NF-κB activity is owed to the inflammatory microenvironment and various oncogenic mutations, and promotes tumor cell proliferation, suppresses apoptosis, and promotes angiogenesis. Molecular genetic studies are ongoing to understand the pathogenesis of PVRL and to target mutations for treatment (4).

A few T-cell intraocular lymphomas are found primarily in the vitreous, with few involving the iris. There is a significantly greater incidence of serous retinal detachment and edema of the macula associated with T-cell lymphoma than the typical B-cell PVRL. Up to 58 percent of the intraocular T-cell lymphomas have systemic involvement, higher than the B-cell PVRL (1,2).

Microscopic Findings. The current approach to diagnosis is via analysis of vitreous biopsy and vitrectomy specimens. This approach is successful in many cases if appropriately handled by the pathologist. A subretinal/RPE biopsy may be necessary because most tumor cells accumulate in this region and the reactive cells are present in the vitreous. Subretinal biopsies may also contain pieces of the retina. The retina is remarkable for perivascular infiltrates of large atypical lymphocytes, but away from the vessels, the lymphocytes and the retina become necrotic (fig. 11-2). The vitreous may have fewer lymphoma cells than the retina, and the choroid usually contains a reactive T-cell infiltrate (2).

Cytopathologic Approach. Vitreoretinal surgeons are aware of the need for sufficient and good quality samples in cases of suspected PVRL. Surgeons are advised to obtain an undiluted vitreous sample (undiluted sample/vitreous biopsy) at the beginning of the procedure. To accomplish this, the vitrector (connected to a syringe) should be positioned into the vitreous.

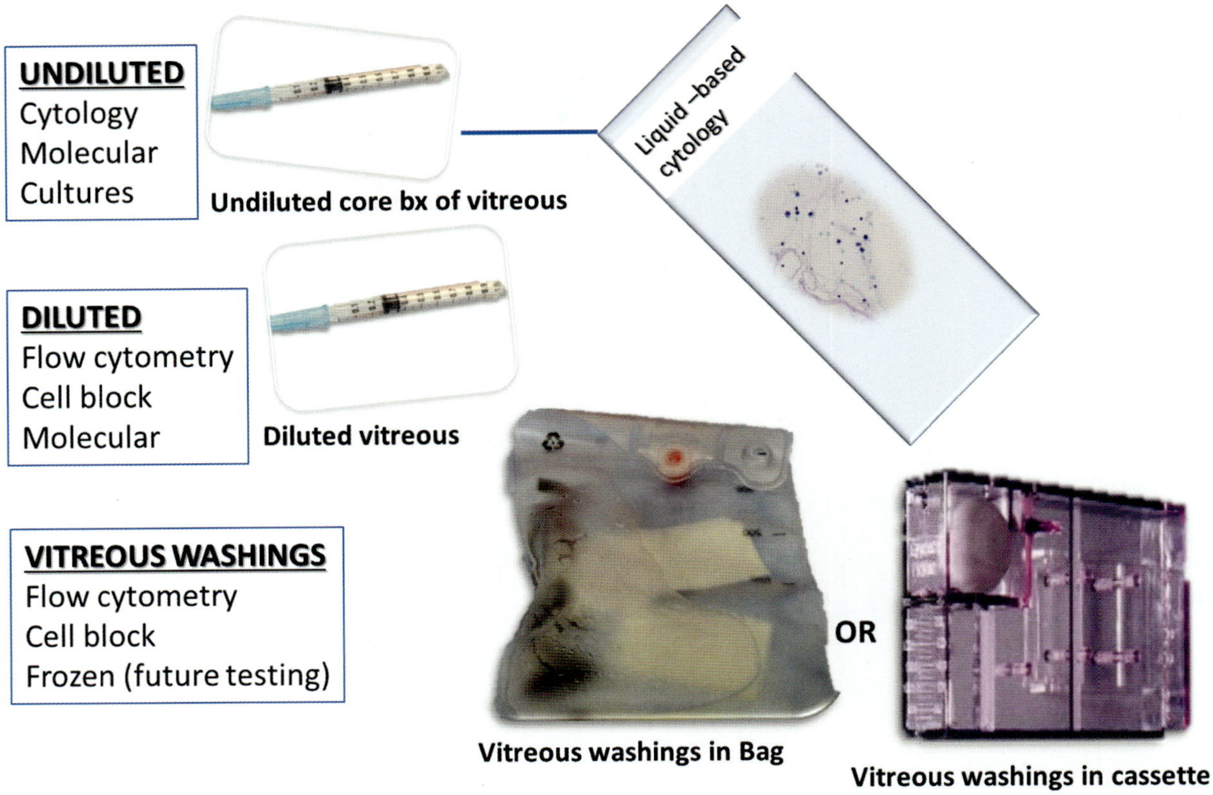

Figure 11-3

SPECIMENS TYPES TO EXCLUDE INTRAOCULAR LYMPHOMA

Undiluted vitreous usually is 1 mm or less, but it is the purest and most concentrated of the specimens. The cytology slide for triaging should be made from this sample because it has the most representative material. If cultures or molecular testing for infectious organisms are needed, this would be the preferred sample since it is the least manipulated. When intraocular lymphoma is suspected, this sample is used for Epstein-Barr virus polymerase chain reaction (EBV PCR) and possibly B- or T- cell PCR. Vitreous (about 1 to 3 mL) is diluted further with some sterile balanced salt solution (BSS) for flow cytometry, generating a cell block, and applying the appropriate molecular techniques to evaluate the sample. Vitreous washings result from the vitrectomy procedure where there is BSS admixed with diluted vitreous (20 to 100 mL); it is the most diluted sample of all. Flow cytometry and preferably cell block are the best tests for this sample. A few milliliters should be frozen for future tests.

The goal is to apply suction without adding any balance salt solution (BSS) to the vitreous. This procedure yields about 1 mL to a maximum of 2 mL. The vitreous contains approximately 4 mL of volume but after removing 1 to 2 mL by suction, the eye begins to collapse, and the retina may detach. Vitrectomy is stopped before the point of collapse. BSS is infused into the vitreous and more sample is obtained immediately, also using a syringe (diluted vitreous sample) and obtaining an average of 2 mL. After this, the vitrectomy proceeds, continuing the infusion of BSS as the vitrector cuts and suctions the diluted vitreous. The vitrectomy instrument collects the diluted vitreous into a bag or container (vitrectomy cassette/vitreous washings) through a tubing system.

In the ideal scenario, the cytology laboratory receives three samples: one syringe with undiluted vitreous, one syringe with diluted vitreous, and the vitrectomy bag/cassette (fig. 11-3). One assesses the type and amount of cellularity from the undiluted sample, a liquid-based slide or cytospin preparation. Based on this quick assessment, the laboratory triages the remainder of the specimens for flow cytometry (if the specimen is cellular and without much necrosis), molecular analysis (viral, clonality, other), and cell block with immunohistochemistry (fig. 11-3) (1,2,4).

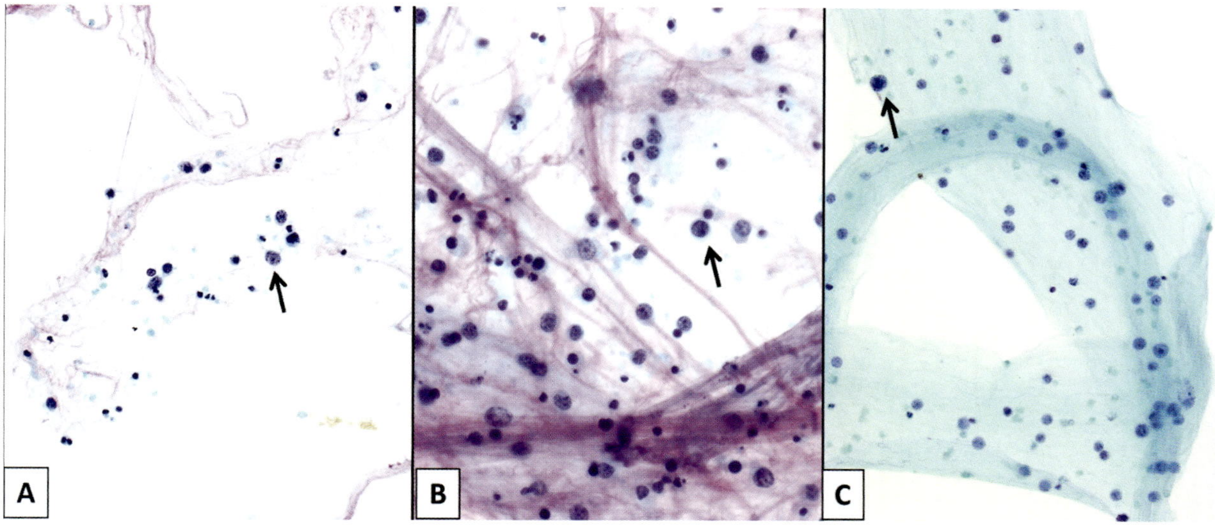

Figure 11-4

PRIMARY VITREORETINAL LYMPHOMA

A: Cytology specimens may be paucicellular, with few atypical large cells (arrow) in a background of apoptotic cells, macrophages, and condensed vitreous.

B: A more cellular specimen with more atypical cells (arrow).

C: Most of the cells in this moderately cellular sample are large atypical lymphocytes (arrow).

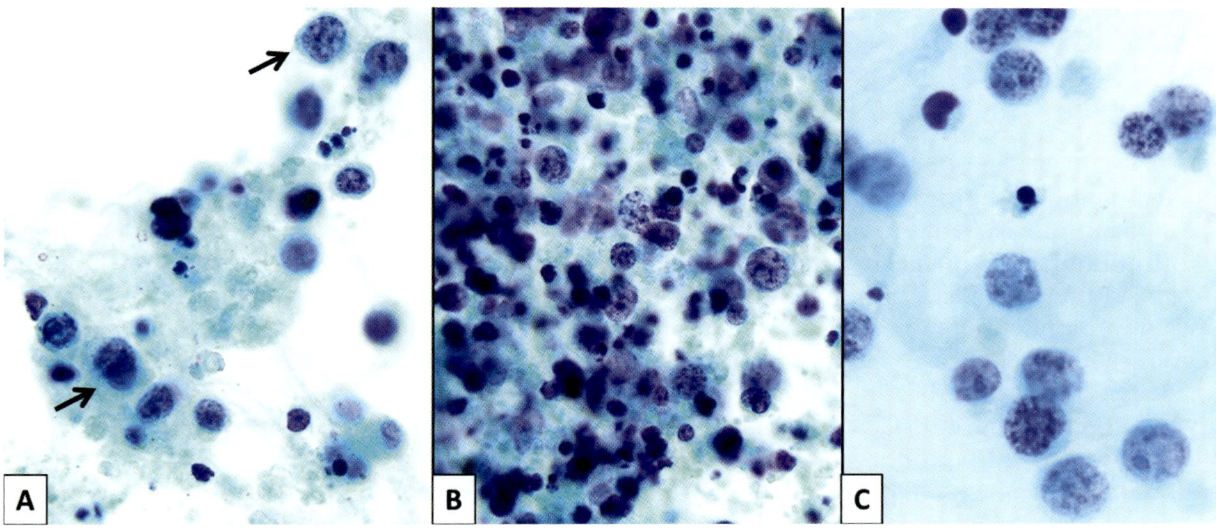

Figure 11-5

PRIMARY VITREORETINAL LYMPHOMA

A: High-power view of a paucicellular specimen shows a focus of atypical large lymphocytes with clumped chromatin and irregular nuclear membranes (arrows) in a background of apoptotic cells and necrotic cytoplasm debris.

B: A cellular specimen with many atypical lymphocytes.

C: Most of the cells in this sample are large atypical lymphocytes.

Cytologic Findings. The specimen may vary from hypocellular to abundant cellularity. There is usually a mixture of cells composed of large atypical cells (scant cytoplasm, irregular nuclear contours, prominent nucleoli, and abnormal chromatin), apoptotic cells, necrotic debris (lymphoglandular bodies), macrophages, and a few small lymphocytes (figs. 11-4, 11-5). These

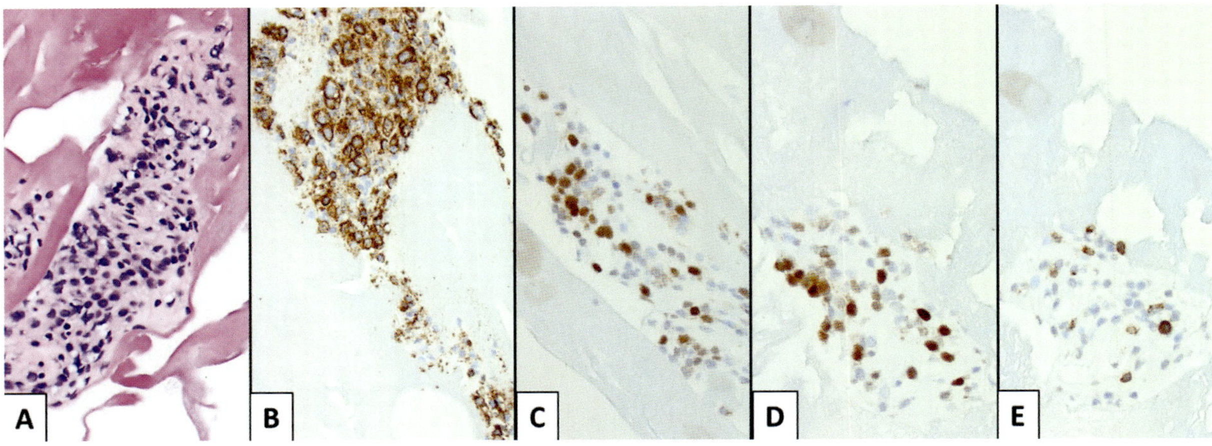

Figure 11-6

DIFFUSE LARGE B-CELL LYMPHOMA, NONGERMINAL TYPE, PRIMARY VITREORETINAL LYMPHOMA

A: Moderate cellularity is seen on a cell block, with few cellular details compared with cytologic samples.
B: CD20 antibody labels the cytoplasm and membranes of large cells and the membranous cellular debris.
C: PAX5 labels the nuclei of the atypical large lymphocytes.
D: MUM1 stains the nuclei of the large cells (coexpressed with PAX5).
E: A few reactive small T lymphocytes are positive for CD3.

cells may be associated with vitreous bands. Cytologic preparations are best for assessing morphology and atypia of the lymphoma cells. The cell block may contain groups of atypical cells, but often with loss of detail (fig. 11-6A) (1,2). The success rate (up to 60 percent) in the diagnosis of PVRL resides mainly in the skills of the technologist in processing the samples and the cytopathologist in interpreting the findings.

Immunohistochemical Findings. The majority of PVRLs are DLBCLs of nongerminal type. The immunophenotype is as follows: positive staining with CD20, PAX5, CD79a; often positive for BCL2, BCL6, and MUM1; and negative for atypical B cells with CD10, CD3, and CD68. CD3 stains the reactive T lymphocytes and CD68 the abundant macrophages that are present in areas of necrotic debris (fig. 11-6B–E) (1,2)

For the rare cases of large T-cell lymphoma, the large cells are positive for CD3. Additional T-cell markers, such as CD30 and ALK1, may be helpful in diagnosis (fig. 11-7) (3).

Flow Cytometric Findings. If enough non-necrotic cells are present, then flow cytometry may be preferable to diluted or vitreous washings. A limited and directed panel is recommended. The results should be interpreted with caution because in hypocellular samples, limited B cells may lead to the misinterpretation of monoclonality because of the lack of other types of B cells in the sample. In these cases, the size of the B cells is essential because in reactive samples there are usually small lymphocytes and most are T cells with a rare small B lymphocyte.

A vitreous reactive infiltrate may have oligoclonal or monoclonal expansions (as a response to viral or other etiologies) of T cells and less frequently, B cells, thus a critical assessment of the results that include the cytologic features should be performed before relying only on flow cytometry results. In the few cases of large T-cell lymphoma, the flow cytometry shows large cells of T-cell origin and aberrant antigen expression (1–3).

Molecular Genetic Findings. Molecular studies are supportive and sometimes the main diagnostic tool in the identification of PVRLs: immunoglobulin gene rearrangements confirm clonality. The sensitivity of clonality studies has been reported to reach 95 percent and depends on the sample quality, laboratory procedures, and primers selected. False positive results occur, for reasons discussed earlier in this chapter. False negative results are due to the variety and amount of somatic mutations and the binding sites of the primers.

The identification of *MYOD88* mutations helps confirm a specific diagnosis of PVRL, because about 70 percent of CNS primary

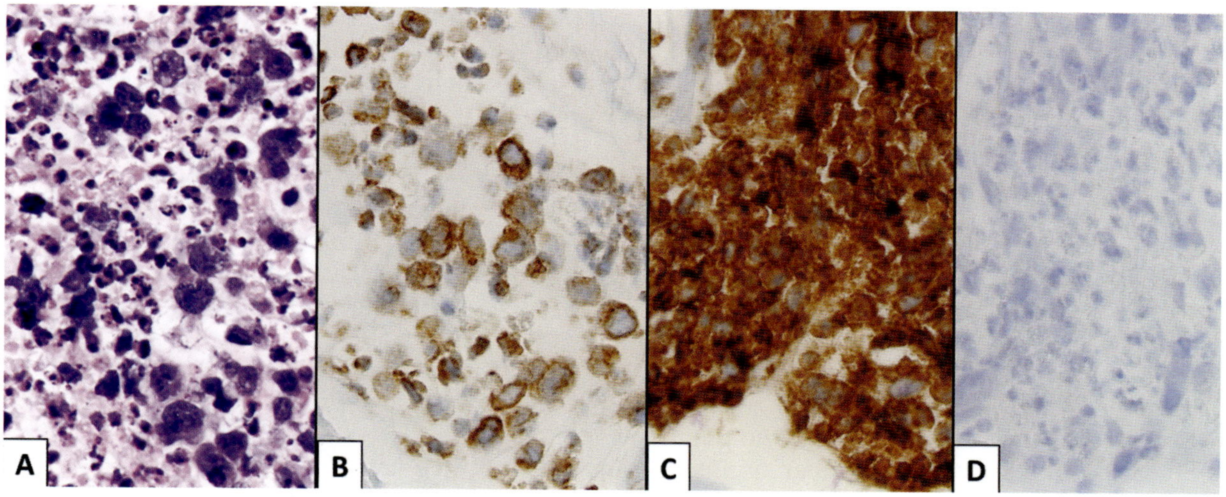

Figure 11-7

LARGE T-CELL LYMPHOMA

A: Cell block with high cellularity and large atypical cells featuring prominent nucleoli, irregular nuclear membranes, and background necrosis.

B: CD3 antibody labels the cytoplasm and membranes of large cells.

C: CD30 labels the atypical large lymphocytes intensely.

D: CD20 is negative.

lymphomas and PVRLs have this unusual mutation. Mutational and PCR *MYOD88* analysis is being substituted for clonality. Next-generation sequencing (NGS) also detects the *MYOD88* mutation and other mutations for diagnosis (4,5).

Cytokine Analysis. The ratio of interleukin (IL)-10 to IL-6 in the aqueous or vitreous fluids is of diagnostic value in patients with suspected PVRL, although the ratio has proven more useful for follow-up after treatment. A ratio of greater than 1 IL-10/IL-6 is supportive of the diagnosis of PVRL (1). Sensitivity is reported to be up to 90 percent.

Differential Diagnosis. The differential diagnosis depends on the precise structure in the eye that is mostly involved by the process, but the most common entity in the differential diagnosis is anterior or posterior uveitis (infectious and noninfectious). If there is involvement predominantly of the retina with necrosis, then acute retinal necrosis (cytomegalovirus [CMV] usually) and retinal vasculitis (infectious or autoimmune) should be considered. The subRPE infiltrates may mimic confluent drusen, RPE detachments, and adult vitelliform dystrophy (yellow cream infiltrates). In the optic nerve, the infiltrates may appear as lesions secondary to sarcoidosis, tuberculosis, leukemic infiltration, and papilledema. The workup should include viral PCR, especially for CMV, herpes simplex (HSV), and zoster virus (HZV) (1,2).

Treatment and Prognosis. Treatment includes local (intravitreous chemotherapy, ocular radiotherapy, CNS radiotherapy, and intrathecal chemotherapy) and systemic (chemotherapy). About 15 percent of the primary vitreoretinal lymphomas present with CNS involvement and about 75 percent of patients develop involvement of CNS at one point in their disease. The survival rate has improved in recent years, with new therapies and the evolution of radiation technology (1,6).

PRIMARY IRIDAL LYMPHOMA

Etiology. The literature records few cases of *primary iridal lymphoma*. There are no reports of primary iris lymphoma in immunocompetent patients. Most primary iris lymphomas are of B-cell lineage, and low-grade extranodal marginal zone type, but few T-cell lymphomas have been studied. More often, iris involvement is secondary to primary vitreoretinal lymphoma, especially when it is recurrent. The few cases reported in the literature of iris lymphoma are either in patients without immune-related cause (patients over 50 years of age) or associated with either EBV-related disease or status post-solid organ transplant, and in these scenarios, patients are younger (7). EBV, PCR, and EBER may detect the *EBV* genome. Patients with status post-organ transplant also have an association with EBV infection (8,9).

Clinical Features. Typically, there is a decrease in visual acuity, redness of the eye, ocular pain associated with varying degrees of anterior chamber cells, and keratic precipitates with or without a visible tan-white infiltrate in the iris stroma. Occasionally, patients have glaucoma, hyphema, vascular congestion in the episclera and the iris, a fibrin clot in the anterior chamber, or pseudohypopyon. If the infiltrate is only present in the iris, the posterior pole is unremarkable. In those cases where the iris is involved secondarily to PVRL, other signs are seen in the vitreous (vitriitis, deposits, cells) and retina (7–9).

Microscopic Findings. The primary iris lymphomas reported in the literature are of low-grade B-cell type, including those associated with status post-organ transplant. Most are extranodal marginal zone type. One of the cases was a follicular type lymphoma with the t(14;18) translocation. There may be small plasmacytic cells and other small lymphocytes in the infiltrate.

Rare cases of large B-cell lymphomas are described in immunocompromised patients. T-cell lymphomas of iris have been reported in cases of systemic/skin primary T-cell lymphoma. Some iris lymphomas extend into the trabecular meshwork and the conjunctival stroma (7,8).

Direct biopsy of the iris and conjunctiva (if there is extraocular extension), or anterior chamber tap with cytologic studies may allow for tissue procurement. Immunohistochemistry and flow cytometry are essential for confirming the diagnosis (versus benign reactive infiltrates) and type of lymphoma (B- or T-cell origin).

Differential Diagnosis. The differential diagnosis is specific to the age group and immune status of the patient. In children, the most frequent entities are juvenile xanthogranuloma, anterior uveitis (rheumatoid arthritis, Bechet disease), diffuse anterior retinoblastoma, resolving hyphema, and leukemic infiltrate. In adults and immunosuppressed patients, the principal entities are anterior and intermediate uveitis (viral, other infectious, sarcoidosis), post-transplant lymphoproliferative disease, metastasis, and leukemic infiltrate (7(9).

Treatment and Prognosis. The few cases reported in the literature have been treated with radiation or intraocular chemotherapy if limited to the iris. The outcome varies with the type of lymphoma: low-grade lymphomas respond best to treatment. In immunosuppressed patients, especially after transplantation, consideration of changes in the immunosuppressive medication and coverage with antivirals and other medications is important during antilymphoma treatment to avoid further complications (7–9).

PRIMARY CILIARY BODY LYMPHOMA

Primary ciliary body lymphoma is a rare lesion, and even less frequent than primary iris lymphoma. There are not enough available data to have an exact incidence. The reported cases are usually of B-cell lineage. Of the four cases reported, half were low-grade extranodal marginal zone (MALT-type) lymphoma, similar to those of the choroid. Two cases were high-grade diffuse large B-cell lymphomas. The lesions present in patients over 50 years of age, without gender preference, and the literature assigns no specific etiology to primary ciliary body lymphomas (10).

The presentation is usually intermediate uveitis of the shallow anterior chamber, with anterior chamber cells, precipitates in the cornea, and anterior vitreous opacities. Ultrasonography shows a dominant ciliary body mass. There may be 360 degree ring-like involvement of the stroma of the ciliary body (10).

The lesion is sampled by fine-needle aspiration biopsy, segmental resection (iridocyclectomy), or enucleation. The lesion is present mainly in the ciliary body stroma and muscle, with extension into the ciliary body epithelium and focal migration of cells into the adjacent vitreous. The type of cells depends on the type of lymphoma but morphology and immunophenotype is that of a low-grade B-cell lymphoma similar to a primary choroidal lymphoma or a DLBCL similar to that seen in PVRL (1–3,7,10).

Similar to the other intraocular primary lymphomas, the treatment depends on the type of lymphoma. The low-grade lymphomas respond to low-dose radiation without recurrence. The high-grade B-cell lymphomas are treated

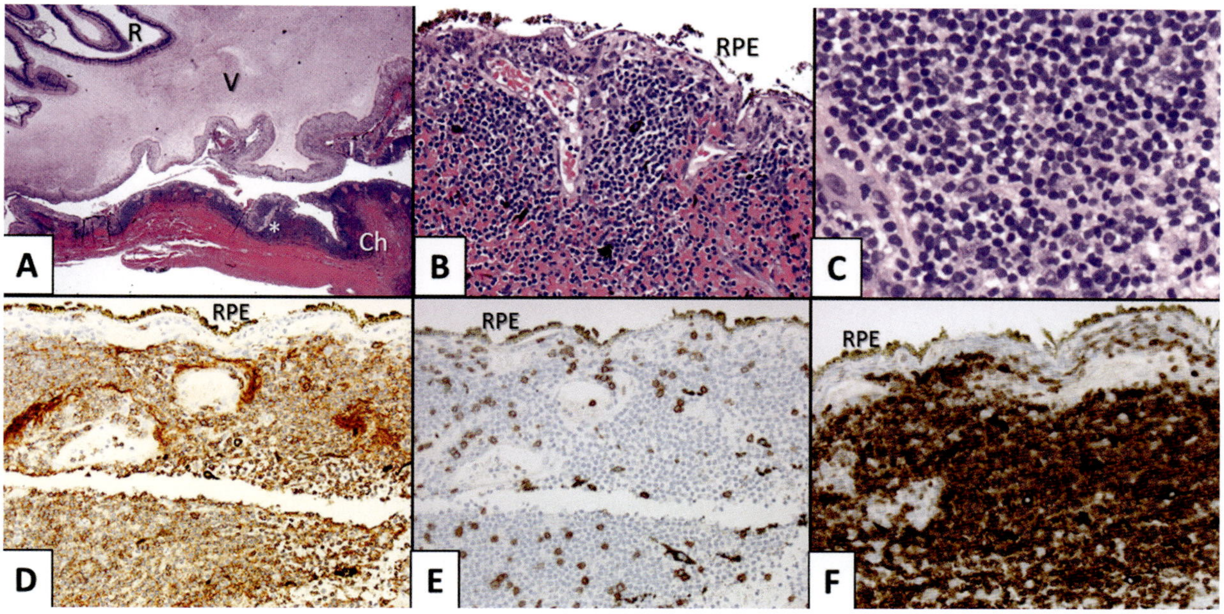

Figure 11-8

PRIMARY CHOROIDAL EXTRANODAL MARGINAL B-CELL LYMPHOMA (MALT TYPE)

A: Low-power view in an evisceration for a perforated eye shows detached retina (R) and cloudy vitreous (V) (acute endophthalmitis secondary to perforation) with a dense hyperchromatic infiltrate in the choroid (Ch) associated with suprachoroidal hemorrhage.
B: Higher-power view of the choroid shows a monotonous infiltrate of small lymphocytes that obliterate choroidal architecture.
C: The lymphocytes are small, and some are plasmacytoid without nucleoli or necrosis.
D: CD20 is positive in the majority of the infiltrating lymphocytes.
E: CD3 is positive in rare lymphocytes.
F: BCL2 is strongly positive in the majority of the infiltrating lymphocytes (co-expressed with CD20).

more aggressively, with combination therapy (intraocular and systemic chemotherapy, and radiation). The prognosis is guarded, similar to that of PVRL (10).

PRIMARY CHOROIDAL LYMPHOMA

Primary choroidal or *uveal lymphoma* is a low-grade lymphoma. In the past, these lesions were classified as reactive processes. Recent studies have shown that primary choroidal lymphoma has a phenotype similar to that of extranodal marginal zone lymphoma of mucosa-associated lymphoid tissue (MALT lymphoma) of the WHO lymphoma classification.

Choroidal lymphoma occurs in adults with no gender preference. Although the exact etiology is unknown, it may develop within a chronic reactive inflammation that transforms into the indolent low-grade lymphoma with clonal B lymphocytes. The presence of background reactive lymphoid follicles in tissue that originally is free of lymphoid tissue (intraocular tissues are devoid of lymphoid or lymphatic tissue) supports the conclusion that this process is reactive or has a reactive component (11–13).

The lesion appears mainly in the choroid and may extend to the ciliary body and rarely into the iris. Infiltrates within the uveal tract appear creamy light brown and these may coalesce to create a thick lesion. Retinal serous detachments may develop over the areas infiltrated by lymphoma. The lymphoma infiltrates may also follow the emissary channels into conjunctiva and form fleshy tan-gray nodules, or they may extend into the posterior episcleral orbital tissue and around the optic nerve. Ultrasonography identifies these lesions (11).

The infiltrates located in the choroidal stroma are composed of small lymphocytes with mildly irregular nuclear contours. These infiltrates are arranged into lymphoid follicles with germinal centers. The monotonous infiltrate consists of clonal B lymphocytes. These cells may infiltrate in sheets into adjacent sclera and conjunctiva.

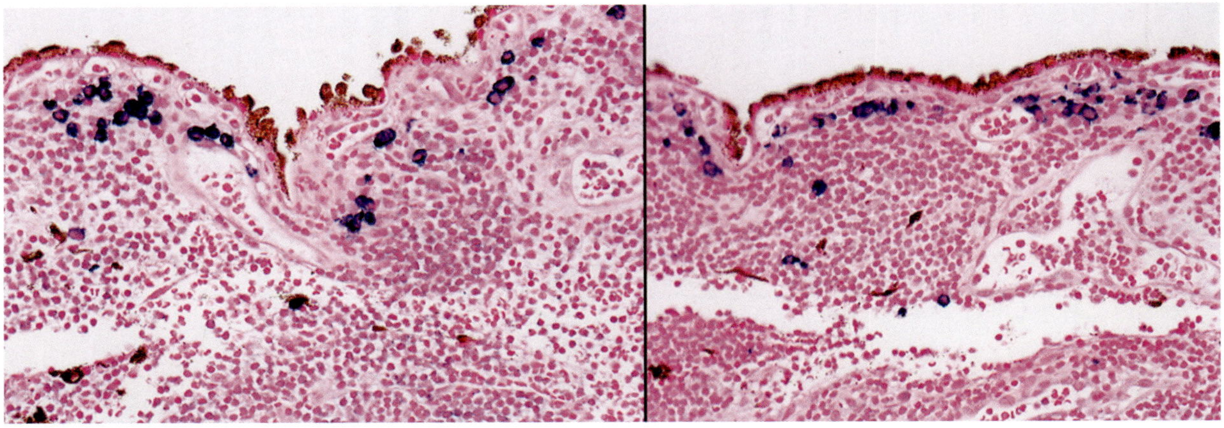

Figure 11-9

PRIMARY CHOROIDAL EXTRANODAL MARGINAL B-CELL LYMPHOMA (MALT TYPE)

Left: In an evisceration for a perforated eye, positive staining of few cells with kappa light chain by in situ hybridization. Right: Positive staining of few cells with lambda light chain by in situ hybridization.

Plasma cells may be present in the lymphocytic infiltrate. The diagnosis is usually made by transscleral biopsy, in enucleated eyes, in evisceration specimens, or less frequently, by transvitreal choroidal biopsy (fig. 11-8) (12,13).

The low-grade lymphoma cells express CD20 and PAX5, and may express CD43, BCL2, and BCL6. The lymphoid follicles may contain IgD- and CD21-positive cells, highlighting the original reactive nature of lymphoid follicles with germinal centers (fig. 11-8D–F). If there is an associated plasmacytoid infiltrate, immunohistochemistry or in situ hybridization for kappa and lambda light chains may prove the clonality of the lesion. Often, however, there is only an altered ratio between kappa and lambda without a clonal pattern (fig. 11-9). The lesion is diagnosed by a combination of a pattern of infiltration, cytomorphology, and immunohistochemistry (12,13).

The entities in the differential diagnosis are sympathetic ophthalmia, sarcoidosis, and Vogt-Koyanagi-Harada disease. All these lesions have an important granulomatous component usually absent in primary choroidal lymphomas (figs. 11-10, 11-11).

Primary choroidal lymphomas respond well to systemic corticosteroids or low-dose radiation. In rare cases, there is systemic involvement and the patient should be treated accordingly (11).

SECONDARY INTRAOCULAR HEMATOLYMPHOID NEOPLASMS

Hematolymphoid neoplasia that involves intraocular tissues secondarily may represent lymphomas or leukemias, and may infiltrate the uveal tissue (predominantly choroid), retina, or vitreous. The choroid is probably the most common intraocular site that is involved by secondary lymphoma or leukemia (11,14).

Secondary Leukemia

General Features. The leukemic infiltrate may be acute or chronic, depending on the type of systemic leukemia. Acute leukemias are five times more common than chronic leukemias. The real incidence is unknown, but reports describe up to 70 percent secondary leukemic involvement in the retina and choroid. There is a slight prevalence in males.

The lesion appears in late disease stages and relapse sites. Studies suggest that the choroid acts as a sanctuary site that allows leukemic cells to evade radiation and chemotherapy. The posterior portion of the eye is the most frequent site of involvement by leukemia: the choroid, retina, and the head of the optic nerve. In children, especially infants, the anterior segment is also involved and may present with pseudohypopyon (14).

Clinical Features. In the retina, leukemic infiltrates present as hemorrhages of all types

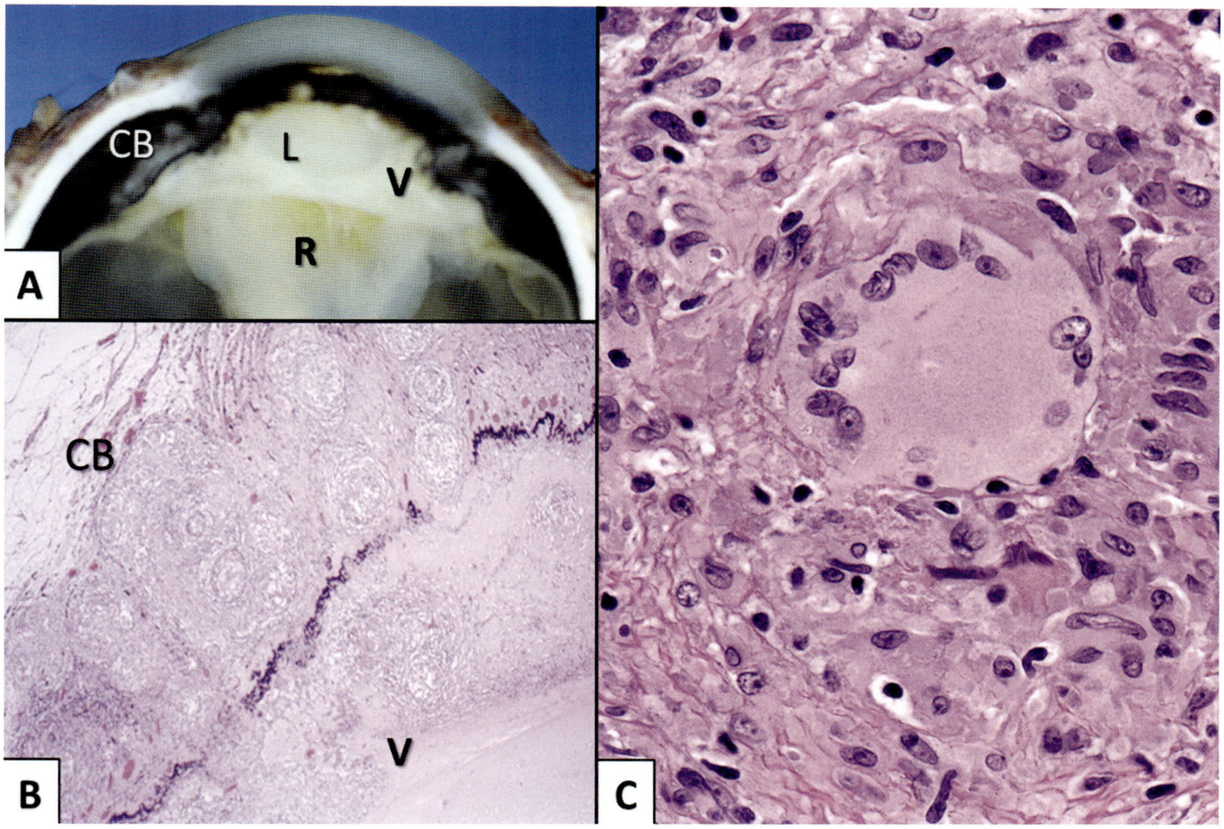

Figure 11-10

ADVANCED SARCOIDOSIS OF EYE

A: Macroscopic view of the anterior segment of the eye with white-tan granulomatous infiltrates in the ciliary body (CB) and choroid extending into the vitreous (V) and pulling the retina (R) in a funnel-shaped retinal detachment.

B: Histologic section of the ciliary body (CB) and anterior vitreous (V) in the same patient. The infiltrates are composed of noncaseating granulomatous inflammation breaking through the ciliary body epithelia into the vitreous.

C: High-power view of the granulomatous infiltrate shows a "naked" type granuloma with a central multinucleated giant cell surrounded by epithelioid histiocytes and concentric fibrosis with a small number of lymphocytes.

that may extend into the vitreous. In chronic leukemia, the vessels of the retina may have aneurysmatic and tortuous vessels with neovascularization. These are secondary to increased blood viscosity. In the choroid, infiltrates may cause serous retinal detachments and creamy infiltrates with secondary RPE changes. Some oncologists propose screening of patients in clinical remission because the choroid may show the infiltrates (11).

Microscopic Findings. Choroidal biopsies, fine-needle aspirates of aqueous humor lesions, and enucleated eyes show choroidal involvement by the same type of leukemia as in the primary bone marrow site (fig. 11-12). Choroidal infiltrates may show blastic elements in cases of chronic systemic leukemia. The leukemic cells populate the perivascular tissue; Bruch membrane prevents cells from involving the retina. Chronic changes in the RPE may accompany the choroidal involvement.

In the retina, the infiltrate is dense and predominantly found in the inner layers around the retinal vessels, with secondary segmental necrosis and hemorrhage. The type of cellularity corresponds to the type of systemic leukemia and immunohistochemistry helps confirm the diagnosis (11,13,14).

Treatment and Prognosis. Treatment of secondary leukemic involvement of intraocular tissues is with systemic chemotherapy. If the dense infiltrate does not respond promptly

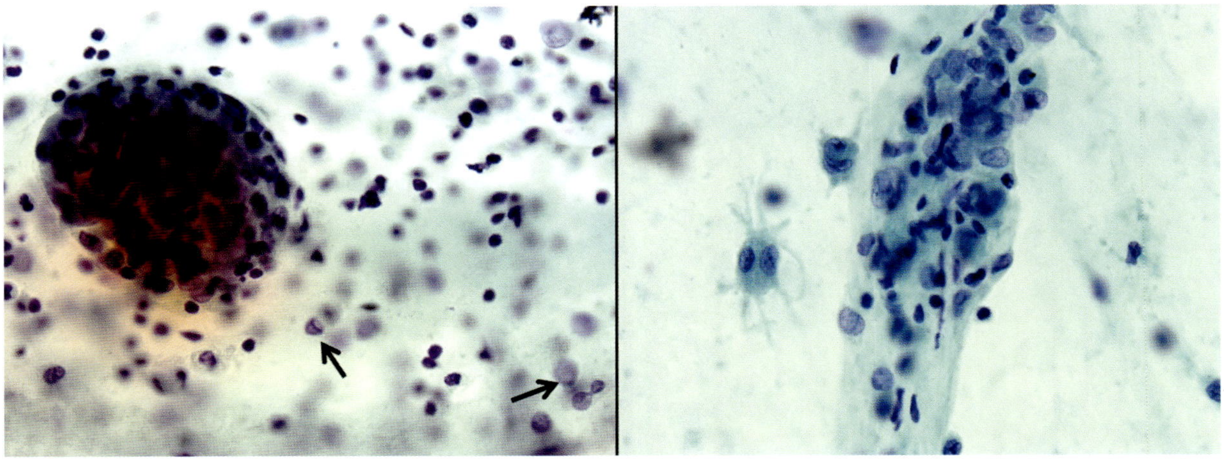

Figure 11-11

SARCOIDOSIS OF UVEAL TRACT AND VITREOUS

Left: The specimen is cellular, with occasional well-formed granulomas (center) and scattered lymphocytes and histiocytes (arrow) in the background.

Right: A poorly formed granuloma with mainly epithelioid histiocytes (center) is in a background of condensed vitreous. Spindle cells are often binucleated and form bands and membranes in the condensed vitreous.

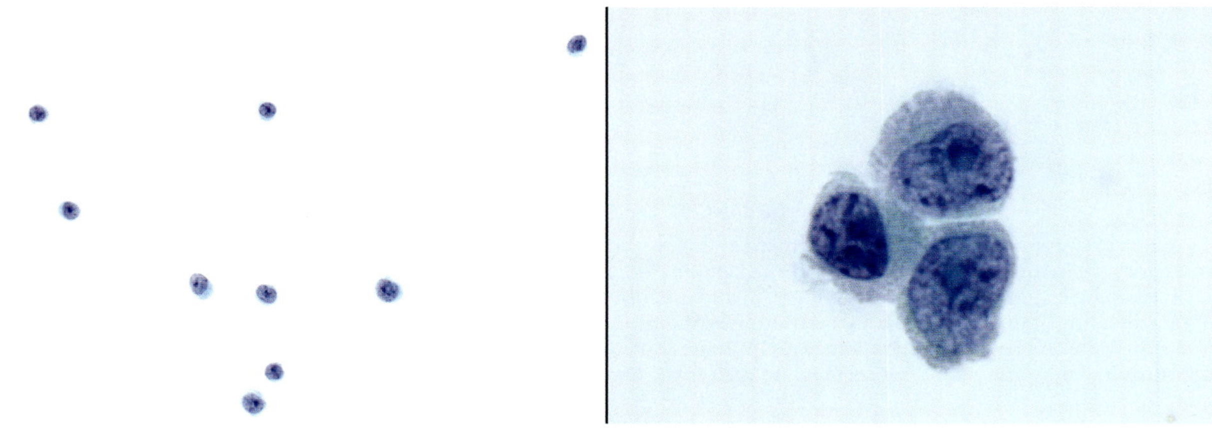

Figure 11-12

ACUTE MYELOGENOUS LEUKEMIA

Left: The anterior chamber in a 2-month-old boy is involved. The clinical differential diagnosis included acute endophthalmitis, retinoblastoma, and recurrence of acute myelogenous leukemia (AML). Cellular specimen in an anterior chamber needle aspiration. Aqueous humor is acellular in normal conditions.

Right: High-power view of the infiltrate discloses large cells with abundant granular cytoplasm and large irregular nuclei with prominent nucleoli, features consistent with the patient's previous diagnosis of AML.

to chemotherapy, local low-dose radiation is used. Some studies associate the presence of hemorrhage and signs of retinal ischemia and necrosis with mortality, and if these signs are present at presentation, patients have a shorter life expectancy. These prognostic features may help select patients that need more aggressive treatment at diagnosis (11).

Secondary Lymphoma

Secondary lymphoma of the uveal tract is non-Hodgkin B-cell lymphoma in most cases (fig. 11-13). T-cell lymphomas may involve the anterior segment, retina, and vitreous. The type of lymphoma is the same as the systemic lymphoma. It presents in patients 50 years or

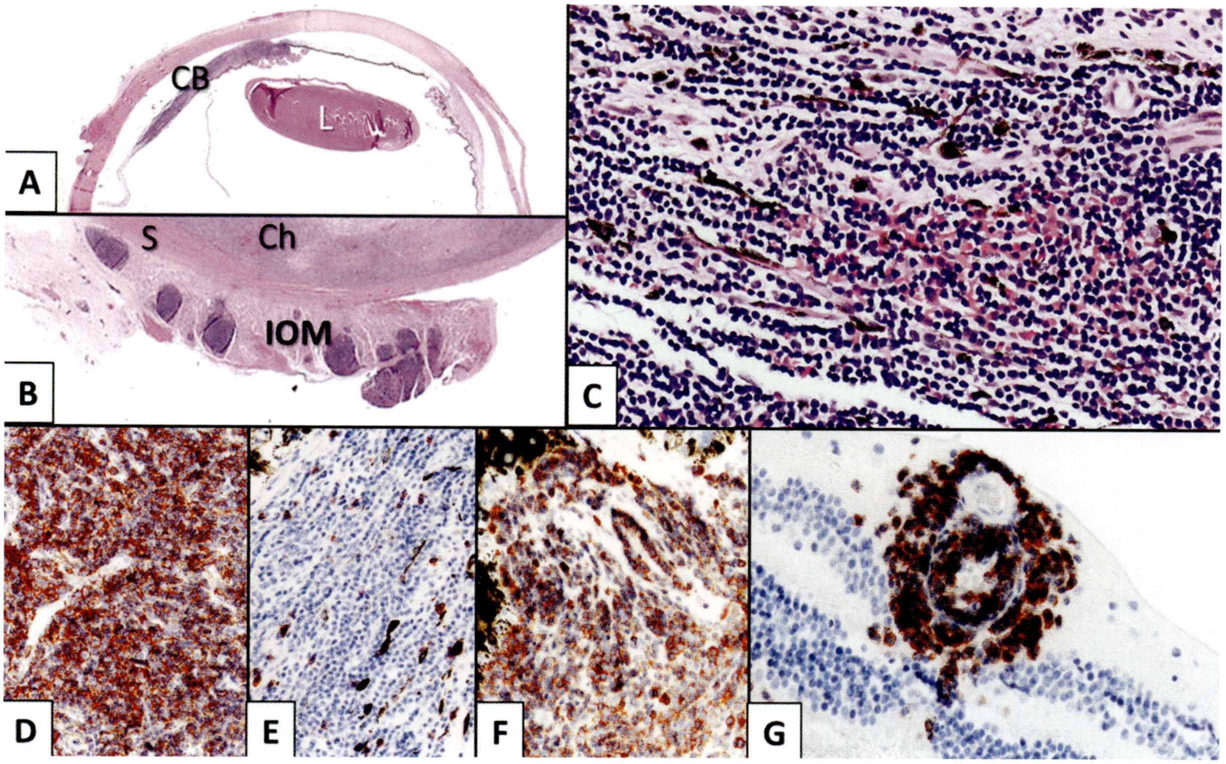

Figure 11-13

SECONDARY INVOLVEMENT OF THE UVEAL TRACT

The ciliary body (CB), choroid (Ch), and retina are involved by non-Hodgkin B-cell chronic lymphocytic leukemia/chronic lymphocytic lymphoma with extraocular extension into the inferior oblique muscle (IOM).

A: Low-power view of a histologic section of the anterior segment of the eye shows the diffuse involvement of the ciliary body (CB).

B: Low-power view of the posterior temporal aspect of the eye shows an infiltrate in the choroid and episcleral tissue, especially the inferior oblique muscle (IOM).

C: The lymphocytes are small and diffusely infiltrate the choroid.

D: CD20 is positive in the majority of the infiltrating lymphocytes.

E: CD3 is positive in rare lymphocytes.

F: CD5 is strongly positive in most of the infiltrating lymphocytes (co-expressed with CD20).

G: The retina shows a perivascular infiltrate of CD20-positive small lymphocytes.

older. Most (80 percent) patients have a known diagnosis of systemic lymphoma; about 30 percent have bilateral uveal involvement. The symptoms and signs are similar to those of primary choroidal lymphoma, including extension into the extraocular tissues (sclera, episclera, conjunctiva, and orbit) (11,14).

Choroidal biopsy, fine-needle aspiration biopsy, or enucleated eyes show the involvement of the choroid by a diffuse infiltrate or by areas of follicular lymphocytic infiltrates. The morphology and immunophenotype correspond to the primary lymphoma. Scleral and extraocular extension occur, but are generally limited by Bruch membrane. There may be associated retinal detachment and vitreous infiltrates of scattered lymphoma cells. Occasionally, there is involvement of ciliary body and iris extending from the choroidal main infiltrate; the iris and ciliary body may be the only uveal structure involved by systemic lymphoma (12,14).

Treatment is usually systemic chemotherapy, with occasional local radiation to prevent local damage. If left untreated there may be secondary neovascularization of iris and retina with glaucoma and loss of vision (14).

REFERENCES

1. Chan CC, Rubenstein JL, Coupland SE, et al. Primary vitreoretinal lymphoma: a report from an International Primary Central Nervous System Lymphoma Collaborative Group symposium. Oncologist 2011;16:1589-99.
2. Grimm SA, Pulido JS, Jahnke K, et al. Primary intraocular lymphoma: an International Primary Central Nervous System Lymphoma Collaborative Group report. Ann Oncol 2007;18:1851-5.
3. Chaput F, Amer R, Baglivo E, et al. Intraocular T-cell lymphoma: clinical presentation, diagnosis, treatment, and outcome. Ocul Immunol Inflamm 2017;25:639-48.
4. Gonzales J, Doan T, Shantha JG, et al. Metagenomic deep sequencing of aqueous fluid detects intraocular lymphomas. Br J Ophthalmol 2018;102:6-8.
5. Kraan W, Horlings HM, van Keimpema M, et al. High prevalence of oncogenic MYD88 and CD79B mutations in diffuse large B-cell lymphomas presenting at immune-privileged sites. Blood Cancer J 2013;3:e139.
6. Kim MM, Dabaja BS, Medeiros J, et al. Survival outcomes of primary intraocular lymphoma: a single-institution experience. Am J Clin Oncol 2016;39:109-13.
7. Mashayekhi A, Shields CL, Shields JA. Iris involvement by lymphoma: a review of 13 cases. Clin Exp Ophthalmol 2013;41:19-26.
8. Yamada K, Hirata A, Kimura A, Tanihara H. A case of primary B-cell type non-Hodgkin lymphoma originating in the iris. Am J Ophthalmol 2003;136:380-2.
9. Chan SM, Hutnik CM, Heathcote JG, Orton RB, Banerjee D. Iris lymphoma in a pediatric cardiac transplant recipient: clinicopathologic findings. Ophthalmology 2000;107:1479-82.
10. Konstantinidis L, Angi M, Coupland SE, Damato B. Primary B-cell lymphoma of the ciliary body with 360° ('ring'-like) growth pattern. Eye (Lond) 2014;28:355-6.
11. Mashayekhi A, Shukla SY, Shields JA, Shields CL. Choroidal lymphoma: clinical features and association with systemic lymphoma. Ophthalmology 2014;121:342-51.
12. Coupland SE, Joussen A, Anastassiou G, Stein H. Diagnosis of a primary uveal extranodal marginal zone B-cell lymphoma by chorioretinal biopsy: case report. Graefes Arch Clin Exp Ophthalmol 2005;243:482-6.
13. Coupland SE, Foss HD, Hidayat AA, Cockerham GC, Hummel M, Stein H. Extranodal marginal zone B cell lymphomas of the uvea: an analysis of 13 cases. J Pathol 2002;197:333-40.
14. Coupland SE, Damato B. Understanding intraocular lymphomas. Clin Exp Ophthalmol 2008;36:564-78.

APPENDICES

Appendix 1

STAGING OF CONJUNCTIVAL CARCINOMA, 8TH ED. AJCC CANCER STAGING MANUAL, 2017[a]

Definition of Primary Tumor (T)	
TX	Primary tumor cannot be assessed
T0	No evidence of primary tumor
Tis	Carcinoma in situ
T1	Tumor ≤5 mm in greatest dimension, which invades through basement membrane without invasion of adjacent structures[b]
T2	Tumor ≥5 mm in greatest dimension, which invades through basement membrane without invasion of adjacent structures
T3	Tumor invades adjacent structures excluding the orbit
T4	Tumor invades the orbit with or without further extension
T4a	Tumor invades orbital soft tissues, without bone invasion
T4b	Tumor invades bone
T4c	Tumor invades adjacent paranasal sinuses
T4d	Tumor invades brain
Definition of Regional Lymph Nodes (N)	
NX	Regional lymph nodes cannot be assessed
N0	No regional lymph node metastasis
N1	Regional lymph node metastasis
Definition of Distant Metastasis (M)	
M0	No distant metastasis
M1	Distant metastasis
Definition of Histopathologic Grade (G)	
GX	Grade cannot be assessed
G1	Well-differentiated
G2	Moderately differentiated
G3	Poorly differentiated
G4	Undifferentiated

[a]Data from reference 1.

[b]Adjacent structures include cornea, forniceal conjunctiva, palpebral conjunctiva, tarsal conjunctiva, lacrimal punctum and canaliculi, plica, caruncle, anterior and posterior eyelid lamellae, eyelid margin, and intraocular compartments.

Appendix 2

STAGING OF CONJUNCTIVAL MALIGNANT MELANOMA, 8TH ED. AJCC CANCER STAGING MANUAL, 2017[a]

Definition of Primary Clinical Tumor (cT)	
TX	Primary tumor cannot be assessed
T0	No evidence of primary tumor
T1	Tumor of the bulbar conjunctiva
T1a	Tumor of bulbar conjunctiva involving <1 quadrant
T1b	Tumor of bulbar conjunctiva involving >1 but <2 quadrants
T1c	Tumor of bulbar conjunctiva involving >2 but <3 quadrants
T1d	Tumor of bulbar conjunctiva involving >3 quadrants
T2	Tumor of nonbulbar (forniceal, palpebral, tarsal) conjunctiva and the caruncle
T2a	Noncaruncular tumor involving <1 quadrant of the nonbulbar conjunctiva
T2b	Noncaruncular tumor involving <1 quadrant of the nonbulbar conjunctiva
T2c	Caruncular tumor involving <1 quadrant of the nonbulbar conjunctiva
T2d	Caruncular tumor involving >1 quadrant of the nonbulbar conjunctiva
T3	Tumor of any size with local invasion
T3a	Tumor of any size invading the globe
T3b	Tumor of any size invading the eyelid
T3c	Tumor of any size invading the orbit
T3d	Tumor of any size invading the nasolacrimal duct, lacrimal sac, or paranasal sinuses
T4	Tumor of any size with invasion of the central nervous system

Definition of Primary Pathological Tumor (pT)	
TX	Primary tumor cannot be assessed
T0	No evidence of primary tumor
Tis	Tumor confined to the conjunctival epithelium
T1	Tumor of bulbar conjunctiva with the invasion of substantia propria
T1a	Tumor of bulbar conjunctiva with the invasion of substantia propria, <2.0 mm in thickness
T1b	Tumor of bulbar conjunctiva with the invasion of substantia propria, >2.0 mm in thickness
T2	Tumor of nonbulbar conjunctiva with the invasion of substantia propria
T2a	Tumor of nonbulbar conjunctiva with the invasion of substantia propria
T2b	Tumor of nonbulbar conjunctiva with the invasion of substantia propria, <2.0 mm in thickness
T3	Tumor of nonbulbar conjunctiva with the invasion of the substantia propria, >2.0 mm in thickness
T3a	Tumor of any size with local invasion
T3b	Tumor of any size invading the globe
T3c	Tumor of any size invading the eyelid
T3d	Tumor of any size invading the orbit
T4	Tumor of any size invading the nasolacrimal duct, lacrimal sac, or paranasal sinuses
	Tumor of any size with invasion of the central nervous system

Regional Lymph Nodes (N)	
NX	Regional lymph nodes cannot be assessed
N0	No regional lymph node metastasis
N1	Regional lymph node metastasis

Distant Metastasis (M)	
M0	No distant metastasis
M1	Distant metastasis

[a]Data from reference 1.

Appendix 3

STAGING OF EYELID CARCINOMA, 8TH ED. AJCC CANCER STAGING MANUAL, 2017[a]

Definition of Primary Tumor (T)	
TX	Primary tumor cannot be assessed
T0	No evidence of primary tumor
Tis	Carcinoma in situ
T1	Tumor ≤10 mm in greatest dimension
T1a	Tumor does not invade the tarsal plate or the eyelid margin
T1b	Tumor invades the tarsal plate or the eyelid margin
T1c	Tumor involves full-thickness of the eyelid
T2	Tumor >10 mm but ≤20 mm in greatest dimension
T2a	Tumor does not invade the tarsal plate or the eyelid margin
T2b	Tumor invades the tarsal plate or the eyelid margin
T2c	Tumor involves full-thickness of the eyelid
T3	Tumor >10 mm but ≤20 mm in greatest dimension
T3a	Tumor does not invade the tarsal plate or the eyelid margin
T3b	Tumor invades the tarsal plate or the eyelid margin
T3c	Tumor involves full-thickness of the eyelid
T4	Any eyelid tumor that invades adjacent ocular, orbital, or facial structures
T4a	Tumor invades ocular or intraorbital structures
T4b	Tumor invades (or erodes through) the bony walls of the orbit or extends to the paranasal sinuses or invades the lacrimal sac/nasolacrimal duct or brain
Definition of Regional Lymph Nodes (N)	
NX	Regional lymph nodes cannot be assessed
N0	No regional lymph node metastasis
N1	Metastasis in a single ipsilateral lymph node based on clinical evaluation or imaging
N1a	Metastasis in a single ipsilateral lymph node based on lymph node biopsy
N1b	Metastasis in a single ipsilateral lymph node based on lymph node biopsy
N2	Metastases in a single ipsilateral lymph node, >3 cm in greatest dimension, or in bilateral or contralateral lymph nodes
N2a	Metastasis in a single ipsilateral lymph node based on lymph node biopsy
N2b	Metastasis in a single ipsilateral lymph node based on lymph node biopsy
Definition of Distant Metastasis (M)	
M0	No distant metastasis
M1	Distant metastasis
Definition of Histopathologic Grade (G) (Applies predominantly to SCC[b] and sebaceous carcinoma; not used for BCC)	
GX	Grade cannot be assessed
G1	Well-differentiated
G2	Moderately differentiated
G3	Poorly differentiated
G4	Undifferentiated

[a]Data from reference 1.
[b]SCC= squamous cell carcinoma; BCC = basal cell carcinoma.

Appendix 4

STAGING OF ORBITAL SARCOMA, 8TH ED. AJCC CANCER STAGING MANUAL, 2017[a]

Definition of Primary Tumor (T)

TX	Primary tumor cannot be assessed
T0	No evidence of primary tumor
T1	Tumor ≤2 cm in greatest dimension
T2	Tumor >2 cm in greatest diameter without invasion of the bony walls or globe
T3	Tumor of any size with invasion of bony walls
T4	Tumor of any size with invasion of the globe or periorbital structures, including the eyelid, conjunctiva, temporal fossa, nasal cavity, paranasal sinuses, and/or central nervous system

Definition of Regional Lymph Nodes (N)

NX	Regional lymph nodes cannot be assessed
N0	No regional lymph node metastasis
N1	Regional lymph node metastasis

Definition of Distant Metastasis (M)

M0	No distant metastasis
M1	Distant metastasis

Definition of Histopathologic Grade (G)

French Federation of Cancer Centers Sarcoma Group (FNCLCC) histologic grading system is adopted for grading of primary orbital sarcomas

[a]Data from reference 1.

Appendix 5

STAGING OF LACRIMAL GLAND TUMORS, 8TH ED. AJCC CANCER STAGING MANUAL, 2017[a]

Primary Tumor (T)

TX	Primary tumor cannot be assessed
T0	No evidence of primary tumor
T1	Tumor 2 cm or smaller in greatest dimension, with or without extraglandular extension into the orbital soft tissue
T1a	No periosteal or bone involvement
T1b	Periosteal involvement only
T1c	Periosteal and bone involvement
T2	Tumor larger than 2 cm but no larger than 4 cm in greatest dimension
T2a	No periosteal or bone involvement
T2b	Periosteal involvement only
T2c	Periosteal and bone involvement
T3	Tumor larger than 4 cm in greatest dimension
T3a	No periosteal or bone involvement
T3b	Periosteal involvement only
T3c	Periosteal and bone involvement
T4	Involvement of adjacent structures, including sinuses, temporal fossa, pterygoid fossa, superior orbital tissue, cavernous sinus, or brain
T4a	Tumor ≤2 cm in greatest dimension
T4b	Tumor >2 cm and ≤4 cm in greatest dimension
T4c	Tumor >4 cm in greatest dimension

Regional Lymph Nodes (N)

NX	Regional lymph nodes cannot be assessed
N0	No regional lymph node metastasis
N1	Regional lymph node metastasis

Distant Metastasis (M)

M0	No distant metastasis
M1	Distant metastasis

[a]Data from reference 1.

Appendix 6

STAGING OF OCULAR ADNEXAL LYMPHOMAS, 8TH ED. AJCC CANCER STAGING MANUAL, 2017[a]

Definition of Primary Tumor (T)

TX	Lymphoma extent not specified
T0	No evidence of lymphoma
T1	Lymphoma involving the conjunctiva alone without eyelid or orbital involvement
T2	Lymphoma with orbital involvement with or without conjunctival involvement
T3	Lymphoma with preseptal eyelid involvement with or without orbital involvement and with or without conjunctival involvement
T4	Orbital adnexal lymphoma and extraorbital lymphoma extending beyond the orbit to adjacent structures, such as bone, maxillofacial sinuses, and brain

Definition of Regional Lymph Nodes (N)

NX	Involvement of lymph nodes not assessed
N0	No evidence of lymph node involvement
N1	Involvement of lymph node region or regions draining the ocular adnexal structures and superior to the mediastinum (preauricular, parotid, submandibular, and cervical nodes)
N1a	Involvement of a single lymph node region superior to the mediastinum
N1b	Involvement of two or more lymph node regions, superior to the mediastinum
N2	Involvement of lymph node regions of the mediastinum
N3	Diffuse or disseminated involvement of peripheral and central lymph node regions

Definition of Distant Metastasis (M)

M0	No evidence of involvement of other extranodal sites
M1a	Noncontiguous involvement of tissue or organs external to the ocular adnexa (e.g., parotid glands, submandibular gland, lung, liver, spleen, kidney, breast)
M1b	Lymphomatous involvement of the bone marrow
M1c	Both M1a and M1b involvement

Definition of Histopathologic Grade (G)

A grade is assigned only to follicular lymphomas, as described by 2017 WHO classification[b]

[a]Data from reference 1.
[b]Data from reference 2.

Appendix 7

DEFINITION OF PRIMARY UVEAL MELANOMA

✔ T Category	T Criteria
TX	Primary tumor cannot be assessed
T0	No evidence of primary tumor
T1	Tumor limited to the iris
T1a	Tumor limited to the iris, not more than 3 clock hours in size
T1b	Tumor limited to the iris, more than 3 clock hours in size
T1c	Tumor limited to the iris with secondary glaucoma
T2	Tumor confluent with or extending into the ciliary body, choroid, or both
T2a	Tumor confluent with or extending into the ciliary body, without secondary glaucoma
T2b	Tumor confluent with or extending into the ciliary body and choroid, without secondary glaucoma
T2c	Tumor confluent with or extending into the ciliary body, choroid, or both, with secondary glaucoma
T3	Tumor confluent with or extending into the ciliary body, choroid, or both, with scleral extension
T4	Tumor with extrascleral extension
T4a	Tumor with extrascleral extension ≤5 mm in largest diameter
T4b	Tumor with extrascleral extension >5 mm in largest diameter

Note: Iris melanomas originate from, and are predominantly located in, this region of the uvea. If less than half the tumor volume is located within the iris, the tumor may have originated in the ciliary body, and consideration should be given to classifying it accordingly.

✔ T Suffix	Definition
(m)	Select if synchronous primary tumors are found in single organ.

Appendix 7, continued

DEFINITION OF PRIMARY UVEAL MELANOMA, continued

Figure 67.1: Classification of ciliary body and choroid uveal melanoma based on thickness and diameter

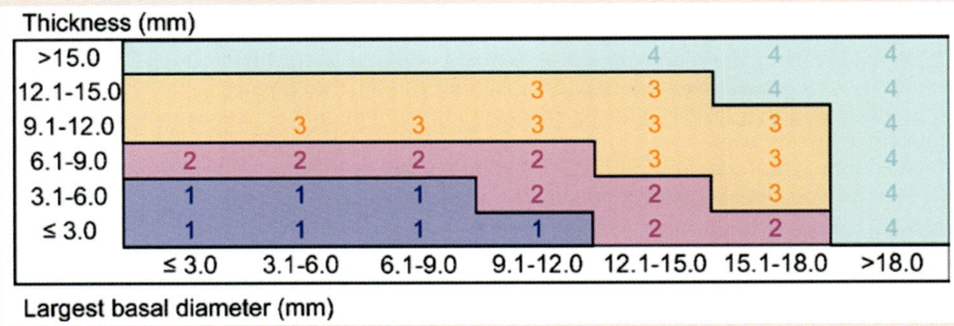

✔ T Category	T Criteria
TX	Primary tumor cannot be assessed
T0	No evidence of primary tumor
T1	Tumor size category 1
T1a	Tumor size category 1 without ciliary body involvement and extraocular extension
T1b	Tumor size category 1 with ciliary body involvement
T1c	Tumor size category 1 without ciliary body involvement but with extraocular extension ≤5 mm in largest diameter
T1d	Tumor size category 1 with ciliary body involvement and extraocular extension ≤5 mm in largest diameter
T2	Tumor size category 2
T2a	Tumor size category 2 without ciliary body involvement and extraocular extension
T2b	Tumor size category 2 with ciliary body involvement
T2c	Tumor size category 2 without ciliary body involvement but with extraocular extension ≤5 mm in largest diameter
T2d	Tumor size category 2 with ciliary body involvement and extraocular extension ≤5 mm in largest diameter
T3	Tumor size category 3
T3a	Tumor size category 3 without ciliary body involvement and extraocular extension
T3b	Tumor size category 3 with ciliary body involvement
T3c	Tumor size category 3 without ciliary body involvement but with extraocular extension ≤5 mm in largest diameter
T3d	Tumor size category 3 with ciliary body involvement and extraocular extension ≤5 mm in largest diameter
T4	Tumor size category 4
T4a	Tumor size category 4 without ciliary body involvement and extraocular extension
T4b	Tumor size category 4 with ciliary body involvement
T4c	Tumor size category 4 without ciliary body involvement but with extraocular extension ≤5 mm in largest diameter
T4d	Tumor size category 4 with ciliary body involvement and extraocular extension ≤5 mm in largest diameter
T4e	Any tumor size category with extraocular extension >5 mm in largest diameter

Notes: 1. Primary ciliary body and choroidal melanomas are classified according to the four tumor size categories defined in Figure 67.1.

2. In clinical practice, the largest tumor basal diameter may be estimated in optic disc diameters (DD; average: 1 DD = 1.5 mm), and tumor thickness may be estimated in diopters (average: 2.5 diopters = 1 mm). Ultrasonography and fundus photography are used to provide more accurate measurements.

3. When histopathologic measurements are recorded after fixation, tumor diameter and thickness may be underestimated because of tissue shrinkage.

✔ T Suffix	Definition
(m)	Select if synchronous primary tumors are found in single organ.

Appendix 7, continued

DEFINITION OF REGIONAL LYMPH NODE (N)

✔	N Category	N Criteria
	NX	Regional lymph nodes cannot be assessed
	N0	No regional lymph node involvement
	N1	Regional lymph node metastases or discrete tumor deposits in the orbit
	N1a	Metastasis in one or more regional lymph node(s)
	N1b	No regional lymph nodes are positive, but there are discrete tumor deposits in the orbit that are not contiguous to the eye. (choroidal and ciliary body melanoma only)

✔	N Suffix	Definition
	(sn)	Select if regional lymph node metastasis identified by SLN biopsy only.
	(f)	Select if regional lymph node metastasis identified by FNA or core needle biopsy only.

DEFINITION OF DISTANT METASTASIS (M)

The terms pM0 and MX are NOT valid categories in the TNM system. Assignment of the M category for clinical classification may be cM0, cM1, or pM1. Any of the M categories (cM0, cM1, or pM1) may be used with pathological stage grouping.

✔	M Category	M Criteria
	cM0	No distant metastasis by clinical classification
	cM1	Distant metastasis
	cM1a	Largest diameter of the largest metastasis ≤3.0 cm
	cM1b	Largest diameter of the largest metastasis 3.1–8.0 cm
	cM1c	Largest diameter of the largest metastasis ≥8.1 cm
	pM1	Distant metastasis, microscopically confirmed
	pM1a	Largest diameter of the largest metastasis ≤3.0 cm, microscopically confirmed
	pM1b	Largest diameter of the largest metastasis 3.1–8.0 cm, microscopically confirmed
	pM1c	Largest diameter of the largest metastasis ≥8.1 cm, microscopically confirmed

AJCC PROGNOSTIC STAGE GROUPS

Always refer to the specific chapter for rules on clinical and pathological classification of this disease.

✔	When T is...	And N is...	And M is...	Then the stage group is...
	T1a	N0	M0	I
	T1b–d	N0	M0	IIA
	T2a	N0	M0	IIA
	T2b	N0	M0	IIB
	T3a	N0	M0	IIB
	T2c–d	N0	M0	IIIA
	T3b–c	N0	M0	IIIA
	T4a	N0	M0	IIIA
	T3d	N0	M0	IIIB
	T4b–c	N0	M0	IIIB
	T4d–e	N0	M0	IIIC
	Any T	N1	M0	IV
	Any T	Any N	M1a–c	IV

REGISTRY DATA COLLECTION VARIABLES

1. Tumor site (ICD code lacks specificity): ☐ ciliary body (use this staging form) ☐ iris (use Uveal Melanoma-Iris Melanoma staging form)
2. Largest basal diameter and thickness of tumor:
3. Ciliary body involvement:
4. Extraocular extension:

Appendix 7, continued

REGISTRY DATA COLLECTION VARIABLES, Continued

5. Histologic type:
6. Chromosome 3 and 8 loss or gain:
7. Gene expression profile:
8. Mitotic count (number of mitoses per 40 HPF, determined by using a 40x objective with a field area of 0.152 mm2):
9. Extravascular matrix patterns (extracellular closed loops and networks, defined as at least three back-to-back closed loops, is associated with death from metastatic disease):
10. Microvascular density:

HISTOLOGIC GRADE (G)

✔	G	G Definition
	GX	Grade cannot be assessed
	G1	Spindle cell melanoma (>90% spindle cells)
	G2	Mixed cell melanoma (>10% epithelioid cells and <90% spindle cells)
	G3	Epithelioid cell melanoma (>90% epithelioid cells)

Note: Because of the lack of universal agreement regarding which proportion of epithelioid cells classifies a tumor as mixed or epithelioid, some ophthalmic pathologists currently combine grades 2 and 3 (nonspindle, i.e. epithelioid cells detected) and contrast them with grade 1 (spindle, i.e. no epithelioid cells detected).

LYMPHOVASCULAR INVASION (LVI)

✔	Component of LVI Coding	Description
	0	LVI not present (absent)/not identified
	1	LVI present/identified, NOS
	2	Lymphatic and small vessel invasion only (L)
	3	Venous (large vessel) invasion only (V)
	4	BOTH lymphatic and small vessel AND venous (large vessel) invasion
	9	Presence of LVI unknown/indeterminate

Appendix 8

CLINICAL CLASSIFICATION OF RETINOBLASTOMA

Group Type	IRCC[a]	AJCC 8th Edition/OOTF[b]
Small tumors	A. Tumor <3mm, distant to the fovea or or optic nerve B. Same as A without vitreous seeds plus subretinal fluid <5mm from the base of tumor or <3mm tumor at the fovea	cT1. Intraretinal tumors(s) with subretinal fluid not more than 5 mm from the base of any tumor cT1a. Tumors <3 mm and further than 1.5 mm from the disc and fovea cT1b. Tumors >3 mm or closer than 1.5 mm from the disc and fovea
Focal vitreous seeds and subretinal fluid	C. Focal vitreous seeds close to the tumor or with subretinal fluid less than a quadrant	cT2. Intraocular tumor(s) with retinal detachment, vitreous seeding or subretinal seeding cT2a. Subretinal fluid more than 5 mm of the base of any tumor cT2b. Tumors with vitreous seeding and/or subretinal seeding
Large tumors, extensive tumor seeds, other ocular changes	D. Massive vitreous seeds or diffuse, extensive subretinal masses E. Neovascular glaucoma, tumor in in contact with lens, tumor in segment, phthisis, diffuse infiltrative tumor, nonsalvable eyes	cT3. Advanced intraocular tumor(s) cT3a. Phthisis or pre-phthisis bulbi cT3b. Tumor involvement of the pars plana, ciliary body, lens, zonules, iris or anterior chamber cT3c. Intraocular pressure greater than 25mm Hg cT3d. Hyphema and/or massive vitreous hemorrhage cT3e. Aseptic orbital cellulitis
Extraocular extension	N/A	cT4. Extraocular tumor(s) involving the orbit, including the optic nerve cT4a. Radiologic evidence of retrobulbar optic nerve involvement or thickening of the optic nerve or involvement of the orbital tissues cT4b. Extraocular tumor clinically evident with proptosis and/or an orbital mass

[a]Murphree AL. Intraocular retinoblastoma: the case for a new group classification. Ophthalmol Clin N Am 2005;18:41-53.
[b]Amram AL, Rico G, Kim JW, et al. Vitreous seeds in retinoblastoma: clinicopathologic classification and correlation. Ophthalmology 2017;124:1540-7.
[c]IIRC = International Intraocular Retinoblastoma Classification; AJCC = American Joint Commission for Cancer; OOTF = Ophthalmic Oncology Task Force.

APPENDICES REFERENCES

1. Amin MB, Edge SB, Greene FL, et al., eds. AJCC cancer staging manual, 8th ed. Switzerland: Springer; 2017.
2. Swerdlow SH, Campo E, Harris NL, et al. WHO classification of tumours of haematopoietic and lymphoid tissues, 4th ed. Lyon: IARC Press; 2017.

Index*

*In a series of numbers, those in boldface indicate the main discussion of the entity.

T

U

V

W

X

Y

Z